T0213875

Health Informatics

This series is directed to healthcare professionals leading the transformation of healthcare by using information and knowledge. For over 20 years, Health Informatics has offered a broad range of titles: some address specific professions such as nursing, medicine, and health administration; others cover special areas of practice such as trauma and radiology; still other books in the series focus on interdisciplinary issues, such as the computer based patient record, electronic health records, and networked healthcare systems. Editors and authors, eminent experts in their fields, offer their accounts of innovations in health informatics. Increasingly, these accounts go beyond hardware and software to address the role of information in influencing the transformation of healthcare delivery systems around the world. The series also increasingly focuses on the users of the information and systems: the organizational, behavioral, and societal changes that accompany the diffusion of information technology in health services environments.

Developments in healthcare delivery are constant; in recent years, bioinformatics has emerged as a new field in health informatics to support emerging and ongoing developments in molecular biology. At the same time, further evolution of the field of health informatics is reflected in the introduction of concepts at the macro or health systems delivery level with major national initiatives related to electronic health records (EHR), data standards, and public health informatics.

These changes will continue to shape health services in the twenty-first century. By making full and creative use of the technology to tame data and to transform information, Health Informatics will foster the development and use of new knowledge in healthcare.

More information about this series at http://www.springer.com/series/1114

Charles P. Friedman • Jeremy C. Wyatt
Joan S. Ash

Evaluation Methods in Biomedical and Health Informatics

Third Edition

 Springer

Charles P. Friedman
Department of Learning Health Sciences
University of Michigan Medical School
Ann Arbor, MI, USA

Joan S. Ash
Department of Medical Informatics and
Clinical Epidemiology, School of Medicine
Oregon Health & Science University
Portland, OR, USA

Jeremy C. Wyatt
Department of Primary Care, Population
Sciences and Medical Education,
School of Medicine
University of Southampton
Southampton, UK

Originally published in the series "Computers and Medicine"

ISSN 1431-1917   ISSN 2197-3741   (electronic)
ISBN 978-3-030-86455-2   ISBN 978-3-030-86453-8   (eBook)
https://doi.org/10.1007/978-3-030-86453-8

This Springer imprint is published by the registered company Springer Nature Switzerland AG
The registered company address is: Gewerbestrasse 11, 6330 Cham, Switzerland

Charles Friedman dedicates this book to his spouse Patti who inspires him every day, and to the consummate professor of physics, Edwin Taylor, who impressed him with the essential service to science that a textbook provides.

Jeremy Wyatt dedicates this book to his spouse Sylvia, to mark her enduring support and love, without which his contribution would not have been possible.

Joan Ash dedicates this book to her husband Paul, who contributed his medical knowledge in addition to unwavering support for the project, and to her daughter Britt and son Deren, who have continuously added their nursing and healthcare software development experience to her informatics knowledge base.

Foreword

With the continuing growth and maturation of biomedical and health informatics, we have seen increasing numbers of novel applications emerging from both academic and industrial settings. Acceptance and adoption of such innovations typically depend on formal studies that meet the rigorous standards that the medical community has come to expect when considering the use of new diagnostic, therapeutic, or management options. By analogy, new pharmaceuticals are typically adopted not only after regulatory certifications (e.g., approval by the Food and Drug Administration) but also formal studies that demonstrate safety and effectiveness – usually documented in peer-reviewed publications. The need for such studies is equally important for informatics innovations, although the nature of software or "smart" devices introduces complexities that often exceed those that have characterized the drug approval process. Developers of informatics systems—seeking to demonstrate not only the safety and efficacy of their products but also their acceptability, utility, and cost-effectiveness—therefore need to anticipate and understand the evaluation requirements that will affect their own innovations as they move from R&D to routine care or public health environments. Drs. Charles "Chuck" Friedman, Jeremy Wyatt, and Joan Ash recognize these needs and offer this new volume, building upon two earlier editions, which nicely meets them.

Drs. Friedman and Wyatt first came together to write the first edition of this volume in the early 1990s. I was on the faculty at Stanford University School of Medicine at the time, directing research and graduate training in biomedical informatics (which we then called "medical information sciences"). I had known both Chuck and Jeremy from earlier visits and professional encounters, but it was coincidence that offered them sabbatical breaks in our laboratory during the same academic year. Knowing that each had strong interests and skills in the areas of evaluation and clinical trial design, I hoped they would enjoy getting to know one another and would find that their scholarly pursuits were both complementary and synergistic. To help stir the pot, we even assigned them to a shared office that we tried to set aside for visitors, and within a few weeks, they were putting their heads together as they learned about the evaluation issues that were rampant in our laboratory. One of the most frequent inquiries from graduate students is, "Although I am happy with my research focus and the work I have done, how can I design and carry out a practical evaluation that proves the value of my contribution?"

Informatics is a multifaceted interdisciplinary field with research that ranges from theoretical developments to projects that are highly applied and intended for near-term use in clinical settings. The implications of "proving" a research claim accordingly vary greatly depending on the details of an individual student's goals and thesis statement. Furthermore, the dissertation work leading up to an evaluation plan is often so time-consuming and arduous that attempting the "perfect" evaluation is frequently seen as impractical or as diverting students from central programming or implementation issues that are their primary areas of interest. They often ask what compromises are possible so that they can provide persuasive data in support of their claims without adding two to three years to their graduate-student life.

The on-site contributions by Drs. Friedman and Wyatt during that year were marvelous. They served as local consultants as we devised evaluation plans for existing projects, new proposals, and student research. By the spring, they had identified the topics and themes that needed to be understood better by those in our laboratory, and they offered a well-received seminar course on evaluation methods for medical information systems. It was out of the class notes formulated for that course that the first edition of this volume evolved.

The book has become the "Bible" for evaluation methods as they relate to the field of biomedical and health informatics. It has had an impact in diverse biomedical informatics training environments both in the USA and abroad, leading to the publication of a revised second edition in 2006.

In the subsequent 15 years, the need for such guidance has been accentuated even more, both in the academic community and in the burgeoning informatics industry. This new third edition has accordingly been eagerly awaited, and I am delighted that Drs. Friedman and Wyatt have taken on the significant update and revision reflected in these pages. I know of no other writers who have the requisite knowledge of statistics and cognition, coupled with intensive study of biomedical informatics and an involvement with creation of applied systems as well. Importantly, they have also recognized that they needed to add a third author who has special expertise in the area of qualitative methods, Dr. Joan Ash. Drs. Friedman, Wyatt, and Ash are scholars and educators, but they are also practical in their understanding of the world of clinical medicine and the realities of system implementation and validation in settings that often defy formal controlled trials. Thus, this edition is of value not only to students of biomedical informatics, but as a key reference for all individuals involved in the implementation and evaluation of basic and applied systems in biomedical informatics.

Those who are familiar with the second edition will recognize that the general structure is very similar in this new version. However, every chapter has been modernized to reflect the substantive changes as the field and cultural attitudes toward computing have evolved. Examples and citations have been updated, and there is a broadening of focus to extend to health issues generally, rather than a former focus largely on clinical systems. This broadening is reflected in the new book title, which includes *health informatics* in addition to *biomedical informatics*. Four new chapters have been added (Chaps. 4 and 5 which discuss study design and planning, plus Chaps. 13 and 17, which discuss correlational studies and mixed-methods studies, respectively).

I commend this updated volume to you, whether your interests are based in academia or in the practical arena of commercial system development. There is no similar volume that presents evaluation detail with a specific focus on both research and practical aspects of informatics systems in biomedicine and health. It also reflects the remarkable expertise and practical experience of the authors, all three of whom are highly regarded for their accomplishments and their ability to present concise but detailed summaries of the key concepts contained in this volume.

Edward H. Shortliffe
Department of Biomedical Informatics
Vagelos College of Physicians and Surgeons
Columbia University in the City of New York
New York, USA

Preface

Introducing the "Tree of Evaluation"

This volume is a substantial revision and modernization of *Evaluation Methods in Biomedical Informatics, Second Edition*, which was originally published in 2006. Much in the world of informatics has changed since the appearance of the second edition, and this volume has changed accordingly. This is signaled by the addition of the important word "health" to the title of the book, to reflect the increasing role that personal health applications play in the informatics world and to emphasize that the methods described in the book include but also transcend the work of the medical profession.

This remains primarily a textbook with a focus on widely applicable evaluation methods, including both quantitative and qualitative approaches. The chapters on quantitative studies retain an emphasis on measurement which is very important in informatics and is not typically covered in statistics courses. The book is written for use in formal courses for students of informatics at all levels, and for use as a reference by informatics researchers and professionals. It makes only modest assumptions about knowledge of basic statistics.

Because the author team spans the Atlantic, readers will find examples and references that reflect experiences and work performed on two continents.

The most significant changes from the second edition include:

- Based on the authors' additional years of teaching experience, a significant attempt at clarifying complex issues of methodology that have proven challenging for students.
- The addition of three important new topics: evidence-based informatics (Chap. 5), methods using "real-world" data (Chap. 13), and mixed-methods studies (Chap. 17).
- Reconstructing the book with a larger number of somewhat shorter chapters.
- An update of examples and references, while respecting the importance of older studies that for decades have been guideposts for work in the field. Several older studies are retained as examples because of their value as teaching cases.

- The use of hyperlinks to enable users of the digital version to navigate quickly between interdependent, and now numbered, chapters and sections.
- Additional self-tests, with answers, and "Food for Thought" questions.

Like the earlier volumes, the third edition uses "evaluation" in the title while recognizing that the methods introduced apply to both evaluation and research. For this reason, this book could have been called "Empirical Methods..." to cover both. "Evaluation" was retained in the title to reflect the nature of informatics as an applied science grounded in practice -- and the fact that many studies of information resources using the methods described in this book are intended to inform practical decisions as opposed to, or in addition to, resulting in publications. To help readers clarify this, the initial chapters devote significant attention to the distinction between evaluation and research.

The Tree of Evaluation

To illustrate the topics covered and how the key ideas of the book relate to one another, we have represented the contents of this volume as a tree in the figure below. In several places, the discussions in the chapters refer back to this Fig. 1.

The progression of topics in the book rises from the base of the tree embracing the full range of approaches to evaluation. The early chapters discuss a range of issues common to all approaches and trace the differences between quantitative and qualitative methods to their philosophical roots. As the discussion progresses up the tree, it follows a major branch on the left to quantitative methods, introducing there an all-important distinction between studies that emphasize measurement and those that emphasize demonstration. This distinction is important because measurement is under-appreciated in informatics. From there, the discussion follows another major branch on the right into the world of qualitative methods, subsequently rising to a central branch of mixed methods. At the highest level of elevation, the tree reaches a canopy of important topics emphasizing communication and ethics.

What Isn't Included

Informatics is an increasingly vast and diverse field, and no single work can cover all aspects of it. In revising this volume, we became increasingly aware of topics to which only passing reference is made or in some cases were not included at all.

Most important, while the methods introduced in this book are widely applicable, the discussions make no attempt to describe all of the information resources and purposes of their use, to which these methods can be applied. The examples illustrated are chosen for their potential to illustrate specific methods. Inevitably, readers of this book will take away from it not so much a deeper sense of "what works and

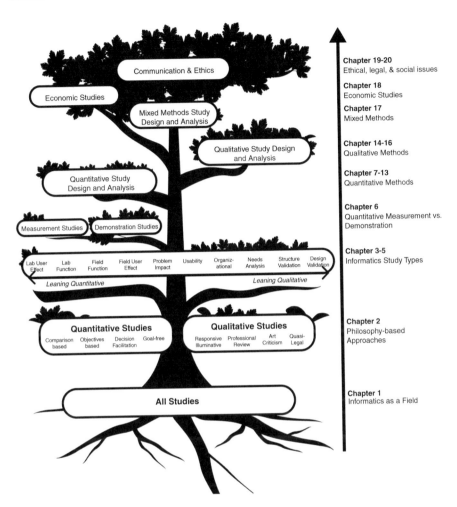

Fig. 1 The tree of evaluation

what doesn't" to improve health with information resources, but rather a much greater ability to determine rigorously what works and what doesn't—and why.

Because this is a basic methods textbook, scholars of evaluation methodology will not find cutting-edge techniques discussed here. We also omitted detailed discussion of what might be called intermediate level methods, such as principal component analysis and generalizability theory. For this reason, doctoral students who wish to employ these methods in their dissertation research will likely need further reading or course work.

While the evaluation methods addressed in this book are applicable to studies of information resources employed in bioinformatics and computational biology, the examples in the text include relatively few applications in these important fields.

In Sum

Returning to the Tree of Evaluation and extending the metaphor, we hope this volume serves for informatics the same purposes that a tree serves in nature, most notably endurance through its emphasis on the basic methods that will change more slowly than the resources to which they are applied. And like a tree that serves many purposes from cleaning the air to a play place for children, this volume will hopefully prove useful to everyone who seeks to improve health through the best possible use of the best possible information resources. Finally, those of us who work intimately with these methods appreciate them as a thing of beauty, much like a tree. And much like a tree that provides shelter for many forms of life, large and small— helping them to thrive and multiply—we hope our tree will both stimulate and sustain the growth of the field of informatics.

Ann Arbor, MI
Southampton, UK
Portland, OR

Charles P. Friedman
Jeremy C. Wyatt
Joan S. Ash

Acknowledgments

We are enormously grateful for the superb assistance of several colleagues in the creation of this book.

Monica Guo produced the artwork for many figures, including the "Tree of Evaluation," and provided vital support to maintain consistency within and across chapters as they each went through several stages of review and revision. It is difficult to imagine how this volume—with authors spanning two continents, two versions of the English language, and eight time zones—could have come together without Monica's creativity, dedication, and attention to detail. Her invariably cheerful "can do" approach to this work was a source of buoyancy for the entire team.

Joshua Rubin, as our work drew to a conclusion, provided enormous assistance in researching policy issues and in ensuring the consistency and completeness of references. We are equally appreciative of Josh's creativity, dedication, and attention to detail.

Professor Mark Roberts of the University of Pittsburgh provided valuable insights and suggestions that supported the revision of the chapter on economic analysis.

We are grateful to Ted Shortliffe for writing a Foreword to this edition, as he so ably did for the previous two editions.

We also thank the students we collectively taught at the University of Michigan, Oregon Health & Science University, and several universities in the UK. Their curiosity and passion for learning helped us recognize which sections of the second edition required clarification and which additional topics should be addressed in this third edition. In particular, we appreciate Keaton Whittaker's assistance with developing the new mixed-methods chapter.

And we are exceedingly grateful to our wonderful spouses—Patti, Sylvia, and Paul—for their encouragement to proceed through this work during a global pandemic, and their patience and understanding as we invaded what otherwise would have been family time to bring this work to conclusion.

Contents

Part I Landscape of Evaluation in Informatics

1 What Is Evaluation and Why Is It Challenging? 3
  1.1 Introduction 3
  1.2 The Place of Evaluation Within Informatics 4
  1.3 Framing Evaluation as Answering Questions 6
  1.4 Examples of Information Resources 9
  1.5 Further Distinguishing Evaluation and Research. 11
  1.6 Reasons for Performing Evaluations 12
  1.7 Evaluation in Informatics as a Complex
      Socio-Technical Endeavor 13
      1.7.1 Studying People 14
      1.7.2 Professional Practice and Culture 14
      1.7.3 Safety and Regulation 15
      1.7.4 Domain Knowledge Complexity 16
      1.7.5 Chains of Effect 17
      1.7.6 Decision Making Processes 18
      1.7.7 Data Availability and Quality. 18
      1.7.8 Performance Standards 19
      1.7.9 Computer-Based Information Resources. 19
  1.8 Concluding Observations. 20
  Answers to Self-Tests 22
    Self-Test 1.1 22
    Self-Test 1.2 22
  References. 22

2 The Panorama of Evaluation Approaches 25
  2.1 Introduction 25
  2.2 Deeper Definitions of Evaluation. 26
  2.3 The Common Anatomy of All Evaluation Studies. 27
  2.4 Diverse Roles in Evaluation Studies 28
  2.5 Some Philosophical Bases of Evaluation. 30

2.6 The Specific Anatomies of Quantitative and Qualitative Studies . . 32
2.7 Multiple Approaches to Evaluation 33
   2.7.1 Approaches Rooted in Quantitative Assumptions 34
   2.7.2 Approaches Rooted in Qualitative Assumptions 35
2.8 Conclusion: Why Are There so Many Approaches? 37
Answers to Self-Tests 40
References. 41

3   **From Study Questions to Study Design: Exploring the Full Range of Informatics Study Types**. 43
3.1 Introduction 43
3.2 The Key Role of Study Questions to Guide Study Design 44
   3.2.1 Why Study Questions Are Important. 45
   3.2.2 The Qualitative Process of Getting to a Final Set of Study Questions 46
   3.2.3 What Is the Right Number of Study Questions? 46
   3.2.4 How Should Study Questions Be Stated? 47
3.3 Ten Informatics Study Types 48
   3.3.1 Needs Assessment Studies. 50
   3.3.2 Design Validation Studies 50
   3.3.3 Structure Validation Studies. 50
   3.3.4 Usability Studies 51
   3.3.5 Laboratory Function Studies 51
   3.3.6 Field-Function Studies. 52
   3.3.7 Laboratory-User Effect Studies 52
   3.3.8 Field User Effect Studies 53
   3.3.9 Problem Impact Studies. 54
   3.3.10 Organizational and System Studies 54
3.4 Specific Features Distinguishing the Ten Informatics Study Types. 55
   3.4.1 The Setting in Which the Study Takes Place 55
   3.4.2 The Status of the Information Resource Employed. 55
   3.4.3 The Sampled Resource Users 55
   3.4.4 The Sampled Tasks 57
   3.4.5 What Is Observed or Measured 57
   3.4.6 Sampling of Users and Tasks. 58
3.5 Conclusions: Approaching a Recipe for Study Design 59
Answers to Self-Tests 60
References. 61

4   **Study Design Scenarios and Examples** 63
4.1 Introduction 63
4.2 A Scenario Invoking a Needs Assessment Study. 64
   4.2.1 Typical Study Questions 64
   4.2.2 Performing Needs Assessments. 65
4.3 A Scenario Invoking a Usability and/or a Function Study 66
   4.3.1 Typical Study Questions 67
   4.3.2 Performing Usability and Function Studies. 67

4.4   A Scenario Invoking a Field User Effect Study   68
     4.4.1   Typical Study Questions   69
     4.4.2   Performing Effect Studies   69
4.5   Scenario Invoking Problem Impact and Organization
     and System Studies   70
     4.5.1   Typical Study Questions   71
     4.5.2   Performing Impact and Organizational and S
           ystem Studies   71
4.6   Exemplary Evaluation Studies from the Literature   72
     4.6.1   Needs Assessment Study: "Using Ethnography to
           Build a Working System: Rethinking
           Basic Design Assumptions"   73
     4.6.2   Usability Study: "Assessing Performance of an
           Electronic Health Record Using Cognitive Task Analysis".   73
     4.6.3   Lab Function Study: "Diagnostic Inaccuracy of
           Smartphone Applications for Melanoma Detection"   73
     4.6.4   Field Function Study: "Evaluation of a Machine Learning
           Capability for a Clinical Decision Support
           System to Enhance Antimicrobial Stewardship
           Programs"   74
     4.6.5   Lab User Effect Study: "Applying Human Factors
           Principles to Alert Design Increases Efficiency
           and Reduces Prescribing Errors in a Scenario-Base
           Simulation"   74
     4.6.6   Field User Effect Study: "Reminders to
           Physicians from an Introspective Computer
           Medical Record: A Two-Year Randomized Trial"   75
     4.6.7   Field User Effect Study: "Electronic Health Records
           and Health Care Quality Over Time in a
           Federally Qualified Health Center"   75
     4.6.8   Problem Impact Study: "Effects of a Mobile
           Phone Short Message Service on Antiretroviral
           Treatment Adherence in Kenya (WelTel Kenya1): A
           Randomised Trial"   76
     4.6.9   Organizational Study: "Electronic Health Record
           Adoption in US Hospitals: The Emergence of a Digital
           'Advanced Use' Divide"   76
4.7   Summary and Conclusions   77
References   77

5   Study Planning   79
5.1   Introduction   80
5.2   Adopting an Evaluation Mindset   81
5.3   Why It May Not Work Out as Planned   83
     5.3.1   Sometimes Stakeholders Would Rather Not Know   84
     5.3.2   Differences in Values   84

5.4 Planning a Study 85
    5.4.1 Assessing Stakeholder Needs 85
    5.4.2 Choosing the Level of Evaluation 87
    5.4.3 Matching What Is Studied to the Stage in the Life Cycle .. 88
    5.4.4 Considering Context: Local Resources and Organizational
           Knowledge.. 89
    5.4.5 The Focus: Matching What Is Evaluated to the Type of
           Information Resource............................. 89
    5.4.6 The Search for Evidence 90
5.5 The Value of Evidence-Based Health Informatics.............. 90
5.6 Searching the Literature................................... 91
    5.6.1 Identifying a Search Topic......................... 92
    5.6.2 Frameworks and Theories 92
5.7 How to Conduct a Literature Search 93
5.8 Systematic Reviews and Narrative Meta-Analyses 95
    5.8.1 Systematic Reviews............................... 95
    5.8.2 Meta-Narrative Reviews 95
5.9 Project Management 96
    5.9.1 The Team.. 96
    5.9.2 Project Management Tools......................... 97
5.10 Summary 98
Answers to Self-Tests 98
References.. 99

Part II Quantitative Studies

6 The Structure of Quantitative Studies 103
6.1 Introduction 103
6.2 Elements of a Measurement Process 105
    6.2.1 Measurement 106
    6.2.2 Attribute.. 106
    6.2.3 Object and Object Class........................... 106
    6.2.4 Attribute–Object Class Pairs 107
    6.2.5 Attributes as "Constructs" 107
    6.2.6 Measurement Instruments 108
    6.2.7 Independent Observations 109
6.3 Key Object Classes and Types of Observations for Measurement in
     Informatics... 110
    6.3.1 Key Object Classes 110
    6.3.2 Key Categories of Observations.................... 111
6.4 Levels of Measurement 112
6.5 Importance of Measurement in Quantitative Studies........... 114
6.6 Measurement and Demonstration Studies 115
6.7 Goals and Structure of Measurement Studies 117
6.8 The Structure and Differing Terminologies of Demonstration
     Studies 120

6.9   Demonstration Study Designs   121
    6.9.1   Prospective and Retrospective Studies.   121
    6.9.2   Descriptive, Interventional, and Correlational Studies   122
6.10   Demonstration Studies and Stages of Resource Development   124
Answers to Self-tests.   125
References.   126

7   Measurement Fundamentals: Reliability and Validity   129
7.1   Introduction   130
7.2   The Classical Theory of Measurement: Framing Error as
    Reliability and Validity   131
7.3   Toward Quantifying Reliability and Validity   132
7.4   A Conceptual Analogy Between Measurement and Archery   133
7.5   Two Methods to Estimate Reliability.   135
7.6   Quantifying Reliability and Measurement Errors   137
    7.6.1   The Benefit of Calculating Reliability   137
    7.6.2   Enabling Reliability Estimation Using Objects-by-
        Observations Matrices   138
    7.6.3   The Standard Error of Measurement   139
7.7   Design Considerations in Measurement Studies to Estimate
    Reliability.   141
    7.7.1   Dependence of Reliability on Populations of Objects and
        Observations   141
    7.7.2   Effects on Reliability of Changing the Number of
        Observations in a Measurement Process   142
    7.7.3   Value of the Prophesy Formula: How Much Reliability Is
        Enough?   142
    7.7.4   Effects on Reliability of Changing the
        Number of *Objects* in a Measurement Process   143
7.8   Validity and Its Estimation.   144
    7.8.1   Distinguishing Validity from Reliability   144
    7.8.2   Content Validity   146
    7.8.3   Criterion-Related Validity   147
    7.8.4   Construct Validity   148
7.9   Concluding Observations About Validity.   150
7.10   Generalizability Theory: An Extension of Classical Theory   151
Answers to Self-Tests   152
References.   154

8   Conducting Measurement Studies and Using the Results.   155
8.1   Introduction   155
8.2   Process of Measurement Studies   156
8.3   How to Improve Measurement Using Measurement
    Study Results   158

8.4   Using Measurement Study Results to Diagnose
        Measurement Problems 159
        8.4.1   Analyzing the Objects-by-Observations Matrix 159
        8.4.2   Improving a Measurement Process 162
8.5   The Relationship Between Measurement
        Reliability and Demonstration Study Results 164
        8.5.1   Effects of Measurement Error in Descriptive and
                Interventional Demonstration Studies 165
        8.5.2   Effects of Measurement Error in Correlational
                Demonstration Studies........................... 166
8.6   Demonstration Study Results and Measurement Error:
        A Detailed Example.................................... 168
        8.6.1   Reliability Estimate 171
        8.6.2   Demonstration Study Correcting the Correlation for
                Attenuation..................................... 171
        8.6.3   Additional Considerations from This Example 172
8.7   Conclusion 174
Answers to Self-Tests 174
References... 176

9   Designing Measurement Processes and Instruments............. 177
9.1   Introduction 178
9.2   When Judges' Opinions Are the Primary Observations. 180
        9.2.1   Sources of Variation Among Judges 181
        9.2.2   Number of Judges Needed for Reliable Measurement 182
        9.2.3   Improving Measurements That Use
                Judges as Observations 182
9.3   When Tasks Are the Primary Observation................... 184
        9.3.1   Sources of Variation Among Tasks 185
        9.3.2   Number of Tasks Needed for Reliable Measurement 187
        9.3.3   Improving Measurements That Use
                Tasks as Observations 188
9.4   When Items Are the Primary Observation................... 194
        9.4.1   Sources of Variation Among Items 194
        9.4.2   Number of Items Needed for Reliable Measurement 196
        9.4.3   Improving Measurements That Use
                Items as Observations 196
        9.4.4   The Ratings Paradox 199
9.5   Conclusion 200
Answers to Self-Tests 201
References... 202

10   Conducting Demonstration Studies........................... 205
10.1   Introduction 206
10.2   Overview of Demonstration Studies 207
        10.2.1 Types of Demonstration Studies.................... 207

10.2.2   Terminology for Demonstration Studies   208
10.2.3   Demonstration Study Types Further Distinguished   210
10.2.4   Internal and External Validity of
         Demonstration Studies   212
10.3   The Process of Conducting Demonstration Studies.   213
10.4   Threats to Internal Validity That Affect All
       Demonstration Studies, and How to Avoid Them   214
       10.4.1   Assessment Biases and the Hawthorne Effect   214
       10.4.2   Incomplete Response and Missing Data.   215
       10.4.3   Insufficient Data to Allow a Clear
                Answer to the Study Question   216
       10.4.4   Inadvertently Using Different Data Items or Codes.   217
10.5   Threats to the External Validity and How to Overcome Them . . .   217
       10.5.1   The Volunteer Effect.   218
       10.5.2   Age, Ethnic Group, Gender and Other Biases   218
       10.5.3   Study Setting Bias   219
       10.5.4   Implementation Bias.   219
       10.5.5   Training Set Bias and Overfitting.   219
10.6   Analysis Methods for Descriptive Studies That Also Apply to All
       Study Types   221
       10.6.1   Graphical Portrayal of the Results   222
       10.6.2   Indices of Central Tendency   222
       10.6.3   Indices of Variability.   222
10.7   Conclusions   224
Answers to Self-Tests   225
References.   227

11   **Design of Interventional Studies**   229
     11.1   Introduction   229
     11.2   Challenges When Designing Interventional Studies to Attribute
            Causation   230
            11.2.1   Secular Trends   231
            11.2.2   Regression to the Mean   232
            11.2.3   Association Is Not Causation   233
     11.3   Control Strategies for Interventional Studies.   233
            11.3.1   Uncontrolled Study.   233
            11.3.2   Historically Controlled or Before-After Studies   234
            11.3.3   Simultaneous External Controls.   235
            11.3.4   Internally and Externally Controlled Before-After
                     Studies   236
            11.3.5   Simultaneous Randomized Controls   237
            11.3.6   Cluster Randomized and Step Wedge Designs.   238
            11.3.7   Matched Controls as an Alternative to
                     Randomization, and the Fallacy of Case-Control Studies   239
            11.3.8   Summary of Control Strategies   240

11.4 Biases and Challenges to the Internal Validity of
Interventional Studies 241
11.4.1 The Placebo Effect 241
11.4.2 Allocation and Recruitment Biases 242
11.4.3 Biases Associated with Data Collection............. 243
11.4.4 Carryover Effect 244
11.4.5 Complex Interventions and Associated Biases........ 244
11.5 Statistical Inference: Considering the Effect of Chance 248
11.5.1 Effect Size and "Effectiveness" 248
11.5.2 Statistical Inference 249
11.5.3 Study Power 250
11.5.4 Data Dredging 251
11.6 Conclusions 251
Answers to Self-Tests 253
References... 256

12 Analyzing Interventional Study Results 259
12.1 Introduction 259
12.2 Grand Strategy for Analysis of Study Results 260
12.3 Analyzing Studies with Discrete Independent and
Discrete Dependent Variables: Basic Techniques 262
12.3.1 Contingency Tables 262
12.3.2 Using Contingency (2 × 2) Tables: Indices of Effect Size 263
12.3.3 Chi-Square and Fisher's Exact Test for Statistical
Significance 265
12.3.4 Cohen's Kappa: A Useful Effect Size Index.......... 266
12.4 Analyzing Studies with Discrete Independent and
Discrete Dependent Variables: ROC Analysis............... 267
12.4.1 Estimating the Performance of a
Classifier Using a Single Cut Point 267
12.4.2 ROC Analysis................................. 268
12.5 Analyzing Studies with Continuous Dependent
Variables and Discrete Independent Variables............... 270
12.5.1 Two Group Studies with Continuous
Outcome Variables 270
12.5.2 Alternative Methods for Expressing Effect Size 271
12.5.3 Logic of Analysis of Variance (ANOVA)............ 272
12.5.4 Using ANOVA to Test Statistical Significance........ 273
12.5.5 Special Issues with ANOVA...................... 275
12.6 Analyzing Studies with Continuous Independent
and Dependent Variables 277
12.6.1 Studies with One Continuous Independent
Variable and One Continuous Dependent Variable 277
12.6.2 Studies with Multiple Independent Variables
and a Continuous Dependent Variable 278

12.6.3 Relationships to Other Methods 279
12.7 Choice of Effect Size Metrics: Absolute Change,
Relative Change, Number Needed to Treat 280
12.8 Special Considerations when Studying Predictive Models 281
12.8.1 Use of Training and Testing Datasets 281
12.8.2 Some Cautionary Tales About Studying Predictive
Models 282
12.9 Conclusions 283
Answers to Self-Tests 284
References 286

13 Designing and Carrying Out Correlational Studies
Using Real-World Data 289
13.1 Introduction 289
13.2 Reasons for Carrying Out Correlational Studies 290
13.3 Types of Data Used in Correlational Studies 292
13.4 Advantages and Disadvantages of Correlational
Studies Using Routine Data 293
13.4.1 Advantages 293
13.4.2 Disadvantages 294
13.5 How to Carry Out Correlational Studies Using Person-Level
Datasets 295
13.5.1 Agreeing to the Study Questions, Cohorts, and Study
Variables 296
13.5.2 Identify the Datasets, Code Sets, and Data Analysis
Routines 296
13.5.3 Develop and Finalize the Data Analysis Plan 297
13.5.4 Applying for Permissions 298
13.5.5 Virtual Data Enclave (Virtual Safe Haven) 298
13.5.6 Data Linkage 298
13.5.7 Data Cleaning 299
13.5.8 Estimate the Size and Direction of Potential Biases 300
13.5.9 Test and Refine the Data Analysis Plan and Carry Out the
Data Analysis 300
13.5.10 Carry Out Sensitivity Analyses 300
13.5.11 Check for Statistical Disclosure Risk 301
13.5.12 Disseminate the Study Results 301
13.6 The Challenge of Inferring Causation from Association 301
13.6.1 Spurious Correlation 302
13.6.2 Bradford Hill's Criteria to Support
Causality in Correlational Studies 303
13.7 Factors Leading to False Inference of Causation 304
13.7.1 Selection Bias 306
13.7.2 Information Bias 307
13.7.3 Confounding 309

13.7.4 Some General Methods to Detect or Control for
Confounding. 313
13.8 Advanced Analysis Methods for Correlational Studies. 314
13.8.1 The Case-Crossover Design 314
13.8.2 Instrumental Variable Analysis. 315
13.8.3 Regression Discontinuity Design 316
13.9 Conclusions 317
Answers to Self-Tests 319
References. 323

Part III   Qualitative Studies

14   An Introduction to Qualitative Evaluation Approaches 329
14.1 Introduction 329
14.2 Definition of the Qualitative Approach 330
14.3 Motivation for Qualitative Studies: What People Really
Want to Know 332
14.4 Support for Qualitative Approaches. 333
14.5 Why Are Qualitative Studies Especially Useful in Informatics?. . 335
14.6 When Are Qualitative Studies Appropriate? 336
14.7 Rigorous, but Different, Methodology. 337
14.8 Qualitative Approaches and Their Philosophical Premises 338
14.9 Natural History of a Qualitative Study 340
14.9.1 Negotiation of the "Ground Rules" of the Study 340
14.9.2 Immersion into the Environment and Initial Data
Collection 340
14.9.3 Iterative Qualitative Loop. 341
14.9.4 Preliminary Report 342
14.9.5 Final Report 343
14.10 Summary: The Value of Different Approaches 343
14.11 Two Example Summaries 344
Answers to Self-Tests 345
References. 346

15   Qualitative Study Design and Data Collection. 347
15.1 Introduction 347
15.2 The Iterative Loop of Qualitative Processes 349
15.3 Theories, Frameworks, and Models that Can Guide the Process. . 350
15.4 Strategies for Study Rigor 353
15.4.1 Reflexivity. 353
15.4.2 Triangulation. 354
15.4.3 Member Checking 356
15.4.4 Data Saturation. 356
15.4.5 Audit Trail 357
15.5 Specifics of Conducting Qualitative Studies 359
15.5.1 Site and Informant Selection 360

15.5.2 The Team 361
15.6 Techniques for Data Collection 362
  15.6.1 Entering the Field 362
  15.6.2 Interviews 363
  15.6.3 Focus Groups 367
  15.6.4 Observations 368
  15.6.5 Field Notes 370
  15.6.6 Gathering Naturally Occurring Data 372
15.7 Toolkits of Complementary Techniques 374
  15.7.1 Case Studies 374
  15.7.2 Action Research 375
  15.7.3 Rapid Assessment Process 375
  15.7.4 Virtual Ethnography Techniques 376
15.8 The Future of Qualitative Data Gathering Methods 377
Answers to Self-Tests 377
References 379

16 Qualitative Data Analysis and Presentation of Analysis Results 381
16.1 Introduction 381
16.2 What Is Qualitative Analysis? 382
16.3 Approaches to Analysis 382
16.4 Qualitative Data Management 385
16.5 Qualitative Analysis Software 386
16.6 How to Code 387
16.7 How to Interpret 391
16.8 Other Analysis Approaches 392
  16.8.1 Narrative Analysis 392
  16.8.2 Analysis of Naturally Occurring Text or Visual Data ... 393
16.9 Using Graphics in the Analysis and Reporting Processes 394
16.10 Evaluating the Quality of Qualitative Analysis 396
16.11 The Future of Qualitative Analysis 397
Answers to Self-Tests 397
References 398

Part IV Special Study Types

17 Mixed Methods Studies 403
17.1 Introduction 403
17.2 The Ultimate Triangulation 404
17.3 Why Use Mixed Methods 405
17.4 Designing Mixed Methods Studies 406
17.5 Data Collection Strategies 410
17.6 Data Analysis Strategies 410
17.7 Presenting Results of Mixed Methods Studies 411
17.8 Usability Studies as Exemplars of Mixed Methods 412

17.8.1 What Is Usability? 412
17.8.2 Importance of Usability 413
17.8.3 Usability Testing Methods 413
17.8.4 Examples of Usability Studies in Informatics 414
17.9 Organizational Systems-Level Studies 416
17.9.1 What Are Organizational Systems-Level Studies?...... 416
17.9.2 Methods Used in Organizational and System-Level
Studies 416
17.10 Special Ethical Considerations in Mixed Method Studies....... 418
17.11 The Future of Mixed Methods Inquiry 418
Answers to Self-Tests....................................... 419
References... 420

**18 Principles of Economic Evaluation and Their Application to
Informatics.. 423**
18.1 Introduction to Economic Studies 423
18.1.1 Motivation for Economic Analysis.................. 424
18.2 Principles of Economic Analysis......................... 425
18.2.1 Types of Cost Studies........................... 426
18.3 Conducting an Economic Analysis 428
18.3.1 The Perspective of the Analysis 428
18.3.2 Time Frame of the Analysis 430
18.4 Definition and Measurement of Costs 431
18.4.1 Why Costs Do Not Equal Charges................. 431
18.4.2 Mechanics of Cost Determinations 433
18.5 Definition and Measurement of Outcomes and Benefits........ 434
18.5.1 Matching the Correct Outcome Measure to the Problem. 434
18.5.2 Adjusting for Quality of Life 435
18.6 Cost-Minimizing Analysis............................... 435
18.7 Cost-Effectiveness Analysis............................. 437
18.7.1 The Incremental Cost-Effectiveness Ratio 438
18.7.2 Cost-Benefit Analyses: The Cost-Benefit Ratio 440
18.7.3 Discounting Future Costs and Benefits 442
18.8 Sensitivity Analysis 444
18.8.1 Structural Sensitivity Analyses.................... 444
18.8.2 Parameter Estimate Sensitivity Analyses 445
18.9 Special Characteristics of Cost-Effectiveness Studies in Biomedical
and Health Informatics................................. 447
18.9.1 System Start Up and Existing Infrastructure Costs 447
18.9.2 Sharing Clinical Information System Costs 448
18.10 Critically Appraising Sample Cost-Effectiveness Studies in
Biomedical and Health Informatics....................... 448
18.11 Conclusion 450
Answers to Self-Tests....................................... 451
References... 452

Part V Ethical, Legal and Social Issues in Evaluation

19 Proposing Studies and Communicating Their Results 457
  19.1 Introduction 457
  19.2 Writing Evaluation Proposals 458
    19.2.1 Why Proposals Are Necessary and Difficult 458
    19.2.2 Format of a Study Proposal 460
    19.2.3 Suggestions for Expressing Study Plans 460
    19.2.4 Special Issues in Proposing Evaluations 464
  19.3 Refereeing Evaluation Studies. 465
  19.4 Communicating the Results of Completed Studies 467
    19.4.1 What Are the Options for Communicating
         Study Results? 467
    19.4.2 What Is the Role of the Study Team: Reporter or Change
         Agent?. 469
    19.4.3 Role of the Evaluation Contract 470
  19.5 Specific Issues of Report Writing 471
    19.5.1 The IMRaD Format for Writing Study Reports 471
    19.5.2 Writing Qualitative Study Reports 472
  19.6 Conclusion. 473
  References. 474

20 Ethics, Safety, and Closing Thoughts 475
  20.1 Introduction 475
  20.2 Broad Ethical Issues in ICT Evaluation. 476
    20.2.1 Standard Practices for the Protection of Human Subjects 476
    20.2.2 The Safety of Implementing ICT 479
    20.2.3 Encouraging Diversity 480
  20.3 Ethical Issues During Each Phase of a Study 481
    20.3.1 Planning for Ethical Evaluation 481
    20.3.2 The Data Gathering and Storage Phase 484
    20.3.3 Analysis and Interpretation Phase 487
    20.3.4 Reporting Results Phase. 487
  20.4 Ethical Considerations for the Future 488
  20.5 Closing Thoughts About the Future of Informatics Evaluation. . . 489
    20.5.1 What Will Be Studied. 489
    20.5.2 Future Trends in Evaluation Methods 492
  Answers to Self-Tests 493
  References. 494

Glossary 497

Index. 521

Part I
Landscape of Evaluation in Informatics

Chapter 1
What Is Evaluation and Why Is It Challenging?

Learning Objectives

The text, examples, self-tests, and Food for Thought questions in this chapter will enable the reader to:

1. Distinguish between evaluation and research and explain why both are examples of empirical studies.
2. Explain why evaluation plays an important role in the field of biomedical and health informatics.
3. Identify the application domain which a given information resource primarily addresses.
4. For a given evaluation study scenario, identify the primary reason or purpose motivating the conduct of the study.
5. Explain why complexities specific to biomedical and health informatics create challenges for teams performing evaluation studies.

1.1 Introduction

The field of biomedical and health informatics addresses the collection, management, processing, and communication of information related to health care, biomedical research, personal health, public and population health, and education of health professionals. Informaticians focus much of their work on "information resources" that carry out these functions (Shortliffe and Cimino 2014).

This chapter introduces many of the topics that are explored in more detail in later chapters of this book. It gives a first definition of evaluation, distinguishes it from research, describes why evaluation is needed, discusses who is involved, and

© Springer Nature Switzerland AG 2022
C. P. Friedman et al., *Evaluation Methods in Biomedical and Health Informatics*, Health Informatics, https://doi.org/10.1007/978-3-030-86453-8_1

notes some of the challenges inherent to evaluation in biomedical and health informatics that distinguish it from evaluation in other areas. In addition, the chapter lists some of the many types of information systems and resources, the questions that can be asked about them, and the various perspectives of those concerned.

1.2 The Place of Evaluation Within Informatics

Biomedical and health informatics draws its methods from many disciplines and from many specific lines of creative work within these disciplines. The fields underlying informatics are what may be called its basic sciences. They include, among others, computer science and engineering, information science, cognitive science, decision science, implementation science, organizational science, statistics, epidemiology and linguistics. One of the strengths of informatics has been the degree to which individuals from these different disciplinary backgrounds, but with complementary interests, have learned not only to coexist, but also to collaborate productively. Formal training opportunities in informatics, drawing on these basic sciences, are now widespread across the world (Florance 2020).

This diverse intellectual heritage for informatics can make it difficult to define creative or original work in the field (Friedman 1995). The "tower" model, shown in Fig. 1.1, asserts that creative work in informatics occurs at four levels that build on one another.

Creative work at every level of the tower can be found on the agenda of professional meetings in informatics and published in journals within the field. The topmost layer of the tower embraces empirical studies—studies based on data—of information resources that have been developed using abstract models and perhaps deployed in settings of ongoing health care, personal health, public health, research, or education. Because informatics is so intimately concerned with the improvement of health of individuals and populations, the benefits and value of resources produced by the field are matters of significant ongoing interest. Studies occupy the topmost layer because they rely on the existence or anticipation of models, systems, and settings where the work of interest is underway. There must be *something* to study. As will be seen later, studies of information resources usually do not await the ultimate deployment of these resources. Even conceptual models may be studied empirically, and information resources themselves can be studied through successive stages of their development.

Studies occupying the topmost level of the tower model are the focus of this book. Empirical studies include measurement and observations of the performance of information resources and the behavior of people who in some way use—or might use—these resources, with emphasis on the interaction between

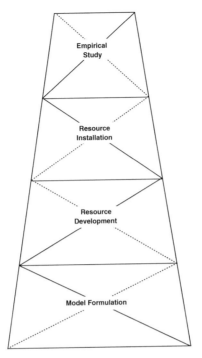

Fig. 1.1 Tower model. (Adapted from the *Journal of the American Medical Informatics Association*, with permission)

the resources and the people and the consequences of these interactions. Included under empirical studies are activities that have traditionally been called "evaluation" and "research". The difference between evaluation and research, as discussed in more detail in a later section, is subtle but important. While both employ methods grounded in the collection and analysis of data, evaluations are shaped by questions of interest to audiences separate from members of the team conducting a study. While almost always influenced by the work of others, research studies are guided by the members of the study team itself. As discussed in the Preface to this book, "evaluation" has been included in the title instead of "empirical methods" or "research" for several reasons: the historical use of the term "evaluation" in the field; the culture of informatics which places high value on pragmatic questions and the use of study results to inform development and improvement of computing artifacts; and societal requirements to know that artifacts deployed in the world are safe and effective. Even though evaluation studies address the needs of specific audiences, their results also contribute to a growing evidence base for the field of informatics, as will be discussed in Chap. 5.

Fig. 1.2 A "fundamental theorem" of biomedical and health informatics

Another way to look at the role of evaluation in biomedical and health informatics is to consider the "inequality" illustrated in Fig. 1.2. This inequality has been proposed as a "Fundamental Theorem" of biomedical and health informatics (Friedman 2009).

The theorem suggests that the goal of informatics is to deploy information resources that help persons—clinicians, students, scientists, and, increasingly, all members of society—do whatever they do "better" than would be the case without support from these information resources. Most individuals who work with information and communication technology in health and biomedicine believe that the inequality in the theorem will hold when the information resources have been properly designed and implemented. Seen in this light, many, but not all, evaluation studies examine whether the fundamental theorem is satisfied. When the theorem is not satisfied, studies can suggest helpful modifications in the information resources themselves or the ways in which they are used. The theorem also reminds study teams and their audiences that informatics is as much, or more, about the people using and organizations deploying information resources as it is about the resources themselves.

1.3   Framing Evaluation as Answering Questions

The famed cultural icons of the Twentieth Century, Gertrude Stein and Alice B. Toklas, may have foreseen the challenge of informatics evaluation in a conversation that occurred when Ms. Stein was on her deathbed. Ms. Stein asked Ms. Toklas: "What is the answer?" Ms. Toklas hesitated, not knowing what to say to her dying friend. Sensing her hesitation, Ms. Stein went on to say to Alice: "In that case, what is the question?" (The New York Times 1967).

Building on this sage advice, from the outset it is useful to think of evaluation in terms of identifying questions that are important to key stakeholders, and then finding the best possible means to answer them. As it applies to biomedical and health informatics, this problem can initially be expressed as the need to answer a basic set of questions. To the inexperienced, these questions might appear deceptively simple.

- A biomedical or health information resource is developed. Is this resource performing as intended? How can it be improved?
- Subsequently, the resource is introduced into a functioning clinical, scientific or educational environment. Again, is it performing as intended, and how can

it be improved? Does it make any difference, in terms of clinical, scientific or educational practice? Are the differences it makes beneficial? Are the observed affects those envisioned by the developers or are there unexpected effects?

Note that we can append "why, or why not?" to each of these questions. And there are of course many more potentially interesting questions than have been listed here.

Out of this multitude of possible questions comes the first challenge for anyone planning an evaluation study: to identify the most appropriate set of questions, of interest to often multiple audiences, to explore in a particular situation. The questions to study in any particular situation are not inscribed in stone. Many more questions about a specific information resource can be stated than can be explored, and it is often the case that the most interesting questions reveal themselves only after a study has commenced. Adding further complexity, evaluation studies are inextricably political, both in their design and in the interpretation and use of their results. There are usually legitimate differences of opinion over the relative importance of particular questions. Before any data are collected, those conducting a study may find themselves in the role of referee between competing views and interests as to what questions should be on the table for exploration.

Even when consensus questions can be stated in advance, they can be difficult to answer persuasively. Some would be easy to answer if a unique kind of time machine, an "evaluation machine," were available.

As imagined in Fig. 1.3, an evaluation machine would somehow enable all stakeholders to see how a work or practice environment would appear if the information resource had never been introduced.[1] By comparing real history with insights from an evaluation machine, accurate conclusions can potentially be drawn about the effects of the resource. Even if an evaluation machine existed, however, it could not answer all questions of possible interest. For example, it could not explain why these effects occurred or how to make the resource better. To obtain this information, it is necessary to communicate directly with the actors in the real history to understand how they used the resource and their views of that experience. There is usually more to a complete evaluation than demonstration of outcomes.

In part because there is no evaluation machine, but also because ways are needed to answer the wider range of important questions for which the machine would not help, there can be no simple solution to the problem of evaluation. There is, instead, an interdisciplinary field of evaluation with an extensive methodological literature that will be introduced in Chap. 2. This literature details many diverse approaches to evaluation, all of which are currently in use, reflecting the full range of study purposes and questions that can be addressed. These approaches will be introduced

---

[1] Readers familiar with methods of epidemiology may recognize the "evaluation machine" as an informal way of portraying the counterfactual approach to the study of cause and effect. More details about this approach may be found in a paper by Maldonado (Maldonado 2016).

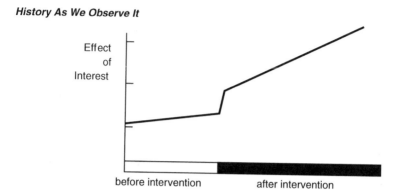

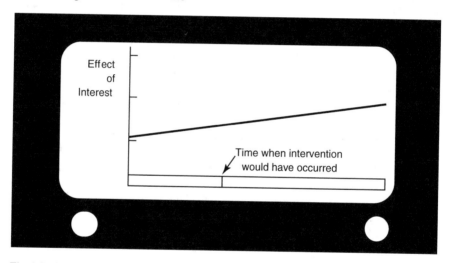

Fig. 1.3 Hypothetical "evaluation machine"

in the next chapter. The approaches differ in the kinds of questions that are seen as primary, how specific questions get onto the agenda, and the data collection and analysis methods used to answer these questions. In informatics, it is especially important that such a range of methods is available because the questions of interest vary dramatically—from the focused and outcome-oriented (Does implementation of this resource improve health?) to very pragmatic, market-oriented questions, such as those famously stated by Barnett in relation to clinical information resources[2]:

1. Is the system used by real people with real patients?

---

[2] These questions were given to the authors in a personal communication on December 8, 1995.

2. Is the system being paid for with real money?
3. Has someone else taken the system, modified it, and claimed they developed it?

Evaluation is challenging in large part because there are so many questions and options for addressing them, and there is almost never an obvious best way to proceed. The following points bear repeating:

1. In any evaluation setting, there are many potential questions to address. What questions are asked shapes (but does not totally determine) the answers that are generated.
2. There may be little or no consensus on what constitutes the best set of questions, or how to prioritize questions.
3. There are often many different ways to address individual questions, each with advantages and disadvantages.
4. There is no such thing as a "perfect study".

As a result, individuals conducting evaluations are in a continuous process of compromise and accommodation. At its root, the challenge of evaluation is to collect and communicate information that is useful to one or more stakeholders while acting in this spirit of compromise and accommodation.

1.4 Examples of Information Resources

As noted in the Preface, the field of biomedical and health informatics centers on "information resources" supporting the collection, management, processing, and communication of information related to health care, biomedical research, personal health, public and population health, and education of health professionals. Information resources take many forms to suit these many purposes. While the information resources of primary interest in this book employ digital technology—and thus consist of computing hardware, software, and communication components—resources that are not digital can also enter the picture. Printed journals and learned colleagues are important examples of information resources that do not employ digital technology.

An information resource can support one or more health-related application domains. A partial list of health-related information resources employing digital technology, for each of five application domains, includes:

- For *health care applications*, such information resources include systems to collect, store, and communicate information about individual patients (electronic health records) as well as systems to reason and generate advice based on computable medical knowledge (decision-support systems).
- For basic, translational, and clinical *research applications*, information resources include databases spanning levels from scale from small molecules to complete organisms; algorithms to analyze these data within and across levels of scale, gener-

ating models that can explain phenomena not previously understood.[3] Other resources supporting research promote the development and execution of clinical trials.

* For *personal health*, pertinent information resources include applications ("apps") accessible from smartphones, watches and other portable devices that collect and manage information necessary to enhance health and well-being.
* For *public and population health* applications, information resources collect and manage information related to vital statistics, as well as information at higher levels of aggregation including communities, jurisdictions (counties, states, or provinces) to enable health status tracking and improvement interventions at these higher levels of geographical scale.
* To support the *education of health professionals*, pertinent resources include multimedia platforms for delivering educational content, learning management systems with associated analytical capabilities, and simulations employing increasingly sophisticated virtual and augmented reality technology.

There is clearly a wide range of biomedical and health information and communication resources to evaluate, and each information resource has multiple aspects that can be studied. The technically minded might focus on the system components, asking such questions as: "Does the resource have an appropriately modular structure?" or "How quickly can the resource recover from a power interruption?" Users of the resources might ask more pragmatic questions, such as: "Is the information in this resource completely up-to-date?" or "How much time must I invest in becoming proficient with this resource, and will it do anything to help me personally?" Those with management responsibilities might wish to understand the impact of these resources on the missions of their organizations, asking questions such as: "How well does this electronic patient record support quality improvement activities?" or "How will sophisticated educational simulations change the role of faculty as teachers?" Thus, evaluation methods in biomedical and health informatics must address not only a wide range of different information resources, but also a wide range of questions about them.

This book does not exhaustively describe how each possible evaluation method can be used to answer each kind of question about each kind of information resource. Instead, it describes the range of available techniques and focuses on those that seem most useful to answer the questions that frequently arise. It describes in detail methods, techniques, study designs, and analysis methods that apply across a range of evaluation purposes. These evaluation methods are applicable to the full range of information resources described in the bulleted list above.

In the language of software engineering, the focus in this book places more emphasis on software validation than software verification. Validation studies explore whether the "right" information resource was built, which involves both determining that the original specification was fit to the problem it was built to address and that the resource as built is performing to specification. By contrast, verification means checking whether the resource was built to specification. For a

---

[3] The authors use the term "bioinformatics" to refer to the use of information resources in support of biological research, and more specifically molecular biology and genomics.

software engineering perspective on evaluation, the reader can refer to several contemporary texts (Somerville 2016).

Self-Test 1.1
Identify each of the following information resources as *primarily* supporting either health care, personal health, research, public/population health, or education.

1. A mobile phone app that measures and reports amounts of daily exercise.
2. A geographic information system that tracks a disease outbreak by census tracts.
3. A system that helps discover optimal drug therapies based on individual genotypes.
4. A system that recommends to clinicians optimal drug therapy based on the genotype of an individual patient.
5. A system that represents human anatomy as three-dimensional digital images.
6. A system that maintains records of vaccination status of all residents of a country.

1.5   Further Distinguishing Evaluation and Research

As discussed previously, evaluation and research use the same empirical methods. The primary difference is in the binding of the evaluation study team to an "audience," who may be one or two individuals, a large group, or several groups of stakeholders who share a "need to know" but may be interested in many different things. It is what the audience wants to know—not what the study team wants to know—that largely determines the questions that an evaluation study examines. Successful evaluation studies illuminate questions of interest to the audience. The relationship between a study team and its audience corresponds in many ways to that between attorneys and their clients.

By contrast, a research study team's allegiance is usually to a focused set of questions or methods that in many ways define the team. The primary audience for a research study team is other teams interested in these same or closely related questions. A research team's scientific agenda builds on the discoveries of related teams, and in turn contributes to the agendas of the other teams.

Although many important scientific discoveries have been accidental, researchers as a rule do not actively seek out unanticipated effects. However, evaluation study teams often do exactly that. Whereas researchers usually value focus and seek to exclude from their study as many extraneous variables as possible, evaluation study teams usually seek to be comprehensive. A complete evaluation of an information resource focuses on developmental as well as in-use issues. In research, laboratory studies often carry more credibility because they are conducted under controlled circumstances and they can illuminate cause and effect relatively unambiguously. During evaluation, field studies often carry more credibility because they illustrate more directly (although perhaps less definitively) the utility and impact of the resource. Researchers can afford to, and often must, lock themselves into a single paradigm for data collection and analysis. Even within a single study, evaluation study teams often employ many paradigms.

1.6   Reasons for Performing Evaluations

Like any complex, time-consuming activity, evaluation can serve multiple purposes. There are at least five major reasons for evaluating biomedical and health information resource (Wyatt and Spiegelhalter 1991).

1. *Promotional*: The use of information resources in health and biomedicine, requires assurance to clinicians, patients, researchers, and educators that these resources are safe and bring benefit to persons, groups, and institutions through improved cost-effectiveness or safety, or perhaps by making activities possible that were not possible before.
2. *Scholarly*: Despite the distinction between evaluation and research, the methods and results of many evaluation studies make them suitable for publication. Publication enriches the evidence base of informatics and can confirm or question the principles that lie at the foundation of the field.
3. *Pragmatic*: Without evaluating the resources they create, developers can never know which techniques or methods are more effective; and very important, how to improve their resources over successive cycles of development; or why certain approaches failed.
4. *Managerial*: Before using an information resource, clinicians, researchers, educators, and administrators must be satisfied that it is functional and be able to justify its use in preference to alternative information resources and the many other innovations that may compete for the same budget.
5. *Medical-legal*: To market an information resource and potentially reduce the risk of liability, developers should obtain accurate information to allow them to label it correctly, safely, and effectively. Clinicians and professional users need evaluation results to inform their professional judgment before and while using these resources. Doing so would support the argument that these users are "learned intermediaries." An information resource that treats the users merely as automatons, without human intervention, risks being judged by the stricter laws of product liability. However, it is worth noting that several US courts have struck down the liability protection afforded by this "learned intermediary doctrine". Nonetheless, providing professional users with evaluation results mitigates the risk of actually causing harm, and thus can work to reduce the potential for liability (O'Connor 2017; Ratwani et al. 2018).

Awareness of the major reasons for conducting an evaluation often helps frame the major questions to be addressed and reduces the possibility that the focus of the study will be misdirected.

Self-Test 1.2

In each of the following scenarios, imagine that a colleague has approached you and described a need for an evaluation study. Based on each description, is your colleague describing a study motivation that is primarily promotional, scholarly, pragmatic, managerial, or medical-legal?

Scenario 1: Your colleague says: "Our researchers are complaining that they can't use the digital laboratory notebook solution we have deployed and are going back to using paper. We need to know how to improve the digital notebook."

Scenario 2: Your colleague says: "We have deployed a sepsis prediction algorithm that has been shown to be highly predictive of sepsis in many studies, but our clinicians are still reluctant to use it. They need to know that it works here."

Scenario 3: Your colleague says: "Our current learning management system runs slowly and "goes down" (becomes non-functional) frequently. We need to convince the administration to invest in a new one."

Scenario 4: Your colleague (from another institution) says: "I'm really glad you published your study of that clinical trials management system. We're deploying the same system and learned a lot about how to do it."

1.7 Evaluation in Informatics as a Complex Socio-Technical Endeavor

The evaluation of biomedical information resources lies at the intersection of three areas, each notorious for its complexity (Fig. 1.4): health and biomedicine as a field of human endeavor, information resources using contemporary technology, and the methodology of evaluation itself. The sections that follow address the challenges deriving from combinations of these three sources of complexity.

Fig. 1.4 Venn diagram summarizing the sources of complexity for evaluation in biomedical and health informatics

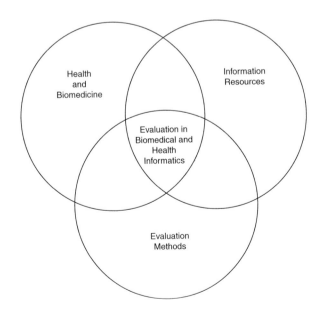

1.7.1   Studying People

Evaluation studies, as envisioned in this book, do not focus solely on the structure and function of information resources; they also address their impact on persons who are customarily users of these resources and on the outcomes of users' interactions with them. To understand users' actions, study teams must confront the gulf between peoples' private opinions, public statements, and actual behavior (Abroms et al. 2014). What is more, there is clear evidence that the mere act of studying human performance changes it, a phenomenon usually known as the Hawthorne effect (Roethligsburger and Dickson 1939). Finally, humans vary widely in their responses to stimuli, from minute to minute and from one person to another. Thus, evaluation studies of biomedical or health information resources require analytical tools from the behavioral and social sciences, biostatistics, and other scientific fields dedicated to "human" problems where high variability is the rule rather than the exception.

1.7.2   Professional Practice and Culture

Health care providers practice under strict legal, regulatory and ethical obligations to give their patients the best care available, to do them no harm, keep them informed about the risks of all procedures and therapies, and maintain confidentiality. These obligations often impinge on the design of evaluation studies. For example, because healthcare workers have imperfect memories and patients take holidays and participate in the unpredictable activities of real life, it is impossible to impose a strict discipline for data recording, and study data are often incomplete. Before any studies of a deployed information resource can begin, healthcare workers and patients are entitled to a full explanation of the possible benefits and disadvantages of the resource (Musen et al. 2021). Even if the study is a randomized trial, since half the patients or healthcare workers will be allocated randomly to the intervention group, all need to be counseled and give their consent prior to being enrolled in the study. Moreover, practice culture in biomedicine and health is characterized by a high degree of autonomy, particularly among those at higher levels of the professional hierarchy. In such a cultural environment, "assigning" anyone to an experimental group that constrains their professional practice may be anathema (Rafferty et al. 2001).

Similar challenges to evaluation apply to the educational domain. For example, students in professional schools are in a "high-stakes" environment where grades and other measures of performance can shape the trajectory of their future careers. Students will be understandably attracted to information resources they perceive as advantageous to learning and averse to those they perceive to offer little or no benefit at a great expenditure of time. Randomization or other means of arbitrary assignment of students to groups, for purposes of evaluative experiments, may also be seen as anathema. Similar feelings may be seen in biological research, where, for example, competitiveness among laboratories may mitigate against controlled studies of information resources deployed in these settings.

Professionals are often more skeptical of new practices than existing ones, and tend to compare an innovation with an idealization of what it is replacing (Mun et al. 2006). The standard required for proving the effectiveness of computer-based information resources may be inflated beyond that required for other methods or technologies. This leads to what may be called an "evaluation paradox": Professional clinicians, scientists, and educators will not adopt a new technology until its effectiveness has been demonstrated on a large scale; but demonstrations on a large scale are not possible in the absence of a significant level of adoption.

1.7.3   Safety and Regulation

Regulations apply to those developing or marketing clinical therapies, medical devices or investigational technology—areas where safety is critical. It is now clear in the United States and the United Kingdom that these regulations apply to any computer-based information resource that fits the descriptions in the box below, and not only to closed loop devices that manage patients directly, without a human intermediary.

The United States Food & Drug Administration (FDA) Definition of a Medical Device (Federal food, drug, and cosmetic act 2021)
The term "device" (except when used in [select other sections of the act]) means an instrument, apparatus, implement, machine, contrivance, implant, in vitro reagent, or other similar or related article, including any component, part, or accessory, which is:

- (A) recognized in the official National Formulary, or the United States Pharmacopeia, or any supplement to them,
- (B) intended for use in the diagnosis of disease or other conditions, or in the cure, mitigation, treatment, or prevention of disease, in man or other animals, or
- (C) intended to affect the structure or any function of the body of man or other animals, and

which does not achieve its primary intended purposes through chemical action within or on the body of man or other animals and which is not dependent upon being metabolized for the achievement of its primary intended purposes. The term "device" does not include software functions excluded pursuant to… [another section of the act].

The US FDA also offers an approach to classifying such medical devices (United States Food and Drug Administration (FDA) 2021).

The United Kingdom Medicines and Healthcare Regulatory Agency Definition of a Medical Device (United Kingdom Medicines and Healthcare products Regulatory Agency (MHRA) 2020)
"A medical device is described as any instrument, apparatus, appliance, software, material or other article used alone or combined for humans to:

- diagnose, prevent, monitor, treat or alleviate disease
- diagnose, monitor, treat, alleviate or compensate for an injury or handicap
- investigate, replace or modify the anatomy or a physiological process
- control conception

Serious consequences, financial and otherwise, may follow from failure by software developers to adequately evaluate and register products that fit within the definition of medical devices (Wyatt 1994). As consumer-focused health and wellness apps proliferate, there is an ever-expanding ethical argument for regulation of such apps. While health and wellness apps not intended for medical use are presently not regulated by the US FDA, bioethicists have recognized that leaving such oversight to the private sector may not be effective in the long term (Kasperbauer and Wright 2020). Regulation in this space serves an important public interest of safeguarding patient safety, while also balancing the importance of continued innovation. With innovation happening at a seemingly ever-increasing pace, curating a regulatory environment that balances competing interests while keeping up with technological advancement will prove to be increasingly challenging and important.

1.7.4 Domain Knowledge Complexity

Biomedicine is well known to be a complex domain, with students in the United States spending a minimum of 7 years to gain professional certification in medicine. It is not atypical for a graduate student in molecular biology to spend five or more years in pursuit of a PhD and follow this with a multi-year postdoctoral training experience. Professional training in other fields, such as nursing and pharmacy, has lengthened in recognition of the nature and complexity of what is to be learned. Biomedical knowledge itself and methods of healthcare delivery and research change rapidly. The number of articles in the "Medline" database was over 30 times larger in 2019 than it had been 50 years previously (National Library of Medicine (NLM) 2021). For this reason, it can be very challenging to keep a biomedical or health information resource up to date, and this may even require revision of the resource during the course of an evaluation study.

The problems that biomedical and health information resources assist in solving are complex, highly variable across regions (Birkmeyer et al. 2013), and difficult to describe. Patients often suffer from multiple diseases, which may evolve over time and at differing rates, and they may undergo a number of interventions over the course of the study period. Two patients with the same disease can present with very different sets of clinical signs and symptoms. There is variation in the interpretation of patient data among medical centers. What may be regarded as an abnormal result or an advanced stage of disease in one setting may pass without comment in another because it is within their laboratory's normal limits or is an endemic condition in their population. Thus, simply because an information resource is safe and effective when used in one center on patients with a given diagnosis, this cannot predict the results of evaluating it in another center or in patients with a different disease profile.

1.7.5   Chains of Effect

The links between introducing a biomedical or health information resource and achieving improvements in the outcomes of interest are long and complex, even in comparison with other biomedical interventions such as drug therapy. As illustrated in Fig. 1.5, this is because the functioning of an information resource and its impact

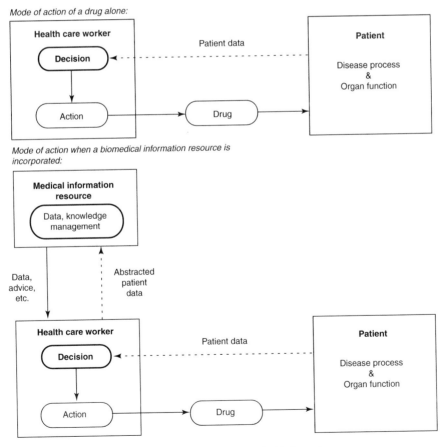

Fig. 1.5 Mode of action of a drug compared to a biomedical information resource

may depend critically on how human "users" actually interact with the resource. These challenges are increasingly relevant with the rapid growth of personal health "apps". In the clinical domain especially, it is probably unrealistic to look for quantifiable changes in health status outcomes following the introduction of many information resources until one has first documented the resulting changes in the structure or processes of health care delivery. For example in their classic study, MacDonald and colleagues showed during the 1980s that the Regenstrief healthcare information system positively affected providers' clinical actions with its alerts and reminders (McDonald et al. 1984). Almost 10 years later, the same study team reported a reduction in the length of hospital stay, which is a health outcome measure (Tierney et al. 1993). Linking the behavior of care providers to health outcomes is the domain of the field of health services research. Because of the proliferation of findings from health services research, in many circumstances it may be sufficient, within the realm of biomedical and health informatics, to limit studies to the effects of an information resource on a clinical process while referring to the literature of health services research to link those process changes to health outcomes. For example, demonstrating that an information resource increases the proportion of patients with heart attacks who are given a clot-dissolving drug can be assumed to improve the survival from heart attacks and patients' functioning thereafter.

1.7.6 Decision Making Processes

In the clinical domain, the processes of decision-making are complex and have been studied extensively (Elstein et al. 1978; Norman and Eva 2010). Clinicians make many kinds of decisions—including diagnosis, monitoring, investigation, choice of therapy, and prognosis—using incomplete and fuzzy data, some of which are appreciated intuitively and not recorded in the clinical notes. If an information resource generates more effective management of both patient data and medical knowledge, it may intervene in the process of medical decision-making in a number of ways; consequently, it may be difficult to decide which component of the resource is responsible for the observed changes. Often this does not matter, but if one component is expensive or hard to create, understanding why and how the resource brings benefit becomes important. Understanding why a resource brings benefit can be much more difficult than determining the magnitude of this benefit.

1.7.7 Data Availability and Quality

In recent years there has been an enormous proliferation of data that can be used by information resources to improve practice across a wide range of health-related domains. These developments are lead, in part, by the work of the U.S. Institute of Medicine to envision a Learning Healthcare System where:

... science, informatics, incentives, and culture are aligned for continuous improvement and innovation, with best practices seamlessly embedded in the delivery process, [with] patients and families active participants in all elements, and new knowledge captured as an integral by-product of the delivery experience. (McGinnis et al. 2007)

Capturing knowledge as an integral by-product of the delivery experience requires that all events of care are completely and accurately captured as data suitable to analysis to generate new knowledge. Unfortunately, this is not generally the case; "missingness" and inaccuracies of data in electronic health record systems are well-documented problems (Haneuse et al. 2021). Learning systems to improve health, as well as education and research, are an important vision. However, these systems are dependent on data to fuel learning.

Chapter 13 of this book will address studies that are based on the accumulation of "real world" data. The results of these studies can be no better than the data that fuel them.

1.7.8 Performance Standards

There is a general lack of "gold standards" in health care: meaning that in many clinical cases, there is often only a weak consensus on what problem(s) a patient has and what are the best courses of action to improve that person's health. For example, diagnoses are rarely known with 100% certainty, partly because it is unethical to do all possible tests in every patient, or even to follow up patients without good cause, and partly because of the complexity of the human biology (Graber 2013). When attempting to establish a diagnosis or the cause of death, even if it is possible to perform a postmortem examination, correlating the observed changes with the patients' symptoms or findings before death may prove impossible. Determining the "correct" management for a patient is even harder, as there is wide variation even in so-called consensus opinions, which is reflected in wide variations in clinical practice even in neighboring areas (Zhang et al. 2010).

1.7.9 Computer-Based Information Resources

From a software engineering perspective, the goals of evaluating an information resource are to verify that the program code faithfully performs those functions it was designed to perform. One approach to this goal is to try to predict the resource's performance from knowledge of the program's structure. However, although software engineering and formal methods for specifying, coding, and testing computer programs have become more sophisticated, programs of even modest complexity challenge these techniques. Since we cannot logically or mathematically derive a program's function from its structure, we often are left with exhaustive "brute-force" techniques for program verification.

The task of verifying a program (obtaining proof that it performs all and only those functions specified) using brute-force methods increases exponentially according to the program's size and the complexity of the tasks it performs. This is an "NP-hard" problem.[4] Put simply, to verify a program using brute-force methods requires application of every combination of possible input data items and values for each in all possible sequences. This entails at least n factorial trials of the program, where n is the number of input data items.

Most information resources such as electronic health records are deliberately configured or tailored to a given institution as a necessary part of their deployment. Hence, it may be difficult to compare the results of one evaluation with a study of the same information resource conducted in another location. It is even reasonable to ask whether two variants of the "same" software application configured for two different environments should be considered the same application. In addition, the notoriously rapid evolution of computer hardware and software means that the time course of an evaluation study may be greater than the lifetime of the information resource itself. While this problem is often exaggerated, and even used by some as a reason not to invest in evaluation, it nonetheless is an important factor shaping how evaluation studies in biomedical and health informatics should be designed and conducted.

Biomedical information resources often contain several distinct components, including interface, database, communication, access control, reasoning, and maintenance programs, as well as data, knowledge, business logic, and dynamic inferences about the users and their current activities. Such information resources may perform a range of functions for users. This means that if study teams are to answer questions such as, "What part of the information resource is responsible for the observed effect?" or "Why did the information resource fail?," they must be familiar with each component of the information resource, their functions, and their potential interactions.

1.8 Concluding Observations

No one could pretend that evaluation is easy. This entire book describes ways that have been developed to solve the many challenges discussed in this chapter. First, study teams should recognize that a range of evaluation approaches are available. What will, in Chap. 5, be called an "evaluation mindset" includes the awareness that every study is to some extent a compromise. To help overcome the many potential difficulties, study teams require knowledge and skills drawn from a range of disciplines, including statistics, measurement theory, psychology, sociology, and anthropology. To avoid committing excessive evaluation resources at too early a stage, the

---

[4] "NP" stands for "Nondeterministic Polynomial-time" and is a term used in computational complexity theory.

intensity of evaluation activity should be titrated to the stage of development of the information resource: It is clearly inappropriate to subject to a multicenter randomized trial a prototype information resource resulting from a 3-month student project (Stead et al. 1994). This does not imply that evaluation can be deferred to the end of a project. Evaluation plans should be appropriately integrated with system design and development from the outset, as further discussed in Chap. 3.

If the developers are able to enunciate clearly the aims of an information resource and the major stakeholder groups, defining the questions to be answered by an evaluation study becomes easier. As will be seen in later chapters, study teams should also watch for adverse or unexpected effects. Life is easier for study teams if they can build on the work of their predecessors; for example, many studies require reliable and valid quantitative ways to measure relevant attitudes, work processes, or relevant outcomes. If these measurement tools already exist, study teams should use them in their studies rather than developing new measures, which would have to undergo a time-consuming process of thorough validation. One valuable role study teams may play is to dampen the often-unbridled enthusiasm of developers for their own systems, focusing the developers' attention on a smaller number of specific benefits it is reasonable to expect.

As illustrated above, there are many potential problems when evaluating biomedical and health information resources, but evaluation is possible, and many thousands of useful evaluations have already been performed. For example, Kwan and colleagues reviewed the results of 108 controlled trials of decision support systems studying the care given to over 1.2 million patients and concluded that most showed clear evidence of an impact on clinical processes and a smaller number showed improved patient outcomes (Kwan et al. 2020). Designing formal experiments to detect changes in patient outcome due to the introduction of an information resource is possible, as will be discussed in a later chapter. It is not the purpose of this book to deter study teams, merely to open their eyes to the complexity of this area.

Food for Thought
1. How would you respond to someone who argued that the three "pragmatic" questions attributed to Barnett (see Sect. 1.3) are, in and of themselves, sufficient to evaluate an information resource—in other words that these relatively easy-to-answer questions provide everything one needs to know?
2. It is perhaps reasonable to believe that, someday, it will be possible to build an evaluation machine for informatics, as described in Sect. 1.3. This might be possible through advanced predictive modeling of the dynamics of organizations and human behavior; and using these models it will be possible to credibly compare the predicted futures of an organization with and without deployment of a specific information resource. Do you believe that such an evaluation machine can replace the need for studies supported by empirical data? Why or why not?
3. Section 1.7 of this chapter provides nine areas of challenges and sources of complexity to evaluation in biomedical and health informatics. Identify the three of

these challenge areas that you consider to be the most challenging and, for each, explain why you believe this challenge area belongs on your shortlist.

4. Suppose you were president of a philanthropic foundation that supported a wide range of projects related to biomedical and health informatics. When investing the scarce resources of your foundation, you might have to choose between funding system/resource development and evaluation studies of resources already developed. Faced with this decision, what proportion of your organization's philanthropic budget would you give to each? How would you justify your decision?

Answers to Self-Tests

Self-Test 1.1

1. Personal health
2. Public/population health
3. Research
4. Health care
5. Education
6. Public/population health

Self-Test 1.2

1. Pragmatic
2. Promotional
3. Managerial
4. Scholarly

References

Abroms LC, Boal AL, Simmens SJ, Mendel JA, Windsor RA. A randomized trial of Text2Quit: a text messaging program for smoking cessation. Am J Prev Med. 2014;47:242–50.

Birkmeyer JD, Reames BN, McCulloch P, Carr AJ, Campbell WB, Wennberg JE. Understanding of regional variation in the use of surgery. Lancet. 2013;382:1121–9.

Elstein A, Shulman L, Sprafka S. Medical problem solving: an analysis of clinical reasoning. Cambridge, MA: Harvard University Press; 1978.

Federal food, drug, and cosmetic act, Amended. 21 USC 301, Amended through PL 117-11 (2021). Available from: https://www.govinfo.gov/content/pkg/COMPS-973/pdf/COMPS-973. pdf. Accessed 9 Jun 2021.

Florance V. Training for research careers in biomedical informatics and data science supported by the National Library of Medicine. In: Berner ES, editor. Informatics education in healthcare: lessons learned. 2nd ed. Cham: Springer Nature; 2020. p. 13–22. https://doi.org/10.1007/978-3-030-53813-2_2.

Friedman CP. Where's the science in medical informatics? J Am Med Inform Assoc. 1995;2:65–7.

Friedman CP. A "fundamental theorem" of biomedical informatics. J Am Med Inform Assoc. 2009 Mar;16:169–70.

Graber ML. The incidence of diagnostic error in medicine. BMJ Qual Saf. 2013;22:21–7.

Haneuse S, Arterburn D, Daniels MJ. Assessing missing data assumptions in EHR-based studies: A complex & underappreciated task. JAMA Netw Open. 2021;4:e210184.

Kasperbauer TJ, Wright DE. Expanded FDA regulation of health and wellness apps. Bioethics. 2020;34:235–41.

Kwan JL, Lo L, Ferguson J, Goldberg H, Diaz-Martinez JP, Tomlinson G, et al. Computerised clinical decision support systems and absolute improvements in care: meta-analysis of controlled clinical trials. BMJ. 2020;370:3216. https://doi.org/10.1136/bmj.m3216.

Maldonado G. The role of counterfactual theory in causal reasoning. Ann Epidemiol. 2016;26:681–2.

McDonald CJ, Hui SL, Smith DM, Tierney WM, Cohen SJ, Weinberger M, et al. Reminders to physicians from an introspective computer medical record: a two-year randomized trial. Ann Intern Med. 1984;100:130–8.

McGinnis JM, Aisner D, Olsen L, editors. The learning healthcare system: workshop summary. Institute of Medicine (IOM). Washington, DC: The National Academies Press; 2007.

Mun YY, Jackson JD, Park JS, Probst JC. Understanding information technology acceptance by individual professionals: toward an integrative view. Inf Manag. 2006;43:350–63.

Musen MA, Middleton B, Greenes RA. Clinical decision-support systems. In: Shortliffe EH, Cimino JJ, editors. Biomedical informatics: computer applications in health care and biomedicine. 5th ed. Cham: Springer Nature; 2021. p. 795–840. https://doi.org/10.1007/978-3-030-58721-5_24.

National Library of Medicine (NLM). MEDLINE citation counts by year of publication (as of January 2021). Bethesda, MD: United States National Institutes of Health (NIH), National Library of Medicine (NLM); 2021. Available from: https://www.nlm.nih.gov/bsd/medline_cit_counts_yr_pub.html. Accessed 19 Jun 2021

Norman GR, Eva KW. Diagnostic error and clinical reasoning. Med Educ. 2010;44:94–100.

O'Connor M. An end to preemptively limiting the scope of a manufacturer's duty: why the Arizona Court of Appeals was right in striking down the learned intermediary doctrine. Ariz State Law J. 2017;49:607.

Rafferty AM, Ball J, Aiken LH. Are teamwork and professional autonomy compatible, and do they result in improved hospital care? BMJ Qual Saf. 2001;10:32–7.

Ratwani RM, Hodgkins M, Bates DW. Improving electronic health record usability and safety requires transparency. JAMA. 2018;320:2533–4.

Roethligsburger F, Dickson W. Management and the worker. Cambridge, MA: Harvard University Press; 1939.

Shortliffe EH, Cimino JJ, editors. Biomedical informatics: computer applications in health care and biomedicine. 4th ed. London: Springer; 2014.

Somerville I. Software engineering. 10th ed. London: Pearson; 2016.

Stead W, Haynes RB, Fuller S, Friedman CP, Travis LE, Beck JR, et al. Designing medical informatics research and library-resource projects to increase what is learned. J Am Med Inform Assoc. 1994;1:28–33.

The New York Times. Alice Toklas, 89, is dead in Paris; literary figure Ran Salon with Gertrude Stein. Archive of The New York Times. 1967 Mar 8; Sect A:45. Available from: https://www.nytimes.com/1967/03/08/archives/alice-toklas-89-is-dead-in-paris-literary-figure-ran-salon-with.html. Accessed 9 Jun 2021.

Tierney WM, Miller ME, Overhage JM, McDonald CJ. Physician inpatient order writing on microcomputer workstations: effects on resource utilization. JAMA. 1993;269:379–83.

United Kingdom Medicines and Healthcare products Regulatory Agency (MHRA). Guidance: medical devices: how to comply with the legal requirements in Great Britain. London: United Kingdom Medicines and Healthcare products Regulatory Agency (MHRA); 2020. Available from: https://www.gov.uk/guidance/medical-devices-how-to-comply-with-the-legal-requirements. Accessed 9 Jun 2021

United States Food and Drug Administration (FDA). How to determine if your product is a medical device. Silver Spring, MD: United States Food and Drug Administration (FDA); 2021. Available from: https://www.fda.gov/medical-devices/classify-your-medical-device/how-determine-if-your-product-medical-device. Accessed 9 Jun 2021

Wyatt JC. Clinical data systems, part 1: data and medical records. Lancet. 1994;344:1543–7.

Wyatt J, Spiegelhalter D. Evaluating medical expert systems: what to test, and how? In: Talmon JL, Fox J, editors. Knowledge based systems in medicine: methods, applications and evaluation: proceedings of the workshop "system engineering in medicine", Maastricht, March 16-18, 1989. Lecture notes in medical informatics. Heidelberg: Springer; 1991. https://doi.org/10.1007/978-3-662-08131-0_22.

Zhang Y, Baicker K, Newhouse JP. Geographic variation in the quality of prescribing. N Engl J Med. 2010;363:1985–8.

Chapter 2
The Panorama of Evaluation Approaches

Learning Objectives
The text, examples, and self-tests in this chapter will enable you to:

1. Explain why there is a need for multiple approaches to evaluation in informatics.
2. Given a description of a study, specify the stakeholders who play specific roles in the study's planning, conduct and/or communication.
3. Relate the methods of quantitative approaches and those of qualitative approaches to their philosophical roots.
4. Compare and contrast the "anatomies" of quantitative studies and those of qualitative studies.
5. Given a description of a study, assign it to one of House's eight classifications.

2.1 Introduction

The previous chapter aimed to establish the purposes of evaluation and how they can be of benefit to biomedical and health informatics and, ultimately, to the health of individuals and populations. Section 1.7 described many aspects of biomedical and health informatics that make evaluation complex and challenging. Fortunately, many good minds—representing an array of philosophical orientations, methodological perspectives, and domains of application—have explored ways to address these difficulties. Many of the resulting approaches to evaluation have met with substantial success. The resulting range of solutions, comprising what a can called a field of evaluation itself, is the focus of this chapter.

This chapter will develop some common ground across all evaluation work while simultaneously appreciating the range of tools available. This appreciation is the initial step in recognizing that evaluation, though difficult, is possible.

© Springer Nature Switzerland AG 2022
C. P. Friedman et al., *Evaluation Methods in Biomedical and Health Informatics*, Health Informatics, https://doi.org/10.1007/978-3-030-86453-8_2

2.2 Deeper Definitions of Evaluation

Not surprisingly, there is no single accepted definition of "evaluation". A useful goal for the reader may be to evolve a personal definition that makes sense, can be concisely stated, and can be publicly defended without embarrassment. The reader is advised not to settle firmly on a definition now, as a personal definition is likely to change many times based on later chapters of this book and other experiences. To begin development of a personal viewpoint, the classic evaluation literature offers three contrasting definitions. All three definitions have been modified to apply more specifically to biomedical and health informatics.

Definition 1 (*adapted from the original text by Rossi and Freeman*): Evaluation is the systematic application of social science research procedures to judge and improve the way information resources are designed and implemented (Rossi and Freeman 1989; Rossi et al. 2018).[1]

Definition 2 (*adapted from Guba and Lincoln*): Evaluation is the process of describing the implementation of an information resource and judging its merit and worth (Guba and Lincoln 1981).

Definition 3 (*adapted from House*): Evaluation is a process leading to a settled opinion that something about an information resource is the case, usually—but not always—leading to a decision to act in a certain way (House 1980).

The first definition ties evaluation to the empirical methods of the social sciences. What this definition actually means depends, of course, on one's definition of the social sciences. At the time they first wrote this definition, the authors would certainly have believed that it includes only methods that depend on quantitative data. Judging from the contents of their book, these authors probably would not have seen the more qualitative methods derived from ethnography and social anthropology as useful to evaluation studies.[2] Their definition further implies that evaluations are carried out in a planned, orderly manner, and that the information collected can lead to two types of results: improvement of the resource and some determination of its value.

The second definition is somewhat broader. It identifies descriptive questions (How is the resource being used?) as an important component of evaluation, while implying the need for a complete evaluation to result in some type of judgment. This definition is not as restrictive in terms of the methods used to collect information. This openness is intentional, as these authors embrace the full gamut of methodologies, from the experimental to the anthropological.

The third definition is the least restrictive and emphasizes evaluation as a process leading to deeper understanding and consensus. Under this definition, an evaluation

---

[1] This classic text has been revised multiple times. An "international edition" was published in 2018 (Rossi et al. 2018).

[2] The authors have stated (p. 265) that "assessing impact in ways that are scientifically plausible and that yield relatively precise estimates of net effects requires data that are quantifiable and systematically and uniformly collected."

could be successful even if no judgment or action resulted, so long as the study resulted in a clearer or better-shared idea by some significant group of individuals regarding the information resource or its context.

When shaping a personal definition, it is important to keep in mind something implied but not explicitly stated in the above definitions: that evaluation is an empirical process, as discussed in Sect. 1.2. This means during an evaluation study, data of varying shapes and sizes are always collected. It is also important to view evaluation as an applied or service activity. Evaluations are tied to and shaped by the specific information resource(s) under study, and the context of the evaluation. Evaluation is useful to the degree that it sheds light on issues such as the need for, function of and utility of those information resources.

2.3   The Common Anatomy of All Evaluation Studies

Even though there is a plethora of evaluation approaches, there are some structural elements common to all evaluation studies (Fig. 2.1). As stated above, evaluations are guided by the need of some individual or group (the stakeholders) to know something about the resource. No matter who that someone is—the development team, funding agency, or other individuals and groups—the evaluation must begin with a process of negotiation. The outcome of these negotiations is a mutual understanding of why and how the evaluation is to be conducted, usually stated in a written contract or agreement. Either as part of the negotiations or in a separate subsequent process, the study team will work with stakeholders to generate a list of the questions the evaluation seeks to answer. The next element of the study is investigation: the collection of data to answer these questions, and, depending on the evaluation approach selected, to answer other questions that arise during the study.

The next element in every evaluation is a mechanism for reporting the results back to those individuals with a need to know. The format of the report must be in line with the stipulations of the contract; the content of the report follows from the questions asked and the data collected. The report is most often a written document, but it does not have to be. The purposes of some evaluations are well served by oral reports, webinars or live demonstrations. It is the study team's obligation to establish a process through which the results of their study are communicated, thus

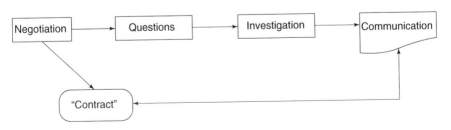

Fig. 2.1 Anatomy of all evaluation studies

creating the opportunity for the study's findings to be put to constructive use. No study team can guarantee a constructive outcome for a study, but there is much that can be done to increase the likelihood of a salutary result. In addition, note that a salutary result of a study is not necessarily one that casts the resource under study in a positive light. A salutary result is one where the stakeholders learn something important or useful from the study findings.

The diagram in Fig. 2.1 may seem unnecessarily complicated to students or study teams who are building their own information resource and wish to evaluate it in a preliminary way. To these individuals, a word of caution: Even when evaluations appear simple and straightforward at the outset, they have a way of becoming complex. Much of this book deals with these complexities and how they can be anticipated and managed.

2.4 Diverse Roles in Evaluation Studies

It is also important to appreciate the multiple roles that are played in the conduct of each study. A review of these roles is useful to help understand the process of evaluation and to help those planning studies to anticipate everything that needs to be done. At the earliest stage of planning a study, and particularly when an evaluation contract is being negotiated, attention to these roles and their relationships helps ensure that the contract will be complete and will serve well in guiding the conduct of the study. It is an excellent idea to make a list of these roles and indicate which individuals or groups occupy each one. Sometimes this exercise requires a few educated guesses, but it should still be undertaken at the outset.

Any of these actors may be affected by a biomedical or health information resource, and each may have a unique view of what constitutes benefit. Each of these individuals or groups may have different questions to ask about the same information resource. The design of evaluation studies requires consideration of the perspectives of all "stakeholders", or at least those considered most important. A major challenge for study teams is to distinguish those persons who must be considered from those whose consideration is optional.

These roles and how they interrelate are illustrated in Fig. 2.2. It is important to note that the same individual may play multiple roles, and some roles may be shared by multiple individuals. In general, the smaller the scope of the study, the greater the overlap in roles.

The first set of roles relates to the individuals who conduct the study. These individuals include the director of the study, who is the person professionally responsible for the study, and any staff members who might work for the director. As soon as more than one person is involved in the conduct of a study, interpersonal dynamics among members of this group become an important factor contributing to the success or failure of the study.

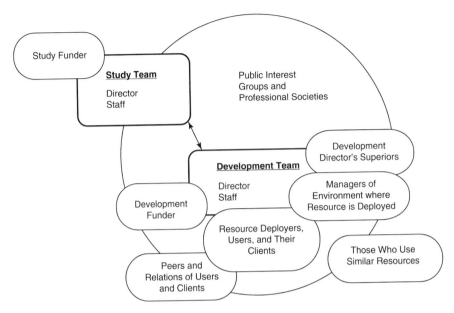

Fig. 2.2 Roles in an evaluation study of an information resource

A second set of roles relates to the deployers and users of the resource and the clients of those users. The identity of these individuals and groups derives from the resource's application domain. For an information resource supporting health care, the resource users are typically healthcare professionals, and their clients are the patients receiving care. For a resource supporting personal health, the users include anyone motivated, for any of a variety of reasons, to use it. For a resource supporting research, the users are those conducting, participating in, or overseeing research studies in the environment where the resource is deployed.

A third set of roles is related directly to the development of the resource under study. Those who fulfill these roles include the director of the resource's development team and their staff. If the resource developer is a commercial vendor, their clients are the organizations where the resource is deployed.

A fourth set of roles includes individuals or groups (or both) who, although they are not developing the resource or participating in an evaluation study, nonetheless may have a profound interest in the study's outcome. These individuals or groups include those who fund the development of the resource, those who fund the evaluation (who may be different from those who fund the resource), supervisors of the director of resource development, managers of the environment where the resource is deployed, those who use similar resources in other settings, peers and relations of the clients, and a variety of public interest groups, regulators and professional societies.

Self-Test 2.1

Nouveau Community Hospital (NCH) is a privately-owned 250-bed hospital, with adjoining outpatient clinics, in a mid-sized American city.[3]

Eighteen months ago, the leadership of NCH decided to develop locally and deploy YourView, a comprehensive "portal" enabling patients to make appointments, to communicate via secure email with staff and care providers, and to view their medical record information that is stored on the hospital's EHR. Immediately following a trial period, YourView went live for all NCH patients.

Based on early experience and anecdotal feedback, the leaders of NCH have grown concerned about two specific aspects of the YourView project and have decided to commission an evaluation study to explore both of these aspects:

A. *There is concern that some patients are finding it difficult to navigate YourView to access the information they are seeking, make appointments, and communicate with their care providers.*

B. *The leadership has heard that patients are concerned about the privacy of their information now that it is accessible electronically through the portal. The leadership wishes to know to what extent these concerns exist, and what is the best way to respond to them.*

The leadership of NCH is soliciting proposals to address the issues listed above.

Put yourself in the role of the leader of a team that will be proposing a study to address the concerns above. With reference to Fig. 2.2, indicate some of the individuals or groups who play the roles described in the figure. For example, you would be the study director and your staff would be the members of the evaluation team.

2.5   Some Philosophical Bases of Evaluation

Several authors have developed classifications (or "typologies") of evaluation methods or approaches. Among the most noteworthy, and useful for designing evaluations in informatics, was that developed in 1980 by Ernest House (House 1980). A major advantage of House's typology is that each approach is elegantly linked to an underlying philosophical model, as detailed in his book. This classification divides evaluation practice into eight discrete approaches, four of which may be viewed as "quantitative" and four as "qualitative." These eight approaches form the first level above the roots of the "Tree of Evaluation" found in the Preface to this book.

Following House's argument, quantitative approaches derive from a logical–positivist philosophical orientation—the same orientation that underlies the classical

---

[3]This example is not based on or intended to depict a specific hospital in the United States or elsewhere.

experimental sciences. The major premises underlying quantitative approaches are as follows:

- Information resources, the people who use them, and the processes they affect, all have attributes that can be measured. All observations of these attributes will ideally yield the same results. Any variation in these results would be attributed to measurement error. It is also assumed that a study team can measure these attributes without affecting how the resource under study functions, or is used.
- Rational persons can and should agree on what attributes of a resource are important to measure and what results of these measurements would be identified as more desirable, correct, or a positive outcome. If an initial consensus does not exist among these rational persons, they can be brought to consensus over time.
- While it is possible to disprove a well-formulated scientific hypothesis, it is never possible to fully prove one; thus, science proceeds by successive disproof of previously plausible hypotheses.
- Because numerical measurement allows precise statistical analysis of performance over time or performance in comparison with some alternative, numerical measurement is *prima facie* superior to a verbal description. Qualitative data may be useful in preliminary studies to identify hypotheses for subsequent, precise analysis using quantitative methods.
- Through these kinds of comparisons, it is possible to prove beyond reasonable doubt that a resource is superior to what it replaced or to some competing resource.

Chapters 6 through 13 of this book address quantitative methods in detail, largely by exploring the issues listed above.

Contrast the above summary of the quantitative mindset with a set of assumptions underlying the "intuitionist–pluralist" worldview that gives rise to qualitative approaches to evaluation:

- What is observed about a resource depends in fundamental ways on the observer. Different observers of the same phenomenon might legitimately come to different conclusions. Both can be "objective" in their observations even if they do not agree. It is not necessary that one is right and the other wrong.
- It does not make sense to speak of the attributes of a resource without considering its context. The value of a resource emerges through study of the resource as it functions in a particular patient care, scientific or educational environment.
- Individuals and groups can legitimately hold different views on what constitutes the desirable outcome of introducing a resource into an environment. There is no reason to expect them to agree, and it may be counterproductive to try to lead them to consensus. One important contribution of an evaluation would be to document the ways in which stakeholders disagree, and the possible reasons for this.
- Verbal description can be highly illuminating. Qualitative data, emphasizing words over numbers, are valuable in and of themselves, and can lead to conclusions as convincing as those drawn from purely quantitative data. Therefore, the

value of qualitative data goes far beyond that of identifying issues for later "precise" exploration using quantitative methods.

- Evaluation should be viewed as an exercise in argument, rather than demonstration, because any study, as House points out, "will appear equivocal when subjected to serious scrutiny" (House 1980, p. 72).

When encountered for the first time, the evaluation approaches that derive from this qualitative philosophical perspective may seem strange, imprecise, and "unscientific". This perhaps stems from widespread acceptance of the quantitative worldview in the biomedical sciences. The importance and utility of qualitative approaches to evaluation are increasingly accepted, however, within biomedical and health informatics (Scott and Briggs 2009; Kaplan and Maxwell 2005; Greenhalgh and Russell 2006). It is therefore important for those trained in classical experimental methods at least to understand, and possibly even also to embrace, the qualitative worldview if they are going to conduct fully informative evaluation studies. Chapters 14 through 16 of this book describe qualitative approaches in detail.

2.6 The Specific Anatomies of Quantitative and Qualitative Studies

Having described a general anatomy of all evaluation studies and then divided the universe of studies into two groups, it is reasonable to ask how this general anatomy differs across the groups, before proceeding to describe all eight approaches in House's typology. Figure 2.3 illustrates the typical anatomy of a quantitative study, and Fig. 2.4 a qualitative study. The differences are seen primarily in the "investigation" aspect of the process.

The investigation component of quantitative studies can be seen as a *Linear Investigative Sequence* proceeding from detailed design of the study, to the

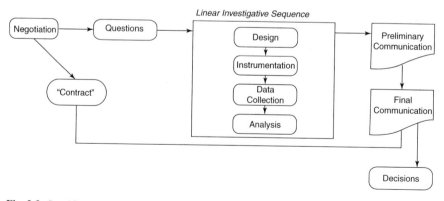

Fig. 2.3 Specific anatomy of quantitative studies

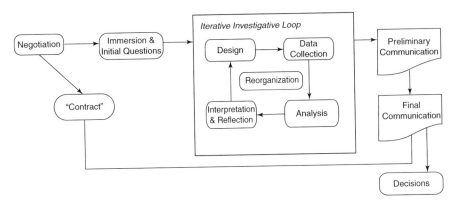

Fig. 2.4 Specific anatomy of qualitative studies

development or selection of instruments for collection of quantitative data, and then to the collection and subsequent analysis of these data. In principle, at least, the study teams go through this investigative sequence once, then proceed from there to report their findings.

By contrast, the qualitative approach includes, as a key step, an initial period of "immersion" into the environment under study—the process of becoming familiar with, and familiar to, those in the environment. This is followed by an *Iterative Investigative Loop*. The investigation proceeds through cycles beginning with a design for initial qualitative data collection, followed by data collection and analysis, and subsequent interpretation of and reflection on what has been learned from the data. This leads to the design for the next iteration. Throughout each cycle, the study team is open to reorganizing its beliefs, which can lead to entirely new evaluation questions for exploration in subsequent cycles.

2.7   Multiple Approaches to Evaluation

House (House 1980) classified evaluation into eight approaches, four of which are quantitative and four as qualitative. Although most evaluation studies conducted in the real world can be unambiguously tied to one of these approaches, some studies exhibit properties of several approaches and cannot be cleanly classified. The label "evaluation approach" has been deliberately chosen, so it is not confused with "evaluation method." In later chapters, an evaluation method will be defined as the procedure for collecting and analyzing data, whereas an evaluation approach is a broader term: the strategy directing the design and execution of an entire study. Following this exposition of eight approaches is an exercise for the reader to classify each of a set of biomedical and health informatics evaluation study scenarios into one of these categories.

2.7.1 Approaches Rooted in Quantitative Assumptions

The first four approaches derive from the quantitative philosophical position.

Comparison-Based Approach In the comparison-based approach, the information resource under study is compared to one or more contrasting conditions, which may include other resources or the absence of the resource entirely. The comparison is based on a relatively small number of "outcome variables" that are assessed in all groups. This approach thus seeks to simulate the "evaluation machine", as described in Sect. 1.3, using randomization or other types of controls, and statistical inference to argue that the information resource was the cause of any differences observed. Early exemplars of comparison-based studies include the work of McDonald and colleagues on physician reminders (McDonald et al. 1984), the work of Evans and colleagues on decision support in antibiotic prescribing (Evans et al. 1998). Many studies, that have been summarized in systematic reviews, have followed (Moja et al. 2014; Bright et al. 2012). The classical Turing "indistinguishability test" (Turing 1950) can be seen as a specific model for a comparison-based evaluation.

Objectives-Based Approach The objectives-based approach seeks to determine if a resource meets its design or performance objectives. These studies are comparative only in the sense that the observed performance of the resource is viewed in relation to the pre-stated objectives. The primary concern is whether the resource is performing up to expectations, not if the resource is outperforming what it replaced. The objectives that are the benchmarks for these studies are typically stated at an early stage of resource development. Although clearly suited to laboratory testing of a new resource, this approach can also be applied to testing a deployed resource as well. Consider the example of a resource to provide "real time" advice to emergency room physicians. The designers might set as an objective that the system's advice be available within 2 min of the time the patient is first seen. An objectives-based evaluation study would measure the time taken for this advice to be delivered and compare it to the stated objective.

Decision-Facilitation Approach With the decision-facilitation approach, evaluation seeks to resolve issues important to developers and managers, so these individuals can make decisions about the future of the resource. The questions posed are those that the decision-makers state, although those conducting the evaluation may help the decision-makers frame these questions so they are more amenable to empirical study. The data-collection methods will follow from the questions posed. These studies tend to be "formative" in focus, meaning that the studies are conducted at early stages of resource development and the results are used to influence further development, which in turn generates new questions for further study. A systematic study of alternative formats for computer-generated advisories, conducted while the resource to generate the advisories is still under development, provides a good example of this approach (Scott et al. 2011).

Goal-Free Approach   With the three approaches described above, the evaluation is guided by a set of goals for the information resource. Therefore, any study is polarized by these manifest goals and is designed to uncover only anticipated, rather than unanticipated, effects. With the "goal-free" approach, those conducting the evaluation are purposefully blinded to the intended effects of an information resource and collect whatever data they can gather to enable them to identify all the effects of the resource, regardless of whether or not these are intended (Scriven 1973). This approach is rarely applied in practice, but it is useful to individuals designing evaluations to remind them of the multiplicity of effects an information resource can engender.

2.7.2   Approaches Rooted in Qualitative Assumptions

There are four main qualitative approaches to evaluation.

Quasi-Legal Approach   The quasi-legal approach establishes a mock trial, or other formal adversarial proceeding, to judge a resource. Proponents and opponents of the resource offer testimony and may be examined and cross-examined in a manner resembling standard courtroom procedure. Based on this testimony, a jury witness to the proceeding can then decide about the merit of the resource. As in a debate, the issue can be decided by the persuasive power of rhetoric, as well as the persuasive power of evidence which is portrayed as fact. There are few published examples of this technique formally applied to informatics, but the technique has been applied to facilitate difficult decisions in other biomedical areas (Arkes et al. 2008; Smith 1992).

Art Criticism Approach[4]   The art criticism approach relies on methods of criticism and the principle of "connoisseurship." (Eisner 2017). Under this approach, an experienced and respected critic, who may or may not be trained in the domain of the resource but has a great deal of experience with resources of this generic type, works with the resource over a period of time. They then write a review highlighting the benefits and shortcomings of the resource. Within informatics, the art criticism approach may be of limited value if the critic is not expert in the subject-matter domain of the biomedical information resource under review. For example, if the resource provides advice to users or automates a task that was previously performed manually, a critic without domain knowledge could offer useful insights about the resource's general functioning and ease of use, but would be unable to judge whether the automated task was carried out properly or whether the advice provided was clinically or scientifically valid. Because society does not routinely expect critics to agree, the potential lack of interobserver agreement does not invalidate this

---

[4]While House chose to call this approach "art criticism", it is derived from the methods and traditions of criticism more generally. Nothing in this approach is specific to art.

approach. Although they tend to reflect less direct experience with the resource than would be the case in a complete "art criticism" study, software reviews that routinely appear in technical journals and magazines are examples of this approach in common practice.

Professional Review Approach The professional review approach includes the well-known "site visit" approach to evaluation. Site visits employ panels of experienced peers who spend several days in the environment where the resource is developed and/or deployed. Site visits often are directed by a set of guidelines specific to the type of project under study but sufficiently generic to accord the reviewers a great deal of control over the conduct of any particular visit. They are generally free to speak with whomever they wish and to ask of these individuals whatever they consider important to know. They may request documents for review. Over the course of a site visit, unanticipated issues may emerge. Site visit teams frequently have interim meetings to identify these emergent questions and generate ways to explore them. As a field matures, it becomes possible to articulate criteria for professional review that are sufficiently specific and detailed to obviate the need for formal site visits, allowing professional reviews to take place asynchronously and off-site. Seen in this light, the process of certifying biomedical software is an example of professional review (Russ and Saleem 2018).

Responsive/Illuminative Approach The responsive/illuminative approach seeks to represent the viewpoints of those who are users of the resource or an otherwise significant part of the environment where the resource operates (Hamilton et al. 1977). The goal is understanding, or "illumination," rather than judgment. The methods used derive largely from ethnography (the scientific description of peoples and cultures with their customs, habits, and mutual differences). The study teams immerse themselves in the environment where the resource is operational. The designs of these studies are not rigidly predetermined, but develop dynamically as the study teams' experience accumulates. The study team begins with a minimal set of orienting questions; the deeper questions that receive thorough ongoing study evolve over time. Many examples of studies using this approach can be found in the literature of informatics (Walker et al. 2019; Ash et al. 2021; Ventres et al. 2006).

Self-Test 2.2

Associate each of the following hypothetical studies with a particular approach to evaluation. For each, try to identify a single approach that is a "best fit."

(a) A comparison of different user interfaces for a genomic sequencing tool, conducted while the resource is under development, to help the tool designers create the optimal interface.

(b) A visit to the work site of the submitters of a competing renewal of a research grant, for purposes of determining whether the grant should be awarded.

(c) Inviting a noted consultant on intelligent tutoring systems design to spend a day on campus to offer suggestions regarding the prototype of a new system.

(d) Conducting patient chart reviews before and after introduction of an information resource, without telling the reviewer anything about the nature of the information resource or even that the intervention was an information resource.

(e) Video recording clinical rounds on a hospital service (ward) where an information resource has been implemented and periodically interviewing members of the ward team.

(f) Determining if a new version of a metabolomics database executes a standard set of performance tests at the speed the designers projected.

(g) Randomizing patients so their medical records are maintained, either by a new system or an older one, and then seeking to determine if the new system influences clinical trial protocol recruitment and compliance.

(h) Staging a mock debate at a health-sciences library staff retreat to decide whether the library should digitize its collection to the extent that no physical books or journals would reside in the library's building.

2.8   Conclusion: Why Are There so Many Approaches?

From the above examples, it should be clear that it is possible to employ almost all of these approaches to evaluation in biomedical and health informatics. Why, though, are there so many approaches to evaluation? The intuitive appeal—at least to those schooled in experimental science—of the comparison-based approach seems unassailable. *Why do evaluation any other way if study teams can definitively demonstrate the value of an information resource, or lack thereof, with a controlled study?*

The goal-free approach signals one shortcoming of comparison-based studies that employ classical experimental methods. Although these studies can appear definitive when proposed, they inevitably rely on an intuitive, arbitrary, or even political choice of questions to explore or outcomes to measure. What is measured is often what *can be* measured with the kind of quantitative precision that the philosophical position underlying this approach demands. It is often the case that the variables that are most readily obtainable and most accurately assessed (e.g., length of hospital stay), and which therefore are employed as outcome measures in studies, are difficult to relate directly to the effects of a biomedical information resource because there are numerous intervening or confounding factors. Studies may have null results not because there are no effects, but because these effects are not manifest in the outcome measures used. In other circumstances, outcomes cannot be unambiguously assigned a positive value. For example, if use of a computer-based tutorial program is found to raise medical students' national licensure examination scores, which are readily obtained and highly reliable, it usually does not settle the argument about the value of the tutorial program. Instead, it may only kindle a new argument about the validity of using the examination as an outcome measure. In the most typical case, a resource will produce several effects: some positive and some negative. Unless the reasons for these mixed effects can somehow be explored further, the impact of a resource cannot be comprehensively understood, or it may be seriously mis-estimated. When there are mixed results, often the resource is judged entirely by the single result of most interest to the group holding the greatest power. For example, a resource that otherwise improves nursing care may be branded a categorical failure because it proved to be more expensive than anticipated.

Comparison-based studies are also limited in their ability to explain differences that are detected or to shed light on why, in other circumstances, no differences are found. Consider, for example, a resource developed to identify "therapeutic misadventures"—problems with drug therapy of hospitalized patients—before these problems can become medical emergencies. Such a resource would employ a knowledge base encoding rules of proper therapeutic practice and would be connected to a hospital information system containing the clinical data about in-patients. When the resource detected a difference between the rules of best practice and the care experience of a specific patient, it would issue an advisory note to the clinicians responsible for the care of that patient. If a comparison-based study of this system's effectiveness employed only global outcome measures, such as length of hospital stay or morbidity and mortality, and the study yielded null results, it would not be clear what to conclude. It may be that the resource is having no beneficial effect, but it also may be that a problem with the implementation of the resource—which, if detected, can be rectified—is accounting for the null results. A failure to deliver the advisories in a visible place in a timely fashion could account for an apparent failure of the resource overall. In this case, incorporating a qualitative evaluation, specifically designed to understand how and when the resource is used and how participants responded to its output, might reveal the problems with the resource and result in a much more valuable study.

The existence of multiple alternatives to the comparison-based approach also stems from features of biomedical information resources and from the challenges, as discussed in Chap. 1, of studying these resources. Specifically:

1. Biomedical information resources are frequently revised; the resource may change in significant ways for legitimate reasons before there is time to complete a comparison-based study.
2. Alternative approaches are well suited to developing an understanding of how the resource functions within a particular environment. The success or failure may be attributable more to match or mismatch with the environment than intrinsic properties of the resource itself. Without such understanding, it is difficult to know how transferable a particular resource is and what factors are important to explore as the resource is considered for adoption by a location other than its development site.
3. There is a need to understand how users employ biomedical information resources, which requires an exercise in description, not judgment or comparison. If the true benefits of information and knowledge resources emerge from interaction of person and machine, approaches to evaluation that take the nature of human cognition into account must figure into a complete set of investigative activities.

4. Alternative approaches offer a unique contribution in their ability to help us understand *why* something happened in addition to *that* something happened. The results of a comparison-based study may be definitive in demonstrating that a resource had a specific impact on research, education, or patient care; but these results may tell us little about what component of the resource made the difference or the chain of events through which this effect was achieved.

The arguments above should be interpreted as another plea for open-mindedness and eclecticism in evaluation. In evaluation studies, the methods should follow from the questions and the context in which the evaluation is set, including the funding and time available. It is shortsighted to give any particular method of study higher status than the problem under study.

Self-Test 2.3
Refer back to Self-test 2.1 and the description of the "YourView" study at Nouveau Community Hospital. Briefly describe how a study might be conducted for each of the combinations of purposes and approaches described below:

(a) A comparison-based study addressing the concern about difficulties navigating the portal.
(b) A responsive-illuminative study addressing the concern about information privacy.
(c) A professional review study addressing the concern about privacy.
(d) An art criticism study addressing the concern about navigation.
(e) An objectives-based study addressing the concern about navigation.
(f) A decision-facilitation study addressing concerns about privacy.

Food for Thought
To deepen understanding of the concepts presented in this and the previous chapter, consider the following questions:

1. Many studies in informatics focus on an information resource's effect on the behavior of persons who are users of the resource: for example, the extent to which the introduction of a digital laboratory notebook improves the way researchers go about doing their research. Consider a study where researchers are observed in their day-to-day research activities before and after introduction of a digital notebook. Do you believe that independent observers would agree as to whether the digital notebook led to improved research in their labs? Can the study still be useful if they don't?
2. Many of the evaluation approaches assert that a single unbiased observer is a legitimate source of information during an evaluation, even if that observer's data or judgments are unsubstantiated by others. Can you offer some examples in wider society where we vest important decisions in a single experienced—and presumably impartial—individual?
3. Do you agree with the statement that all evaluations appear equivocal when subjected to serious scrutiny? Why or why not?

Answers to Self-Tests

Self-Test 2.1

The information resource is *YourView*. A partial list of pertinent roles includes:

- Users of *YourView*: Patients, Staff, Care Providers
- Peers and Relations of Users: Primarily family members of patients.
- Development funder: NCH
- Development team: Technical staff of the *YourView* team.
- Director the development team: Supervisor of the technical staff.
- Managers of the Environment: Hospital administrators at NCH
- Those who use similar resources: Other community hospitals
- Public interest groups and professional societies: Patient advocacy groups, hospital associations, informatics societies

Self-Test 2.2

(a) Decision-facilitation
(b) Professional review
(c) Art criticism
(d) Goal-free
(e) Responsive/illuminative
(f) Objectives-based
(g) Comparison-based
(h) Quasi-legal

Self-Test 2.3

There are many possible answers to this question. Some examples include:

(a) Once patients' primary usability concerns have been identified, a prototype new interface for YourView could be developed. A comparison-based study would compare the patients' ability to compete specific tasks with the old and new interface.

(b) Patients, staff, and clinicians could be interviewed over a period of time, individually or in groups, by a study team to better understand the sources of these concerns and understand what measures might address these concerns.

(c) A panel of privacy experts is engaged by NCH management to spend 2–3 days at the hospital meeting with various groups and issuing a report.

(d) An expert in usability is asked to work through the interfaces in *YourView* and offer suggestions as to how they can be improved.

(e) A group of typical users is given a series of structured tasks to complete using YourView, with the objective that the average user will complete 90% of the tasks within 1 min.

(f) Three alternative policies for privacy protection in *YourView* are developed. A survey is administered to a large sample of users to determine which policy would be most acceptable.

References

Arkes HR, Shaffer VA, Medow MA. The influence of a physician's use of a diagnostic decision aid on the malpractice verdicts of mock jurors. Med Decis Making. 2008;28:201–8.

Ash JS, Corby S, Mohan V, Solberg N, Becton J, Bergstrom R, et al. Safe use of the EHR by medical scribes: a qualitative study. J Am Med Inform Assoc. 2021;28:294–302.

Bright TJ, Wong A, Dhurjati R, Bristow E, Bastian L, Coeytaux RR, et al. Effect of clinical decision-support systems: a systematic review. Ann Int Med. 2012;157:29–43.

Eisner EW. The enlightened eye: qualitative inquiry and the enhancement of educational practice. New York: Teachers College Press; 2017.

Evans RS, Pestotnik SL, Classen DC, Clemmer TP, Weaver LK, Orme JF Jr, et al. A computer-assisted management program for antibiotics and other antiinfective agents. N Engl J Med. 1998;338:232–8.

Greenhalgh T, Russell J. Reframing evidence synthesis as rhetorical action in the policy making drama. Healthcare Policy. 2006;1:34–42.

Guba EG, Lincoln YS. Effective evaluation: improving the usefulness of evaluation results through responsive and naturalistic approaches. San Francisco, CA: Jossey-Bass; 1981.

Hamilton D, MacDonald B, King C, Jenkins D, Parlett M, editors. Beyond the numbers game: a reader in educational evaluation. Berkeley, CA: McCutchan; 1977.

House ER. Evaluating with validity. Beverly Hills, CA: Sage; 1980.

Kaplan B, Maxwell JA. Qualitative research methods for evaluating computer information systems. In: Anderson JG, Aydin CE (eds). Evaluating the organizational impact of healthcare information systems. Health informatics series. New York: Springer; 2005:30–55.

McDonald CJ, Hui SL, Smith DM, Tierney WM, Cohen SJ, Weinberger M, et al. Reminders to physicians from an introspective computer medical record: a two-year randomized trial. Ann Intern Med. 1984;100:130–8.

Moja L, Kwag KH, Lytras T, Bertizzolo L, Brandt L, Pecoraro V, et al. Effectiveness of computerized decision support systems linked to electronic health records: a systematic review and meta-analysis. Am J Pub Health. 2014;104:12–22.

Rossi PH, Freeman HE. Evaluation: a systematic approach. Newbury Park, CA: Sage Publications; 1989.

Rossi PH, Lipsey MW, Henry GT. Evaluation: a systematic approach (International Edition). Newbury Park, CA: Sage; 2018.

Russ AL, Saleem JJ. Ten factors to consider when developing usability scenarios and tasks for health information technology. J Biomed Inform. 2018;78:123–33.

Scott PJ, Briggs JS. A pragmatist argument for mixed methodology in medical informatics. J Mixed Methods Res. 2009;3:223–41.

Scott GP, Shah P, Wyatt JC, Makubate B, Cross FW. Making electronic prescribing alerts more effective: scenario-based experimental study in junior doctors. J AM Med Inform Assoc. 2011;18:789–98.

Scriven M. Goal-free evaluation. In: House ER, editor. School evaluation. Berkeley, CA: McCutchan; 1973.

Smith R. Using a mock trial to make a difficult clinical decision. BMJ. 1992;305:1284–7.

Turing AM. Computing machinery and intelligence. Mind Q Rev Psychol Philos. 1950;59:433–60.

Ventres W, Kooienga S, Vuckovic N, Marlin R, Nygren P, Stewart V. Physicians, patients, and the electronic health record: an ethnographic analysis. Ann Fam Med. 2006;4:124–31.

Walker RC, Tong A, Howard K, Palmer SC. Patient expectations and experiences of remote monitoring for chronic diseases: systematic review and thematic synthesis of qualitative studies. Int J Med Inform. 2019;124:78–85.

Chapter 3
From Study Questions to Study Design: Exploring the Full Range of Informatics Study Types

Learning Objectives

The text, examples, self-tests and Food for Thought questions in this chapter will enable the reader to:

1. Explain why clear statements of study questions are important to the design and successful execution of a study.
2. Express the purposes of a study as one or more study questions, using the PICO format where appropriate.
3. Distinguish between study questions that are too global and those that are too granular.
4. Describe the characteristics of each of the ten informatics study types.
5. Explain why each study type has the specific features described in Sect. 3.4 and Table 3.2.
6. Given a description of a study, assign it to one of the ten informatics-specific study types.
7. For a given set of study questions, select study type(s) that are suited to these questions and justify this choice.

3.1 Introduction

This chapter takes us one level higher in the "Tree of Evaluation" as introduced in the Preface to this book.

Chapter 1 of this book introduced the challenges of conducting evaluation studies in biomedical and health informatics, along with the specific sources of complexity that give rise to these challenges. Chapter 2 elaborated this discussion by introducing the range of approaches that are useful to conduct evaluations, not only in biomedical informatics but also across many areas of human endeavor. Chapter 2

© Springer Nature Switzerland AG 2022
C. P. Friedman et al., *Evaluation Methods in Biomedical and Health Informatics*, Health Informatics, https://doi.org/10.1007/978-3-030-86453-8_3

also stressed that study teams can address many of these challenges by viewing each evaluation study as anchored by specific purposes. Each study is conducted for some identifiable client group, often to inform specific decisions that must be made by members of that group. Studies become possible by focusing on the specific purposes motivating the work, often framing these purposes as a set of questions and choosing the approach or approaches best suited to those purposes. A study is then considered successful if it provides credible information that is helpful to members of the client group, often but not always to make important decisions.

In this chapter, the focus returns to informatics per se. This chapter explores the specific purposes of evaluation studies of information resources in biomedical and health settings. While Chap. 2 provided a tour of the approaches available to study almost anything using the methods of science, this chapter provides a tour of informatics-specific evaluation purposes that range from exploration of the need for an information resource that has not yet be built to exploration of a resource's impact on health, education, or research outcomes.

The approach in this chapter is to provide a comprehensive listing of what is possible in designing and conducting evaluation studies. By providing virtually all of the options and breaking them down into logical groupings, this "catalogue" approach can simplify the design process by allowing study teams to make choices from this list. This strategy also helps ensure that study teams do not overlook important options for conducting a study. While there is no such thing as a "cookbook" for evaluation study design, the information in this chapter makes what can be called a fair approximation to a cookbook intended to be helpful to study teams.

Chapter 4, to follow, is a companion to this one. Chapter 4 provides examples of informatics studies that represent each study type introduced here. Chapter 4 also introduces four prototypical evaluation scenarios that provide more-detailed guidance about choosing what to study and what study types to employ in specific situations.

Even experienced study teams wrestle with the problem of deciding what to study and how. Even with a list of options to give structure to the process, every study must, to some significant degree, be custom designed. In the end, decisions about what evaluation questions to pursue and how to pursue them are exquisitely sensitive to each study's special circumstances and constrained by the resources that are available for it. Evaluation is very much the art of the possible.

3.2   The Key Role of Study Questions to Guide Study Design

Sect. 2.3 of the previous chapter provided a description of the stages of all evaluation studies. The first stage in any study is negotiation resulting in a contract. Most important, this negotiation process will establish the overall purpose of the study: Does it encompass technical issues (e.g., the reliability of the hardware that powers the resource), issues relating to people (such as the training required to use the resource)—or more far-reaching issues such as the impact of resource on the quality and efficiency of work within the organization?

The contract will also identify the deliverables that are required when the study is completed, who has interests in or otherwise will be concerned about the study results, where the study personnel will be based, the resources available, the time-line for reports and other deliverables, any constraints on what can be studied or where, and budgets and funding arrangements. The details of the negotiation process are beyond of the scope of this book but are discussed in several references on evaluation (Posovac 2011; Yarbrough et al. 2011; Grembowski 2016; Owen 2020). Signers of the contract should include at minimum: an individual representing the funder of the evaluation, the director of the study team, and senior managers of any sites where the information resource is deployed and/or where data will be collected.

The more specific study questions that will guide the detailed design of the study may be included as part of the initial negotiations and thus included in the contract, or they may be determined through a separate process that follows the contract negotiation. How to determine the study questions is the subject of the following sub-sections.

3.2.1   Why Study Questions Are Important

Once the study's purpose has been established, the next step is to convert the per-spectives of the concerned parties, and what these individuals or groups want to know, into a finite set of study questions. It is important to recognize that, for any setting that is interesting enough to merit formal evaluation, the number of potential study questions is infinite. The process of identifying a tractable number of ques-tions is therefore essential and has a number of benefits:

- It helps to crystallize thinking of both study teams and key members of the audi-ence who are the "stakeholders" in the evaluation.
- It guides the study teams and clients through the critical process of assigning priority to certain issues, thus productively narrowing the focus of a study.
- It converts broad statements of purpose (e.g., "to evaluate a new app for diabetes control") into specific questions that can potentially be answered (e.g., "Will users of the app persist in their use sufficiently to generate a significant health benefit?").
- It allows different stakeholders in the evaluation process—patients, professional groups, managers—to see the extent to which their own concerns are being addressed and to ensure that these feed into the evaluation process.
- Most important, it enables the team to proceed to the next step of actually design-ing their study. This avoids the pitfalls of inappropriately applying the same set of methods to any study, irrespective of the questions to be addressed, or, equally inappropriately, choosing to emphasize only those questions compatible with the methods a study team prefers.

The granularity of the study questions envisioned here resides in a space between the broad purpose of a study and the very specific questions that might be asked directly to end users of an information resources on a questionnaire or via an interview. For example, a study's broad purpose might be to "evaluate a weight reduction mobile app". One of very, very many possible questions that could be asked to users of the app is: "What specific features of the app do you find difficult to use?" The type of study questions discussed here, that provide essential guidance to the designers of a study, have an in-between level of specificity: for example: "What patterns and levels of app use are associated with greater or lesser degrees of weight loss?"

3.2.2 The Qualitative Process of Getting to a Final Set of Study Questions

In formal evaluations, the questions that are addressed are those that survive a narrowing process that begins with discussions involving stakeholder groups and must ultimately be ratified by those funding the study or otherwise in a position of authority. Therefore, when starting a formal evaluation, a major decision is whom to consult to establish the questions that will get "on the table," how to log and analyze their views, and what weight to place on each of these views. There is always a range of interested parties to any study; a mapping of the individuals and groups that play specific roles in the setting under study will provide important guidance as to whom to consult or in what order. (See Sect. 2.4.) Beyond that, the study teams will apply their common sense and learn to follow their instincts. For studies with a large scope and multi-year timeframe, it is often useful to establish an advisory group to the study team, so the final decisions regarding the questions guiding a study have legitimacy and are not perceived as biased or capricious.

The discussions with various stakeholder groups inform the hard decisions regarding the scope of the study. Even for a study likely to use quantitative approaches exclusively, the process of shaping the study questions is fundamentally qualitative in nature, so the methods and tools for qualitative data analysis discussed in Chap. 16, are applicable at this stage of study design. To manage through the process, it is useful to reflect on the major interests identified after each round of discussions with stakeholders and then identify potential study questions that might address these interests. The tools used for qualitative data analysis are also of potential value to organize and narrow into a more tractable form what is, almost inevitably, a large set of study questions at the outset. What is important at this stage is to keep a sense of perspective, distinguishing the issues as they arise and organizing them into some kind of hierarchy; for example, classifying them as low-, medium-, and high-level. Inter-dependencies among the questions should be noted; and care should be taken to avoid intermingling more global issues with more focused ones.

3.2.3 What Is the Right Number of Study Questions?

Clearly, there is no cut and dried answer to this question; but based on experience, a range of 3–7 questions is a useful guideline. When the number of study questions falls below 3, their level of specificity approaches that of the overall study scope and will not be helpful to inform the study design. When the number of study questions exceeds 7, in general they become too granular and do not map neatly to the design decisions study teams must make. Those familiar with research grant applications might equate evaluation study questions to the specific aims that are typically required for research grants. The number of specific aims in successful research grant applications typically falls into this 3–7 range.

3.2.4   How Should Study Questions Be Stated?

Again, there are no hard and fast rules for this, but some guidelines may be helpful:

- First and foremost, questions should be formulated grammatically as questions.
- Second, the questions should be phrased in the space between expressions of broad purpose and the very specific questions that might be asked of study participants.
- Third, the expression of the study question should not include a description of anticipated study methods.

For example, an appropriately formulated study question, might be:

"What patterns and levels of app use are associated with varying degrees of weight loss?"

A "question" inappropriately phrased as a statement that also invokes study methods, might be:

"Conduct a survey of users and analyze user log files to determine what patterns and levels of app use are associated with varying degrees of weight loss."

A question that is an expression of broad purpose might be:

"How successful is the app?"

The PICO format (Roever 2018), appropriate to many types of studies, can be a helpful guide for framing study questions. The PICO model, as adapted for informatics studies, expresses four elements comprising a complete study question:

P = Participants in the study—who are typically resource users such as students, patients, members of the public, or health care professionals—and the setting from which the participants are drawn;

I = Information resource or other **Intervention** being studied;

C = Comparison between the information resource under study with alternative resource or the absence of the resource;

O = Outcomes of interest to the stakeholders and the study team.

An example of an informatics study question in PICO format, would be:

In a sample of elderly diabetic persons (P), do those who use a specially designed app in conjunction with diet counseling (I) exhibit greater weight loss (O) than those who receive counseling only (C)?

Note that the complete PICO format applies only to study questions that are comparison-based. However, in most study settings there will be identified participants (P), some kind of information resource (I), and some envisioned outcome (O).

As discussed in the following section, there are many options for study design. By co-mingling study question formulation with study design, a team may close prematurely on a study design that is not optimal, impractical, or mismatched to the question. Study teams should begin the process of designing a study only after the questions have been formulated. In evaluation, design follows from purpose.

Self-Test 3.1

1. Classify each of the study questions below as either: (1) a broad statement of purpose, (2) an appropriately specific study question, (3) an overly-specific question that might be directed to a participant:

   (a) Are researchers who use the digital laboratory notebook better able than non-users to identify salient information in their recorded observations?
   (b) Are you able to understand what the app is telling you to do?
   (c) Was the implementation of the digital laboratory notebook successful?
   (d) For the third simulated case (the patient with acute chest pain), how helpful were the diagnoses suggested by the resource?
   (e) For a set of general internal medicine cases, to what extent does the resource improve the diagnostic hypotheses of medical residents?

2. From the study questions above:

   (a) Identify those that are appropriately specific for use in an evaluation study.
   (b) For the questions that are appropriately specific, identify the elements of the PICO format in each question.

3.3 Ten Informatics Study Types

Following the identification of study questions, study teams can begin to make key decisions about the actual design of their study. The design process begins by associating the study questions, and features of the broader context in which the study will be conducted, with one (or very often more than one) study type. Covering the set of study questions comprehensively may require two or more study types that are implemented in harmony with one another. The ten study types introduced in Table 3.1 are specific to informatics and form the next level of the "Tree of Evaluation". As shown in the table, each study type is characterized by the focus of the study, the aspect of the information resource on which it focuses, the stakeholders who will be primarily interested in the results, and whether study methods used tend to be quantitative, qualitative, or both.

Table 3.1 Classification of informatics study types by broad study foci, the stakeholders most concerned and the method(s) most commonly employed

Study type	Study focus	Resource aspect primarily studied	Audience/ stakeholders primarily interested in results	Family of methods most commonly employed
1. Needs assessment	The problem to be solved	Need for the resource	Resource developers, funders of the resource	Qualitative
2. Design validation	Conformance of the development process to accepted standards	Design and development process	Funders of the resource, resource developers, certification agencies	Qualitative
3. Structure validation	Design of the resource in relation to its intended function	Resource static structure	Insurers, resource developers, certification agencies	Qualitative
4. Usability	Ability of users to navigate the resource to carry out key functions	Resource user interfaces	Resource developers, resource users	Both
5. Laboratory function	The potential of the resource to be beneficial	Resource performance under controlled conditions (efficacy)	Resource developers, funders, users	Quantitative
6. Field function	The potential of the resource to be beneficial in the real world	Resource performance in actual use	Resource developers, funders, users	Quantitative
7. Lab user effect	Likelihood of the resource to change user behavior	Resource performance under controlled conditions (efficacy)	Resource developers, funders, users	Quantitative
8. Field user effect	Impact on user behavior in the real world	Resource effectiveness	Resource users and their clients, resource purchasers and funders	Both
9. Problem impact	Effect of the resource on the health problem it was designed to solve	Resource effectiveness	The universe of stakeholders	Both
10. Organization and system	Relationships between the resource and the organizational context in which it is deployed	Broader implications of the resource	Members of the organization where study is conducted, and similar organizations and policy makers	Both

The following sub-sections describe the ten study types. The reader may wish to navigate to relevant sections of Chap. 4 to see relevant scenarios and examples.

3.3.1 Needs Assessment Studies

Needs assessment seeks to clarify the information problem the resource is intended to solve, as discussed further in Sect. 4.2. These studies ideally take place before the resource is designed. They usually take place in the setting where the resource is to be deployed, although simulated settings may sometimes be used to study problems in communication or decision-making, as long as the potential users of the resource are included. Ideally, these potential users will be studied while they work with realistic problems or tasks to clarify how information is used and communicated, as well as identify the causes and consequences of inadequate information flows. The study team seeks to understand users' skills, knowledge, and attitudes, as well as how they make decisions or take actions. To ensure that developers and funders have a clear model of how a proposed information resource will fit with working practices and the overall organizational culture, study teams also may need to study health care or research processes, team functioning, and the cultural norms and values of the larger organization in which work is done. While qualitative methods are used more frequently in needs assessments, the consequences of the current challenges may be quantified in terms of costs or adverse outcomes. Needs assessments tend not to be published, because they are inward-looking and focused on problems specific to an identified deployment site. More details on how to conduct needs assessments are available in several books and articles (Witkin and Altschuld 1995; Yen et al. 2017; Weigl et al. 2012). An example study is described in Sect. 4.6.1.

3.3.2 Design Validation Studies

Design validation focuses on the quality of the processes of information resource design and development, most likely by asking an expert to review these processes using the qualitative "art criticism" approach to evaluation described in Sect. 2.7.2. The expert may review documents, interview the development team; compare the suitability of the technical standards, software-engineering methodology and programming tools used to others available; and generally apply their expertise to identify potential flaws in the approach used to develop the software, as well as constructively suggest how these might be corrected.

3.3.3 Structure Validation Studies

Structure validation addresses the static form of the software, often after the first prototype has been developed. This type of study is most usefully performed using the "art criticism" approach and is conducted by an expert or a team of experts with experience in developing software for the problem domain and concerned users. For these purposes, the study team needs access to both summary and detailed documentation about the system architecture, the structure and function of each module, and the interfaces between them. The expert might focus on the appropriateness of the algorithms that have been employed and check that they have been correctly implemented. Experts might also examine the data structures (e.g., whether they are appropriately normalized) and knowledge bases (e.g., whether they are evidence-based, up-to-date, and modeled in a format that will support the intended analyses or reasoning). Most of this will be done by inspection and discussion with the development team, without actually running the software.

Note that the study types listed up to this point do not require a functioning information resource. However, beginning with usability testing below, the study types require the existence of at least a functioning prototype.

3.3.4   Usability Studies

Usability studies address how well intended users can operate or navigate the resource to carry out tasks to meet their needs. Usability studies can focus either on a resource under development prior to deployment, or a resource that has been deployed. Section 4.3 in the following chapter provides a scenario related to usability studies. Usability studies employ both quantitative and qualitative methods. Quantitative data can be collected by the resource itself through automatically-generated logs of user actions, and by more sophisticated technical methods, such as eye-tracking tools. Qualitative data result from direct observation and/or recording of the interactions, and debriefing interviews of users after interactions with the resource. Many software developers have usability testing labs equipped with one-way mirrors and sophisticated measurement systems staffed by experts in human–computer interaction to carry out these studies—an indication of the importance increasingly attached to this type of study (Middleton et al. 2013; Abbott and Weinger 2020). An example study is described in Sect. 4.6.2.

3.3.5   Laboratory Function Studies

Laboratory-function studies go beyond usability of the resource to explore whether whatever the resource does has the potential to be beneficial. Users' abilities to manipulate the resource so it does what they want it to do addresses the usability question, but a function study addresses whether the results of these interactions also hold promise to achieve health-, science-, or education-related goals. Function studies relate directly to how the resource performs in relation to what it is trying to achieve for the user or the organization. In the common case where the resource provides information, advice or guidance to a user or groups of users, laboratory function studies will explore the extent to which the information or advice is scientifically accurate and appropriate to the circumstances. If the resource is dependent on "input" from the user, to what extent do the users provide the kinds of input the resource requires to generate useful results? To what extent do users understand the information, advice or guidance that the resource generates? Laboratory function studies engage samples of users who complete tasks under controlled conditions. Section 4.3 also addresses function studies and an example laboratory function study is found is Sect. 4.6.3.

3.3.6   Field-Function Studies

Field-function studies are variants of laboratory-function studies. In field-function studies, the resource is not ready for use in the real world, but the goal is to test it using real world problems or challenges. So the resource is "pseudo-deployed" in a real workplace and employed by real users, up to a point. As real users interact with the resource over real tasks, the results of these interactions—for example, the information or advice the resource might generate—goes to the study team instead of back to the users. There is no immediate access by the users to the output or results of interactions with the resource that might influence their real-world decisions or actions. The output of the resource is recorded for later review by the study team, and perhaps also by the users themselves. In a field-function study of a weight loss app for diabetics, users would be given access to the app for a fixed period of time and asked to use the app as they normally would. Any advice for diet or exercise modification would, however, be sent only to the study team for analysis. The study team might also choose to share the advice with the users engaged in the study, for their opinions as an additional source of data. An example study is described in Sect. 4.6.4.

3.3.7   Laboratory-User Effect Studies

With the transition from function studies to user-effect studies, attention turns from the resource's potential to how the resource might shape users' behavior. In laboratory-user effect studies, the intended users of the resource—practitioners, patients, researchers, students, or others—interact with the resource and are asked what they *would do* with the advice or other output that the resource generates, but no actions are taken as a result. Laboratory user effect studies are conducted under controlled conditions, with the users assigned to complete sets of tasks created by study team. Although such studies involve individuals who are representative of the end-user population, no actual health care, research or educational practices are affected by a study of this type. An example study is described in Sect. 4.6.5.

To illustrate the subtle difference between a usability, function and user-effect study, consider a resource that allows users to conduct searches of the biomedical literature or other biomedical database. The examples below illustrate typical study questions that differentiate usability, function, and effect studies:

- Usability study question: Can the user navigate the system as well as enter and improve their search terms?
- Function study question: Can the user conduct a search that yields relevant references as judged by others?
- Effect study question: Can the users conduct a search yielding results that alter their assessment of a problem?

As the types progress from studies conducted under controlled conditions to those conducted in the real world, the difference parallels the contrast, in clinical research, between efficacy and effectiveness. "Efficacy can be defined as the performance of an intervention under ideal and controlled circumstances, whereas effectiveness refers to its performance under 'real-world' conditions" (Singal et al. 2014). In general, the studies discussed above address efficacy. The studies discussed below address effectiveness.

3.3.8 Field User Effect Studies

Field-user effect studies focus on real actions involving health promotion and care, the conduct of research, education or other modes of practice. This requires an information resource that is stable and usable, restricting this kind of study to information resources that have been deployed for routine use. This type of study provides an opportunity to test whether the informational services provided through interaction with the resource affect users' decisions and actions in significant ways. In user effect studies, the emphasis is on the behaviors and actions of users, and not the consequences of these behaviors. There are many rigorous studies of carefully developed information resources revealing that the information resource is either not used, despite its seeming potential, or that when used, no decisions or actions are improved as a result (Eccles et al. 2002; Murray et al. 2004). These findings demonstrate that, however promising the results of previous evaluations, an information resource with the potential to be beneficial requires at least an effect study to determine its value once it is rolled out into routine practice.

Section 4.4 offers a scenario related to field user effect studies. Example studies are described in Sects. 4.6.6 and 4.6.7.

3.3.9 Problem Impact Studies

Problem impact studies are similar to field user effect studies in many respects, but differ profoundly in what is being explored. Problem impact studies examine whether the original problem or need that motivated creation or deployment of the information resource has been addressed in a satisfactory way. This often requires an investigation that looks beyond the behaviors and actions of care providers, researchers, or patients to examine the consequences of these actions. For example, an information resource designed to reduce medical errors may affect the behavior of some clinicians who employ the resource, but for a variety of reasons, leave the error rate unchanged. This can occur if the causes of errors are multi-factorial and the changes induced by the information resource address only some of these factors. In other domains, an information resource may be widely used by researchers to access biomedical information, as determined by a user effect study, but a subsequent problem impact study may or may not reveal effects on scientific productivity. New educational technology may change the ways students learn but may or may not increase their performance on standardized examinations. An example study is described in Sect. 4.6.8.

3.3.10 Organizational and System Studies

All the of study types discussed up to this point focus on the information resource itself, the intended users of the resource, the interactions between them, and the effects of these interactions. However, information resources and their users exist in social and organizational settings that can shape these interactions, and the deployment of an information resource can profoundly change the organization or larger system in which the resource is deployed. The pervasive culture of a social or organizational setting shapes the extent to which an information resource is used, will be accepted, and will prove to be beneficial. These relationships are critical to the ultimate success, and oftentimes the survival, of a deployed resource in an environment. With the advent of regional and national health information infrastructures (Yasnoff et al. 2004; Friedman et al. 2017), the outcomes of greatest interest may exist only at these large levels of scale. An example study is described in Sect. 4.6.9.

Field user-effect studies, problem impact studies, and organizational studies can also be sensitive to all-important unintended consequences. Oftentimes, deployment of an information resource intended to solve a target problem creates other unintended and/or unanticipated problems. Electronic health records intended to improve the quality and safety of health care had an unintended and unanticipated consequence of making care providers into "coders", greatly extending their workdays and causing significant resentment (Colicchio et al. 2019; Harrison et al. 2007).

The scenario in Section 4.5 of Chap. 4 elaborates on impact and organizational studies. Economic studies, discussed in detail in Chap. 18, represent a key subset of both problem impact studies and organizational studies.

3.4   Specific Features Distinguishing the Ten Informatics Study Types

Table 3.2 introduces a set of features that further describe and differentiate the ten study types introduced in the previous section. In addition to adding detail to the description of the study types, these features directly support study design by specifying the design options available once a study team has elected to perform a study of a particular type.

3.4.1   The Setting in Which the Study Takes Place

Studies of the design process, the resource structure, and many resource functions are typically conducted outside the work environment - in a "laboratory" setting where studies can take place under carefully controlled conditions. Studies to elucidate the need for a resource and studies of its effect and impact would usually take place in ongoing usage settings—known generically as the "field". The same is true for studies of the impact of a resource on organizations. These studies can take place only in a setting where the resource is routinely available for use and where these activities occur and/or important decisions are made.

3.4.2   The Status of the Information Resource Employed

Study types 1–3 do not require a working version of the resource. For some other kinds of studies, a prototype version of the resource is often sufficient, whereas for studies in which the resource is employed by intended users to support real decisions and actions, a deployed version that is robust and reliable is required.

3.4.3   The Sampled Resource Users

The information resources that are the foci of this book function through interaction with one or more "users". These "users" bring to every interaction their personal knowledge, attitudes and values, as well as their understanding of how best to interact with the resource.[1] The results of all studies hinge on these characteristics of the users who participate. The users who participate in a study are always, in some way, sampled from a larger population. It is never possible to study a resource by

---

[1] That said, studies of closed-loop devices that function with no direct human intermediation or interaction can employ many of the study types described here, with "users" in a passive role.

Table 3.2 Factors distinguishing the ten informatics study types

Study type	Study setting	Status of the resource	Sampled users	Sampled tasks	What is observed or measured
1. Needs assessment	Field	None, or pre-existing resource to be replaced	Anticipated resource users	Naturally occurring	User skills, knowledge, decisions, or actions; care processes, costs, team function or organization; patient outcomes
2. Design validation	Lab	Plans for the resource	None	None	Quality of design method or team
3. Structure validation	Lab	Prototype version	None	None	Quality of resource structure, components, architecture
4. Usability	Lab or field	Prototype or deployed version	Proxy or intended users	Authored, simulated, abstracted, naturally occurring	Speed of use, user comments, completion of tasks
5. Laboratory function	Lab	Prototype or deployed version	Proxy or intended users	Authored, simulated, abstracted	Speed and quality of data collected or displayed; accuracy of advice given
6. Field function	Field	Prototype or deployed version	Proxy or intended users	Naturally occurring	Speed and quality of data collected or displayed; accuracy of advice given
7. Lab user effect	Lab	Prototype or deployed version	Intended users	Authored, simulated, abstracted	Impact on user knowledge, simulated/ pretended decisions or actions
8. Field user effect	Field	Deployed version	Intended users	Naturally occurring	Extent and nature of resource use. Impact on user knowledge, real decisions, real actions
9. Problem impact	Field	Deployed version	Intended users	Naturally occurring	Care processes, costs, team function, cost effectiveness
10. Organization and system	Field	Deployed version	Intended users	Naturally occurring	Structural and functional aspects of organizations, culture and values

engaging all possible intended users. In some cases, the users of the resource in the study are not the actual intended users for whom the resource is ultimately designed. Instead, they may be members of the development or evaluation teams, or other individuals, who can be called "proxy users", who are chosen because they are conveniently available or because they are more affordable than, for example, licensed professionals.

3.4.4   The Sampled Tasks

In studies that employ a prototype or deployed version of the resource, the sampled users typically employ the resource to complete tasks that are meaningful to their professional or personal lives. (The resource is actually "run".) The precise nature of these tasks depends on what the resource is designed to do. For example, if the resource is patient-facing app, the tasks might be healthy food recipes to be prepared with assistance from the app. If the resources is an analytical tool for scientists, the tasks may be datasets for which the tool has been designed to be helpful.

For studies conducted in the field, the information resource is employed to complete tasks as they occur naturally in the experience of these users. For studies conducted in more controlled laboratory settings, the tasks are assembled in advance of the conduct of the study. They can be authored by experts to suit specific study purposes, or they can be abstracted versions of cases or problems that previously occurred in the real word, shortened to suit the specific purposes of the study. As with the resource users discussed previously, the tasks employed in a study carry serious implications for the study results, and tasks are always sampled in some way.

3.4.5   What Is Observed or Measured

All evaluation studies entail observations that generate data that are subsequently analyzed to address the study questions. Table 3.2 portrays the wide range of observations and measures that are possible, and how these differ by study type. More specifically, what is measured or observed in a study will follow directly from previously stated evaluation questions. Returning to the example study question presented in Sect. 3.2.4, the question:

What patterns and levels of app use are associated with varying degrees of weight loss?

would call for measurement of the patterns and levels of app use along with weight, assessed over time, for the individuals who use the app.

3.4.6 Sampling of Users and Tasks

In the paragraphs above, the term "sampling" has been introduced for both tasks and users. In evaluation studies, tasks and users are always sampled from some real or hypothetical population. Sampling occurs, albeit in different ways, in both quantitative and qualitative studies. Sampling of users and tasks are major challenges in evaluation study design. It is never possible, practical, or desirable to try to study everyone doing everything possible with an information resource. By choosing representative samples of users and tasks, study teams can be confident of challenging the resource with a reasonable spectrum of what is expected to occur in routine use. Under some circumstances, it is also important to know what will happen to the resource's usability, functions, or effects if the resource encounters extremes of user ability, data quality, or task difficulty. For these purposes, techniques to produce representative samples of users and/or tasks can be supplemented by purposive samples that create extreme atypical circumstances or provide a stress test of the resource. The quantitative and qualitative methods chapters of this book will address sampling issues in greater detail. The very specific challenge of sampling tasks in quantitative studies is addressed in Sect. 9.3.

Self-Test 3.2

For each of the following hypothetical evaluation scenarios, list which of the ten types of studies listed in Table 3.1 they include. Some scenarios may include more than one type of study.

1. A state makes a major investment in health information exchange to give emergency departments across the state access to medical histories of all state residents in the event of medical emergencies. A study of the number of allergic drug reactions before and after introduction of the exchange system is undertaken.
2. The developers of an app for persons with congestive heart failure recruit five potential users to help them assess how readily each of the main functions of the app can be accessed from the opening screen.
3. A study team performs a thorough analysis of the information required by psychiatrists to whom patients are referred by a community social worker.
4. A biomedical informatics expert is asked to review a Ph.D. student's dissertation project which entails creation of a prototype information resource. The expert requests copies of the student's code and documentation for review.
5. A new intensive care unit clinic information system is implemented alongside manual paper charting for a month. At the end of this time, the quality of the data generated by the new system and data recorded on the paper charts is compared. A panel of intensivists is asked to identify, independently, episodes of hypotension from each data set.
6. A resource is developed to enable patients to ask health related questions in their own words and obtain factually correct answers. The resource is tested by asking patients to state three health questions of primary interest to them. The questions

are recorded and then "spoken" to the resource. Panels of judges rate the correctness of the resource's answers.

7. A large health care delivery system wishes to understand the effects on job satisfaction of all health system personnel following the introduction of a comprehensive vendor-provided electronic health record system.

8. A program is devised that generates a predicted 24-h blood glucose profile using seven clinical parameters. Another program uses this profile and other patient data to advise on insulin dosages. Diabetologists are asked to prescribe insulin for the patient given the 24-h profile alone, and then again after seeing the computer-generated advice. They also are asked their opinion of the value of the advice.

9. A resource to identify patients eligible for clinical trials is implemented across a large health system. The rates of accrual to trials, before and after implementation, are compared.

3.5 Conclusions: Approaching a Recipe for Study Design

The selection of one, and often more, study types is a vital first step in evaluation study design. As noted previously, each study design always follows from the study questions. The ideas presented in this chapter approach as closely as possible something akin to a cookbook recipe for study design. The steps in this "recipe" are:

1. Define the broad study features and memorialize them in a contract.
2. Develop the more detailed (3–7) questions to guide the study.
3. Select one or more study types, as described in Tables 3.1 and 3.2, that, together, enable the team to address all of the study questions.
4. For each study type selected, and with reference to Table 3.2, begin more detailed study design by noting how the choice of study types guides, and in some cases constrains, many design options.
5. Depending on the study types chosen, quantitative and/or qualitative methods will be used. Chapters 6, 7, 8, 9, 10, 11, 12, and 13 provide detailed design guidance for quantitative studies. Chapters 14, 15, and 16 do the same for qualitative studies. Chapter 17 addresses mixed methods studies which use both.

The "Tree of Evaluation" illustrates the relationship between the eight generic evaluation approaches introduced in Sect. 2.7, which are applicable to any field, and the ten informatics-specific study types introduced in this chapter. While it is possible in principle to use any of the eight approaches in any of the ten study types, the "Tree" illustrates that some informatics study types align with the approaches that lean to the quantitative side, and other types align with other approaches that lean to qualitative side. For example, design and structure validation studies in informatics invite use of the art criticism and professional review approaches. The evaluation scenarios and selected studies presented in Chap. 4 will make these relationships more explicit.

Food for Thought

1. Consider the statement offered in Sect. 3.2.2 that even the most quantitative studies begin with a qualitative process of engaging stakeholders to arrive at a set of agreed-upon study questions. In other words, a study striving to be maximally objective is grounded in a subjective exercise. Is this strange or disconcerting? Why or why not?
2. Refer to the Nouveau Community Hospital scenario within Self-test 2.1. For either of the areas of concern motivating the study, articulate three or more study questions at the "just right" level of specificity between broad statements of purpose and questions that might be asked directly to participants. Where appropriate, express these in PICO format.

Answers to Self-Tests

Self-Test 3.1

1a. ii
1b. iii
1c. i
1d. iii
1e. ii

2a. Questions (a) and (e) are appropriately specific study questions.
2b. For Question a:

- P = researchers
- I = digital lab notebook
- C = users vs. non-users of notebook
- O = identifying salient information

  For Question e:

- P = medical residents
- I = information resource designed to improve diagnosis
- C = (implied) before use and after use comparison
- O = quality of diagnostic hypotheses

Self-Test 3.2

1. Problem impact
2. Usability test
3. Need assessment
4. Design validation
5. Field function
6. Lab function

7. Organizational
8. Field user effect
9. Problem impact.

References

Abbott PA, Weinger MB. Health information technology: fallacies and sober realities - Redux A homage to Bentzi Karsh and Robert wears. Appl Ergonom. 2020;82:102973.

Colicchio TK, Cimino JJ, Del Fiol G. Unintended consequences of nationwide electronic health record adoption: challenges and opportunities in the post-meaningful use era. J Med Internet Res. 2019;21:13313.

Eccles M, McColl E, Steen N, Rousseau N, Grimshaw J, Parkin D, et al. Effect of computerised evidence based guidelines on management of asthma and angina in adults in primary care: cluster randomised controlled trial. BMJ. 2002;325:941–8.

Friedman CP, Rubin JC, Sullivan KJ. Toward an information infrastructure for global health improvement. Yearb Med Inform. 2017;26:16–23.

Grembowski D. The practice of health program evaluation. 2nd ed. Thousand Oaks, CA: Sage Publications; 2016.

Harrison MI, Koppel R, Bar-Lev S. Unintended consequences of information technologies in health care - an interactive sociotechnical analysis. J Am Med Inform Assoc. 2007;14:542–9.

Middleton B, Bloomrosen M, Dente MA, Hashmat B, Koppel R, Overhage JM, et al. Enhancing patient safety and quality of care by improving the usability of electronic health record systems: recommendations from AMIA. J Am Med Inform Assoc. 2013;20:2–8.

Murray MD, Harris LE, Overhage JM, Zhou XH, Eckert GJ, Smith FE, et al. Failure of computerized treatment suggestions to improve health outcomes of outpatients with uncomplicated hypertension: results of a randomized controlled trial. Pharmacotherapy. 2004;24:324–37.

Owen JM. Program evaluation: forms and approaches. 3rd ed. London: Routledge; 2020.

Posovac EJ. Program evaluation: methods and case studies. London: Routledge; 2011.

Roever L. PICO: model for clinical questions. Evid Based Med Pract. 2018;3:1–2.

Singal AG, Higgins PD, Waljee AK. A primer on effectiveness and efficacy trials. Clin Transl Gastroenterol. 2014;5:45.

Weigl BH, Gaydos CA, Kost G, Beyette FR Jr, Sabourin S, Rompalo A, et al. The value of clinical needs assessments for point-of-care diagnostics. Point Care. 2012;11:108–13.

Witkin BR, Altschuld JW. Planning and conducting needs assessments: a practical guide. Thousand Oaks, CA: Sage Publications; 1995.

Yarbrough DB, Shulha LM, Hopson RK, Caruthers FA. The program evaluation standards: a guide for evaluators and evaluation users. 3rd ed. Thousand Oaks, CA: Sage Publications; 2011.

Yasnoff WA, Humphreys BL, Overhage JM, Detmer DE, Brennan PF, Morris RW, et al. A consensus action agenda for achieving the national health information infrastructure. J Am Med Inform Assoc. 2004;11:332–8.

Yen PY, McAlearney AS, Sieck CJ, Hefner JL, Huerta TR. Health information technology (HIT) adaptation: refocusing on the journey to successful HIT implementation. JMIR Med Inform. 2017;5:28.

Chapter 4
Study Design Scenarios and Examples

Learning Objectives
Since this chapter is based on the content of Chap. 3, it supports achievement of the Chap. 3 learning objectives.

4.1 Introduction

This chapter is a companion to Chap. 3 and supports the learning objectives in that chapter. It offers scenarios and illustrations of the informatics study types previously introduced in that chapter. As such, this chapter has no specific learning objectives as it is designed to enrich and deepen the understanding of specific concepts rather than presenting new ones. Readers of this chapter will likely find themselves navigating back and forth between Chaps. 3 and 4. The provided hyperlinks will assist in this process.

The initial sections of this chapter offer four scenarios that collectively capture many of the opportunities for formal evaluation studies to provide valuable information to a range of stakeholders. Each scenario maps to one or two specific study types. The discussion of each scenario illustrates the multiplicity of evaluation questions that arise naturally in informatics.

- A health problem or opportunity has been identified that seems amenable to an information/communication resource, but there is a need to define the problem in more detail.
- A prototype information resource, or a new component of an existing resource, has been developed, but its usability and potential for benefit need to be assessed prior to deployment.
- A locally developed information resource has been deployed within an organization, but no one really knows how useful it is proving to be.

© Springer Nature Switzerland AG 2022
C. P. Friedman et al., *Evaluation Methods in Biomedical and Health Informatics*, Health Informatics, https://doi.org/10.1007/978-3-030-86453-8_4

- A commercial resource has been deployed across a large enterprise, and there is need to understand what impact it has on users, as well as on the organization.

The second section of the chapter illustrates many of the informatics study types with brief descriptions of published papers.

4.2 A Scenario Invoking a Needs Assessment Study

A health problem or opportunity has been identified that seems amenable to an information/communication resource, but there is a need to define the problem in more detail.

The emphasis here is clearly on understanding the nature and cause of a problem, and particularly whether the problem is due to poor information collection, management, processing, analysis, or communication. An information resource is unlikely to help if the cause lies elsewhere. It is likely that careful listening to those who have identified or "own" the problem will be necessary, as will some observation of the problem when it occurs, interviews with those who are affected, and assessment of the frequency and severity of the problem and its consequences to patients, professionals, students, organizations, and others. The emphasis will be on assessing needs as perceived across a range of constituencies, studying potential resource users in the field, assessing user skills, knowledge, decisions, or actions, as well as work processes, costs, team function, or organizational productivity. If there are existing information resources in the setting, studies of their use and the quality of data they hold can provide further valuable insights into the nature of the problem and reason it occurs. If no information problem is revealed by a thorough needs assessment, there is probably no need for a new information resource, irrespective of how appealing the notion may seem from a technical point of view.

4.2.1 Typical Study Questions

This scenario typically raises several questions that can be addressed using a variety of evaluation methods. These questions include:

- What is the problem, why does it matter, and how much effort is the organization likely to devote to resolving it?
- What is the history of the problem, and has anyone ever tackled it before? How, and with what outcome?
- Where, when, and how frequently does the problem occur? What are the consequences for staff, other people, and the organization?
- Is the problem independent of other problems, or is it linked to, or even a symptom of, other problems somewhere else?

- What are all the factors leading to the problem, and how much might improvements in information handling ameliorate it?
- Who might generate better information to help others resolve the problem; how do these alternative information providers perceive the problem and their potential part in resolving it?
- What kinds of information need to be obtained, processed, and communicated for the information resource users and others concerned in solving the problem? From where, to where, when, how, and to whom?
- What other information resources exist or are planned in the environment?
- What quality and volume of information are required to ameliorate the problem enough to justify the effort involved?
- How, in general, can these information/communication needs be met? What specific functions should be built into an information resource to meet the identified needs?
- Is there anything else that a potential resource developer would like to know before spending their time on this problem?

4.2.2   Performing Needs Assessments

The success of any biomedical or health information resource depends on how well it fulfills a health care, research, or educational need. Usually, before developers begin the design of an information resource, someone—often a representative of the potential user community—has identified a problem amenable to a solution via improved utilization of biomedical information or knowledge. Sometimes, however, and particularly for information resources employing cutting-edge technology, the project is initiated by the developers without careful conceptualization of the real-world problem the resource is intended to address. Often, the project deteriorates from this point as the developer tries to persuade the increasingly mystified professionals in that site of the potential of a "breakthrough" that is, in fact, a solution in search of a problem. Thus, defining the need for an information resource before it is developed is an important precursor to any developmental effort.

Let us say, for the sake of argument, that a clinician notices that the postoperative infection rate on a certain hospital ward is high, and that, for unknown reasons, patients may not be receiving the prophylactic antibiotics that are known to be effective. The clinicians uncovering the problem may merely note it, or they may try to define and understand it more, which sets the stage for an evaluation study. In increasing order of complexity, a study team that forms to conduct the study may discuss the problem with colleagues, conduct a staff survey to explore the nature and extent of the perceptions of the problem, collect actual healthcare data to document the problem, and perhaps compare the locally collected data with published results based on data collected elsewhere. Such careful study would require a definition of what constitutes a postoperative infection (e.g., does it include chest infections as well as wound infections?) and then an audit of postoperative infections and the drugs prescribed before and after surgery.

Defining the problem is always necessary to guide the choice of a specific solution. While professionals in informatics may be prone to believe that all problems can be addressed with advanced information- or knowledge-based technologies, problems and potentially effective solutions to address them come in many shapes and sizes. A new or revised information resource may not be what is needed. For example, the root cause of the example problem above may be that a confusing interface leads to mis-entered orders and the wrong drug is administered. Particularly in the short term, this etiology recommends a very different kind of solution from what would be obtained if the source of the problem were out-of-date antibiotic knowledge on the part of the prescribers themselves. Once the mechanisms of the problem are uncovered, it may prove most efficient to address them with educational sessions, wall posters, or financial incentives instead of new information technology. If an information resource is the chosen solution, the resource users and developers need to choose the appropriate kind of information to provide (advice to prescribe an appropriate antibiotic from the formulary), the appropriate time it is to be delivered (6 h before to 2 h after surgery), and the appropriate mode of delivery (incorporation into an order set, advice on-screen to the resident or intern, or a printed reminder affixed to the front of the case record).

An example of a needs assessment study is found in Sect. 4.6.1 below.

4.3   A Scenario Invoking a Usability and/or a Function Study

A prototype information resource, or a new component of an existing resource, has been developed, but its usability and potential for benefit need to be assessed prior to deployment.

For this scenario, it is important to recognize that the target resource could take many shapes and sizes. The prototype resource could be a relatively self-contained system, such as a new patient facing mobile app; or it could be a new functional module designed to add value to an existing resource, such as an advisory system to improve drug prescribing based on pharmacogenomics. In this scenario, the study team must first clearly identify the resource under study and place clear boundaries around it, since in modern information technology, any resource will be connected to a larger ecosystem of resources. If the resource entails a high level of interactivity with the end-users, and especially if the users are entering information, the need for a formal usability study always arises. If the resource places end-users in a relatively passive role, such as a machine learning early-warning system for deterioration of patients in critical care, a function study conducted in the lab or the field may be more appropriate. It is essential to recognize that usability of a resource by those who designed it is a necessary but insufficient condition for usability by those for whom it is intended.

4.3.1   Typical Study Questions

For this scenario, typical study questions will include:

- Who are the target users, and what are their background skills and knowledge?
- Does the resource make sense to target users?
- Following a brief introduction, can target users navigate themselves around important parts of the resource?
- Can target users select relevant tasks in reasonable time and with reasonable accuracy using the resource?
- What user characteristics correlate with the ability to use the resource and achieve fast, accurate performance with it?
- What other kinds of people can use it safely?
- How to improve the screen layout, design, wording, menus, etc.
- Is there a long learning curve? What user training needs are there?
- How much ongoing help will users require once they are initially trained?
- What concerns do users have about the information resource—usability, accuracy, privacy, effect on their jobs, other side effects?
- Based on the performance of prototypes, does the resource generate clinically valid and scientifically valid information as output?
- Do different stakeholders have different views about the validity of this information?
- In what ways does the validity of the output depend on how end-users employ the resource?

4.3.2   Performing Usability and Function Studies

These questions fall within the scope of the usability and laboratory-function testing approaches listed in Tables 3.1 and 3.2. Usability studies engage a range of techniques, largely borrowed from the human–computer interaction field and employing both quantitative and qualitative approaches. These include:

- Seeking the views of potential users after both a demonstration of the resource and a hands-on exploration. Methods such as focus groups may be very useful to identify not only immediate problems with the software and how it might be improved, but also potential broader concerns and unexpected issues that may include user privacy and long-term issues around user training and working relationships.
- Studying users while they carry out a list of predesigned tasks using the information resource. Methods for studying users include watching over their shoulder; video observation (sometimes with several video cameras per user); use of eye-tracking technology; think out-loud protocols (asking the user to verbalize their

impressions as they navigate and use the system); and automatic logging of key-strokes, navigation paths, and time to complete tasks.

- Use of validated questionnaires to capture user impressions, often before and after an experience with the system, with one example being the mHealth App Usability Questionnaire (Zhou et al. 2019).
- Specific techniques to explore how users might improve the layout or design of the software. For example, to help understand what users think of as a logical menu structure for an information resource, study teams can use a card-sorting technique (United States General Services Administration (GSA) 2021). This entails listing each function available on all the menus on a separate card and then asking users to sort these cards into several piles according to which function seems to go with which.

Depending on the aim of a usability study, it may suffice to employ a small number of potential users. Nielsen has shown that, if the aim is only to identify major software faults, the proportion identified rises quickly up to about five or six users, then much more slowly to plateau at about 15 to 20 users (Nielsen and Landauer 1993). Five users will often identify 80% of software problems. These small sample usability studies serve, more often than not, to suggest ways to improve the resource. For this reason, usability studies are often performed iteratively to test the value of improvements made in response to studies done previously.

Function studies often employ larger samples and the focus shifts from interaction with users to what the information resource generates as output. This output can take many forms, depending on the purpose of the resource. For many resources, this output takes the form of advice: for patients, how to change their health-related behavior; for clinicians, how to alter a patient care plan; for students, what learning resources might address their knowledge deficits; and for researchers, what analytical techniques are best suited to their study design. In all cases, the emphasis in a function study is on the validity and the *potential* utility of this output.

Exemplary usability and function studies are described in Sects. 4.6.2, 4.6.3, and 4.6.4 below.

4.4   A Scenario Invoking a Field User Effect Study

A locally developed information resource has been deployed within an organization, but no one really knows how useful it is proving to be.

The situation here is quite different from the preceding scenario. Here, the system is deployed in one part of the organization, so it has moved beyond the prototype stage of development. The key issues are whether the resource is being used, by whom, whether this usage is appropriate, and what behavior and attitude changes are actually occurring in the environment where the resource is deployed. With reference to Tables 3.1 and 3.2, this scenario typically calls for field-user effect studies.

4.4.1   Typical Study Questions

These might include:

- Is the resource being used as intended?
- Is the information being captured or communicated by the resource of good quality (accurate, timely)?
- To what extent and in what ways has user behavior changed as a result of interactions with the resource?
- Are these changes in line with what was intended?
- Are there secondary changes in the behavior of others in the organization who are not direct users of the resource?
- What are users' attitudes and beliefs about using the resource in day-to-day work? Do these beliefs depend on the users' background, work assignments, or role in the organization?
- Does the information resource appear to be causing any specific problems for users? What are these problems, and how often do they occur? What are their causes?
- Do the effects of the resource derive from characteristics unique to a limited set of users or the specific areas where it has been deployed?
- Is the resource having other effects that were unanticipated?

4.4.2   Performing Effect Studies

The evaluation approach under this scenario can employ quantitative or qualitative methods, or a mix of methods as described in Chap. 17. When the intended behavior changes are well defined, such as compliance with specific care guidelines or daily monitoring of personal health habits, these can be measured using quantitative methods to be described in Chapters 6–9, often using already existing measurement tools. When deeper understanding of why the desired effects have or have not occurred, qualitative studies can be particularly beneficial.

Effect studies can be carried out in both laboratory settings and field settings; but, more often, they are performed in the field. Studies in the field are important because study teams cannot assume that users are making best use of the resource, even if its usability has been verified in the lab. It is possible that the resource is being used minimally or not at all. If usability studies and function studies have not been performed, it may be necessary to circle back to these types of studies to understand why limited or unanticipated effects are occurring and where the resource needs improvement.

For effect studies, it is often important to know something about how the users carried out their work prior to the introduction of the information resource, so many effect studies employ a "before and after" design. As will be discussed in detail in Chap. 11, study teams must be careful, in designing such studies, to be able to distinguish between effects due to the resource itself as opposed to other factors. In particular, the effects of a resource that provides information or advice should be studied in relation to the advice provided. If, for example, the resource is intended to improve user decisions, the resource must tell the users something they do not already know. The resource may be providing accurate or otherwise useful advice, and in that sense is a high performing resource, but it will have no benefit unless it augments the users' knowledge. This reasoning aligns closely with the "Fundamental Theorem" introduced in Sect. 1.2.

If the results of effect studies are promising, it may be beneficial to study whether the need for the resource exists in other parts of the organization where it is currently not available or is not being used. If the results are positive, more widespread deployment of the system across the organization can take place.

Exemplary effect studies are described in Sects. 4.6.5, 4.6.6, and 4.6.7 below.

4.5   Scenario Invoking Problem Impact and Organization and System Studies

A commercial resource has been deployed across a large enterprise, and there is need to understand what impact it has on users, their clients, as well as on the organization as a whole.

This evaluation scenario, suggesting a problem impact or organizational study, is often what people think of first when the topic of evaluation is introduced to them. However, it is only one of many types of evaluation studies, and it aligns with a late stage in the life cycle of an information resource. It is important to avoid undertaking an impact or organizational study if no prior formal studies have occurred. This is primarily because a null result of an impact study (a finding of no impact) could be attributable to wide range of causes that would have been revealed, and possibly corrected, by the findings of studies conducted earlier.

Impact-oriented evaluations target the ultimate goals of the information resource. For a resource designed to improve health care, impact can be defined in terms of the health of patients who receive that care. For a resource designed to support research, impact can be defined in terms of the research productivity of the organization where it is deployed.

Impact-oriented evaluations often take place in large organizations that have invested in large-scale information resources and need to understand the returns on the investments they have made and the larger investments that are likely needed in the future. These stakeholders will vary in the outcome measures that, if assessed in a study, will convince them that the resource has been effective. Many such stakeholders will wish to see quantified indices of benefits or harms stemming from the resource—for example, the amount the resource improves productivity or reduces costs, or perhaps other benefits such as reduced waiting times to perform key tasks or procedures, lengths of hospital stay, or occurrence of adverse events. Governmental agencies take interest in impact- and organization-level evaluations to justify the expenditure of public funds in pursuit of improved health of the populations they serve. For example, the U.S. Office of the National Coordinator for Health IT invested significant resources in the evaluation of the government's major investments in health information technology that began in 2009 (United States Office of the National Coordinator for Health Information Technology (ONC) 2018; Gold and McLaughlin 2016).

4.5.1 Typical Study Questions

Study questions that arise out of this scenario include:

- In what fraction of occasions when the resource could have been used was it actually used?
- Who uses it, why, are these the intended users, and are they satisfied with it?
- For information resources addressing health care or personal health, does using the resource change health status or outcomes for patients?
- For research resources, does the resource affect research quality or productivity?
- For educational resources, does the resource affect learning outcomes?
- Does using the resource influence organizational cohesion or communication?
- Does using the resource have persistent effects on user knowledge or skills?
- Does using the resource affect job satisfaction?
- How does the resource influence the whole organization and relevant subunits?
- Do the overall benefits and costs or risks differ for specific groups of users, departments, and the whole organization?
- How much does the resource really cost the organization?
- Should the organization keep the resource as it is, improve it, or replace it?
- What have been the unintended consequences of introducing the resource?
- How can the resource be improved, at what cost, and what benefits would result?

To each of the above questions, one can add "Why or why not?" to gain a broader understanding of what is happening because of use of the resource.

4.5.2 Performing Impact and Organizational and System Studies

If the study team is pursuing quantitative methods, deciding which of the possible outcome variables to include in an impact study and developing ways to measure them can be the most challenging aspect of an evaluation study design. Study teams usually wish to limit the number of measures employed in a study. This is for many reasons: to make optimal use of limited evaluation resources, to minimally burden the users with data collection activities, and to avoid statistical analytical challenges that can result from a large number of outcome measures.

Impact studies can use qualitative approaches to allow the most relevant issues to emerge over time and with increasingly deep immersion into the study environment. This emergent feature of qualitative work obviates the need to decide in advance which impact variables to explore and is one of the major advantages of qualitative approaches.

Because problem impact studies in healthcare settings can be very expensive and time-consuming, study teams can sometimes save much time and effort by measuring impact in terms of certain healthcare processes and behaviors rather than patient outcomes. This is tantamount to using an effect study as a proxy for an impact study. For example, measuring the mortality or complication rate in patients with heart attacks requires data collection from hundreds of patients, as severe complications and death are (fortunately) rare events. However, if large, rigorous trials or meta-analyses have determined that a certain procedure (e.g., giving heart-attack patients streptokinase ideally within 6 h of symptom onset) correlates closely with the desired patient outcome, it may suffice to measure the rate of performing this procedure as a valid surrogate for the desired outcome (RxList 2021). Mant and Hicks demonstrated that measuring the quality of care by quantifying a key process in this way might require one-tenth as many patients as measuring outcomes (Mant and Hicks 1995).

The landscape for impact studies is also changing with the advent of large repositories, containing routinely collected data, that can often be "mined" for relationships between the introduction of information resources and outcomes of interest. Studies conducted using repository data are observational or "real world evidence" studies discussed in Chap. 13. Studies using repository data can be carried out rapidly but have several important limitations, including limitation of what can be learned from the data routinely collected, concerns about quality of the data, and above all, the concern that a statistical relationship cannot be interpreted as a causal relationship. If the research productivity of a university increased 1 year after introduction of a new comprehensive research information system, does that mean that this information resource was the cause of the increase? Later chapters will explore this question in depth.

Organizational and system studies can be conducted at levels of scale ranging from small hospitals and research agencies up to entire nations.

Exemplary problem impact studies and organizational studies are found in Sects. 4.6.8 and 4.6.9 below.

4.6 Exemplary Evaluation Studies from the Literature

This section summarizes studies that align with many of the study types described in Tables 3.1 and 3.2, and illustrate published studies aligning with the four scenarios discussed above.

4.6.1 Needs Assessment Study: "Using Ethnography to Build a Working System: Rethinking Basic Design Assumptions"

Forsythe published in 1992 the archetype for needs assessment studies, as well as a compelling case for performing them (Forsythe 1992). This paper describes how the developers of a patient-facing information resource for individuals suffering from migraine were on the verge of developing the "wrong system", one that did not address the needs of the patients it was intended to serve. A qualitative needs assessment inserted into the project plan generated a more accurate portrayal of user needs and the system's design was redirected accordingly. Even though this paper was written several decades ago, its message still rings true and remains an important resource for everyone in the field of informatics.[1]

4.6.2 Usability Study: "Assessing Performance of an Electronic Health Record Using Cognitive Task Analysis"

Saitwal published a usability testing study that evaluates the Armed Forces Health Longitudinal Technology Application (AHLTA) EHR using a cognitive task analysis approach, referred to as Goals, Operators, Methods, and Selection rules (GOMS) (Saitwal et al. 2010). Specifically, the study team evaluated the system response time and the complexity of the graphical user interface (GUI) when completing a set of 14 prototypical tasks using the EHR. The team paid special attention to inter-rater reliability of the two study team members who used GOMS to analyze the GUI of the system. Each task was broken down into a series of steps, with the intent to determine the percent of steps classified as "mental operators". Execution time was then calculated for each step and summed to obtain a total time for task completion.

4.6.3 Lab Function Study: "Diagnostic Inaccuracy of Smartphone Applications for Melanoma Detection"

Wolf et al. (2013) conducted an evaluation study of smartphone applications capable of detecting melanoma and sought to determine their diagnostic inaccuracy. The study is exemplary of a lab function study and complements the Beaudoin et al. study described below because the study team paid special attention to measuring

---

[1] A more recent example of a needs assessment in the field of rehabilitation is found in a paper by Moody et al. (Moody et al. 2017).

application function in a lab setting using digital clinical images that had a confirmed diagnosis obtained via histologic analysis by a dermatopathologist.

The study team employed a comparative analysis between four different smartphone applications and assessed the sensitivity, positive predictive value, and negative predictive value of each application compared to histologic diagnosis. Rather than focus on the function in a real health care setting with real users, the study team was interested in which applications performed best under controlled conditions.

4.6.4 Field Function Study: "Evaluation of a Machine Learning Capability for a Clinical Decision Support System to Enhance Antimicrobial Stewardship Programs"

Beaudoin et al. conducted an observational study to evaluate the function of a combined clinical decision support system (antimicrobial prescription surveillance system (APSS)) and a learning module for antimicrobial stewardship pharmacists in a Canadian university hospital system. The study team developed a rule-based machine learning module designed from expert pharmacist recommendations which triggers alerts for inappropriate prescribing of piperacillin–tazobactam. The combined system was deployed to pharmacists and outputs were studied prospectively over a 5-week period within the hospital system.

Analyses assessed accuracy, positive predictive value, and sensitivity of the combined system, the individual learning module alone, and the decision support system alone, in comparison with pharmacists' opinions. This is an exemplary field function study because the study team examined the ability of the combined rule-based learning module and APSS to detect inappropriate prescribing in the field with real patients.

4.6.5 Lab User Effect Study: "Applying Human Factors Principles to Alert Design Increases Efficiency and Reduces Prescribing Errors in a Scenario-Base Simulation"

Russ et al. studied the redesign of alerts using human factors principles and their influence on prescribing by providers (Russ et al. 2014). The study is exemplary of a lab user effect study because it analyzed frequency of prescribing errors by providers, and it was conducted in a simulated environment (the Human-Computer Interaction and Simulation Laboratory in a Veterans Affairs Medical Center). The study team was particularly interested in three types of alerts: drug-drug interactions, drug-allergy, and drug disease.

Three scenarios were developed for this study that included 19 possible alerts. These alerts were intended to be familiar and unfamiliar to prescribers. The study team focused on the influence of original versus redesigned alerts on outcomes of perceived workload and prescribing errors. The study team also employed elements of usability testing during this study, such as assessing learnability, efficiency, satisfaction, and usability errors.

4.6.6 Field User Effect Study: "Reminders to Physicians from an Introspective Computer Medical Record: A Two-Year Randomized Trial"

McDonald et al. conducted a two-year randomized controlled trial to evaluate the effects of a computer-based medical record system which reminds physicians about actions needed for patients prior to a patient encounter (McDonald et al. 1984).[2] This study most closely aligns with a field user effect study for the attention to behavior change in preventive care delivery associated with use of the information resource, and is exemplary because its rigorous design accounts for the hierarchical nature of clinicians working in teams without having to manipulate the practice environment. Randomization occurred at the team level and analyses were performed at both the team and individual levels. The study did include problem impact metrics that addressed patients' health status, however no significant changes were observed in these outcomes.

4.6.7 Field User Effect Study: "Electronic Health Records and Health Care Quality Over Time in a Federally Qualified Health Center"

Kern et al. conducted a three-year comparative study across six sites of a federally qualified health center in New York to analyze the association between implementation of an electronic health record (EHR) and quality of care delivery as measured by change in compliance with Stage 1 Meaningful Use quality measures

---

[2] Although there have been hundreds of field user effect studies (Scott et al. 2019) conducted since the work by McDonald et al. in 1984, this study is specifically called out as an archetype for them all.

(Kern et al. 2015). This study is an exemplary field user effect study for its attention to measures of clinician behavior in care delivery through test/screening ordering using the EHR and explicit use of statistical analysis techniques to account for repeated measures on patients over time. The study also includes two problem impact metrics (change in HbA1c and LDL cholesterol) analyzed over the study period; however, the study intent was primarily focused on clinician behavior.

4.6.8 Problem Impact Study: "Effects of a Mobile Phone Short Message Service on Antiretroviral Treatment Adherence in Kenya (WelTel Kenya1): A Randomised Trial"

The study team (Lester et al. 2010) conducted a randomized controlled trial to measure improvement in patient adherence to antiretroviral therapy (ART) and suppression of viral load following receipt of mobile phone communications with health care workers. The study randomized patients to the intervention group (receiving mobile phone messages from healthcare workers) or to the control group (standard care). Outcomes were clearly identified and focused on behavioral effects (drug adherence) and the extent that improvements in adherence influenced patient health status (viral load). The special attention to randomization and use of effect size metrics for analysis are critical components to measuring the overall impact of mobile phone communications on patient health.

4.6.9 Organizational Study: "Electronic Health Record Adoption in US Hospitals: The Emergence of a Digital 'Advanced Use' Divide"

This study team examined whether "hospitals use EHRs in advanced ways that are critical to improving outcomes, and whether hospitals with fewer resources – small, rural, safety-net – are keeping up." (Adler-Milstein et al. 2017). Their data source was a survey conducted annually by the American Hospital Association. They found that 81% of hospitals overall had adopted, at minimum, what was previously defined as a basic EHR system. So-called critical access hospitals, that are smaller and in rural areas, were less likely to have adopted 16 distinct EHR functions. This work pointed to a distinct disparity in use of electronic health records in pursuit of improve health across the United States.

4.7 Summary and Conclusions

The studies cited above are an extremely small sample of the informatics evaluation studies that have been published in the past six decades. For example, a literature review covering the period 1997–2017 revealed 488 published empirical studies addressing the focused topic of clinical decision support as used by qualified practitioners (Scott et al. 2019).

Moreover, the studies described here represent an even smaller fraction of the studies actually conducted. For example, of the 42 evaluation study reports focused on the 2009–2014 electronic health record initiatives of the U.S. Office of the National Coordinator for Health IT (United States Office of the National Coordinator for Health Information Technology (ONC) 2018), an informal review revealed that less than 10 resulted in publications. Looking back on the definitions and functions of evaluation described in Sect. 2.2, this is how it should be. Many studies are directed by purposes that are internal to, or have findings that are useful to, only the organization commissioning them. In these more numerous cases, publication of the methods and results of even an enormously successful and useful study is likely unnecessary and may in certain cases be impermissible.

As will be seen in the following chapter, publication of studies is nonetheless very important through its contribution to "evidence-based health informatics". Among other benefits, published studies, especially those describing their methods in detail, are valuable in helping study teams learn from each other how to improve their approaches and techniques.

References

Adler-Milstein J, Holmgren AJ, Kralovec P, Worzala C, Searcy T, Patel V. Electronic health record adoption in US hospitals: the emergence of a digital "advanced use" divide. J Am Med Inform Assoc. 2017;24:1142–8.

Beaudoin M, Kabanza F, Nault V, Valiquette L. Evaluation of a machine learning capability for a clinical decision support system to enhance antimicrobial stewardship programs. Artificial Intelligence Med. 2016;68:29–36.

Forsythe DE. Using ethnography to build a working system: rethinking basic design assumptions. In: Frisse ME, editor. Proceedings of the 16th annual symposium on computer applications in medical care. New York: McGraw-Hill; 1992. p. 510–4.

Gold M, McLaughlin C. Assessing HITECH implementation and lessons: 5 years later. Milbank Q. 2016;94:654–87.

Kern LM, Edwards AM, Pichardo M, Kaushal R. Electronic health records and health care quality over time in a federally qualified health center. J Am Med Inform Assoc. 2015;22:453–8.

Lester RT, Ritvo P, Mills EJ, Kariri A, Karanja S, Chung MH, et al. Effects of a mobile phone short message service on antiretroviral treatment adherence in Kenya (WelTel Kenya1): a randomised trial. Lancet. 2010;376:1838–45.

Mant J, Hicks N. Detecting differences in quality of care: the sensitivity of measures of process and outcome in treating acute myocardial infarction. BMJ. 1995;311:793–6.

McDonald CJ, Hui SL, Smith DM, Tierney WM, Cohen SJ, Weinberger M, et al. Reminders to physicians from an introspective computer medical record: a two-year randomized trial. Ann Intern Med. 1984;100:130–8.

Moody L, Evans J, Fielden S, Heelis M, Dryer P, Shapcott N, et al. Establishing user needs for a stability assessment tool to guide wheelchair prescription. Disability Rehab Assist Technol. 2017;12:47–55.

Nielsen J, Landauer TK. A mathematical model of the finding of usability problems. In: Proceedings of the INTERACT '93 and CHI '93 conference on human factors in computing systems. New York: Association for Computing Machinery; 1993. p. 206–13.

Russ AL, Zillich AJ, Melton BL, Russell SA, Chen S, Spina JR, et al. Applying human factors principles to alert design increases efficiency and reduces prescribing errors in a scenario-based simulation. J Am Med Inform Assoc. 2014;21:287–96.

RxList. Streptase drug description. Rancho Santa Fe, CA: RxList of WebMD; 2021. Available from https://www.rxlist.com/streptase-drug.htm#description. Accessed 21 Mar 2021.

Saitwal H, Feng X, Walji M, Patel V, Zhang J. Assessing performance of an electronic health record (EHR) using cognitive task analysis. Int J Med Inform. 2010;79:501–6.

Scott PJ, Brown AW, Adedeji T, Wyatt JC, Georgiou A, Eisenstein EL, et al. A review of measurement practice in studies of clinical decision support systems 1998-2017. J Am Med Inform Assoc. 2019;26:1120–8.

United States General Services Administration (GSA). Card sorting. Washington, DC: United States General Services Administration, Technology Transformation Services; 2021. Available from https://www.usability.gov/how-to-and-tools/methods/card-sorting.html. Accessed 18 June 2021.

United States Office of the National Coordinator for Health Information Technology (ONC). HITECH evaluation reports. Washington, DC: United States Office of the National Coordinator for Health Information Technology (ONC); 2018. Available from https://www.healthit.gov/topic/onc-hitech-programs/hitech-evaluation-reports. Accessed 16 June 2021.

Wolf JA, Moreau JF, Akilov O, Patton T, English JC, Ho J, et al. Diagnostic inaccuracy of smartphone applications for melanoma detection. JAMA Dermatol. 2013;149:422–6.

Zhou L, Bao J, Setiawan IM, Saptono A, Parmanto B. The mHealth APP usability questionnaire (MAUQ): development and validation study. JMIR. 2019;7:e11500.

Chapter 5
Study Planning

Learning Objectives

The text, examples, Food for Thought questions, and self-tests in this chapter will enable the reader to:

1. Outline what is meant by an evaluation mindset and describe nine components of the mindset.
2. Describe what might go wrong early in the evaluation process that could threaten the project.
3. Describe the process of option appraisal and its uses.
4. Explain why the stage in the life cycle of the information resource is important when planning the evaluation.
5. Define evidence-based health informatics (EBHI) and describe its value.
6. Explain why, even before searching for literature, a study team should seek local information.
7. Given a general topic of interest, conduct a literature search to find evidence about that topic.
8. Describe how an already established framework or theory helps in conducting a literature search.
9. Describe how systematic reviews differ from standard literature searches.
10. Explain what a meta-narrative review is.
11. Describe the value of using a project management process while conducting an evaluation.

© Springer Nature Switzerland AG 2022
C. P. Friedman et al., *Evaluation Methods in Biomedical and Health Informatics*, Health Informatics, https://doi.org/10.1007/978-3-030-86453-8_5

5.1 Introduction

This chapter forms a natural transition between the previous four chapters that describe the "what" of evaluation studies to the multiple following chapters that describe the "how". The chapter does this by addressing practicalities related to planning and managing studies. The chapter describes what to do once evaluation questions are outlined and interest in conducting an evaluation project has deepened. The chapter begins with a description of a realistic evaluation mindset and warning of what might undermine the project during the planning stage. It then describes the role of stakeholders, the process of seeking evidence by searching for and reviewing the literature, the importance of Evidence Based Health Informatics (EBHI), and the necessity of organizing and conducting an evaluation study as a formal project.

Once preliminary evaluation study questions have been defined and the team has discussed different types of studies that might be useful for this particular evaluation study, what are additional choices that need to be considered before moving forward? Study teams must:

- Take the time to adopt an evaluation mindset and be aware of problems that may interfere with evaluation efforts even at early stages.
- Seek knowledge about stakeholder needs.
- Consider the level of evaluation that will be most useful, and identify the stage or stages in the life cycle of the information resource under study that most need evaluation.
- Capitalize on relevant work that has been done before by reviewing prior studies of information resources similar to the one being evaluated.
- Organize the study as a formal project, utilizing standard project management tools and techniques.

As informatics has become a more mature discipline, its published knowledge base has expanded rapidly. As it has grown into an academic as well as an applied discipline, the number of journals, monographs, and theses and dissertations has increased concomitantly. For this reason, the investment of some time and energy to look back at related work done by others will likely save resources and avoid frustration. For example, organizations often want to survey the users of ICT to assess satisfaction. This is an area where many survey instruments have already been developed, validated, and published. For example, the questionnaire for user interface satisfaction (QUIS) has been used in numerous studies (UMD n.d.; Lesselroth et al. 2009; Carayon et al. 2009). Re-use of a respected and available instrument can improve the study's rigor and save time. As discussed in more detail in Chap. 6, even if the instrument is not a perfect fit for the proposed project, the team members may be able to re-use parts of it, saving valuable time and effort and likely resulting in a better tool.

5.2 Adopting an Evaluation Mindset

Earlier chapters of this book addressed differences, sometimes subtle, between evaluation and research: the former emphasizing the goal of answering questions posed by an identified "audience" and the latter emphasizing the pursuit of open scientific questions of general interest to the field. This differentiation suggests a distinctive "mindset" that evaluation study teams should bring to their work. Nine components of this evaluation mindset are:

1. *Tailor the study to the problem:* Every evaluation is custom made, depending on both the problem at hand and the needs of the stakeholders. Evaluation differs from mainstream views of research in that an evaluation derives importance from the needs of an identified set of stakeholders or "clients" (those with the "need to know") rather than the open questions that inform progress in an academic discipline. Many evaluations also contribute welcome new knowledge of general importance to an academic discipline in addition, however.

2. *Collect data useful for making decisions:* There is no theoretical limit to the number of questions that can be asked or to the data that can be collected in an evaluation study. What is done is primarily determined by the decisions that ultimately need to be made about the resource and the information seen as useful to inform these decisions. Study teams must be sensitive to the distinction between what is necessary (to inform key decisions) versus what is not clearly needed but is readily available. Data collection consumes resources; collecting extraneous data is a waste of resources. On the other hand, identifying what is extraneous is not always easy, especially when conducting qualitative studies. At times, results of qualitative studies can be unexpected and the extraneous data might be important for decision making. In addition, data must come from reputable sources if they are to be useful. The quality of the data must be assessed at the outset: a poorly worded question on a survey will produce poor and misleading data (Leedy and Ormrod 2016).

3. *Plan for both intended* and *unintended effects:* Whenever a new information resource is introduced into an environment, there will be consequences, both positive and negative. Only some of them relate to the stated purpose of the resource, so the others are unintended. During a complete evaluation, it is important to look for and document effects that were intended, as well as those that were not, and continue the study long enough to allow these effects to appear. The literature of innovation is replete with examples of unintended consequences. Rogers' seminal monograph, Diffusion of Innovations describes the unintended consequences of implementation of mechanical tomato pickers which not only did not do as skillful a job as humans, but also generated financial and cultural hardships among laid off human pickers (Rogers 2003). Another example of an unintended consequence is the QWERTY keyboard, the keyboard design that has become a universal standard even though it was actually designed

to *slow* typing out of concern about jamming the keys of a manual typewriter, a mechanical device that has long since vanished from general use (Rogers 2003). Nine types of unintended consequences related specifically to adoption of ICT have been described in the literature, including (1) workflow issues, (2) new kinds of errors, (3) changes in communication patterns and practices, (4) more/ new work for clinicians, (5) never ending system demands, (6) changes in the power structure, (7) overdependence on the technology, (8) emotions, and (9) paper persistence (Campbell et al. 2006; Ash et al. 2007).

4. *Study the resource both while it is under development or being modified and after it is deployed:* In general, the decisions that evaluation can facilitate are of two types. *Formative or constructive* decisions are made for the purpose of improving an information resource. These decisions usually are made while the resource is under development or being locally modified by the ICT group, but they can also be made after the resource is deployed. *Summative* decisions made after a resource is deployed in its envisioned context deal explicitly with the effect or impact of the resource in that context (Brender 2006). It can take many months, and sometimes years, for the impacts of a deployed resource to stabilize within an environment. Before conducting the most useful summative studies, it may therefore be necessary for evaluation teams to allow this amount of time to pass.

5. *Study the resource in both the laboratory and in the field:* Completely different questions arise when an information resource is still in the laboratory and when it is in the field. *In vitro* studies, (Brender 2006) conducted in a usability laboratory, and *in situ* studies, conducted in an ongoing clinical, research, or educational environment, are both important aspects of evaluation.

6. *Go beyond the ICT point of view:* The developers or modifiers of an information resource usually are empathic only up to a point and often are not predisposed to be detached and objective about *their* system's performance. Those conducting the evaluation usually see it as part of their job to get close to the end-users and portray the resource as the users experience it (Brender 2006). The ICT group may not want to hear negative comments, but the feedback is needed if the resource is to be improved.

7. *Take the context into account:* Anyone who conducts an evaluation study must be aware of the entire context or environment within which the information resource will be used. The function of an information resource must be viewed as an interaction between the resource, a set of "users" of the resource, and the social/organizational/cultural context, which does much to determine how work is carried out in that context. Whether a new resource functions effectively is determined as much by its fit with the work environment as by its compliance with the resource designers' operational specifications as measured in the laboratory.

8. *Let the key issues emerge over time:* Evaluation studies represent a dynamic process. The design for an evaluation, as it might be stated in a project proposal, is typically just a starting point. Rarely are the important questions known with

total precision or confidence at the outset of a study. In the real world, evaluation designs must be allowed to evolve as new, important issues come into focus.

9. *Be methodologically eclectic*: It is important to base study design and data-collection methods on the questions to be explored, rather than bringing some predetermined methods or instruments to a study. Some questions are better addressed with qualitative data collected through open-ended interviews and observation. Others are better addressed with quantitative data collected via structured questionnaires, patient chart audits, or logs of user behavior. For evaluation, quantitative data are not clearly superior to qualitative data. Many studies now use data of both types, which are "mixed method" designs (see Chap. 16). Accordingly, those who conduct evaluations should at least know about rigorous methods for collection and analysis of both types.

Food for Thought
An informatics student is developing a phone app so that patients can keep track of their migraine headaches. The student has no funding to hire an evaluation team and will therefore need to evaluate the app with no outside assistance.

1. Which of the nine components of the evaluation mindset should the student pay most attention to while playing the role of both developer and study team?
2. What unintended consequences might arise during the course of this project that the student's evaluation should try to uncover?

5.3   Why It May Not Work Out as Planned

If an evaluation team did possess the "evaluation machine", described earlier in Sect. 1.3, life would be easier, but still not perfect. The team members would design and implement the information resource, let the machine tell them what would have happened had the resources not been implemented, and then compare the two scenarios. The difference would, of course, be a measure of the "effect" of the resource, but there may be many other factors not detectable by the machine that are important to investigate. The unavailability of the evaluation machine, and other factors, have led many creative individuals to devise an assortment of evaluation approaches. Because of the richness and diversity of these approaches, it is safe to say that an informative study probably can be designed to address any question of substantive interest in informatics.

However, even the best-designed studies may not work out as planned. One of the worst failure scenarios occurs when apparently meritorious studies—studies of important issues that are well designed—are never carried out. Resistance to the conduct of a study can develop either before the study is begun or during its progress. There are two principal reasons why this occurs. In both cases, attention to the various roles in an evaluation (see Fig. 2.5) and the importance of advance negotiation of an evaluation contract can both signal problems and help the study designers navigate through them.

5.3.1 Sometimes Stakeholders Would Rather Not Know

Some, perhaps many, resource developers and decision makers purchasing commercial products believe they have more to lose than to gain from a thorough study of their information resource. This belief is more likely to occur in the case of a resource perceived to be functioning very successfully or in the case of a resource that generates a great deal of interest because of some novel technology it employs. There are three logical counterarguments to those who might resist a study under these circumstances: (1) the perception of the resource's success will likely be confirmed by the study; (2) the study, if it supports these perceptions, can show how, and perhaps why, the resource is successful; and (3) a study would generate information leading to improvement of even the most successful resource. With reference to Fig. 2.2, stakeholders outside the ICT team can bring pressure on the ICT developers and decision makers to support or tolerate a study, but studies imposed under these circumstances tend to progress with great difficulty because of a lack of trust between the evaluation and ICT teams.

5.3.2 Differences in Values

Performing an evaluation study adds an overhead to any technology implementation project. It often requires the resource developers or those customizing a product to engage in tasks they would not otherwise undertake (such as programming the resource to function in several different ways; at a minimum, it usually requires some modifications in the project's implementation timeline). If the ICT group does not value the information they obtain from a study, they may, for example, be unwilling to await the results of a study before modifying some aspect of the resource that could be shaped by these results, or they may be reluctant to freeze the production version of a system long enough for a study to be completed. Underlying differences in values also may be revealed as a form of perfectionism. The ICT team or other stakeholders may argue that less-than-perfect information is of no utility because it cannot be trusted. Because everyone makes important decisions based on imperfect information every day, such a statement should not be taken literally. It is more likely revealing an underlying belief that the effort entailed in a study will not be justified by the results that will be generated. Some of these differences in belief between study teams and ICT teams can be reconciled, but others cannot.

The potential for these clashes of values to occur underscores the importance of the negotiations leading to an evaluation contract. If these negotiations come to a complete halt over a specific issue, it may be indicative of a gap in values that cannot be spanned, making the study impossible. In that case, all parties are better off when this is known early. For their part, study teams must respect the values of the ICT group by designing studies that have minimal detrimental impact on the project's developmental activities and timeline. It must be stressed that it is the responsibility of the evaluation team and study director to identify these potential value differences and initiate a collaborative effort to address them. The evaluation team should not expect the ICT team to initiate such efforts, and under no circumstances should they defer resolution of such issues in the interest of getting a study started. When the ICT and evaluation teams overlap, the problem is no less sticky and requires no less attention. In this case, individuals and groups may find themselves with an internal conflict unless they engage in conversations among themselves, alternating between the ICT and the study teams' positions, about how to resolve value differences imposed by the conduct of a study in the context of resource development.

When evaluation studies meet their objectives, the ICT and evaluation teams perceive each other as part of a common enterprise, with shared goals and interests. Communication is honest and frequent. Because no evaluation contract can anticipate every problem that may arise during the conduct of a study, problems are resolved through open discussion, using the evaluation contract as a basis. Most problems between people and organizations are resolved through compromise, and the same is true for evaluation.

5.4   Planning a Study

Figure 2.1 illustrates the entire evaluation process. Figure 5.1 depicts the components needed during the evaluation planning component of the process. These components are shown as a series of decisions about methods based on input from stakeholders, on information about context, and on the evidence base. Each component is described below.

5.4.1   Assessing Stakeholder Needs

As described above, a recurrent and troublesome issue in many evaluation studies is not choosing what to measure, or even which methods to use in a specific study, but how to balance the often-competing demands of the different stakeholders involved in commissioning the study, as well as how to allocate resources fairly to answer the many questions that typically arise.

To resolve this strategic problem, it is very important for the study team to understand what, in the stakeholders' minds, are the key purposes of the study. When study questions are formulated as described in Sect. 3.2.2 it is likely that some questions will have higher priority than others, or that there is a single overarching question that reflects a primary decision to be made. For example, of the many reasons for performing evaluations, very often the overarching purpose is to collect information to support a choice between two or more options: to purchase vendor resource A or B or to develop interface design A or B. Therefore, the more a

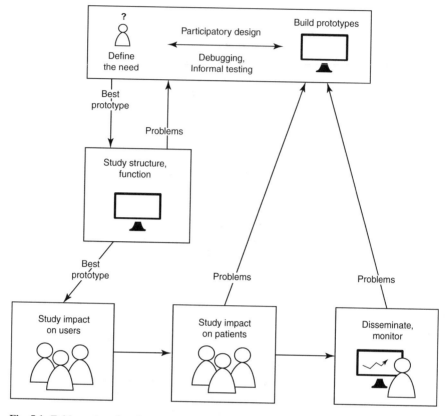

Fig. 5.1 Evidence-based evaluation planning steps

proposed study generates information that is interesting, but does not actually inform this choice, the less value to the stakeholders it will have.

When a study is to inform real-world choices of these types—especially those with significant financial or political consequences—study teams can give structure to the study by using the technique called "option appraisal." (Whitfield 2007). In an option appraisal, a choice is ultimately made by scoring and ranking several options (e.g., to purchase resource A, B, or C), often including the null option (no purchase). The person or group responsible for the choice then selects the criteria they wish to use and how they will score them.

When option appraisal is applied as a planning tool for a study, the criteria used in the option appraisal are reflected in the major questions that need to be answered in the evaluation study. The score assigned against each criterion in the option appraisal depends on the results of evaluation studies that attempt to answer the relevant question. Sometimes, it is relatively easy to assign the scores. For example, if the primary criterion is cost, the "evaluation study" may consist of discussions with the suppliers of each candidate information resource. Often, however, significant data collection is required before the score for each option can be assigned—if the scoring criterion is user experience or the ratio of costs to benefits as discussed in Chap. 18, for example. By focusing on criteria that will be used by decision-makers in an option appraisal, study teams can usually identify the key questions that their evaluation studies need to address and how the results will be used. When criteria emerge from an option appraisal as particularly important—for example, those listed first or weighted more heavily—the study team knows that the results of the study addressing these criteria are crucial, and that extra attention to detail and a higher proportion of the evaluation budget are warranted for studies addressing these criteria.

5.4.2 Choosing the Level of Evaluation

One of the fundamental choices required when planning any evaluation study is the level or scale at which to focus the study. For example, this level can be one or more of the following: a message sent by a patient or another component of the information being processed or communicated through a patient portal, the portal as a whole, a patient or case, a health professional such as a nurse or medical assistant who receives messages, a multidisciplinary clinical team, or part or all of a healthcare delivery organization such as a clinic or group of clinics. To illustrate this, Table 5.1 shows some sample questions that might prompt study teams to study each of these levels.

It is important to realize that logistical factors often require studies to be conducted at higher levels of scale. This occurs when individual objects such as patients or health professionals interact and thus cannot be separated out for study. An example would be studying the impact of an antenatal information kiosk by providing a password to half the women attending a clinic. Because those women without a

Table 5.1 Sample questions at differing levels of evaluation

Level of evaluation	Sample question
A patient message or other component of the information being processed through a patient portal	If messages have a word limit, what is the effect of the limit on the content of messages that are sent?
The information resource as a whole	How long does it take for patient messages to get the attention of the initial clinician-recipient?
A patient	Are patients satisfied when they send messages through the portal?
A clinical team	How are communication patterns within the team influenced by the introduction of the portal?
Part of a healthcare delivery organization (e.g., clinic)	How is the clinic workflow affected by the portal?
The entire healthcare delivery organization	How has the introduction of the portal affected the rate of no-shows across all the clinics, preventing lost revenue?

password could either "borrow" one or look over the shoulders of women with a password using the kiosk, it would be logistically difficult to restrict usage to the intended group. Even if the "control" women failed to share passwords, women typically share their experiences and information in an antenatal clinic; therefore, it would be naïve to assume that if the kiosk were only made available to half the women, the other half would be completely ignorant of its contents. Similar arguments apply to studying the effect of an educational course for health professionals or the impact of a new set of reference databases. These interactions require that the study team raise the level of the evaluation to focus on groups rather than individuals.

5.4.3   Matching What Is Studied to the Stage in the Life Cycle

Evaluation, defined broadly, takes place throughout the resource development cycle: from defining the need to monitoring the continuing impact of a resource once it is deployed. Figure 5.2 outlines the different issues that can be explored, with different degrees of intensity, at each stage in the information resource development or modification life cycle.

Prior to any resource development or modification, as discussed earlier, there may be very active formal evaluation to establish needs. During the early phases of actual resource development or modification, informal feedback and exploration of prototypes is associated with code development and debugging. A single prototype then emerges for more formal testing, with problems being fed back to the development team. Eventually, it passes preset criteria of adequacy, and its effects on users can be tested in a more formal way—though often still under controlled "laboratory" conditions. Once safety is ensured and there is reason to believe that the information resource is likely to bring benefit, its impact can be studied in a limited field test prior to wider dissemination. Once disseminated, it is valuable to monitor the

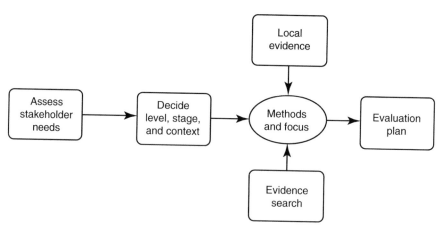

Fig. 5.2 Changing evaluation issues during different stages in the life cycle

effects of the resource on the institutions that have installed it and evaluate it for potential hazards that may only come to light when it is in widespread use—a direct analogy with post marketing surveillance of drugs for rare side effects.

Evaluation is integral to information resource development and modification, and adequate resources must be allocated for evaluation when time and money are budgeted for a development effort. Evaluation cannot be left to the end of a project. However, it is also clear that the intensity of the evaluation effort should be closely matched to the resource's maturity (Brender 2006). For example, one would not wish to conduct an expensive field trial of an information resource that is barely complete, is still in prototype form, may evolve considerably before taking its final shape, or is so early in its development that it may fail because simple programming bugs have not been eliminated. Equally, once information resources are firmly established in practice settings, it may appear that no further rigorous evaluations are necessary. However, key questions may emerge only after the resource has become ubiquitous.

5.4.4   Considering Context: Local Resources and Organizational Knowledge

Prior to searching the literature to discover what has been done elsewhere, it is wise to conduct a search of local organizational knowledge. Study teams should find out about similar project evaluations that have already been conducted in the organization. Building on what has already been discovered makes sense in several ways. It might offer guidance about methods that have already worked well, identify skills of other staff members whose expertise might be tapped, describe processes that were used for managing the evaluation project, and offer results upon which a new evaluation project might build. This organizational knowledge might be the result of quality assurance projects, so it could exist in local files in the form of unpublished reports. These might only be discovered by interviewing staff who either worked on the project or were aware of it at the time. Staff members with knowledge of the history of prior efforts, who have been with the organization for years, can contribute their historical organizational knowledge to planning the upcoming evaluation if they can be identified.

5.4.5   The Focus: Matching What Is Evaluated to the Type of Information Resource

There are many types of information resources containing numerous functional components. Not every function can or should be measured for every resource, and it often requires much thought about the purpose of the evaluation itself to produce a relevant list of issues to pursue. Because resources for evaluation are always limited, it may be helpful to rank the items listed in the order of their likely contribution to answering the questions the evaluation is intended to resolve. Often, as discussed in Chap. 2, priorities are set not by the study teams, but by the stakeholders in the evaluation. The study team's role is then to initiate a process that leads to a consensus about what the priority issues should be.

5.4.6 The Search for Evidence

Since study teams are trying to improve whatever is being evaluated, a literature review can identify HOW the evaluation can be conducted and WHAT is already known. A study team may want to find out if others have already conducted similar evaluations. If so, information about prior evaluations can help the team make decisions about whether to conduct a similar evaluation, whether to duplicate a prior study, or whether to do something entirely new and different. For evaluating internally, the team might want to learn how it has been done before and simply do it the same way. For a more research-oriented approach, the team may want to identify gaps in what is known globally and contribute to the body of knowledge. The team should take the time to search for best practices, identify research questions, help scope an evaluation project and/or search for survey instruments and methods during the planning process. All of these activities are included in the practice of Evidence-Based Health Informatics (EBHI) as discussed in Sect. 5.5 below.

5.5 The Value of Evidence-Based Health Informatics

After the early choices and considerations outlined towards the left in Fig. 5.1 have been addressed, the evaluation team will be better prepared for conducting the next critical step in planning the project: searching for evidence that exists outside the organization. In light of the expanding evidence base that exists in informatics, there has been a movement to recognize the importance of what has been called evidence-based health informatics (EBHI). EBHI is defined as "the conscientious, explicit, and judicious use of the current best **evidence** when making decisions about the introduction and operation of IT in a given healthcare setting." (Rigby et al., 2016, p. 255). EBHI emphasizes reviewing results of others' evaluation/research projects to provide better evidence for development, policy, and purchasing decisions.

The rationale for increasing the use of EBHI is economic as well as academic: ICT is expensive, so all decision making should be based on robust evidence. Conducting an evaluation project is also expensive, so finding prior evidence that can facilitate the evaluation project being planned will save money. Evaluation of internal projects builds the evidence base, but only if the results are shared. The rationale is also ethical: informaticians have an obligation to show benefit, when it exists, and to share potentially harmful situations when they are discovered. If decision makers do not have evidence available to them, the decisions they make might be dangerous. For example, for many years the only source of information about EHR downtime was through the public media, with newspapers headlining computer crashes in large hospital systems. When large healthcare systems have suffered unexpected downtime, it made the front page of some newspapers. Sometimes unintended consequences like this are never evaluated because lawsuits are underway, and often it is because organizations do not want negative experiences harming their reputations. At other times, vendor contracts prohibit discussion of system problems (Koppel and Kreda 2009). However, there is now some rigorous published literature about downtime and hospital systems written by objective researchers, which is more credible and an example of the value of EBHI (Sax et al. 2010; Larsen et al. 2020).

EBHI presents unique challenges because informatics evidence exists in the literature of a very wide breadth of fields including that of all health professions and the sciences underlying them as well as the literature of computer and information sciences—in addition of course to the literature of informatics itself. And, as has already been established in earlier chapters of this book, the literature of informatics draws from a very wide range of methods.

Another unique aspect of informatics is the growing recognition that mixed methods—studies combining quantitative and qualitative methods—are often necessary to address what stakeholders really want to know. This is because of the large numbers of different stakeholders whose interests must be accommodated and because of the wide range of sociotechnical factors that shape the experience and outcomes of ICT development and deployment.

There is a growing number of excellent sources of information about EBHI which provide much more detail for those interested in learning more. The 2013 IMIA Yearbook of Medical Informatics is devoted to the topic of EBHI (Jaulent 2013), and a freely available comprehensive book on EBHI was published in 2016 (Ammenworth and Rigby 2016).

5.6 Searching the Literature

Section 5.4.4 has described how to search for local knowledge. The next step is literature searching for knowledge developed elsewhere. It is important to recognize that there exists a hierarchy of extensiveness of searching for evidence, from a brief simple search to an extensive comprehensive review. Students often say that they are going to conduct comprehensive systematic reviews when they start planning evaluation projects. They often do not understand what is involved in a true systematic review, including the extent of resources needed. The alternative is to start with a plan for doing a brief simple search to get a sense of what already exists and then develop a detailed plan for broader and more in-depth searching if it is needed. The study team should by all means consult with local health sciences librarians for

Fig. 5.3 Continuum of literature review comprehensiveness

assistance: librarians like nothing better than to encourage and teach the use of high-quality searching techniques. See Fig. 5.3 for a rendering of the continuum of literature review comprehensiveness.

5.6.1   Identifying a Search Topic

A simple search using health-related databases is a reasonable first step to get a sense of what is available. Depending on the information resource and kind of evaluation you have in mind, the team can then move on to other databases that might include non-medical literature, more foreign publications, or different formats such as dissertations.

One obvious search topic might be identification of evaluation methods for studying information resources similar to the resource of interest to the evaluation team. For example, if the team wants to assess satisfaction levels among users of a patient portal supplied by an organization, team members might first look for papers that have done evaluations of patient portals elsewhere so that the team might be able to replicate them. This way, the study team can learn about how to do the evaluation as well as what others have discovered. Study teams should not be discouraged if nothing is found. The team's job will be harder in the absence of relevant literature, but on the other hand, the team will have identified a gap in the knowledge base, which will make the study more fundable or publishable. Even if this evaluation will be done as a quality improvement project, the team can aim for publishing it and contributing to the body of knowledge in addition, but only if stakeholders agree and appropriate human subjects permissions are in place. Much depends on the career goals of stakeholders and whether or not publication is rewarded in their circumstances.

5.6.2   Frameworks and Theories

The early evaluation questions already outlined by the team, even if they are not yet finalized, should drive selection of topic areas for the literature search. The authors recommend using the well-known PICO framework, described in Chap. 3, with PICO standing for Problem, Intervention, Comparison or Control, and Outcome (Eriksen and Frandsen 2018), as a checklist for some of the areas about which to search. For example, if the study team plans to evaluate a patient portal, the search might include papers about (1) the problems of accessibility and usability of patient portals, (2) descriptions of similar interventions implemented elsewhere, (3) methods used in those studies, and (4) results of prior studies.

Science progresses by identifying and disproving theories. Health informatics, as a scientific discipline, is developing its evidence base of tested theories. Even if an evaluation project is not going to result in publications that contribute to the evidence base, planning the project can benefit from a review of relevant theories and frameworks. EBHI, which entails carrying out studies designed to improve the evidence base of health informatics, can help in providing background information for any informatics study. EBHI studies often test the effectiveness of incorporating a generic system design principle or theory that has wide potential application across systems and clinical contexts. An example generic principle is whether to present alerts from an e-prescribing system in a modal dialogue box (which requires the user to click OK or Cancel) or in the ePrescribing interface. Using case scenarios and a simulated ePrescribing system, Scott et al. (Scott et al. 2011) carried out a study with junior physicians as participants and found that, while prescribing alerts that appeared on the e-prescribing interface were only about half as effective at reducing prescribing errors as modal dialogue boxes, they were much more acceptable. This makes it more likely that the alert system users will tolerate the alerts and not switch off the alert system altogether, which currently often happens in UK primary care systems. This study was relatively small and limited in scope. However, this is the kind of study that can generate insights that will help others designing alert systems to make them more acceptable, or at least to consider the tradeoff between user acceptance and effectiveness. If an evaluation team is planning on assessing development and use of a set of CDS alerts, the team would benefit from learning about this piece of evidence.

5.7 How to Conduct a Literature Search

After a broad review has been accomplished and the evaluation questions initially identified, the team needs to decide how in-depth and how broad the literature search needs to be. The team may want to limit the review to a quick overview search appropriate for a small internal evaluation, do a search as described below for conducting a standard literature search, or do a full systematic review.

Figure 5.4 outlines basic steps in the search process. Many decisions should be made at the outset about scope, what media to look at (books, journals, web),

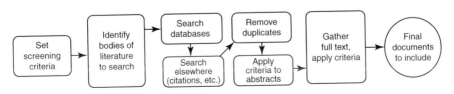

Fig. 5.4 Literature searching process

databases to search, search strategy, selection criteria, format of the product of the searching (to inform stakeholders or to write a formal proposal, for example).

Often it is a good idea to start with looking for books, if the topic is broad enough, and for review articles. Review articles provide summaries of literature on specific topics. For example, a review paper by Hussain et al. reviews all qualitative papers published in the Journal of the American Medical Informatics Association from its inception through 2017. It provides a list, summary, and analysis of all of the papers (Hussain et al. 2020). When a review paper exists in an area of interest, it is a good starting place when reviewing the literature of that area. Next, a search using appropriate databases of peer-reviewed papers is recommended. Librarians are excellent resources and can provide suggested search strategies if the search is at all difficult. For example, a librarian can help identify terms, preferably using a controlled vocabulary rather than keywords to enhance precision, and make the team aware of databases that are appropriate. While doing the search, when a perfect article is located, look at its index terms to find more papers that are similar or search in a citation index database for that purpose.

Once citations are found, the study team can screen them by reading abstracts, finding items that meet preset criteria so that the team can look at the full text, and finally selecting those that are relevant and of high enough quality to meet selection standards. The team might want to venture beyond academic literature to find unpublished information, what is termed grey literature, if needed. It is helpful to use reference software for keeping track of citations and for formatting them for different reports and publications. The most important and difficult step then is to synthesize, interpret, and summarize/make sense of what has been found. This synthesis will help to identify gaps in the literature and to help the team make decisions about how to move forward. Be very wary of what is found online, however, unless it is peer reviewed and indexed in a reputable database (Leedy and Ormrod 2016).

Self-Test 5.1

An evaluation team has been asked to design a study assessing the adoption and use of tablets in waiting rooms in outpatient settings in a large system of clinics. Patients have the option of using them instead of filling out forms by hand, so use is voluntary. The team members believe that Diffusion of Innovations Theory (Rogers 2003) might assist with designing the study, but they do not know much about the theory or about prior studies about patient entered data.

1. What sources would a team member want to look for first and for what topics?
2. What would you do if you did a literature search and found 1000 papers about Diffusion of Innovations in health care?
3. Once you found a few especially helpful papers, what would you do?

5.8 Systematic Reviews and Narrative Meta-Analyses

These categories of reviews result from searching broadly for evidence, screening carefully for inclusion of only the best papers, and evaluating selected papers using a careful process conducted by several experts.

5.8.1 Systematic Reviews

Systematic reviews differ from standard literature searches: they are more rigorous, comprehensive, may include gray literature, and involve only experts as authors. They pay particular attention to the quality of the evidence and assess the validity of each piece of evidence included to answer the review question—which need not always be about effectiveness. They may include a meta-analysis, which is a statistical procedure for summarizing and comparing the data in several or many studies so that more definitive results can be reported (Leedy and Ormrod 2016). The Cochrane Collaboration (Cochrane.org n.d.) is the chief international organization that conducts rigorous systematic reviews and in the U.S. the Agency for Healthcare Research and Quality funds a network of Evidence-Based Practice Centers that produce such reviews under contract (AHRQ n.d.).

A set of guidelines for presentation of the results of systematic reviews, titled Preferred Reporting Items for Systematic Reviews and Meta-Analyses (PRISMA), such as those conducted by the Cochrane Collaboration, offers detailed guidance for publishing these reviews (Moher et al. 2009). This list of guidelines is more precise and comprehensive than needed for many literature reviews done by study teams preparing to assess an intervention and researchers writing papers for publication. However, PRISMA guidelines offer a rigorous standard for doing a high-quality search, even if some steps may not be applicable.

5.8.2 Meta-Narrative Reviews

One example of a newer type of systematic review is the meta-narrative review process developed by Greenhalgh et al. (Greenhalgh 2008; Otte-Trojel and Wong 2016). This is the qualitative equivalent of meta-analyses, which are statistical. An example would be a narrative review of review papers that have been published about the implementation of alerts for drug-drug interactions. A search may find that 20 reviews have been published about different aspects of this topic. A meta-narrative analysis would look across these 20 papers and synthesize the results, most likely reporting meta themes in table form.

5.9 Project Management

Another vital step in planning an evaluation study is to take the time to use a formal project management approach to planning. Informatics evaluation efforts usually meet the definition of projects: a project is a sequence of tasks with a beginning and end, bounded by time, resources, and desired results. Therefore, project management tools and techniques should be applied to help assure that the evaluation (1) is timely, (2) has appropriate resources available, and (3) is of high quality (Ludwick and Doucette 2009; Davies et al. 2020). These are the three cornerstones of project management, and if one is insufficient, the others suffer as well. It is important to negotiate the goals and questions to be addressed, time scale, and budget in advance. See Sect. 5.3 to review what can go wrong if these negotiations do not take place. Regarding resources, it is often difficult to apportion the budget between what is spent on development and what is spent on evaluation activities. According to generally accepted standards, the starting point for the evaluation activity should be at least five percent of the total budget, but a larger percentage is often appropriate if this is a demonstration project or one where reliable and predictable resource function is critical to patient safety. In a closed-loop drug-delivery system, for example, a syringe containing a drug with potentially fatal effects in overdose is controlled by a computer program that attempts to maintain some body function (e.g., blood glucose, depth of anesthesia) constant or close to preprogrammed levels. Any malfunction or unexpected dependencies between the input data and the rate of drug delivery could have serious consequences. In these cases, the ratio of the budget allocated to evaluation rather than development must be larger.

5.9.1 The Team

Evaluation studies are usually conducted by teams. A study manager for the team generally has experience in managing projects and ideally may also be an informatics specialist or a specialist in the biomedical domain of the work. One of the authors once interviewed the professional project manager for implementation of an EHR across a large health system. When the interviewee shared that they had been a project manager for NASA for rocket launches, the interviewer asked them to compare managing an informatics project with a NASA project. Informaticians will not be surprised to learn that the EHR project, with all of its human components, was more difficult. Knowledge of the healthcare domain would have helped this person. In informatics, projects have often been conducted without the benefit of a project manager at all and without the use of proven project management tools, often with difficulty (Davies et al. 2020).

At the first stage of planning a study, it is an excellent idea to make a list of the potential project roles listed in Sect. 2.4, such as project funder, resource developer, user, and community representative, and indicate which stakeholders occupy each role. Sometimes this task requires educated guesses, but the exercise is still useful. This exercise will particularly serve to identify those who should be consulted during the critical early stages of designing a study.

Those planning evaluation projects should be aware of the need to include a diverse and balanced membership on the evaluation team. While the "core" evaluation team may be relatively small, many evaluation projects need occasional access to the specialist skills of computer scientists, ethnographers, statisticians, health professionals and other domain experts, managers, and health economists.

Although many evaluations are carried out by an internal team of individuals already paid by the organization, placing some reliance on external study team members may help to uncover unexpected problems—or benefits—and is increasingly necessary for credibility. Recalling from Chap. 2 the ideal of a completely unbiased "goalfree" evaluation, it can be seen that excessive reliance on evaluation carried out by an internal ICT team can be problematic. Gary et al. showed that studies carried out by resource developers are three times as likely to show positive results as those published by an external team (Gary et al. 2005).

Project advisory groups often are appointed to oversee the quality and progress of the overall effort from the stakeholders' perspective. Such a group typically has no direct managerial responsibility in that the team does not report to that group. This advisory group should be composed of individuals who can advise the study team about priorities and strategy, indemnify them against accusations of bias, excessive detachment or meddling in the evaluation sites, and monitor progress of the studies. Such a group can satisfy the need for a multidisciplinary advisory group and help to ensure the credibility of the study findings.

Self-Test 5.2
Referring back to the scenario described in Self-test 5.1 about patient entered data:

1. The team member has only found three good papers about patients entering data and only one was about the use of tablets in waiting rooms. The team wants to emulate the methods used in the latter study, which combined usage data and interview data. What skills would the present team need to make sure were represented on an internal evaluation team?
2. What skills might the team want in an advisory group?

5.9.2   Project Management Tools

Project management activities and tools include stakeholder analyses, plans including activities and milestones, timelines, assignment of resources, formal agreements that include agreed-on operating procedures, delineation of communication channels and decision-making processes, role assignments, and team building activities. Professional project managers usually take advantage of sophisticated project management software applications, but small projects can exploit simpler and often freely available software. The processes with which software can help are (1) sharing with stakeholders and team members alike what the plan is and the progress being made, and (2) generating graphical depictions of progress using timelines and charts so that the timeline and dependencies on different parts of the project can be easily communicated (Leedy and Ormrod 2016). Small evaluation teams do not need professional project managers with expensive software, but a staff member who knows enough about project management and the availability of simple tools should be assigned to manage the project.

5.10 Summary

Planning an evaluation study takes time and it is often tempting for study team members to launch a study without sufficient planning. This can be a mistake, however, in that knowledge of prior evidence and adherence to the tenets of solid project management can help immensely in assuring that a study will be useful. The authors would like to advocate for and promote the use of evidence in informatics evaluation studies. All too often studies that are conducted have neglected to build on prior work, primarily because evaluation team members do not know that the body of evidence is as rich as it is. In the past, it was not as productive searching the literature because few informatics journals existed and the publication of papers reporting applied research, as opposed to academic research, related to informatics in medical journals was scant. The publication patterns have changed, much more evidence is available, and the future of EBHI is encouraging. Other positive trends related to planning involve increased use of the project management process, which is being recognized as a core competency of informaticians and increasingly used in the field of informatics (Gary et al. 2005).

Answers to Self-Tests

Self-Test 5.1
1. What sources would a team member want to look for first and for what topics?
   The literature about Diffusion of Innovations Theory is vast. The team knows there is a classic book about it, so first a copy of that book should be located. It would be a good starting point if it included up to date references, but the book is old, so references are old. Therefore, the team should next locate papers in the health science literature about DOI and informatics interventions, so the team asks a librarian for guidance searching appropriate databases for papers at the intersection of those two broad topics.
2. What would you do if you find 1000 papers about Diffusion of Innovations and ICT in health care?

You would narrow your search. You could try finding papers about DOI and patient entered data, but there might be other relevant papers of interest as well about DOI and patient facing technologies.
3. Once you found a few especially helpful papers, what would you do?
   Look at the references in those papers. Look at the index terms for those papers and use those terms when doing another search. Search a citation index database to find recent papers that have cited those key papers.

Self-Test 5.2
1. What skills would the present team need to make sure were represented on an internal evaluation team?
   Assuming the existing team members have informatics training, they are likely a mix of clinicians and ICT specialists. They probably know how to gather and analyze usage data, but they might want to make sure they have a usability specialist on the team and a clinician practicing in the outpatient environment. They should add a qualitative methodologist if they do not already have one on the team to plan the interview process. In addition, they will need a project manager.
2. What skills might the team want in an advisory group?
   To help with access to the usage data, a data specialist from the organization could become an advisor. To assist with the qualitative part, both an outside social scientist and a clinic manager who could help with building rapport in the clinics would be useful. The addition of a patient representative to gain that perspective would be a good idea as well.

References

AHRQ (n.d.). https://www.ahrq.gov/research/findings/evidence-based-reports/index.html. Accessed 9 June 2021.

Ammenworth E, Rigby M. Evidence-based health informatics. Amsterdam: IOS Press; 2016.

Ash JS, Sittig DF, Poon EG, Guappone K, Campbell E, Dykstra RH. The extent and importance of unintended consequences related to computerized physician order entry. J Am Med Inform Assoc. 2007;14:415–23.

Brender J. Handbook of evaluation methods for health informatics. Amsterdam: Elsevier; 2006.

Campbell E, Sittig DF, Ash JS, Guappone K, Dykstra R. Types of unintended consequences related to computerized provider order entry. J Am Med Inform Assoc. 2006;13(5):547–56.

Carayon P, et al. Implementation of an electronic health records system in a small clinic: the viewpoint of clinic staff. Behaviour Inform Tech. 2009;28(1):5–20.

Cochrane.org (n.d.). https://www.cochrane.org/about-us. Accessed 9 June 2021.

Davies A, Mueller J, Moulton G. Core competencies for clinical informaticians: a systematic review. Int J Med Inform. 2020;141:104237.

Eriksen MB, Frandsen TF. The impact of patient, intervention, comparison, outcome (PICO) as a search strategy tool on literature search quality: a systematic review. J Med Lib Assoc. 2018;106:420–31.

Gary A, Adhikan N, McDonald H, et al. Effect of computerized clinical decision support systems on practitioner performance and patient outcomes: a systematic review. JAMA. 2005;293:1223–38.

Greenhalgh T. Meta-narrative mapping: a new approach to the systemati review of complex evidence. In: Hurwitz B, Greenhalgh T, Skultans V, editors. Narrative research in health and illness. Boca Raton, FL: Wiley; 2008.

Hussain MI, Figuerredo MC, Tran BD, Su Z, Molldrem S, Eikey EV, Chen Y. A scoping review of qualitative research in JAMIA: past contributions and opportunities for future work. J Am Med Inform Assoc 00(0) 2020, 1–12.

Jaulent MC. Evidence-based health informatics. Yearbook of medical informatics 2013. Stuttgart: Schattauer; 2013.

Koppel R, Kreda D. Health care information technology vendors' "hold harmless" clause: implications for patients and clinicians. JAMA. 2009;301(12):1276–8.

Larsen EP, Rao AH, Sasangohar F. Understanding the scope of downtime threats: a scoping review of downtime-focused literature and news media. Health Inform J. 2020;26(4):2660–72.

Leedy PD, Ormrod JE. Practical research: planning and design. 11th ed. Pearson: Boston, MA; 2016.

Lesselroth BJ, Felder RS, Adams SM, et al. Design and implementation of a medication reconciliation kiosk: the Automated Patient History Intake Device (APHID). J Am Med Inform Assoc. 2009;16(3):300–4.

Ludwick DA, Doucette J. Adopting electronic medical records in primary care: lessons learned from health information systems implementation experience in seven countries. Int J Med Inform. 2009;78:22–31.

Moher D, Liberati A, Tetzlaff J, Altman DG, The PRISMA Group. Preferred reporting items for systematic reviews and meta-analyses: the PRISMA Statement. PLoS Med. 2009;6(7):e1000097.

Otte-Trojel T, Wong G. Going beyond systematics reviews: realist and meta-narrative reviews. In: Ammenwoerth E, Rigby M, editors. Evidence-based health informatics. Amsterdam: IOS Press; 2016.

Rigby M, Magrabi F, Scott P, Doupi P, Hypponen H, Ammenwerth E. Steps in moving evidence-based health informatics from theory to practice. Healthc Inform Res. 2016;22(4):255–60.

Rogers EM. Diffusion of innovations. 5th ed. New York: Free Press; 2003.

Sax U, Lipprandt M, Rohrig R. The rising frequency of IT blackouts indicates the increasing. Yearb Med Informatics. 2010:130–7.

Scott GPT, Shah P, Wyatt JC, Makubate B, Cross FW. Making electronic prescribing alerts more effective: scenario-based experimental study in junior doctors. J Am Med Inform Assoc. 2011;18(6):789–98.

UMD (n.d.). http://www.cs.umd.edu/hcil/quis/. Accessed 9 June 2021.

Whitfield D. Options appraisal criteria and matrix. ESSU Research Report No. 2, June 2007. https://www.researchgate.net/publication/228383838_Options_Appraisal_Criteria_and_Matrix. Accessed 9 June 2021.

Part II
Quantitative Studies

Chapter 6
The Structure of Quantitative Studies

Learning Objectives

The text, examples, and self-tests in this chapter will enable the reader to:

1. For a given measurement process, identify the attribute being measured, the object class, the instruments being employed, and what constitutes an independent observation.
2. For a given measurement process, identify of the level of measurement (nominal, ordinal, interval, ratio) of the attribute(s) included in the measurement process.
3. Explain why measurement is fundamental to the credibility of quantitative studies and why, according to the fundamental precepts of quantitative methods, even the most abstract constructs can be measured objectively.
4. Given a study design, classify it as primarily a measurement or demonstration study.
5. Given a demonstration study design, identify the categories of subjects/participants as well as the independent and dependent variables.
6. Given a demonstration study design, classify the study as prospective or retrospective and also as descriptive, interventional, or correlational.

6.1 Introduction

What the famed epidemiologist Alvin Feinstein wrote in 1987 still rings true today and sets the tone for this chapter.

> Important human and clinical phenomena are regularly omitted when patient care is . . . analyzed in statistical comparisons of therapy. The phenomena are omitted either because they lack formal expressions to identify them or because the available expressions are regarded as scientifically unacceptable. (Feinstein 1987)

© Springer Nature Switzerland AG 2022
C. P. Friedman et al., *Evaluation Methods in Biomedical and Health Informatics*, Health Informatics, https://doi.org/10.1007/978-3-030-86453-8_6

This chapter begins the exploration of quantitative studies in detail. Chapters 6 through 13 address the design of studies, along with how to develop measurement procedures to collect data and how subsequently to analyze the data collected. The methods introduced relate directly to the comparison-based, objectives-based, and decision-facilitation approaches to evaluation described in Chap. 2. They are useful for addressing most of the purposes of evaluation in informatics, the specific questions that can be explored, and the types of studies that can be undertaken—all as introduced in Chap. 3.

More specifically, this chapter introduces a conceptual framework to guide the design and conduct of quantitative studies. The framework employs terminology; and the terminology introduced in this chapter is used consistently in the chapters on quantitative studies that follow. Much of this terminology may already be familiar, but some of these familiar terms are likely used here in ways that are novel. Unfortunately, there is no single accepted terminology for describing the structure of quantitative studies. Epidemiologists, behavioral and social scientists, computer and information scientists, and statisticians have developed their own unique variations. The terminology introduced here is a composite that makes sense for the hybrid field of informatics. Importantly, the terms introduced here are more than just labels. They represent concepts that are central to understanding the structure of quantitative studies and, ultimately, to their design and conduct.

A major theme of this chapter, and indeed all eight chapters on quantitative studies, is the importance of measurement. Three of the chapters in this group focus explicitly on measurement because many of the major problems to be overcome in qualitative study design are, at their core, problems of measurement. Measurement issues are also stressed here because they are often overlooked in research methods courses based in other disciplines.

After introducing the concept of measurement along with some related terminology, this chapter formally establishes the distinction between:

- *measurement studies* designed to explore with how well "quantities of interest" in informatics can actually be measured; and
- *demonstration studies*, which apply these measurement procedures to address evaluation questions of substantive and practical concern.

This discussion will make clear that demonstration studies are what study teams ultimately want to do, but measurement studies sometimes are an initial step required before successful demonstration studies are possible.
The distinction between measurement and demonstration studies is more than academic. In the informatics literature, it appears that measurement issues usually are embedded in, and often confounded with, demonstration issues. For example, a recent review of quantitative studies of clinical decision support systems revealed that only 47 of 391 published studies paid explicit attention to issues of measurement (Scott et al. 2019). This matter is of substantial significance because, as will

be seen in Chap. 8 (Section 8.5), deficiencies in measurement can profoundly affect the conclusions drawn from demonstration studies. The quotation that begins this chapter alerts us to the fact that ability to investigate is shaped by ability to measure. Unless study teams possess or can develop ways to measure what is important to know about information resources and the people who use them, their ability to conduct evaluation studies—at least those using quantitative approaches—is substantially limited.

This chapter continues with a discussion of demonstration studies, seeking to make clear how measurement and demonstration studies differ. This discussion introduces some differences in terminology that arise between the two types of studies. The discussion continues by introducing categories of demonstration studies that are important in matching study designs to the study questions that guide them.

6.2 Elements of a Measurement Process

This section introduces some general rules, definitions, and synonyms that relate to the process of measurement. As noted earlier, these definitions may use some familiar words in unfamiliar ways. The process of measurement and the interrelations of the terms to be defined are illustrated in Fig. 6.1.

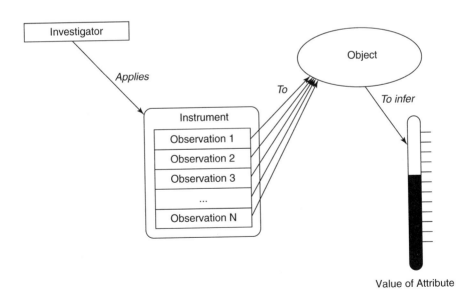

Fig. 6.1 The process of measurement

6.2.1   Measurement

Measurement is the process of assigning a value corresponding to the presence, absence, or degree of presence, of a specific attribute in a specific object. The terms "attribute" and "object" are defined below. Measurement results in either: (1) assignment of a numerical score representing the extent to which the attribute of interest is present in the object, or (2) assignment of an object to a specific category. Taking the temperature (attribute) of a person (object) is an example of the process of measurement.

6.2.2   Attribute

An attribute is what is being measured. Speed (of an information resource), blood pressure (of a person), the correct diagnosis (of a diseased patient), the number of new patient admissions per day (in a hospital), the number of kilobases (in a strand of DNA), and computer literacy (of a person) are examples of attributes that are pertinent within informatics.

6.2.3   Object and Object Class

The object is the entity on which the measurement is made. In the examples of attributes in the section above, the object that pairs with an attribute is given in parentheses. It is useful to think of an attribute as a property of an object, and conversely, as an object as the "holder" or "carrier" of an attribute.

In some cases, it is possible to make measurements directly on the object itself; for example, when measuring the height of a person or the weight of a physical object. In other cases, the measurements are made on some representation of the object. For example, when measuring quality of performance (an attribute) of a procedure (the object), the actual measurement may be made via a recording of the procedure rather than a direct, real-time observation of it.

It is also useful to think of each object as a member of a class. Each act of measurement is performed on an individual object, which is a member of the class. Information resources, persons (who can be patients, care providers, students, and researchers), groups of persons, and organizations (healthcare and academic) are important examples of object classes in informatics on which measurements are frequently made.

6.2.4 Attribute–Object Class Pairs

Having defined an attribute as a property of a specific object that is a member of a class, measurement processes can be framed in terms of *paired* attributes and object classes. Table 6.1 illustrates this pairing of attributes and object classes for the examples discussed above. It is important to be able to analyze any given measurement process by identifying the pertinent attribute and object class. To do this, certain questions might be asked. To identify the *attribute*, the questions might be: What is being measured? What will the result of the measurement be called? To identify the *object class*, the question might be: On whom or on what is the measurement made? There is a worked example followed by Self-test 6.1, later in this chapter, immediately after Sect. 6.2.7. Together, these will expand your knowledge of these important concepts.

6.2.5 Attributes as "Constructs"

All attributes—that is, all things that are measured--are abstractions. This may be difficult to imagine because some of the attributes routinely measured in day-to-day life, especially properties of physical objects such as size and weight, are so intuitive that they become almost labels permanently attached to these objects. Seeing a tall person, we instinctively say, "That person must be 2 meters tall". But in point of fact, the height (attribute) of that person (object) is not known until it is measured. Using imagination, it is possible to conceive of a society where height was unimportant, so it went unmeasured; and moreover, the concept of "height" might not even exist in the minds of the people of that society.

For this reason, attributes are sometimes referred to as "constructs" in part as a reminder that attributes of objects, that are measured, are not an indigenous part of the physical world but rather products of human ingenuity (Binning 2016). Over time, each scientific field develops a set of constructs, "attributes worth measuring,"

Table 6.1 Examples of attribute–object class pairs

Attribute	Object class
Speed	Information resources
Blood pressure	Persons
Correct diagnosis	Patients with a disease
New admissions per day	Institutions that provide health care
Computer literacy	Persons
Number of base-pairs	DNA strands

that become part of the culture and discovery methods of that field. Study teams may tend to view the attributes that are part of their field's research tradition as a persistent part of the landscape and fail to recognize that, at some earlier point in history, these concepts were unknown. Blood pressure, for example, had no meaning to humankind until circulation was understood. Computer literacy is a more recent construct stimulated by contemporary technological developments. Indeed, many of the most creative works of science propose completely new constructs, develop methods to measure them, and subsequently demonstrate their value in describing or predicting phenomena of interest.

Because informatics is a "people science", many studies in informatics address human behavior and the abstract states of mind (knowledge, attitudes, beliefs) that are presumed to shape this behavior. To perform quantitative studies, these abstract states of mind must be formulated as constructs and measured "objectively" in accord with the philosophical precepts discussed in Sect. 2.5. The behavioral, social, and decision sciences have contributed specific methods that enable measurement of such attributes and use of them in studies. *A key idea for this chapter, that may be counterintuitive to those trained in the physical sciences, is that "states of mind" which cannot be directly observed with human senses, can still be measured objectively and employed with rigor in quantitative studies.*

6.2.6   Measurement Instruments

An instrument is the technology used for measurement. The "instrument" encodes and embodies the procedures used to determine the presence, absence, or extent of an attribute in an object. The instrument used in a measurement process follows from the attribute being measured and the object on which the measurement is being made. A diverse array of instruments exists to support studies in informatics. These include questionnaires to measure attitudes of people, image acquisition devices to determine the presence or absence of disease, forms to record appraisals of the observed performance of individuals and groups, and software that automatically records aspects of information resource performance. Software can also function as instruments by computing the values of attributes, such as care quality measures, from EHRs and other health data repositories.

While many types of measurements in informatics can be made by electro-mechanical and computational instruments, other types of measurements require humans "in the loop" as an integral part of the measurement process. For example, performance of complex tasks by people using information resources is appraised by observers who are called "judges". In such instances, a human "judge," or perhaps a panel of judges, are an essential part of the instrumentation for a measurement process. Humans who are part of the instrumentation should not be confused with humans whose performance is being measured.

6.2.7 Independent Observations

An observation is one independent element of measurement data. "Independent" means that, at least in theory, the result of each observation is not influenced in any way by the results of other observations. In some measurement processes, the independence of observations is easily recognized. For example, if a person steps onto a scale (the instrument) on three successive occasions, each of the three measurements (the attribute is weight, the object is a person) is readily assumed to be independent. When quality of a task performance (the attribute) by a surgical team (the object) is rated by a panel of judges (the instrument), each judge's rating can be considered an independent observation as long as the judges do not communicate before offering their assessments. When a questionnaire is the measurement instrument, we assume that a person's response to each question (often called an "item") on the questionnaire is a response to that question only and is uninfluenced by their responses to other questions.

As measurement is customarily carried out, multiple independent observations are employed to estimate the value of an attribute for an object. As seen in the next chapter, this is because multiple independent observations produce a better estimate of the "true" value of the attribute than any single observation. Use of multiple observations also allows for the determination of how much variability exists across observations, which is necessary to estimate the error inherent in the measurement.

An Example of a Measurement Process
Consider an information resource, such as *Isabel*, designed to improve medical diagnosis (Graber and Mathew 2008). Such a system would take patient signs, symptoms, and tests as input. The resource's algorithms would then suggest diseases that are consistent with these findings. A study team is interested in how reasonable the diagnoses suggested by the resource are, even when these diagnoses are not exactly correct. They conduct a study where the "top-five" diagnoses generated by the resource for a sample of test cases are referred to a panel of five experienced physicians for review. The panelists independently review, in the context of the case, each diagnosis set and rate the set on a 1–5 scale of "reasonableness".

The measurement aspects of this process can be divided into its component parts: the attribute being measured, the relevant object class, the instrument, and what constitutes the multiple independent observations.

- To identify the *attribute*, one might ask: What is being measured and what should the result of the measurement be called? The result of the measurement is the "reasonableness" of the top five diagnoses, so that is the attribute. This is a highly abstract construct because "reasonableness" is a state of mind. Nothing like it exists in the physical world.

- To identify the *object class*, one might ask: On whom or on what is the actual measurement made? The measurement is made on the diagnosis set generated by the resource for each case, so the "diagnosis set for each case" is the object class of measurement.
- The *instrument* in this example is a human judge coupled to a form on which the judgements are recorded.
- An *independent observation* is the appraisal of one diagnosis set for one case by one judge.

Self-test 6.1

1. To determine the performance of a computer-based reminder system, a sample of alerts generated by the system (and the patient record from which each alert was generated) is given to a panel of physicians. Each panelist rates each alert on a four-point scale from "highly appropriate to the clinical situation" to "completely inappropriate." Focusing on the measurement aspects of this process, name the attribute being measured, the class of measurement objects, the instrument used, and what constitutes an independent observation.
2. Staff members of a large community hospital undergo training to use a new administrative information system. After the training, each staff member completes a "test," which is comprised of 30 questions about the system, to help the developers understand how much knowledge about the system has been conveyed via the training. Name the attribute being measured, the class of measurement objects, and the instrument used. Describe how this measurement process employs multiple independent observations.
3. In a test of a prototype mobile health app, four testers rated "ease of use" for each of 10 tasks the app was designed to accomplish, using a 5-point scale for each rating. A rating of 5 means "extremely easy to use for this task" and at the other extreme, a rating of 1 means "nearly impossible to use for this task". What are the attribute and object class for this measurement, and what are the independent observations?

6.3 Key Object Classes and Types of Observations for Measurement in Informatics

6.3.1 Key Object Classes

Turning now to the range of measurement issues encountered in the real world of informatics, there are four specific categories of object classes that are often of primary interest in informatics studies: (1) professionals who may be health care providers, researchers, or educators; (2) clients of these professionals, usually patients or students; (3) biomedical information resources themselves; and (4) work groups or organizations that conduct research, provide health care, or provide education.

Among the classes of objects, *professionals* are important in informatics because attributes of these individuals influence whether and how information resources are used. Attributes of professionals that are important to measure include their domain-specific biomedical knowledge, their attitudes toward information technology and their work environment, their experience with information technology, among many others.

Clients emerge as objects of interest for many reasons. When clients are patients receiving health care, their health problems are complex and the attributes of these problems, central to the conduct of evaluation studies of information resources designed to improve their care, are difficult to assess. Important attributes of patients that often require measurement are diagnosis, prognosis, appropriateness of actual or recommended management, the typicality of their disease presentation, as well as their own beliefs and attitudes about health and disease. As patients increasingly access health information and some health services directly from the Internet, many of the attributes of professionals, as listed above, assume increased importance for patients as well. When clients are students receiving training in the health professions or biomedical research, measured attributes about them can be important determinants of what they will learn and what they are capable of learning.

Information resources have many attributes (e.g., data quality, speed of task execution, ease of use, cost, reliability, and degradation at the limits of their domain) that are of vital interest to informatics study teams.

Finally, *work groups and organizations* have many attributes (e.g., mission, age, size, budget structure, complexity, and integration) that determine how rapidly they adopt new technology and, once they do, how they use it.

6.3.2   Key Categories of Observations

The four categories of observations of frequent interest for measurement in informatics are:

1. *Tasks:* In many studies, measurements are made by giving the members of an object class something to do or a problem to solve. Different information resources (objects) may be challenged to process sets of microarray data (tasks) to determine speed of processing or usefulness of results (measured attributes). Alternatively, health care professionals or students (objects) may be asked to review sets of clinical case summaries (tasks) to develop a diagnosis or treatment plan (measured attributes). Within these kinds of performance-based assessments, which occur often in informatics, the challenges or problems assigned to

objects are generically referred to as tasks. The credibility of quantitative studies often hinges on the way the study team manages tasks in measurement and demonstration study design.

2. *Judges:* Many measurement processes in informatics employ judges—humans with particular expertise who provide their informed opinions, usually by completing a rating form, about behavior they observe directly or review, in retrospect, from some record of that behavior. Judges are necessary to measurement in informatics when the attribute being assessed is complex or where there is no clear standard against which performance may be measured.

3. *Items:* These are the individual elements of a form, questionnaire, or test that is used to record ratings, knowledge, attitudes, opinions, or perceptions. On a knowledge test or attitude questionnaire, for example, each individual question would be considered an item.

4. *Logistical factors:* Many measurement processes are strongly influenced by procedural, temporal, or geographic factors, such as the places where and times when observations take place.

6.4 Levels of Measurement

In the process of quantitative measurement, a value of an attribute is assigned to each object. Attributes differ according to how their values are naturally expressed or represented. Attributes such as height and weight are naturally expressed using continuous numerical values whereas attributes such as "marital status" are expressed using discrete values. An attribute's level of measurement denotes how its values can be represented. As discussed in later chapters, an attribute's level of measurement directs the design of measurement instruments and the statistical analyses that can be applied to the measurement results.

There are four such levels of measurement:

1. *Nominal:* Measurement on a nominal attribute results in the assignment of each object to a specific category. The categories themselves do not form a continuum or have a meaningful order. Examples of attributes measured at the nominal level are ethnicity, medical specialty, and the base-pairs comprising a nucleotide. To represent the results of a nominal measurement quantitatively, the results must be assigned arbitrary codes (e.g., 1 for "internists," 2 for "surgeons," 3 for "family practitioners"). The only aspect of importance for such codes is that they be employed consistently. Their actual numerical or alphanumerical values have no significance.

2. *Ordinal:* Measurement at the ordinal level also results in assignment of objects to categories, but the categories have some meaningful order or ranking. For example, physicians often use a "plus" system of recording clinical signs ("++ edema"), which represents an ordinal measurement. The staging of cancers is another clinical example of an ordinal measurement. The status ranking of uni-

versities is another example of ordinal measurement. When coding the results of ordinal measurements, a numerical code is typically assigned to each category, but no aspect of these codes except for their numerical order contains interpretable information.

Note that both nominal and ordinal measurements result in discrete or categorical values and are often referenced using those terms. However, use of the term "categorical" as an umbrella descriptor for nominal and ordinal measures conceals the important difference between them.

3. *Interval:* Results of measurements at the interval level take on continuous numerical values that have an arbitrarily chosen zero point. The classic examples are the Fahrenheit and Celsius scales of temperature. This level of measurement derives its name from the "equal interval" assumption, which all interval measures must satisfy. To satisfy this assumption, equal differences between two measurements must have the same meaning irrespective of where they occur on the scale of possible values. On the Fahrenheit scale, the difference between 50 and 40 degrees has the same meaning as the difference between 20 and 10 degrees. An "interval" of 10 degrees is interpreted identically all along the scale.

4. *Ratio:* Results of measurements at the ratio level have the additional property of a true zero point. The Kelvin scale of temperature, with a zero point that is not arbitrarily chosen, has the properties of ratio measurement. Most physiological attributes (such as blood pressure) and physical measures (such as length) have ratio properties. This level of measurement is so named because one can assign meaning to the ratio of two measurement results in addition to the difference between them.

Just as nominal and ordinal measures are often grouped under the heading of "categorical" or "discrete", interval and ratio measures are often grouped under the heading of "continuous".

In quantitative measurement, it is often desirable to collect data at the highest level of measurement possible for the attribute of interest, with ratio measurement being the highest of the levels. In other words, if the attribute allows measurement at the interval or ratio level, the measurement should be recorded at that level. Doing this ensures that the measured results contain the maximum amount of information. For example, in a survey of healthcare providers, a study team may want to know each respondent's years of professional experience which is, naturally, a ratio measure. Frequently, however, such attributes are assessed using discrete response categories, each containing a range of years. Although this measurement strategy provides some convenience and possibly some sense of anonymity for the respondent (which may generate more complete data with fewer missing values), it reduces to ordinal status what is naturally a ratio variable, with inevitable loss of information. Even if the data are later going to be categorized in service to privacy by preventing re-identification, collecting the data initially at the highest level of measurement is the preferred strategy. Data can always be converted from higher to lower levels of measurement, but it is not possible to go in the other direction.

Self-test 6.2

Determine the level of measurement of each of the following:

1. A person's serum potassium level.
2. A health sciences center's national ranking in research grant funding.
3. The distance between the position of an atom in a protein, as predicted by a computer model, and its actual position in the protein.
4. The "stage" of a patient's cancer diagnosis.
5. The hospital unit to which a patient is assigned following admission to the hospital.
6. A person's marital status.
7. A person's score on an intelligence test, such as an IQ test.

6.5   Importance of Measurement in Quantitative Studies

The major premises underlying the quantitative approaches to evaluation, originally introduced in Sect. 2.5, highlight why measurement is so important. These premises are re-stated here in a somewhat revised form to take advantage of the new concepts and terminology introduced in this chapter.

In quantitative studies, the following are assumed:

- *Attributes* are inherent in the *object* under study. Merit and worth are part of the object and can be measured unambiguously. A study team can measure these attributes without affecting the object's structure or function.
- All rational persons agree (or can be brought to consensus) on what attributes of an object are important to measure and what measurement results would be associated with high merit or worth.
- Because measurement of attributes allows precise statistical portrayals and comparisons across groups and across time, numerical measurement is *prima facie* superior to verbal description.
- Through comparisons of measured attributes across selected groups of objects, it is possible to rigorously address evaluation questions of importance to informatics.

These premises make clear that the proper execution of quantitative studies requires careful and specific attention to methods of measurement. Accurate and precise measurement cannot be an afterthought.[1] Measurement is of particular importance in informatics because, as a relatively young and still evolving field (Hasman et al. 2011), informatics does not have a well-established tradition of "things worth measuring" or proven methods for measuring them. By and large, those planning studies are faced with the task of first deciding what to measure and

---

[1] Terms such as "accuracy" and "precision" are used loosely in this chapter. They will be defined more rigorously in Chap. 7.

then developing their own measurement methods. For most study teams, this task proves more difficult and more time-consuming than initially anticipated. In some cases, informatics study teams can adapt the measures used by others, but they often need to apply these measures to a different setting, where prior experience may not apply.

The choice of what to measure, and how, is an area where there are few prescriptions and where sound judgment, experience, and knowledge of methods come into play. Decisions about *what* and, above all, *how* to measure require knowledge of the study questions, the intervention and setting, and the experience of others who have done similar work. A methodological expert in measurement is of assistance only when teamed with others who know the terrain of biomedical and health informatics.

6.6 Measurement and Demonstration Studies

This section establishes a formal distinction between studies undertaken to develop and refine methods for making measurements, which are called *measurement studies*, and the subsequent use of these methods to address questions of direct importance in informatics, which are called *demonstration studies*.[2]

Measurement studies seek to determine with how much error an attribute of interest can be measured in a population of objects, often also indicating how this error can be reduced. In an ideal quantitative measurement, all independent observations yield identical results. Any disagreement is due to measurement error. For example, measuring the air pressure (attribute) in an automobile tire (object) done three consecutive times at 15 s intervals would be expected to yield identical results and any discrepancies would be attributed to "measurement error" and would call into question some aspect of the measurement process. The developers of a new tire pressure gauge (an instrument) would be well advised to conduct measurement studies, testing their new instrument by determining whether it generates reproducible results over a range of conditions such as different types of inflation valves.

Measurement procedures developed and vetted through measurement studies provide study teams with what they need to conduct *demonstration studies*. Once it is known with how much error an attribute can be measured using a particular procedure, the measured values of this attribute can be employed to answer questions of importance in the world. Continuing our example, if tire manufacturers have confidence in the new pressure gauge as a measurement instrument, they can use it with confidence to measure things of importance to them, such as how quickly a tire, when punctured, loses air.

---

[2] The concept of the measurement study in informatics can be traced to the work of Michaelis et al. (1990).

The bottom line is that study teams must know that their measurement methods are fit for purpose—"fitness for purpose" will be defined more rigorously in the next chapter—*before* collecting data for their demonstration studies. As shown in Fig. 6.2, it is necessary to perform a measurement study to establish the adequacy of all measurement procedures unless the measurement methods to be used have an established "track record."

Example Contrasting Measurement and Demonstration
Return to the example of measuring the "reasonableness" (attribute) of diagnostic hypothesis sets (object) generated by an AI system, where the observation type is human judges. In a perfect quantitative measurement, all judges would give identical ratings to the hypothesis set generated by the system for each case, but this is very unlikely to occur. A *measurement study* would explore the amount of disagreement that exists among the judges across a set of cases. This is a necessary step in evaluating the system, but not what the developers and potential users of the program are primarily interested in knowing. However, once the performance of the measurement process is established, and the amount of error is considered to be acceptable, then a wide range of *demonstration studies* becomes possible. Important demonstration studies might explore the "reasonableness" of the diagnoses the AI system generates, whether the hypotheses generated by the AI system are more "reasonable" than those generated by human clinicians without aid from the system, and whether exposure to the diagnoses generated by the system increases the "reasonableness" of the human clinicians' diagnoses.

The attribute addressed here reflects a highly abstract construct, what may be seen as a "state of mind" of the judges. Even so, quantitative methods can be applied. The abstractness of the attribute does not require use of qualitative methods.

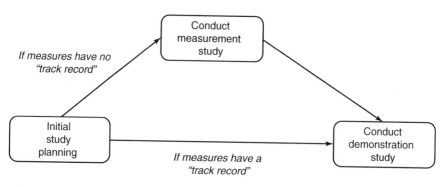

Fig. 6.2 Measurement and demonstration studies

A track record for a measurement procedure develops from measurement studies undertaken and published by others. Even if the measurement procedures of interest have a track record in a particular health care or research environment, they may not perform equally well in a different environment. In these cases, measurement studies may still be necessary even when apparently tried-and-true measurement approaches are being employed. Study teams should always ask themselves—"Are the measurement methods we want to use fit for *our* purpose?—whenever they are planning a study and before proceeding to the demonstration phase. The "box" below offers several examples of measurement studies from the literature.

Examples of Published Measurement Studies

The results of measurement studies can and should be published. When published, they establish a "track record" for a measurement process that enables study teams to employ it with confidence in demonstration studies. Finding existing measurement methods prior to embarking on a study is an example of "Evidence-based Health Informatics" as discussed in Chap. 5. There is a published algorithm to aid study teams in identifying measurement studies and measurement instruments (Terwee et al. 2009).

- The most numerous measurement studies explore the reliability and validity of questionnaires to assess such attributes as health ICT and mobile health app usability (Yen et al. 2014; Zhou et al. 2019); mobile device proficiency (Roque and Boot 2018) as well as more general attitudes toward health ICT (Dykes et al. 2007).
- Other published measurement studies explore methods to measure the impact of clinical decision support (Ramnarayan et al. 2003).
- The literature also includes measurement studies that address the automated computation of quality measures (Kern et al. 2013).

6.7 Goals and Structure of Measurement Studies

The overall goal of a measurement study is to estimate with how much error an attribute of interest can be measured for a class of objects, ultimately leading to a viable measurement process for later application to demonstration studies.

Chapter 7 describes in more detail how measurement errors are estimated. Chapters 8 and 9 address the development of measurement methods in greater technical detail.

One specific goal of a measurement study is to determine how many independent observations are necessary to reduce error to a level acceptable for the demonstration studies to follow. In general, the greater the number of independent observations comprising a measurement process, the smaller the measurement error. (The following chapter will explain why.) This relationship suggests an important trade-off because each independent observation comes at a cost. Returning to the example of the AI diagnostic advisory system, the more judges who rate each hypothesis set, the "better" the measurement will be. However, the time of expert judges is both expensive and a scarce resource. Experts are busy people and may have limited time to devote to participating in this role. Without a measurement study conducted in advance, there is no way to quantify this trade-off and determine an optimal balance point between quality of measurement and availability of judges.

Another goal of measurement studies is to verify that measurement instruments are well designed and functioning as intended. Even a measurement process with an ample number of independent observations--judges, in our running example--will have a high error rate if there are fundamental flaws in the way the process is conducted. For example, if the judges do not discuss, in advance, what "plausibility of a hypothesis set" actually means, the results will reveal unacceptably high error. Fatigue may be a factor if the judges are asked to do too much, too fast. As another example, consider a computer program developed to compute patients' medication costs (attribute) automatically from a computer-based patient record (object). If this program has a bug that causes it to fail to include certain classes of medications, it will not return accurate results. Only an appropriate measurement study can detect these kinds of problems.

Teams designing a measurement study also should try to build into the study features that challenge the measurement process in ways that might be expected to occur in the demonstration study to follow. Only in this way is it possible to determine if the results of the measurement study will apply when the measurement process is put to use. For example, in the diagnostic hypothesis rating task mentioned above, the judges should be challenged in the measurement study with a range of cases typical of those expected in the demonstration study. The program to compute medication costs should be tested with a representative sample of cases from a variety of clinical services in the hospital where the demonstration study will ultimately be conducted. Additionally, a measurement technique may perform well with individuals from one particular culture, but perform poorly when transferred to a different culture. For example, the same questionnaire administered to patients in two hospitals serving very different ethnic groups may yield different results because respondents are interpreting the questions differently. This issue, too, can be explored via an appropriately designed measurement study that includes the full range of sociocultural settings where questionnaire might be used.

Measurement studies are also important to ensure that measurement processes are not trivialized. A measurement process might seem straightforward until it is actually necessary to use it. As an example, consider a study of a new admission–discharge–transfer (ADT) system for hospitals. Measuring the attribute "time to process admission" for the object class of patients is central to the success of this study. Although, on the surface, this attribute might seem trivial to measure, many potential difficulties arise on closer scrutiny. For example, when did the admission process for a patient *actually* begin and end: when each patient entered a waiting room or when they began speaking to an admissions clerk? Were there interruptions, and when did each of these begin and end? Should interruptions be counted as part of processing time?

If human observers are going to be the measurement instrument for this process, a measurement study might include three or four such observers who simultaneously observe the same set of admissions. The discrepancies in the observers' estimates of the time to process these admissions would determine how many observers (whose individual results are averaged) are necessary to obtain an acceptable error rate. The measurement study might reveal flaws in the form on which the observers are recording their observations, or it might reveal that the observers had not been provided with adequate instructions about how, for example, to deal with interruptions. The measurement study could be performed in the admissions suite of a hospital similar in many respects to the ones in which the demonstration study will later be performed. The demonstration study, once the measurement methods have been established, would then explore whether the hospital actually processes admissions faster with the new system than with its predecessor.

Self-test 6.3

Clarke and colleagues developed the TraumAID system (Clarke et al. 1994) to advise on initial treatment of patients with penetrating injuries to the chest and abdomen. Measurement studies of the utility of TraumAID's advice required panels of judges to rate the adequacy of management of a set of "test cases"—as recommended by TraumAID and as carried out by care providers. To perform this study, case data were fed into TraumAID to generate a treatment plan for each case. The wording of TraumAID's plans was edited carefully in hope of ensuring that judges performing subsequent ratings would not know whether the described care was performed by a human or recommended by a computer.

Two groups of judges were employed: one from the medical center where the resource was developed, the other a group of senior physicians from across the country. The purposes of this study were to determine the extent to which judges' ratings of each plan were in agreement, whether judges could detect plans generated by computer merely by the way they are phrased, and whether the home institution of the judges affected the ratings. Is this a measurement or a demonstration study? Explain.

6.8   The Structure and Differing Terminologies of Demonstration Studies

Demonstration studies differ from measurement studies in several respects. First, they aim to say something meaningful about an information resource or address some other question of substantive interest in informatics. With measurement studies, the primary concern is the error inherent in assigning a value of an attribute to each individual object, whereas with demonstration studies, the concern is different. Demonstration studies are concerned with determining the actual magnitude of that attribute in a group of objects, determining if certain groups of objects differ in the magnitude of that attribute, or if there is a relationship between that attribute and other attributes of interest. For example, in a study of an information resource to support management of patients in the intensive care unit:

- A measurement study would be concerned with how accurately and precisely the "optimal care" (attribute) of patients (object class) can be determined.
- A subsequent demonstration study might explore whether care providers supported by the resource deliver care more closely approximating optimal care.

In a study of an "app" to support patients' control their blood pressure, study teams might be interested in whether users are more satisfied with a new version of the app compared to an older version:

- A measurement study may focus on how well satisfaction with the app (attribute) as perceived by its users (object class) can be measured with a short questionnaire.
- A subsequent demonstration study would compare satisfaction among users of the new of old versions.

The terminology used to describe the structure of demonstration studies also differs from that used for measurement studies. Most notably:

- When the *object class* of measurement in a measurement study is people, they are typically referred to as *subjects or participants* in a demonstration study.
- An *attribute* in a measurement study is typically referred to as a *variable* in a demonstration study.

In describing the design of quantitative studies, variables can be of two types: dependent and independent. The *dependent variables* are a subset of the variables in the study that capture outcomes of interest to the study team. For this reason, dependent variables are also called "outcome variables." The *independent variables* are those included in a study to explain the measured values of the dependent variables. Returning to some examples introduced earlier in this chapter, a demonstration study might contrast the time to admit new patients to a hospital before and after a new information resource designed to facilitate the process is introduced. In this example, "time to admit" would be the dependent variable and "before vs. after" the introduction of the new system would be the independent variable.

6.9 Demonstration Study Designs

There are many variations of demonstration study designs that suit many different purposes. Figure 6.3 illustrates these variations.

6.9.1 Prospective and Retrospective Studies

Demonstration studies vary according to the time orientation of data collection.

Prospective Studies: A prospective study is planned and designed in advance, in anticipation of data collection that will occur primarily in the future. The planning of the study precedes the collection of the most important data, and in this sense, prospective studies look forward in time.

Examples of prospective studies include: needs assessment surveys, usability studies of a prototype resource conducted in laboratory conditions, and field user-effect studies that randomize individuals to groups that do and do not employ a patient-facing app.

It is common for teams conducting prospective studies to include in their study some data that pre-existed the planning of the study. For example, a team

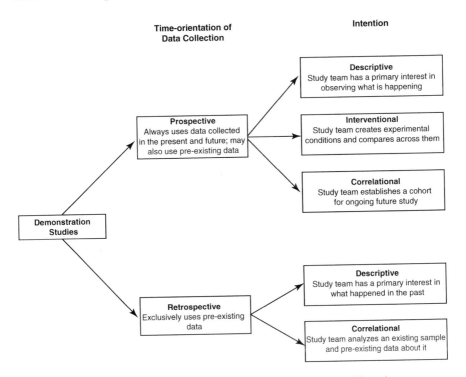

Fig. 6.3 Demonstration study designs differentiated by time-orientation and intention

conducting a field user effect study of a patient-facing app might, with consent, bring into the study dataset some information about each participant's health behavior prior to any participants' use of the app.

Retrospective Studies: In a retrospective study, the data exist prior to the planning and design of the study. Retrospective studies look backward upon data already collected. The data employed in retrospective studies are often collected routinely as a by-product of practice and stored in clinical, research or educational data repositories which makes the data available for a wide range of studies. In a retrospective study, the team selects which data to employ in their analyses out of a larger set of data that is available to them. Retrospective studies are also called "ex post facto" studies.

Examples of purely retrospective studies include needs assessments for a new patient portal that exclusively examines records of previous use of the existing portal, a lab function study to develop and test the accuracy of a machine learning model to predict sepsis based on data accumulated in a clinical data repository, and a problem impact study tracking research productivity following introduction of a research information management system 3 years in the past.

6.9.2   Descriptive, Interventional, and Correlational Studies

Demonstration studies also differ according to intention.

Descriptive Studies: Descriptive studies seek primarily to estimate the value of a dependent variable or set of dependent variables in a selected sample of subjects. *Purely descriptive studies have no independent variables.* If a group of nurses were given a rating form (previously verified through a measurement study to be an acceptable measurement tool) to ascertain the "ease of use" of a nursing information system, the mean value of this variable in a sample of nurses would be the key result of this descriptive study. If this value were found to be toward the low end of the scale, the study team might conclude from this descriptive study that the resource was in need of substantial revision. Although they seem deceptively simple, descriptive studies can be highly informative. Studies of the quality of health information on the Internet illustrate the importance of well-conducted descriptive studies (Daraz et al. 2019).

Referring back to the categorization of evaluation approaches introduced in Sect. 2.7, descriptive studies also can be tied to the "objectives based" approach. When a study team seeks to determine whether a resource has met a predetermined set of performance objectives--for example, whether the response time meets predetermined criteria for being "fast enough"--the logic and design of the resulting demonstration study is descriptive. There is no independent variable. The dependent variable in this example is "speed of execution", usually averaged over a range of tasks executed by the resource.

As illustrated in Fig. 6.3, descriptive studies can be either prospective or retrospective. In the example above, testing the speed of a resource in executing a set of tasks is a prospective study. An example of a retrospective descriptive study would entail inspection of usage log files to determine the extent and nature of use of an information resource.

Interventional Studies: Here, the study team creates a contrasting set of conditions that enable the exploration of the questions of interest. For example, if the study question of primary interest is: "Does a particular patient-facing app promote weight loss?", the study team would typically create one group of users and another group of non-users of the app. After identifying a sample of subjects for the study, the team then assigns each subject to either the "user" or "non-user" condition. In this example, "use" vs. "non-use" of the app is the independent variable. A dependent (outcome) variable of interest, in this case weight loss, is then measured for each subject, and the mean values of this variable for subjects in each condition are compared. Interventional studies align with the "comparison-based" approach to evaluation introduced in Sect. 2.7. They are often referred to as experiments or quasi-experiments.

The classic field user-effect study of reminder systems by McDonald and colleagues (McDonald et al. 1984) is an example of an interventional study applied to informatics. In this study, groups of clinicians (subjects/participants) either received or did not receive computer-generated reminders to carry out recommended aspects of care. Receipt or non-receipt of the reminders were the conditions comprising the independent variable. The study team compared the extent to which clinical actions consistent with recommended care (the dependent variable) took place in each group.

When the study conditions are created prospectively by the study team and all other differences are eliminated by random assignment of subjects, it is possible to isolate and explore the effect due solely to the difference between the conditions. This allows the study team to assert that these effects are causal rather than merely coincidental.

As illustrated in Fig. 6.3, all interventional studies are prospective.

Correlational Studies: Here, study teams explore the relationships among a set of variables, the values of which are routinely generated as a by-product of ongoing life experience, health care, research, educational, or public health practice. Correlational studies are increasingly called "real world evidence" studies and may also be called datamining or quality assessment studies. These studies are becoming increasingly common in both clinical and biological application domains.

Correlational studies may be either prospective or retrospective, as shown in Fig. 6.3. The best-known example of a prospective correlational study is the Framingham Study of heart disease, where a cohort of participants was first identified and then followed over many decades (Dawber 2013). This is a prospective study because the effort was planned before any data were collected, but there were no interventions and all of the data originated from the life and health care experiences of the participants. An informatics example of a prospective correlational study is a field function study which "pseudo-deploys" a machine learning algorithm in a care environment and sends the predictions of the algorithm back to the study team to determine the predictive power of the model. An informatics example of a retrospective correlational study examines use of an online portal by a health system's patients. Looking back several years following the introduction of the portal, a study team might examine the extent and nature of portal use in relation to demographic characteristics, disease experience and other characteristics (Palen et al. 2012).

In correlational studies, there will almost always be more data available than are needed to address a particular set of study questions. Study teams select the variables to include in their analysis from this larger set, and must choose which variables to label as "independent" and "dependent", associating the independent variables with presumed predictors or causes and associating the dependent variables with presumed outcomes or effects. Correlational studies are linked most closely to the "decision facilitation" approach to evaluation discussed in Sect. 2.7.1. Because there is no deliberate creation of experimental conditions or randomization in correlational studies, attributions of cause and effect among the variables selected for analysis require careful consideration, as will be discussed in Chap. 13.

6.10   Demonstration Studies and Stages of Resource Development

This concluding section calls attention to some important relationships between the three demonstration study types, the lifecycle of resource development, and the settings in which demonstration studies can be conducted.

• Descriptive studies are useful over the entire lifecycle and can be conducted in the lab or in the field. Descriptive studies are particularly valuable in early stages of resource development to determine if the resource is achieving pre-determined design specifications; but they may also be valuable to document the extent and nature of use of a resource after it is deployed.
• Interventional studies require a prototype or a deployed version of a resource. Study teams can carry out interventional studies in laboratory settings—for example, to compare two interfaces of a resource under controlled conditions prior to deployment. Alternatively, and as discussed above, study teams can conduct an interventional field study to understand the effects of a deployed resource.
• Correlational studies require a deployed resource and are carried out in field settings that generate real-world data.

Returning to the themes of earlier chapters, study teams can almost always choose among several options for demonstration studies. This chapter emphasizes that, irrespective of the selected demonstration study design, all quantitative studies rely on sound methods of measurement.

Self-test 6.4

Classify each of the following demonstration study designs as descriptive, interventional, or correlational. In each study, who or what are the subjects/participants? Identify the independent and dependent variables.

1. An information resource is developed to aid in identifying patients who are eligible for specific clinical protocols. A demonstration study of the resource is implemented to examine protocol enrollment rates at sites where the resource was and was not deployed.
2. A new clinical workstation is introduced into a network of medical offices. Logs of 1 week of resource use by nurses are studied to document to what extent and for what purposes this resource is being used.
3. A number of clinical decision support resources have been deployed to support care in a large health system. A study team examines which characteristics of patients are predictive of the extent of use of these resources.
4. A study team compiles a database of single nucleotide polymorphisms (SNPs), which are variations in an individual's genomic sequences. The team then examines these SNPs in relation to diseases that these individuals develop, as reflected in a clinical data repository.
5. Students are given access to a database to help them solve problems in a biomedical domain. By random assignment, half of the students use a version of the database emphasizing hypertext browsing capabilities; half use a version emphasizing Boolean queries for information. The proficiency of these students at solving problems is assessed at the beginning and again at the end of the study period.

Answers to Self-tests

Self-test 6.1

1. The attribute is "appropriateness" of each alert. "Alerts" comprise the object class. (Although the panelists need access to the cases to perform the ratings, cases are *not* the object class here because each alert is what is directly rated—the attribute of "appropriateness" is a characteristic of each alert—and because each case may have generated multiple alerts related to its different clinical aspects.) The instrumentation is the rating form as completed by a human judge. Each individual judge's rating of the appropriateness of an alert constitutes an independent observation.
2. The attribute is "knowledge about the administrative information system." Staff members are the object class. The instrument is the written test. Each question on the test constitutes an independent observation.
3. The attribute is "ease of use" of the app. Tasks are the object class. The independent observations are the ratings by each tester.

Self-test 6.2
1. Ratio
2. Ordinal
3. Ratio
4. Ordinal
5. Nominal
6. Nominal
7. Interval (In IQ testing, the average score of 100 is completely arbitrary)

Self-test 6.3

This is a measurement study. The stated purposes of the study have nothing to do with the actual quality of TraumAID's advice. The purposes are exclusively concerned with how well this quality, whatever it turns out to be, can be measured. The quality of TraumAID's advice would be the focus of a separate demonstration study.

Self-test 6.4

1. It is an interventional study because the study team presumably had some control over where the resource was or was not deployed. The site is the "subject" for this study. (Note that this point is a bit ambiguous. Patients could possibly be seen as the subjects in the study; however, as the question is phrased, the enrollment rates at the sites are going to be the basis of comparison. Because the enrollment rate must be computed for a *site*, then site must be the "subject.") It follows that the dependent variable is the protocol enrollment rate; the independent variable is the presence or absence of the resource.
2. It is a descriptive study. Nurses using the system are the subjects. There is no independent variable. Dependent variables are the extent of workstation use for each purpose.
3. It is a correlational study. Patients are the subjects. The independent variables are the characteristics of the patients; the dependent variable is the extent of use of information resources.
4. This is also a correlational study. Patients are the subjects. The independent variable is the genetic information; the dependent variable is the diseases they develop. There is, however, no manipulation or purposeful intervention.
5. Interventional study. Students are the subjects. Independent variable(s) are the version of the database and time of assessment. The dependent variable is the score on each problem-solving assessment.

References

Binning JF. Construct. In: Encyclopedia Britannica. 2016. https://www.britannica.com/science/construct. Accessed 30 Dec 2020.
Clarke JR, Webber BL, Gertner A, Rymon KJ. On-line decision support for emergency trauma management. Proc Symp Comput Applications Med Care. 1994;18:1028.

Daraz L, Morrow AS, Ponce OJ, Beuschel B, Farah MH, Katabi A, et al. Can patients trust online health information? A meta-narrative systematic review addressing the quality of health information on the internet. J Gen Intern Med. 2019;34:1884–91.

Dawber TR. The Framingham study: the epidemiology of atherosclerotic disease. Harvard University Press; 2013.

Dykes PC, Hurley A, Cashen M, Bakken S, Duffy ME. Development and psychometric evaluation of the Impact of Health Information Technology (I-HIT) scale. J Am Med Inform Assoc. 2007;14:507–14.

Feinstein AR. Clinimetrics. New Haven: Yale University Press; 1987. p. viii.

Graber ML, Mathew A. Performance of a web-based clinical diagnosis support system for internists. J Gen Intern Med. 2008;23:37–40.

Hasman A, Ammenwerth E, Dickhaus H, Knaup P, Lovis C, Mantas J, et al. Biomedical informatics–a confluence of disciplines? Methods Inf Med. 2011;50:508–24.

Kern LM, Malhotra S, Barrón Y, Quaresimo J, Dhopeshwarkar R, Pichardo M, et al. Accuracy of electronically reported "meaningful use" clinical quality measures: a cross-sectional study. Ann Int Med. 2013;158:77–83.

McDonald CJ, Hui SL, Smith DM, et al. Reminders to physicians from an introspective computer medical record: a two-year randomized trial. Ann Intern Med. 1984;100:130–8.

Michaelis J, Wellek S, Willems JL. Reference standards for software evaluation. Methods Inf Med. 1990;29:289–97.

Palen TE, Ross C, Powers JD, Xu S. Association of online patient access to clinicians and medical records with use of clinical services. JAMA. 2012;308:2012–9.

Ramnarayan P, Kapoor RR, Coren M, Nanduri V, Tomlinson AL, Taylor PM, et al. Measuring the impact of diagnostic decision support on the quality of clinical decision making: development of a reliable and valid composite score. J Am Med Inform Assoc. 2003;10:563–72.

Roque NA, Boot WR. A new tool for assessing mobile device proficiency in older adults: the mobile device proficiency questionnaire. J Appl Gerontol. 2018;37:131–56.

Scott PJ, Brown AW, Adedeji T, Wyatt JC, Georgiou A, Eisenstein EL, et al. A review of measurement practice in studies of clinical decision support systems 1998–2017. J Am Med Inform Assoc. 2019;26:1120–8.

Terwee CB, Jansma EP, Riphagen II, de Vet HC. Development of a methodological PubMed search filter for finding studies on measurement properties of measurement instruments. Qual Life Res. 2009;18:1115–23.

Yen PY, Sousa KH, Bakken S. Examining construct and predictive validity of the Health-IT Usability Evaluation Scale: confirmatory factor analysis and structural equation modeling results. J Am Med Inform Assoc. 2014;21:e241–8.

Zhou L, Bao J, Setiawan IM, Saptono A, Parmanto B. The mHealth APP usability questionnaire (MAUQ): development and validation study. JMIR. 2019;7:e11500.

Chapter 7
Measurement Fundamentals: Reliability and Validity

Learning Objectives

The text, examples, and practice problems in this chapter will enable you to:

1. Distinguish between the concepts of reliability and validity of a measurement process.
2. Explain why measurement processes reveal their own reliability and why determination of validity requires appeal to a source of information external to the measurement process.
3. Represent the result of a measurement study as an "objects-by-observations" matrix.
4. Given the results of a measurement process expressed in an objects-by-observations matrix, use a spreadsheet or other program to compute a reliability coefficient for that measurement process.
5. Describe the effect of changing the number of observations on the reliability of a measurement process and apply the "Prophecy Formula" to estimate the effect of change the number of observations.
6. Describe the relationship between measurement error and reliability, and apply the formula that relates the standard error of measurement to reliability.
7. Given a description of a validity study, identify it as primarily a content, criterion-related, or construct validity study.
8. Outline the structure of a content, criterion-related, or construct validity study for a given measurement process.

Supplementary Information The online version of this chapter (https://doi.org/10.1007/978-3-030-86453-8_7) contains supplementary material, which is available to authorized users.

7.1 Introduction

*There is growing understanding that all measuring instruments must be critically and empirically examined for their reliability and validity. The day of tolerance of inadequate measurement has ended (*Kerlinger 1986).

This "call to action" from a famed psychologist in 1986 remains true to this day and has been echoed more recently other scholars in a range of health-related fields (Boateng et al. 2018; Moriarity and Alloy 2021; Thanasegaran 2009; Kimberlin and Winterstein 2008). It motivates the very significant attention paid in this volume to the theory and practice of measurement. Section 6.6 established the distinction between *measurement studies* that determine how well (with how much error) an attribute of interest can be measured, and *demonstration studies* that use the results of these measurement to address important evaluation questions. For example, a measurement study might reveal that a particular measurement process can reveal the "speed" of a resource in executing a certain family of tasks to a precision of ±10%. By contrast, a demonstration study might reveal that Resource A completes these tasks with greater speed than Resource B. Measurement and measurement studies are the foci of this chapter as well as Chaps. 8 and 9.

Recall from Sect. 6.6 that measurement studies, when they are necessary, are conducted before any related demonstration studies. All measurement studies have a common general structure and employ a family of analytical techniques, which are introduced in this and the following chapter. In measurement studies, the measurement of interest is undertaken with a sample of objects under conditions similar to those expected in the demonstration study to follow.

The data generated by a measurement study are analyzed to *estimate* what can, in general terms, be characterized as the amount of error inherent in the measurement process. The estimated amount of error derives from a *sample* of objects and sometimes also from a sample of circumstances under which the measurement will be made. In general, the greater the sample size used in a measurement study, the greater the confidence the study team can place in the estimate of the magnitude of the measurement error. Oftentimes the results of a measurement study suggest that the error is too large and that the measurement methods must be refined before a demonstration study can be undertaken with confidence. The results of the measurement study may also suggest the specific refinements that are needed to reduce the error.

A demonstration study usually yields, as a by-product, data that can be analyzed to shed some light on the performance of the measurement methods used in the study. For example, the patterns of responses to a questionnaire by subjects in a demonstration study can be analyzed to estimate the measurement error attributable to the questionnaire in that specific sample of subjects. It is often useful to carry out these analyses to confirm that measurement errors are in accord with expected levels; however, this cannot substitute for measurement studies conducted in advance when the measurement methods do not have a "track record".

When a demonstration study employs measures with a "track record", especially if it is based on published results of measurement studies, the study team can proceed with confidence. For this reason, carefully designed and conducted measurement studies are themselves important contributions to the literature. These publications are a service to the entire field of informatics, enabling the entire community to conduct demonstration studies with greater confidence, without having to conduct measurement studies of their own.

7.2   The Classical Theory of Measurement: Framing Error as Reliability and Validity

All measurements have some degree of error. The value of an attribute in a particular object is never known with absolute confidence and for that reason a branch of science has developed around the theory of error and methods to quantify it. This chapter develops the concept of error according to the so-called classical theory of measurement, which quantifies measurement error using the metrics of reliability and validity (Tractenberg 2010).

In classical theory, reliability is the degree to which measurement is consistent or reproducible. A measurement that is at least acceptably reliable is measuring *something*. Validity is the degree to which that *something* being measured is what the study team intends to measure. Validity of measurement becomes a particular challenge when the attribute being measured is highly abstract. When properties of physical objects, such as height of people, are the attributes of interest, the attribute is not particularly abstract. Human senses provide reassurance that a rigid pole calibrated in inches or centimeters is measuring height and not something else. By contrast, consider measurement of users' *perceptions* of "ease of use" of an information resource via a questionnaire. This attribute is a state of mind not observable with human senses. Responses to a questionnaire could be measuring ease of use of this resource but might also be measuring, to some degree, computer literacy and a more general attitude toward all information technology and not just this resource in particular.

Reliability is a logical precursor to validity. No meaningful exploration of the validity of a measurement process is possible until that process is demonstrated to be acceptably reliable. What constitutes "acceptable" reliability will be discussed later in this chapter. Measurements of height are likely to be very, but not perfectly, reliable. The reliability of "ease of use" assessments is less clear. Both reliability and validity are properties of the measurement process as a whole. Changing any aspect of the complete measurement process may introduce error or change the nature of the error. For example, an "ease of use" questionnaire developed for and shown to be reliable when completed by highly educated care providers may be much less reliable and valid when administered to patients for whom the language of the questionnaire is a second language.

In other disciplines, terms very closely related to reliability and validity are used to describe and quantify measurement error. Clinical epidemiologists and clinical researchers typically use *precision* as a term corresponding to reliability and *accuracy* as a term corresponding to validity (Weinstein et al. 1980).

7.3 Toward Quantifying Reliability and Validity

The previous discussion developed reliability and validity as concepts. The next step is to frame reliability and validity as quantified properties of measurement processes. These properties, themselves, can be estimated to provide a useful indication of whether a measurement process is fit for use in a demonstration study. To this end, classical theory makes several important assumptions that are illustrated in Fig. 7.1.

Base Assumption: All measurements have errors. All measurements are in search of an unknowable exact result that can only be estimated.

Assumption 1: A perfectly reliable measurement will produce the same result every time an independent measurement is conducted under identical conditions. The result of a single measurement (an "observed score") for any object can be represented as the result of a perfectly reliable measurement, the "true score", plus "measurement error".

Assumption 2: Measurement errors are randomly distributed "noise", so over a series of a measurements, the best estimate of an object's true score is the mean value of the observed results. Also, the greater the number of repeated measurements, the more closely the mean of the observed results will approximate the "true score".

Assumption 3: The "true score", while completely reliable, is not completely valid. The true score thus has two components: a completely valid component, capturing the extent to which it reflects the attribute the measurement process is intending to measure; and an invalid component, reflecting the extent to which the measured value also reflects some other "unwanted" attribute(s).

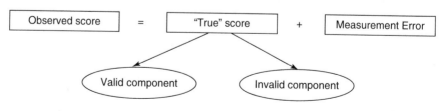

Fig. 7.1 Components of an observed (measured) score in classical theory

Returning to the earlier examples, when measuring the attribute of height in people as objects, random factors such as slouching and stretching can contribute to measurement error. Also, when the instrument is a ruler calibrated in centimeters or inches, random error can be introduced around a determination of the top of an individual's head. These random factors create a common-sense conclusion that averaging repeated measurements will produce a better estimate of an individual's actual height than one observation. As discussed previously, the validity of the process can in this case be verified by direct observation confirming that the rod is measuring height and nothing else.

When measuring the perceived ease of use of an information resource via a questionnaire, matters become a bit more challenging. There are good reasons to believe that an unknown number of uncontrollable factors related to the person's biological and mental state might affect responses to the questionnaire in essentially random ways. Attempting to converge on an estimate of a person's "true" perception of the resource's ease of use by repeatedly administering the questionnaire, following the procedure used when measuring height, will not be effective because people responding to the questionnaire on successive occasions might simply remember and record on the form what they responded on the previous occasion, so these successive observations would not qualify as independent measurements. Moreover, other factors such as computer literacy and general attitudes toward technology might be "polluting" the measured results and affecting their validity. Much of remainder of this chapter, and the next, will introduce ways of addressing these challenges.

7.4   A Conceptual Analogy Between Measurement and Archery

The analogy of a measurement process to an archer shooting at a set of irregularly shaped targets, as shown in Fig. 7.2, may offer insight into the concepts of reliability and validity. In this analogy, each arrow shot by the archer corresponds to a single observation, the result of one act of measurement to assign a value of an attribute to an object. Each target corresponds to one object, with a hidden "bull's-eye" corresponding to an unknowable exact value of the attribute for the object. In Fig. 7.2, the fuzzy dots on the target suggest the location of the bull's-eye. Each target has its bull's-eye in a different place, corresponding to differing exact values of the attribute in each object. The shape of the target suggests to the archer the location of its hidden bull's-eye, but does not reveal it. So for each object, the archer directs multiple arrows at an aim point where the archer *believes* the bull's-eye to be.

Looking now at where these successively-fired arrows land on the target (see Fig. 7.2), there is likely to be enough consistency to the archer's shooting to discern the central point at which the archer was aiming, even if no arrows strike it exactly, and the location of that central point is the best estimate of where the

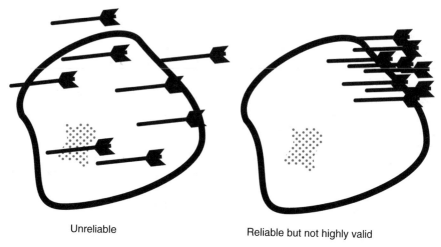

Unreliable   Reliable but not highly valid

Fig. 7.2 Measurement as archery

archer was actually aiming. The amount of scatter of the arrows around this central point provides a sense of the archer's reliability. The more arrows the archer directs at that same point of aim, the more apparent that point of aim will be from inspection of where the arrows land. If the archer shoots an infinite number of arrows at each target, the central point about which these arrows cluster would be exactly equal to the point the archer was aiming for. This central point is analogous, in terms of the classical theory of measurement, to an object's measured score.

If and only if there is enough consistency to the archer's shooting to discern a central point on each target around which the arrows cluster, the archer has enough reliability to prompt a meaningful discussion of validity. The archer has taken cues from the shape of the target to make an educated guess at the location of the invisible bull's eye. Over a series of targets, the smaller the distance between the archer's aim point and the actual location of the hidden bull's-eye on each target, the greater the archer's validity. So archers are "reliable" to the extent that they can land a succession of arrow near the aim point, and "valid" to the extent that they can take cues from the shape of the target as to where the hidden bull's-eye actually is.

A key inference from this analogy: It is possible to estimate the reliability of a measurement process directly by inspecting the results of the measurement process itself, by analogy with the relatively easy task of calculating the scatter of the arrows over a series of targets. Determining the validity of the measurement entails a much more indirect and uncertain process of trying to infer the actual location of the bull's-eyes in relation to where the archer believed them to be. There is no direct way to calculate this from the where the arrows land.

7.5   Two Methods to Estimate Reliability

In general, the reliability of a measurement can estimated by examining the results of multiple independent observations performed on a set of objects. There are two ways of making multiple observations that can be construed to be independent.

Moving from the hypothetical world of archers, how is it possible to undertake multiple observations in the real world of measurement? One logical, and often useful, way is to repeat the same measurement over short time intervals. The differences between these successive measurements provide an estimate of what is known as *test-retest reliability*. The more consistent the results of repeated measurements over a series of objects, the greater the reliability of the measurement process. This approach works well in situations where the objects of study are insensate and will not remember the results of previous measurements. So, for example, the test-retest approach is well suited to measurements of the performance of machines or the measurements of the weight of people or physical objects.

However, the test-retest approach does not work well for the measurement problems that often arise in informatics; for example, studying human states of mind or their performance in interaction with machines. When measuring a state of mind such as perceived ease of use of an information resource, asking people the same questions again within a short time interval will not generate independent observations. People are likely to give the same responses because they will remember and repeat what they did the previous time. In these cases, human memory disqualifies the repeated measurements from being independent. Repeating the measurement over a longer period of time, to defeat the effect of memory, requires reproducing the exact circumstances of the previous measurement, which is often impossible. Moreover, when measuring the performance of people on tasks undertaken in interaction with machines, which is usually time consuming and mentally taxing, it may not be possible to convince people to return for a repeat measurement.

An alternative approach, usually necessary when making measurements of abstract states of mind, employs multiple observations conducted at approximately the same time. *The observations can be crafted in ways that are different enough for each observation to create a unique challenge for the object yet similar enough that they all measure essentially the same attribute.* The agreement among the results of these multiple co-occurring observations, over a series of objects, provides an estimate of internal consistency reliability. Use of multiple judges to rate a performance or multiple questions on questionnaire are examples of the method of multiple co-occurring observations. Using the archery metaphor, the method of multiple co-occurring observations is analogous to having a set of archers, each placed at a slightly different angle to the target, so each archer has a slightly different view of it. Taking cues from the shape of the target, each archer presumes the location of the

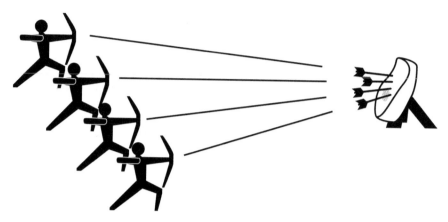

Fig. 7.3 Archery and the method of multiple co-occurring observations

invisible bull's eye. On command, each archer shoots one arrow simultaneously with the other archers (Fig. 7.3), each aiming at where they believe the bull's-eye to be.

To see the contrast between the approaches, reconsider the TraumAID system, originally introduced in the previous chapter, to advise on the care of patients with penetrating injuries of the chest and abdomen (Clarke et al. 1988). This information resource might be studied by asking expert judges to rate the appropriateness of the medical procedures suggested by the system over a series of trauma cases. Thus, the objects are the cases (and more specifically the set of procedures recommended for each case), the judges' ratings comprise the observations, and the attribute is the "appropriateness" of the recommended care for each case. If one or more judges rated each case and then, 3 months later, rated the same set of cases again, the agreement of each judge with him/herself from one occasion to the next would assess test-retest reliability. If all judges rated each case on only one occasion, the agreement among the different judges would be a measure of internal consistency reliability. The appeal of the test-retest approach is limited because the judges may recall the cases, and their ratings of them, even after an interval as long as 3 months, or they may be unwilling to carry out the ratings twice. Lengthening the time interval between ratings increases the risk that changes in the prevailing standard of care against which TraumAID's advice is judged, or the personal experiences of the judges that alter their perceptions of what constitutes appropriate care, might change the context in which these judgments are made. Under these circumstances, disagreements between test and retest judgments could be attributable to sources other than measurement error.

When the method of multiple co-occurring observations is employed in measurement, a set of differing observations designed to measure the same attribute, may be called a *scale, panel* or an *index*.[1] Examples include sets of items on a written test or questionnaire, a panel of human judges or observers, or a related set of performance indicators. The Dow Jones index is computed from the market values of a selected set of securities, and is accepted as a single number representing the more abstract attribute of the performance of the economy as a whole. Approaching measurement via multiple simultaneous observations—forming a scale, panel or index to assess a single attribute—avoids the practical problems associated with the test-retest approach but brings some challenges of its own. Most notably, there is no way to be certain that the multiple observations, that are designed to be assessing a common attribute, are functioning as intended. Judges on a panel, for example, may have different ideas of what a "good performance" is. (By analogy, the archers who release their arrows simultaneously may have different perceptions of where the bull's-eye on each target is.) Chapter 8 will introduce methods for examining how well a scale, index, or panel is performing as part of a measurement process.

7.6   Quantifying Reliability and Measurement Errors

7.6.1   The Benefit of Calculating Reliability

Applying either the test-retest method or the method of multiple co-occurring observations, the best estimate of the true value of an attribute for each object is the average of the independent observations. If the reliability of a measurement process is known, it is then possible estimate the amount of error due to random or unsystematic sources in any individual object's score. This error estimate is known as the *standard error of measurement* and is defined more precisely in the next section. Also, as discussed in Sect. 8.5, knowledge of the reliability of a measurement process can reveal to what degree errors of measurement are contributing to misleading results of demonstration studies. It is important to remember that any measurement process consisting of multiple observations can reveal the magnitude of its own reliability. This contrasts with estimation of validity, that, as we will see later, requires collection of additional data.

---

[1] Technically, there is a difference between a *scale* and an *index* (Crossman 2019) but for purposes of this discussion the terms can be used interchangeably. Also note that the term *scale* has two uses in measurement. In addition to the definition given above, *scale* can also refer to the set of response options from which one chooses when completing a rating form or questionnaire. In popular parlance, one might say "respond on a scale of 1–10" of how satisfied you are with this information resource. Usually, it is possible to infer the sense in which the term "scale" is being used from the context of the statement.

7.6.2 Enabling Reliability Estimation Using Objects-by-Observations Matrices

To estimate the reliability of a measurement process, the first step is to represent the results of the multiple observations made for each object. This can be done using a matrix of objects-by-observations as illustrated in Table 7.1 below. This approach applies to measurement processes where observations are made sequentially (test-retest) and when the observations are co-occurring. In the objects-by-observations matrix in Table 7.1, the rows correspond to the measurement results for each of four objects of measurement, the columns correspond to the results for each of five repeated or co-occurring observations. Each cell expresses the result of one observation for one object. In this example, the result of the measurement process is expressed as integer values 1 through 5. Consistent with classical measurement theory, the result of the measurement, the object score, is computed as the mean of the individual observations for that object.[2]

In addition to portraying the structure of objects-by-observations matrices, Table 7.1 portrays a special case of perfect reliability. Each object displays an identical result for all five observations. There is no scatter or variability in the results from the individual observations; metaphorically, all of the measurement "arrows" landed in the exact same place on each target. Each object's score is the average of the five observations, but in this special case, for all objects this is the average of identical values.

In a much more typical example (Table 7.2), the results of the observations vary for each object. Each object's score is now the average of individual observations whose results are not identical. The degree to which a measurement process is reliable is quantified by computing a reliability coefficient, commonly represented as ρ, that takes on values from 0 to 1, with $\rho = 1$ representing the case of perfect reliability illustrated in Table 7.1. Intuitively, the reliability coefficient will be directly related to the amount of consistency across the observations made on each object, and inversely related to the amount of disagreement among the observations for each object. The reliability coefficient (ρ) for the matrix in Table 7.2 is 0.81. A measurement with this level of reliability is consistent enough to be a measurement

Table 7.1 An objects-by-observations matrix illustrating perfectly reliable measurement ($\rho = 1$)

Object	Observations					Object score
	1	2	3	4	5	
A	3	3	3	3	3	3
B	4	4	4	4	4	4
C	2	2	2	2	2	2
D	5	5	5	5	5	5

---

[2]The object score can also be computed as the summed score of all observations. Using the mean and summed scores will yield the same reliability results as long as there are no missing observations.

Table 7.2 Matrix depicting a more typical measurement result ($\rho = .81$)

Object	Observations					Object score
	1	2	3	4	5	
A	4	5	3	5	5	4.4
B	3	5	5	3	4	4.0
C	4	4	4	4	5	4.2
D	3	2	3	2	3	2.6

of *something*. As seen in the next section, value of ρ associates the measurement with a quantified amount of error.

A spreadsheet available to online users as supplementary material will calculate a reliability coefficient from an objects-by-observations matrix. The reliability coefficient generated by these methods is known as Cronbach's alpha (Cronbach 1951). Other reliability coefficients exist, but coefficient alpha is commonly used and is widely applicable. Most statistical packages will also perform this calculation.

7.6.3   The Standard Error of Measurement

The classical theory of measurement enables the estimation of the amount of random (also called unsystematic) error associated with the result of a measurement process. This is accomplished by computing the *standard error of measurement* (SE_{meas}) using the formula:

$$SE_{meas} = SD\sqrt{1-\rho}$$

where ρ is the reliability coefficient and SD is the standard deviation of all object scores in the measurement process In Table 7.1, the standard deviation of the object scores is 1.29, but because the reliability is perfect ($\rho = 1$), the standard error of measurement is 0. In the more realistic example of Table 7.2, the standard deviation of the objects scores is 0.82. The standard error of measurement, applying the above formula with $\rho = .81$, is 0.36, which can be rounded-off to .4.

Recalling that the mean of the observations for each object is the best estimate of the value of its "true score", the standard error of measurement quantifies the uncertainty of that estimate. More specifically, computation of the standard error of measurement generates an error bracket around the result of a measurement for *each object*, reflecting the uncertainty in that result due to random measurement error. For Object A in Table 7.2, then, the measured score of 4.4 should be roughly interpreted as 4.4 ± 0.4, and the measured score for Object B should be interpreted as 4.0 ± 0.4. The unknowable "true score" has a 68% probability of falling within the measurement result ± SE_{meas} (Fig. 7.4).[3]

---

[3] Those familiar with the concept of confidence intervals might think of observed score ±1 standard error of measurement as the 68% confidence interval for the true score. The 95% confidence interval would be approximately the observed score plus or minus two standard errors of measurement.

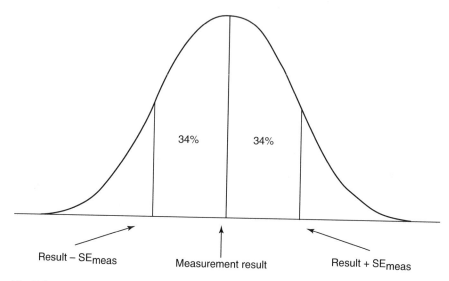

Fig. 7.4 Distribution of measurement results illustrating the standard error of measurement

Self-Test 7.1

1. By inspection of the formula for computing the standard error of measurement, what are the effects on SE_{meas} of (a) adding a constant to each measurement result and (b) multiplying each measurement result by a constant?
2. The measurement result given in the table below has a reliability of 0.88.

Object	Observations			
	1	2	3	4
A	2	4	4	3
B	3	3	3	4
C	2	3	3	2
D	4	3	4	4
E	1	2	1	2
F	2	2	2	2

(a) Compute the score for each object and the standard deviation of these scores. (Standard deviations can be computed by most spreadsheet programs. The formula can be found in any basic statistics text.)
(b) Compute the standard error of measurement.
(c) Would changing the result of Observation 1 on Object A from 2 to 4 increase or decrease the reliability of the measurement process represented in the matrix?
(d) Would changing the result of Observation 4 on Object F from 2 to 4 increase or decrease the reliability of the measurement process represented in the matrix?

7.7 Design Considerations in Measurement Studies to Estimate Reliability

7.7.1 Dependence of Reliability on Populations of Objects and Observations

Table 7.2 above offers the results of a generic measurement study. What constitutes the attribute, the objects, and the observations are all unspecified. In any fully specified example, the attribute and the populations from which the objects and observations are drawn will be fully described. The reliability estimate generated by the measurement study is specific to those populations. Changing either one would be expected to change the magnitude of the reliability.

Table 7.3 recasts Table 7.2 as a specific measurement process. The attribute to be measured is perception of ease of use of search engine interfaces (using a "1–5" rating scale); the objects are database search engine interfaces; and the observations are the ratings by five panelists. The calculated reliability for this objects-by-observations matrix ($\rho = 0.81$ as in Table 7.2) suggests that this panel of raters is measuring with reasonably high reliability the attribute of ease of use. However, it cannot be assumed that a different family of search engines, based on different information retrieval algorithms, will yield the same reliability when judged by the same set of panelists. Similarly, the same degree of reliability cannot be assumed if the same set of search engines is tested using panelists with very different professional backgrounds.

Observations and objects sampled from the same domain are said to be "equivalent". The results of measurement studies can be generalized only to equivalent populations. All too frequently, questionnaires developed for patients with high language proficiency and found to be reliable and valid in those populations are administered to patients with lower language proficiency. A study team cannot assume that the questionnaire will be equally reliable and valid in this non-equivalent population.

Table 7.3 A more specific measurement process (each cell is the rating by one panelist)

Search engine interfaces (Objects)	Expert panel of ratings of "ease of use" (Observations)					Average rating for each interface
	1	2	3	4	5	
A	4	5	3	5	5	4.4
B	3	5	5	3	4	4.0
C	4	4	4	4	5	4.2
D	3	2	3	2	3	2.6

7.7.2 Effects on Reliability of Changing the Number of Observations in a Measurement Process

Increasing the number of equivalent observations in a measurement process typically increases the magnitude of the reliability. Decreasing the number of observations has the reverse effect. This can be seen intuitively by returning to the archery metaphor. For each target, the archer is aiming at a particular point. The greater the number of arrows the archer shoots, the more accurately the central point of the arrow cluster estimates the location of the point of aim.

The Spearman-Brown "prophecy formula" provides a way to estimate the effect on reliability of adding equivalent observations to or deleting observations from a measurement process. If one knows the reliability ρ_k of a measurement process with k observations, an estimate of its reliability ρ_n with n observations is given by

$$\rho_n = \frac{q\rho_k}{1 + (q-1)\rho_k}$$

where $q = n/k$.

To illustrate the prophecy formula's estimated effect of changing the number of observations, consider a hypothetical situation where four judges are asked to independently assess the quality of care for each of a set of 30 clinical cases. In this situation, judges are the observations, cases are the objects, and quality of care is the attribute. Let's assume that the reliability of the four judges, calculated directly from an objects-by observations matrix, is 0.65. Table 7.4 shows the prophesied effects, using the above formula, of adding equivalent judges to or deleting judges from the set of four originally in the study. The result in boldface is the result of the measurement study as actually conducted with four judges. The prophecy formula generates all other values. Table 7.4 shows, for example, that increasing the number of judges from 4 to 6 is prophesied to increase the reliability from .650 to .736.

7.7.3 Value of the Prophesy Formula: How Much Reliability Is Enough?

The prophesy formula is particularly valuable to help designers of measurement processes achieve pre-determined target levels of reliability. As a practical matter, measurements consume resources of time and money in proportion to the number of

Table 7.4 Application of the prophecy formula

	Number of judges							
	1	2	**4**	6	8	10	20	100
Adjustment factor	0.25	0.5	**1.0**	1.5	2.0	2.5	5.0	25.0
Reliability	0.317	0.481	**0.650**	0.736	0.788	0.823	0.903	0.979

observations they include. This in turn raises the question of how much reliability is "good enough": what should the target values of reliability actually be? The degree of reliability that is necessary depends primarily on how the results of the measurement are to be used. If the measurement results are to be used for demonstration studies where the focus is comparisons of groups of objects, a reliability coefficient of 0.7 or above is usually adequate, although higher reliability is always desirable when the resources to achieve it are available. By contrast, a reliability coefficient of 0.9 or above is often necessary when the concern is assignment of scores to individual objects with a very small standard error of measurement—for example, when a decision has to be made as to whether a particular information resource has achieved a pre-stated performance specification, or whether a clinician has attained a level of proficiency necessary to safely perform a procedure.

With reference to Table 7.4, if the purpose of the measurement is for use in a demonstration study to compare the quality of care for groups of cases, the collective opinions of at least six judges would be required in this hypothetical example to achieve a target reliability of more than 0.7. However, if our concern is to assign a very precise "quality of care" score to each case, more than 20 judges would be needed to reach a reliability of 0.9.

7.7.4   Effects on Reliability of Changing the Number of Objects in a Measurement Process

Increasing the number of equivalent objects typically increases confidence in the estimate of the reliability but will not bias the magnitude of the estimate in one direction or another. Measurement studies with small numbers of objects will yield estimates of reliability that are unstable. Tables 7.1, 7.2 and 7.3 in this chapter included only four objects to compactly portray the concept and structure of an objects-by-observation matrix. Both examples contain too few objects for confident estimation of reliability. In both cases, the addition of a fifth object could profoundly change the reliability estimate. However, if the measurement study included 50 objects, the effect of a 51st object would be much less profound.

For measurement study design in situations where an almost unlimited number of objects is available, a rule of thumb would be to include at least 100 objects for stable estimates of reliability. For example, in a measurement study of an attitude questionnaire to be completed by registered nurses, the number of available "objects" (nurses in this case) is large and a minimum sample of 100 should be employed. In situations where the number of available objects is limited, the designer of a measurement process faces a difficult challenge, since it is not desirable to include the same persons in measurement and demonstration studies. So, for example, if patients with a rare disease were the objects of measurement and 500 cases are all that exist, a measurement study using 50 of these persons would be a pragmatic choice while still providing a reasonably stable estimate of reliability.

Self-Test 7.2

1. For the data in Question 2 of Self-Test 7.1, what is the predicted reliability if the number of observations were (a) increased to 10 or (b) decreased to 1?
2. Human testers are asked to assess the ease of use of an information resource to complete a wide range of tasks. In a measurement study, four testers rated ease of use for 20 separate tasks, on a 5-point scale. A rating of 5 means "extremely easy to use for this task" and at the other extreme, a rating 1 means "nearly impossible to use for this task".

    (a) What are the objects and what are the independent observations for this measurement process?
    (b) If Tester 1 gave Task 1 a rating of "4", and the measurement process is perfectly reliable, what score would Tester 2 have given for Task 1? Explain your answer.
    (c) All other things being equal, if the reliability of this measurement process is .7 as carried out, what would be prognosticated reliability if two additional testers were added to the process?

3. The Critical Assessment of Techniques for Protein Structure Assessment (CASP) is an annual competition that tests methods for predicting the structure of proteins. Alternative prediction methods are challenged by sets of test amino acid sequences. Each sequence corresponds to a three-dimensional protein structure that is unknown to the competitors but has been determined experimentally. Quality of the prediction methods is based on the percentage of carbon atoms in the predicted structure that are within some specified distance of their actual location in the protein structure, as determined experimentally. In one of the CASP competitions, 14 test sequences were used with 123 prediction methods. Assume, although this was not the case, that in the competition all methods were applied against all test sequences.

    (a) Frame this as a measurement problem. Name the attribute, observations, and objects.
    (b) What would be the dimensionality of the objects-by-observations matrix for this CASP competition?
    (c) In order to designate one method as the winner, by classical theory what level of reliability would the organizers of CASP likely wish to achieve?

7.8 Validity and Its Estimation

7.8.1 Distinguishing Validity from Reliability

Reliability estimates indicate the degree of random, unsystematic "noise" in a measurement. The other important aspect of a measurement process is its validity, which indicates the degree to which the measurement results reflect the "polluting" effects of attributes other than the attribute of interest. To the extent that a measurement process is reliable, the results have meaning. To the extent that a measurement process is valid, the results mean what they are intended to mean. More rigorously, the validity of a measurement process is the fraction of the perfectly reliable true score that reflects the attribute of interest (see Fig. 7.1). As previously established, a measurement process can reveal its own reliability by portraying the results in an objects-by-observations matrix and analyzing the consistency of the observations. *Estimation of validity is a different process requiring the collection of additional information.* Even though this chapter presents validity after reliability, reliability and validity of measurement are of equal importance. Study teams should not assume that a measurement process is valid, solely because it appears to be reliable.

Returning to the archery metaphor (Fig. 7.2), validity corresponds to the extent of concordance, over a series of targets, between the point at which the archer was aiming on each target and the unknown location of each target's bull's-eye. The reliability of the archers can be determined directly from the scatter of the arrows on each target. However, determining how close to the invisible bull's-eyes the archers' points of aim really were requires additional information. Again metaphorically, one way to do this is to ask the archers to verbalize how they decided where to aim and then ask some knowledgeable judges whether this reasoning made sense. This procedure, while providing some useful information about the archers' "validity", is not likely to be definitive. This analogy illustrates the fundamental challenge of estimating validity. When collecting collateral information to establish validity, no single source of such information is completely credible, and often multiple different sources must be used.

To take an example from the real world, some smartphone apps allow drivers to report the presence of vehicles stopped on the shoulder of the road on which they are moving. This is an example of a measurement process with "presence of vehicle on shoulder" as the attribute, a road location as the object, and drivers' reports as the observations. These reports are reliable to the extent that multiple drivers report a "stopped vehicle's presence" at the same place and at approximately the same time. However, in order for these reports to be fully valid, there must *really* be a stopped vehicle where and when the drivers report it. And there are many reasons to suspect that these reports may not be fully valid. For example, some drivers may incorrectly report a hazard in the road itself as stopped vehicle, or they may report a vehicle as being on the shoulder when in fact it is parked further from the roadway itself. Validation thus requires information from other sources, which can take many forms. One approach would be to safely park on the shoulder a mix of vehicles and other objects, and then study the reports of drivers using the app to determine if they accurately distinguish between the vehicles and other objects.

It is useful to think of measurement validation in much the same way that attorneys approach their work. The goal of attorneys is to assemble evidence that makes a compelling case on behalf of their clients, a case that will be convincing to a judge and/or jury. So it is with validity of measurement. The goal, in a measurement study, is to make a compelling case for validity by assembling the strongest possible evidence. Often multiple studies are used to argue for a measurement process's

Table 7.5 Types of validation studies

Study types	Related names of sub-types	Driving question	Strengths	Weaknesses
Content validation	Face validation	Does the measurement process appear to be valid?	Relatively easy to study	Makes a weak case for validity
Criterion-related validation	Predictive validation Concurrent validation	Do the measurement results correlate with an external standard?	Can make a strong case for validity, when an external standard exists	Standards may not exist or may be controversial
Construct validation	Convergent validation Discriminant validation	Do the measurement results reproduce a hypothesized pattern of correlations with other measures?	Makes the strongest case for validity	Requires much additional data collection

validity, and the strongest case for the validity of a measurement process results from accumulating evidence across a complementary set of studies. More rigorously, studies of validity are of three general types: content, criterion-related, and construct (VandenBos 2007). These study types are summarized in Table 7.5 and discussed in the following sections.

7.8.2   Content Validity

This is the most basic notion of validity, also known more loosely as "face" validity. The driving question is whether the measurement process, by inspection or observation, appears to be valid. Estimation of content validity addresses questions such as: By inspection of the instruments, do the observations appear to address the attribute that is the measurement target? Does the measurement process make sense or do the results defy common sense?

Assessment of content validity requires the instrument and procedures to be inspected. It does not necessarily require measurements actually to be taken. For example, content validity of a questionnaire can be determined by asking a panel of experts to review the questions being asked on the questionnaire and offering an opinion as to whether each question addresses the attribute it is intended to measure. The opinions of the informed individuals comprise the additional data that are always required in validation studies.

For example, a study team may be constructing a questionnaire to measure an information resource's "ease of use" (attribute) as perceived by its users. The independent observations are the questions on the questionnaire. The items on the questionnaire are intended to comprise a scale, with each item seeking to measure "ease of use" from a slightly different perspective. In a content validity study, a group of judges may be asked: Which of the following items belong on the "ease of use" scale on this questionnaire, assuming respondents are asked to agree or disagree with each of the following statements?

1. The displayed information was clearly formatted
2. I was able to get the system to do what I wanted it to do
3. I was able to get help when I needed it
4. The system responded quickly when I made an entry

There might be little disagreement that Items 1 and 2—addressing information displays and the user interface—belong on the scale. However, some inspectors might argue that the "system response time" item (Item 4) addresses the resource's performance rather than its ease of use, and that "accessibility of help" (Item 3) is not a characteristic of the resource itself but rather of the environment in which the resource is installed. These inspectors would argue that Items 3 and 4 bring unwanted "polluting" factors into the measurement, making it less valid.

Content validity is relatively easy to assess, but it provides a relatively weak argument, grounded in subjective judgment, for the validity of a measurement process. Content validity is rarely adequate as the sole approach to arguing for the validity of a measurement process. Nonetheless, the content validity of a measurement process cannot be assumed and must be verified through an appropriately designed component of a measurement study.

7.8.3 Criterion-Related Validity

For criterion-related validity the driving question is different: Do the results of a measurement process correlate with some external standard or predict an outcome of particular interest? For example, a study team developing a scale to measure computer literacy may do a criterion validity study by examining whether those who score highly on the computer literacy scale also learn more quickly to navigate an unfamiliar piece of software. Similarly, does a scale that rates the quality of radiotherapy treatment plans identify treatment plans associated with better therapy outcomes? In the earlier example of a smartphone app to crowdsource the presence of stopped vehicles, the strategy of parking vehicles and other objects on the shoulder would provide evidence of criterion-related validity.

Unlike content validity, which can be assessed through inspection or observation of the measurement instruments and processes themselves, determination of criterion-related validity requires a study where measurements are made, on a representative sample of objects, using the instrument being validated as well as the instruments needed to assess the external criteria. Using a previous example, estimating the criterion-related validity of a computer literacy scale in a measurement study requires that the scale be completed by a sample of health professionals who also try their hand at using an unfamiliar piece (or pieces) of software. The correlation between their scores on the literacy scale and the time taken to master the unfamiliar software would generate an estimate of criterion-related validity. In this case, the polarity of the relationship would be expected to be negative: lower mastery times associated with higher literacy scores.

Determination of criterion-related validity depends on the identification of specific criteria that will be accepted as reasonably definitive standards, and for which reliable and valid measurement methods already exist. Also, the measurement properties of the external criteria must themselves established, and found to be acceptable, before a criterion-related validity study using these criteria can be performed.

There are no hard and fast standards for interpreting the results of a criterion-related validity study. In general, the greater the correlation with the external criterion, the stronger the case for validity—and the polarity of the correlation must be in the anticipated direction.

As a very general rule of thumb, correlations between the measurement process that is being validated and the external criterion should exceed 0.5.

When the external criterion used in a validity study is an attribute that can only be known in the future, criterion-related validity is often referred to as "predictive" validity. It is often possible to identify predictive criteria for validation of new measurement methods, but predictive validation studies can take months or years to complete. For example, a criterion-related validity study might employ subsequent survival rates of patients as a criterion to validate methods for measuring the quality of radiotherapy treatment plans. However, assessment of the criterion in a sample of patients could take many years. When the criterion is an attribute that can be measured at the same time as the measurement process under study, this is often termed a "concurrent" validation. The discussion above of a computer literacy scale, with time to master unfamiliar software as the criterion, is illustrative of concurrent validation.

7.8.4 Construct Validity

Construct validity resembles criterion-related validity in many ways, but the approach is more sophisticated. When exploring construct validity, the driving question is: Does the measurement of the attribute under study correlate with measurements of other attributes (also known as constructs) in ways that would be expected theoretically? This method is the most complex, but in many ways the most compelling, way to estimate validity. Assessment of construct validity in a measurement study typically requires measurement of multiple additional attributes, in contrast with criterion-related validity which typically involves only a single external standard.

The focus in construct validity is on verifying a set of relationships that have been hypothesized in advance. Some of the relationships between the attribute under study and the additional attributes included in the measurement study will be hypothesized to be high; others will be hypothesized to be small or zero. In contrast to criterion-related validation, where the more desirable result is always higher

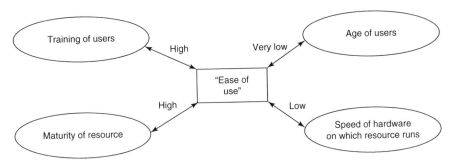

Fig. 7.5 Hypothesized relationships in a construct validity study of an "ease of use" questionnaire

correlation, for construct validity the desirable result is the replication of a hypothesized pattern of correlations, some positive and others negative, however theory dictates that pattern to be.

Consider once again the example of a questionnaire to assess users' perceptions of the "ease of use" of a biomedical information resource. As shown in Fig. 7.5, the attribute of "ease of use" might be expected to correlate highly, but not perfectly, with some other attributes, such as "computer literacy" of each user and the "extent of the training" each user had received. At the same time, ease of use would *not* be expected to correlate highly with the ages of the resource users, and only a low correlation might be expected with the speed of each user's computer on which the software runs. To assess construct validity of the "ease of use" questionnaire, data on all four additional attributes shown in Fig. 7.5 would be collected, using previously validated procedures, along with the ease-of-use data collected via the questionnaire to be validated. Correlation coefficients between these measures, as computed from the measurement data on all five attributes, would be computed and the results compared with those hypothesized. If the expected pattern is reproduced, the study team can make a strong case for the validity of the new measure. When the hypothesized pattern of correlations is not reproduced, the deviations often can cue the study teams to which "polluting" attributes are being tapped by the new questionnaire—and thus point the way to modifications that will increase its validity.

Additional terms are often used in relation to construct validity. In construct validity studies, the correlations that are hypothesized to be high are referred to as indicators of convergent validity; the correlations that are hypothesized to be low are referred to indicators of discriminant validity. Statistical methods such as principal components analysis or factor analysis, which are beyond the scope of this text, are often employed to estimate a measurement instrument's construct validity (Watson 2017; Dykes et al. 2007). In cases where the observations of interest are items on a form, these methods allow determination of the degree to which items that are expected to be highly intercorrelated (convergent validity) and minimally intercorrelated (discriminant validity) behave as expected.

7.9 Concluding Observations About Validity

In sum, concern about validity of measurement extends throughout all of science. When astrophysicists detect a signal (some nonrandom pattern of waveforms) in their radio telescopes and try to determine what it means, their concern is with validity. In general, concern with validity increases when the attributes being measured are more abstract and only indirectly observable, but this concern never vanishes. For example, the concept of walking is relatively concrete, and there exists a robust consensus about what "taking a walking step" means. Yet a smartphone app to measure a person's distance walked in fixed time interval may very well be measuring movements other than walking distance, threatening the validity of the measurement result (Orr et al. 2015). The picture becomes more complex as one moves into the realm of human states of mind and their associated attributes. When attitudes and other states of mind, such as "satisfaction," become the focus of measurement, the need for formal measurement studies that address validity becomes self-evident.

Self-Test 7.3

1. In the example above of the smartphone app designed to measure walking distance in a fixed time interval, name some "unwanted" attributes that may threaten the validity of the measurement results.
2. Is each of the following studies primarily concerned with content, criterion-related, or construct validity?

    (a) A study team develops knowledge test related to research regulation and ethics. As part of the process of validating the test, the team studies the correlation between scores on the test and the number of comments Institutional Review Boards generate on research protocols submitted by people who have taken the test.

    (b) A study team developing a questionnaire to measure computer literacy hypothesizes that computer literacy will be moderately positively correlated with years of formal education, highly positively correlated with holding a position that requires computer use, uncorrelated with gender, and weakly correlated with age.

    (c) A study team developing a computer literacy questionnaire convenes a panel of experts to identify the core competencies defining computer literacy.

3. A new information resource "DIATEL" has been developed to assist diabetic patients to find information to help them manage their disease. The developers of DIATEL have asked you to develop a measurement instrument, in the form of a questionnaire, to administer to samples of patients who use the resource. The instrument is designed to ascertain patients' perceptions of the "value" to them of the information they have obtained from DIATEL. The initial request is that the instrument consist of ten questions, so it can be completed quickly. Patients completing this questionnaire will respond by answering "yes" or "no" to each question.

(a) Provide an example of what one question on the instrument might look like.
(b) Describe a measurement study that will ascertain the reliability of the instrument using the method of multiple simultaneous observations. Include in your description what data will be collected, from whom, and how you will analyze the data leading to an estimation of the instrument's reliability.
(c) After trials of the questionnaire, which reveal a reliability of .7 but suggest that the instrument is taking too long to complete, you are asked to create a "short form" of the questionnaire by deleting five of the questions. You object. What would be the prognosticated reliability of the "short form"?
(d) Suggest some kinds of "external" data that, if collected, could be used to assess the validity of this questionnaire.

7.10   Generalizability Theory: An Extension of Classical Theory

Although the classical theory of measurement serves the purposes and level of this book, this chapter concludes with a brief exposition of generalizability theory which modernizes and extends classical theory. Using generalizability theory (Briesch et al. 2014; Shavelson et al. 1989), study teams can conduct measurement studies with multiple observations of interest. For example, if a measurement requires multiple judges, each of whom complete a multiitem rating form, use of generalizability theory is required to determine the reliability of this process. Study teams can analyze multiple potential sources of error and model the relations among them. This additional power is of potential importance in biomedical informatics because it can mirror the complexity of almost any measurement problem addressed in the field. Generalizability theory (G-theory) allows computation of a generalizability coefficient, which is analogous to a reliability coefficient in classical theory.

Although the specific computational aspects of G-theory are beyond the scope of this discussion, the theory has great value as a heuristic for measurement study design. If a study team can conceptualize and formulate a measurement problem in terms appropriate to study via G-theory, a psychometrician or statistician with appropriate experience can handle the details. The basic idea is the same as in classical theory but with extension to multiple observations of interest, whereas classical theory is limited to one observation of primary interest. Analytical methods derived from the analysis of variance (ANOVA) are used to decompose the total variability of the measurements made on the objects included in the study.[4] The total variability is decomposed into variability due to objects, variability due to the multiple observations and their statistical interactions, and variance due to other sources not explicitly modeled by the observations included by the team in the measurement study.

---

[4] The basics of ANOVA are discussed in Chap. 12 of this book.

The generalizability coefficient is represented by:

$$\rho = \frac{V_{objects}}{V_{total}} = \frac{V_{objects}}{V_{objects} + V_{facets} + V_{other}}$$

where "facets" is a shorthand term for the types of observations specifically included in the measurement study. In the example above, there would be two facets: items and judges.

In statistical terms, the subscripted V's in the above formula are variance components, which can be computed using methods derivative from ANOVA. V_{facets} and V_{other} are taken to represent sources of measurement error. Expressions for V_{facets} explicitly involve the number of values each facet can assume, which makes it possible to use G-theory to model the effects on reliability of changing the number of values for any of the facets that are part of the measurement process. For example, for the basic one-facet models that have been developed throughout this chapter using classical theory, the formula for the generalizability coefficient is:

$$\rho = \frac{V_{objects}}{V_{objects} + \left(V_{objects \times observations} / N_{observations} \right)}$$

This equation is exactly equivalent to the Spearman-Brown prophecy formula.

A major value of G-theory derives from its applicability to more complex measurement problems than classical theory allows. In addition to measurement studies with multiple types of observations, more complex measurement designs can be analyzed, including so-called nested designs of the type described in Sect. 9.3.3.

For most readers of this book, G-theory is something good to know about, but not necessarily to be proficient in.

The following chapter discusses how all of the measurement theory and concepts introduced in this chapter can be put to work to improve measurement in informatics studies.

Answers to Self-Tests

Self-Test 7.1

1. (a) Adding a constant has no effect on the standard error of measurement, as it affects neither the standard deviation nor the reliability.
   (b) Multiplication by a constant increases the standard error by that same constant.
2. (a) The scores are 13, 13, 10, 15, 6, 8 for Objects A–F. The standard deviation of the six scores is .86.
   (b) .30.

(c) The reliability would increase because the scores for Object 1, across observations, become more consistent. The reliability in fact increases to 0.92.

(d) A decrease in reliability.

Self-Test 7.2

1. (a) 0.95.
   (b) 0.65.
2. (a) In this case, the tasks are the objects and the testers are the observations.
   (b) In a perfectly reliable measurement, all observations have the same value for a given object. So Tester 2 would also give a rating of "4" for Task 1.
   (c) By the Prophecy Formula, the estimated reliability would be .78.
3. (a) The attribute is, for a given method and test sequence, the percentage of carbon atoms within the threshold distance. The observations are the test sequences. The objects are the prediction methods.
   (b) The matrix would have 14 columns corresponding to the test sequences as observations and 123 rows corresponding to prediction methods as objects.
   (c) A very high reliability, on the order of .9 would be sought. The demonstration study seeks to rank order the objects themselves, as opposed to comparing groups of objects. This suggests the use of a large number of test sequences.

Self-Test 7.3

1. It is possible that other types of movements, beside whatever the study team defines as "walking" would count as walking steps. These movements might include climbing stairs, standing up or sitting down, or cycling. Also, computation of distance walked will be influenced by stride length which is in turn related to a person's height. So the person's height could be a "polluting" construct.
2. (a) Criterion-related validity. Number of comments generated by the IRB, which would indicate a person's ability to produce a highly compliant research protocol, could be a considered a criterion for validation in this case. The direction of the relationship between scores on the test and number of comments would be negative: greater knowledge should generate smaller numbers of comments.
   (b) Construct validity. The validation process is based on hypothesized relationships, of varying strength, between the attribute of interest (computer literacy) and four other variables.
   (c) Content validity.
3. (a) Some examples:
   On a scale of 1–5 (5 being highest), rate the effect DIATEL has had on your HbA1c levels.
   Rate from strongly agree to strongly disagree:
   "I am healthier because of my use of DIATEL"
   "I plan to continue to use DIATEL"
   (b) This study can be done by using each item of the questionnaire as an "independent" observation. A representative sample of diabetic patients who have used DIATEL would be recruited. The patients would complete the questionnaire. The data would be displayed in an objects (patients) by observations (items) matrix. From this, the reliability coefficient can be computed.

(c) Using the Prophecy Formula, the reliability with five items is predicted to be .54.

(d) Examples of external data include number of DIATEL features patients use (Do patients who use more features give higher ratings on the questionnaire?), whether patients use DIATEL in the future (Do patients' ratings on the questionnaire predict future use?), whether patients mention DIATEL to their clinicians (Do patients who rate DIATEL higher mention it more often?)

References

Boateng GO, Neilands TB, Frongillo EA, Melgar-Quiñonez HR, Young SL. Best practices for developing and validating scales for health, social, and behavioral research: a primer. Front Public Health. 2018;149:1–18.

Briesch AM, Swaminathan H, Welsh M, Chafouleas SM. Generalizability theory: a practical guide to study design, implementation, and interpretation. J Sch Psychol. 2014;52:13–35.

Clarke JR, Cebula DP, Webber BL. Artificial intelligence: a computerized decision aid for trauma. J Trauma. 1988;28:1250–4.

Cronbach LJ. Coefficient alpha and the internal structure of tests. Psychometrika. 1951;16:297–334.

Crossman A. The differences between indexes and scales: definitions, similarities, and differences. ThoughtCo. 2019. https://www.thoughtco.com/indexes-and-scales-3026544. Accessed 10 Jun 2021.

Dykes PC, Hurley A, Cashen M, Bakken S, Duffy ME. Development and psychometric evaluation of the impact of health information technology (I-HIT) scale. J Am Med Inform Assoc. 2007;14:507–14.

Kerlinger FN. Foundations of behavioral research. New York: Holt, Rinehart and Winston; 1986.

Kimberlin CL, Winterstein AG. Validity and reliability of measurement instruments used in research. Am J Health-Syst Pharmacy. 2008;65:2276–84.

Moriarity DP, Alloy LB. Back to basics: the importance of measurement properties in biological psychiatry. Neurosci Biobehav Rev. 2021;123:72–82.

Orr K, Howe HS, Omran J, Smith KA, Palmateer TM, Ma AE, et al. Validity of smartphone pedometer applications. BMC Res Notes. 2015;8:1–9.

Shavelson RJ, Webb NM, Rowley GL. Generalizability theory. Am Psychol. 1989;44:922–32.

Thanasegaran G. Reliability and validity issues in research. Integr Dissemin. 2009;4:35–40.

Tractenberg RE. Classical and modern measurement theories, patient reports, and clinical outcomes. Contemp Clin Trials. 2010;31:1–3.

VandenBos GR. Validity. APA Dictionary of Psychology. Washington, DC: American Psychological Association; 2007. https://dictionary.apa.org/validity. Accessed 10 Apr 2021.

Watson JC. Establishing evidence for internal structure using exploratory factor analysis. Measure Eval Counsel Develop. 2017;50:232–8.

Weinstein MC, Fineberg HV, Elstein AS, Frazier HS, Neuhauser D, Neutra RR, et al. Clinical decision analysis. Philadelphia: W.B. Saunders; 1980.

Chapter 8
Conducting Measurement Studies and Using the Results

Learning Objectives
The text, examples, and self-tests in this chapter will enable the reader to:

1. Given a description of a measurement study, assign each aspect of the study to a particular "stage", as described in Sect. 8.2.
2. Explain the difference between correlation and calibration of observations in a measurement process.
3. Use the method of part-whole correlations ("corrected" or "uncorrected") to diagnose problems with a measurement process and recommend steps to improve the process.
4. Describe the difference between standard errors of measurement and standard errors of the mean.
5. Explain why measurement error affects the results of demonstration studies.
6. Given the result of a demonstration study expressed as a correlation of two variables, and the reliabilities of the measurements of the two variables, estimate the effect of attenuation due to measurement error on the demonstration study result.

8.1 Introduction

This chapter moves from the theoretical emphasis of Chap. 7 to actual measurement practice. Chapter 7 introduced the concepts of reliability and validity as ways of thinking about and estimating the errors in a measurement process. This chapter begins by addressing the process of actually conducting measurement studies and then introduces methods to put measurement study results to use in improving measurement processes. The chapter continues by exploring the relationship between measurement errors and the results of demonstration studies. Chapter 9 will conclude the discussion of measurement by examining the three most common types of observations in informatics studies—tasks, judges, and items—and introducing the

© Springer Nature Switzerland AG 2022
C. P. Friedman et al., *Evaluation Methods in Biomedical and Health Informatics*, Health Informatics, https://doi.org/10.1007/978-3-030-86453-8_8

specifics of how to design measurement processes for each type. Greater detail on the methods discussed in this chapter are available in books focused on measurement methods (Streiner et al. 2015; Frey 2018).

8.2 Process of Measurement Studies

Recall from Sect. 6.2 that measurement studies require multiple independent observations on each of a set of objects. The data collected during a measurement study take the form of an objects-by-observations matrix, as illustrated in Table 7.1. In the quantitative view of the world, all independent observations of the same phenomenon should yield the same result. The closer the observations approach agreement for each object, the more reliable, and therefore objective and trustworthy, the measurement process can be considered to be. Disagreement reflects unwanted "subjectivity" in the measurement process.

A perfectly reliable measurement process, never seen in the real world, is one where all observations of the same object are in perfect agreement. Section 7.6 introduced methods for quantifying reliability of a measurement process, through calculations made directly from the objects-by-observations matrix. The previous chapter also emphasized that perfect reliability does not guarantee a perfect measurement of the desired attribute. Even if the measurement is perfectly reliable, the results may be to some degree invalid, reflecting attributes other than the attribute intended for measurement. Estimating the validity of a measurement process requires collection and analysis of data external to the measurement process itself.

Because the conduct of a complete measurement study, exploring both reliability and validity, is complex and time-consuming, it is vitally important that results of measurement studies be published so other informatics study teams can reuse the measurement methods developed and documented by their colleagues. Many examples of published measurement studies exist in the informatics literature; Sect. 6.7 cites just a few. Many more are needed.

With this brief review as background, the specific steps for conducting a complete measurement study are:

1. Design the measurement process to be studied. Precisely define the attribute(s), object class, instrumentation, measurement procedures, and what will constitute the multiple independent observations. Recall that the object class is an expression of who or what the measurements are made on.
2. Decide from which hypothetical population the objects in the measurement study will be sampled. It may also be necessary to decide from which population the observations will derive. (For example, if the observations are to be made by human judges, what real or hypothetical group do the selected judges represent?) This step is key because the results of the measurement study cannot be generalized beyond these populations.

3. Decide how many objects and how many independent observations will be included in the measurement study. This step determines the dimensionality of the objects-by-observations matrix for the data collected.

4. Collect data using the measurement procedures as designed, along with any additional data that may be used to explore validity. It is often useful to conduct a pilot study with a small number of objects before undertaking the complete measurement study.

5. Analyze the objects-by-observations matrix to estimate reliability. Cronbach's alpha can be computed using any of several computer programs for statistical analysis.

6. Conduct any content, criterion-related, or construct validity studies that are part of the measurement study. This will make use of the additional data collected in Step 4.

7. If the reliability or validity proves to be too low, attempt to diagnose the problem using methods described in Sect. 8.3.

8. If the results of the measurement study are sufficiently favorable, proceed directly to a demonstration study. Otherwise, repeat the measurement study, with revised measurement procedures.

Example

In a realistic but hypothetical situation, suppose that a study team is interested in the performance of a decision support system based on a machine learning model, and so seeks to assess the attribute "accuracy of advice" for each patient case evaluated by this information resource.

Patient cases are the object class of measurement. Abstracts of each patient's history and reports of the system's advice are the available information about each case (object) to enable the assessments. Human judges and any forms used to record the assessments comprise the instrumentation. A rating of the advice for one case by one judge is an independent observation.

For the measurement study, the study team elects to use 50 cases and 6 judges, each of whom will rate the accuracy of advice for each of the cases. The dimensions of the resulting objects-by-observations matrix will be 50 by 6. The choice of cases and judges is nontrivial because the results of the study cannot be generalized beyond the characteristics of populations from which these cases and judges are selected.

To increase the generalizability, the study team selects 50 cases from a citywide network of hospitals and six expert clinician judges from across the country. Conducting the study requires the resource to generate its advice for all 50 cases, and for each of the judges to review and rate the advice for all of the cases as well. The reliability of the ratings is estimated from the resulting objects-by-observations matrix to be 0.82. Using the Prophesy Formula, it is predicted that four judges will exhibit a reliability of 0.70. Given the time and effort required for the demonstration study to follow, the study team decides to use only four judges in the demonstration study.

Self-Test 8.1

With reference to the example described in the box above:

1. Which of the eight stages of conducting a measurement study are explicitly represented in the example above? Which are not represented?
2. What is the predicted reliability of this measurement process using one judge only? Would you consider this figure acceptable?
3. How might validity be explored in this hypothetical measurement study?

8.3 How to Improve Measurement Using Measurement Study Results

Two courses of action exist for a study team if a measurement study reveals suboptimal reliability. In this discussion, the term *observations* can refer to judges observing some kind of performance, items on a questionnaire, or tasks carried out by machines or people.

- *Option 1: Modify the number of independent observations in the measurement process.* When a measurement study reveals a low reliability (typically a coefficient of less than 0.70), the study team can improve it in a "brute force" way by increasing the number of independent observations drawn from the same population. In a measurement study employing test-retest reliability, this would simply entail increasing the number of repeated observations. In a measurement study employing multiple simultaneous observations, this would entail adding more observations drawn from the same populations: for example, adding more judges to performance ratings or more items to a questionnaire. Increasing the number of observations increases the work involved in conducting each measurement, increasing the time and expense incurred when conducting any subsequent demonstration studies. In some situations, as seen in the example above, a measurement study can yield higher-than-needed reliability, and the results of the measurement study can help the study team streamline the study by reducing the number of independent observations per object.
- *Option 2: Diagnose and correct specific problems with the measurement.* Alternatively, the study team can try to understand what is going wrong in the measurement process and take specific steps to correct it. In a measurement process with judges for example, they might identify and replace judges whose ratings seem unrelated to the ratings of their colleagues, or give all judges an improved form on which to record the ratings. This more elegant approach can increase reliability without increasing the resources required to conduct the measurement.

When a study team responds to a measurement study result by changing the number of observations (Option 1 above), it is often not necessary to repeat the measurement study because the impact of the change can be estimated from the Prophesy

Formula. When employing Option 2, the changes are more fundamental (e.g., a change in the format of a rating instrument or altering the composition of a panel of judges), and it may be necessary to repeat the measurement study, possibly going through several iterations until the measurement process reaches the required level of performance.

Also important, a well-intentioned attempt to improve only the reliability of a measurement can also affect its validity. When using Option 1, if the observations added to or deleted from the measurement process differ in some important way from the retained observations, this can inadvertently shift the attribute the measurement is addressing. Use of Option 1 requires at least a defensible assumption that the observations added or deleted are representative of the same population as the retained observations. If judges are added to a panel, for example, they should be similar in background and training to the pre-existing members of the panel. As discussed below, use of Option 2 requires a study team to carefully consider whether the changes they have made will affect the attribute that is the target, and thus affect the validity of the measurement process.

8.4   Using Measurement Study Results to Diagnose Measurement Problems

The following section illustrates how to identify which elements of a measurement process may need modification.

8.4.1   Analyzing the Objects-by-Observations Matrix

Analysis of the objects-by-observations matrix can reveal which of the observations, if any, are sources of unreliability and/or invalidity by behaving "differently" from the other observations. An observation that is "well behaved" should be both: 1) correlated with the other observations, and 2) calibrated with the other observations. For an observation to be correlated with other observations, an object with a high score on that observation should also tend to have a high score for the other observations. For an observation to be calibrated with other observations, the results of the observations should have similar values. An observation that is correlated with the other observations may not necessarily be calibrated with them. For example, a judge who is part of a panel rating a series of performances may agree with the other judges about which performances are better or worse, but may consistently give more (or less) favorable ratings than the other judges to all performances. The ratings of a correlated but uncalibrated judge will not affect the rank order of the performance ratings but will skew the values of these ratings upward or downward (Walsh et al. 2017).

The process of diagnosing a measurement problem is essentially a process of looking for observations that are either uncorrelated, uncalibrated, or both. The process begins with the objects-by-observations matrix. To examine the degree to which each observation is correlated with the others, first compute a *corrected part–whole correlation* between each observation and the other observations as a group (Mourougan and Sethuraman 2017).[1] Pearson product-moment correlation coefficients are customarily used for this purpose.[2] Each correlation may be computed using the following for two attributes, denoted *x* and *y*, with measurements of both attributes performed on *i* objects:

$$ r = \frac{\sum \left(x_i - \bar{x} \right)\left(y_i - \bar{y} \right)}{\sqrt{\sum_i \left(x_i - \bar{x} \right)^2 \sum_i \left(y_i - \bar{y} \right)^2}} $$

where x_i and y_i are values of the individual observations of x and y, and $\bar{x}$ and $\bar{y}$ are the mean values of x and y over all objects in the study sample. This formula looks imposing but is a built-in function of all spreadsheet and statistical programs.

Tables 8.1, 8.2 and 8.3 illustrate how this works. Table 8.1, identical to Table 7.2, portrays an objects-by-observations matrix resulting from a measurement process with four objects and five observations. The reliability coefficient for this measurement is 0.81. (As discussed in the previous chapter, real measurement studies will employ more than four objects.)

To compute the corrected part–whole correlation for each observation, it is necessary to create ordered pairs of numbers representing the score for each observation and the total score for each object, *excluding that observation*. Table 8.2 illustrates computation of the corrected part–whole correlation for Observation 1 from Table 8.1. The ordered pairs used to compute the correlation consist of each

Table 8.1 Objects-by-observations matrix

Objects	Observations					Total object score
	1	2	3	4	5	
A	4	5	3	5	5	22
B	3	5	5	3	4	20
C	4	4	4	4	5	21
D	3	2	3	2	3	13
Total observation score	14	16	15	14	17	

---

[1] Some texts refer to part-whole correlations as item-total correlations.

[2] When the purpose of computing the coefficients is to inspect them to determine if the observations are "well behaved," the Pearson coefficient is widely used and is the only coefficient discussed explicitly here. The Pearson coefficient assumes that the variables are both measured with interval or ratio properties and normally distributed. Even though both assumptions are frequently violated, the Pearson coefficient provides useful guidance to the study team performing measurement studies.

object's score for Observation 1 paired with that object's total score *but excluding the score for Observation 1*. Because Object A has a total score of 22, excluding Observation 1 yields a corrected total score of 18.

Repeating this process for each object creates the ordered pairs shown on Table 8.2. The corrected part-whole correlation can be computed directly from values in this table.

This process continues, observation by observation, to compute each of the corrected part-whole correlations. The counterpart to Table 8.2 used to compute the corrected-part whole-correlation for Observation 2 would share the structure of Table 8.2, but with the middle column containing the scores for Observation 2 and the rightmost column containing the total scores excluding the value of Observation 2.

Table 8.3 portrays the objects-by-observations matrix with the corrected part-whole correlation for each observation included as the bottom row shaded in grey. These results suggest that Observation 3, with a corrected part-whole correlation of 0.11, is not well correlated with the others. The other four part-whole correlations are large.

Although somewhat cumbersome, it can be helpful to also compute and inspect the matrix of correlations across all pairs of observations in a measurement process. This is more cumbersome than the method of part-whole correlations because a measurement process with N observations has $N(N-1)/2$ pairwise correlation coefficients, as opposed to N part-whole coefficients. Analysis of a measurement process with 10 observations, for example, involves inspection of 45 coefficients. Table 8.4

Table 8.2   Ordered pairs for computing the corrected part–whole correlation for Observation 1

Object	Score for Observation 1	Corrected total score, *excluding the score for Observation 1*
A	4	18
B	3	17
C	4	17
D	3	10

Table 8.3   Objects-by-observations matrix with corrected part–whole correlations

Objects	Results of five observations					Total score
	1	2	3	4	5	
A	4	5	3	5	5	22
B	3	5	5	3	4	20
C	4	4	4	4	5	21
D	3	2	3	2	3	13
Corrected part–whole correlations	0.62	0.83	0.11	0.77	0.90	

Table 8.4   Correlations between observations for the "typical measurement result" in Table 8.1

Observations	1	2	3	4	5
1	—	0.41	−0.30	0.89	0.91
2		—	0.49	0.73	0.74
3			—	−0.13	0.09
4				—	0.94

displays these correlations for the measurement results given in Table 8.1. Overall, there are 5(5 − 1)/2, or 10, correlation coefficients to inspect. The correlation between observations i and j is found at the intersection of the ith row and jth column. Because the correlation between Observations i and j will be the same as the correlation between Observations j and i, the redundant correlations are not shown.

Table 8.4 also shows that Observation 3 is not well behaved. The correlations between Observation 3 and the other observations, seen in grey background, show no consistent pattern: two are negative, and two are positive.

Observations that are "well behaved" should be at least modestly and positively correlated with all other observations. Typically, a well-behaved observation will have a corrected part-whole correlation above 0.3. Each well-behaved observation works to increase the reliability of the measurement of the attribute. A badly behaved observation can have the opposite effect. When a measurement study reveals that a specific observation is not well behaved, and thus does not belong with the others in the group of observations, it should be revised or deleted from the measurement process. In the example above, deleting Observation 3 actually increases the reliability of measurement from 0.81 to 0.89 even though the number of observations in the measurement process is decreased.

As discussed in Sect. 7.5, a set of observations that is well behaved as a group can be said to comprise an index, or scale.

8.4.2   Improving a Measurement Process

This section describes how problems with a measurement process can be diagnosed using part–whole correlations. For brevity's sake, the material that follows refers to corrected part–whole correlations more simply as part–whole correlations:

1. *If all part–whole correlations are reasonably large but the reliability is lower than required:* Employ Option 1 by adding defensibly equivalent observations to the measurement process.
2. *If most part–whole correlations are low:* Something affecting all observations is fundamentally amiss. Check aspects of the measurement process that relate to all observations; for example, if human judges are using a rating form, they may have been inadequately trained. Or, if the observations are items on a questionnaire, they may all be phrased at a reading level that is too sophisticated for the people completing the questionnaire.
3. *If one (or perhaps two) observations display low part–whole correlations:* First try deleting the misbehaving observation(s). The reliability may be higher and the entire measurement process more efficient if so pruned. Alternatively, try modifying or replacing the misbehaving observation(s), but always keep in mind that selectively deleting observations can affect validity: what is actually being measured by the observations in the set.
4. *If two or more observations display modest part–whole correlations while the others are high:* This situation is ambiguous and may indicate that the observations as a group are measuring two or more different attributes. In this case,

each subset displays high intercorrelation of its member observations, but the observations from different subsets are not correlated with each other. This possibility cannot be fully explored using part–whole correlations and requires either careful inspection of the full intercorrelation matrix or use of more advanced statistical techniques, such as principal component or factor analysis as discussed below. If the study team expected the observations to address a single attribute and in fact they address multiple discrete attributes, the entire measurement process is not performing as intended and should be redesigned. (To work out some examples in detail, complete Self-Test 8.2, below.)

If a specific observation is not well behaved, as in Outcome 3 above, several things may be happening. It will be necessary to pinpoint the problem in order to fix it. For example, consider items on a questionnaire as a set of observations. A misbehaving item may be so poorly phrased that it is not assessing anything at all, or perhaps the questionnaire respondents employed in the measurement study lack some specific knowledge that enables them to respond to the item. Alternatively, the item may be well phrased but, on logical grounds, does not belong with the other items on the scale. This situation can be determined by inspecting the content of the item, or, if possible, talking to the individuals who responded to the item to see how they interpreted it. Because development of reliable and valid measurement processes can be painstaking and time-consuming, a study team should employ existing measurement tools and instruments whenever possible.

Objects-by-observation matrices resulting from measurement studies may also be analyzed using more advanced statistical techniques such as principal components analysis or factor analysis (Watson 2017). These techniques analyze the matrix of intercorrelations as illustrated in Table 8.4, and will reveal whether the observations as a group address a single shared attribute, or possibly more than one. These methods can also reveal which observations are not well-behaved.

Self-Test 8.2

1. Refer to Table 8.1. If the scores for Observation 5 were {4,3,4,2} instead of {5,4,5,3} as shown in the table, would this primarily affect the correlation or the calibration of Observation 5?
2. Consider the following measurement result, with a reliability of 0.61. What is your diagnosis of this result? What would you do to improve it?

| Objects | Results of six observations | | | | | |
	1	2	3	4	5	6
A	4	3	5	2	1	4
B	2	4	5	3	2	2
C	3	4	3	4	4	3
D	2	3	1	2	1	2
E	3	3	2	2	4	3
Part–whole correlations						
	0.49	0.37	0.32	0.37	0.32	0.49

3. Consider the following measurement result, for which the reliability is 0.72.

Objects	Results of six observations					
	1	2	3	4	5	6
A	4	5	4	2	2	3
B	3	3	3	2	2	2
C	4	4	4	4	5	4
D	5	5	4	2	2	1
Part–whole correlations						
	0.21	0.13	0.71	0.76	0.68	0.51

The matrix of correlations among items for these observations is as follows.

Items	1	2	3	4	5	6
1	—	0.85	0.82	0	0	−0.32
2			0.87	−0.17	−0.17	−0.13
3				0.33	0.33	0.26
4					1	0.78
5						0.78

How would you interpret these results? What would you do to improve this measurement process?

8.5   The Relationship Between Measurement Reliability and Demonstration Study Results

The previous section of this chapter has addressed methods to improve reliability of measurement. This section addresses a more fundamental question: Why is reliability important? More specifically, what effects on a demonstration study will result from sub-optimal reliability?

The previous chapter, and specifically Sect. 7.6.3, introduced the relationship between reliability and the "error bracket" around a measurement result *for each individual object*. The magnitude of this error bracket, the *standard error of measurement,* varies inversely with the reliability of the measurement process. The higher the reliability, the smaller the standard error of measurement.

By contrast, in demonstration studies, the focus shifts from the measured value of an attribute on individual objects to the distribution of the measured values of that attribute (now called a variable in demonstration study parlance) across groups of objects (now called participants and subjects). A simple example, using demonstration study parlance, illustrates this difference:

- **Measurement result for one participant:** *Ms. Lopez's satisfaction with a diabetes app is 8.5 ± 0.5 where the measured value of the "satisfaction" attribute is 8.5 (on a scale of 10) and 0.5 is the standard error of measurement.*
- **Measurement result for a group of participants:** *After a sample of 100 diabetics used the app for 3 months, their mean satisfaction score was 7.2 with a standard deviation of 1.6.*

In demonstration studies involving groups of objects, measurement errors add unexplained random variability, what can be thought of as "noise", to the results of measurements. The greater the *standard error of measurement attached to each object*, the greater the *standard deviation (SD) of the distribution of the scores for a group of objects*. Importantly, because the effects of unreliability are random, they do not bias the actual estimate of the mean value of an attribute for the object group.

Crudely, it may be helpful to think of random measurement errors as a "silent factor" eroding the informativeness of demonstration studies. Random errors do their work of injecting noise into the demonstration study results without their presence being recognized unless the study team knows, from previously conducted measurement studies, the reliability of the measurement methods employed.

The following sections discuss the effects of measurement error in more detail.

8.5.1   Effects of Measurement Error in Descriptive and Interventional Demonstration Studies

In many descriptive and interventional demonstration studies, as discussed in Sect. 6.9, the result of primary interest is expressed as the mean value of a measured variable in a sample of participants. The expression of the mean is usually accompanied by an expression of the uncertainty, or standard error, around the mean (SE_{mean}). For a sample of size N, the standard error of the mean is computed as:

$$SE_{mean} = \frac{SD}{\sqrt{N}}$$

where SD is the standard deviation of the variable for N participants. In terms of confidence intervals, the mean value of a variable plus or minus the SE_{mean} generates the 68% confidence interval around the mean, and the mean value plus or minus twice the SE_{mean} generates the 95% confidence interval.[3]

---

[3] This is a useful approximation for computing the 95% confidence interval. A more exact formula is 95%CI = Mean ± (1.96 × SE_{mean}).

As discussed above, randomly distributed measurement errors have the straight-forward effect of increasing the uncertainty around the computed mean values of variables (attributes) of interest. The greater the measurement error, the greater the standard error of the mean and confidence intervals, and thus the greater the uncertainty. In the above example of a satisfaction questionnaire, with a standard deviation of 1.6 in a sample of 100 people, the standard error of the mean is 0.16.

In many interventional demonstration studies the question of primary interest is whether the mean of some attribute differs among specified groups of objects: often a group receiving an intervention vs. a control group. Because measurement error adds imprecision to the estimates of the means for each group, the lack of reliability decreases the probability that a measured difference between the mean values in the groups will be interpreted as a "significant" effect as opposed to an effect due to chance. For example, a demonstration study might compare the costs of managing hypertensive patients with and without support from a "care navigator" using a clinical decision support tool. With samples of 100 patients in each group, the observed mean ($\pm SE_{mean}$) of these costs might be \$595 ($\pm$\$2.0) per patient per year in the "care navigator" group and \$600 ($\pm$\$2.0) in the group not using it. Applying methods of statistical inference to be introduced in Chap. 12, this difference is not statistically significant. Suppose, however, it was found that the measurement methods used to determine these costs were so unreliable that measurement error was contributing a substantial part of the variability reflected in the standard error of the mean. Through use of improved measurement methods in a repeat of the demonstration study, it may be possible to reduce the standard error of the mean by 25%. If this were done, the results of a hypothetical replication of the study might then be \$595 ($\pm$\$1.6) for the navigator group and \$600 ($\pm$\$1.6) for the control group. This difference *is* statistically significant.[4]

8.5.2 Effects of Measurement Error in Correlational Demonstration Studies

In correlational studies, the focus shifts from comparing the mean values of variables (measured attributes) for two or more samples, to computing the correlations between two or more different variables as measured in a single sample. For example, a demonstration study might examine the correlation between perceived ease of use of a clinical workstation with scores on a knowledge test administered at the end of a training session in a single sample of care providers. In this type of study, the "silent" effect of measurement error will reduce the observed magnitude of the correlation between the two attributes. This effect of measurement error on correlation

---

[4] For those experienced in inferential statistics, a *t*-test performed on the case with the larger standard errors reveals $t = 1.77$, $df = 198$, $p = .08$. With the reduced standard errors, $t = 2.21$, $df = 198$, $p = .03$.

is known as *attenuation*. An approximate correction for attenuation of the correlation between two variables can be used if the reliabilities of the measurements of the variables are known:

$$r_{corrected} = \frac{r_{observed}}{\sqrt{\rho_1 \rho_2}}$$

where $r_{corrected}$ = correlation corrected for measurement error (attenuation)

$r_{observed}$ = observed or actually measured correlation
ρ_1 = reliability of measurement 1
ρ_2 = reliability of measurement 2.

To see why this attenuation happens, first consider two variables (Y_{true} and X_{true}), each measured with perfect reliability ($\rho = 1$) and between which the correlation is high. The relation between Y_{true} and X_{true} is shown in the upper part of Fig. 8.1. The values of Y_{true} and X_{true} fall nearly on a straight line and the correlation coefficient is

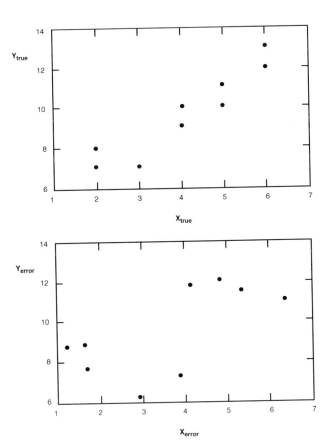

Fig. 8.1 Effect of measurement error on the correlation between two variables

0.95. Applying a small, randomly distributed error to Y_{true} and X_{true}, generates representative values of these variables as they might be measured with less than perfect reliability: Y_{error} and X_{error}. The plot of Y_{error} versus X_{error} in the lower part of Fig. 8.1 shows the degradation of the relationship and the corresponding attenuation of the correlation coefficient in this case from 0.95 to 0.68.

This correction must be applied with caution because it makes the assumption central to classical measurement theory that all measurement errors are randomly distributed. If this assumption turns out not be valid, it is possible to obtain corrected correlations that are overinflated. An example of the effect of attenuation is discussed in detail later in the chapter.

Self-Test 8.3

1. In a demonstration study, the diagnostic assessments of the acuity of skin lesions by a pool of expert dermatologists are found to correlate (coefficient of correlation = 0.45) with assessments created by a machine learning system that interprets images of these lesions. The reliability of the assessments by the dermatologists is found, in a measurement study, to be 0.7.

   (a) Focusing on the measurement process for obtaining the dermatologists' assessments, what is the attribute being assessed; what are the objects of measurement, and what constitute independent observations?
   (b) Make an argument that the reliability of the automated assessments of the machine learning system can be assumed to be perfect.
   (c) What is the predicted "true" correlation between the machine learning assessments and the dermatologists' assessments, correcting for the less than perfect reliability of the dermatologists' assessments?

2. Assume that the assumptions underlying the correction for attenuation hold. Use the correction formula to show that $r_{observed} \leq \sqrt{\rho_1 \rho_2}$. (*Hint:* Use the fact that the corrected correlation must be ≤ 1.0.) This equation is important because it points out that the reliability of a measure sets an upper bound on the correlation that can be measured between it and other measures.

8.6   Demonstration Study Results and Measurement Error: A Detailed Example

This section takes a classic example from the informatics literature to illustrate how measurement error can lead a study team to underestimate the true value of the correlation between the resource and the standard. This example is based the work of van der Lei and colleagues (Van der Lei et al. 1991), who explored the performance

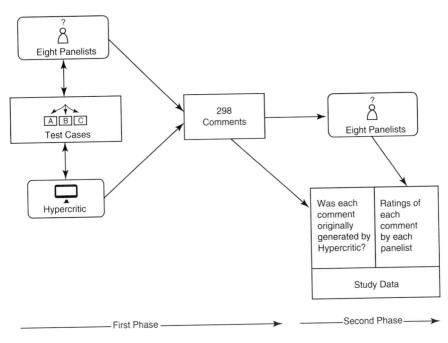

Fig. 8.2 Structure of the Hypercritic study

of Hypercritic, a knowledge-based system that offers critiques of the treatment of hypertensive patients through computational examination of the electronic records of these patients.[5] This example presents a modified version of the original study, which explored the question: To what extent are the critiques generated by Hypercritic in agreement with the critiques offered by human experts?

Figure 8.2 illustrates a simplified structure of the study. In the first phase of the study, Hypercritic and eight members of an expert panel independently examined the care records of a set of hypertensive patients for the purpose of generating comments about the care of each patient. Each comment identifies a way in which the care of that patient might have been improved. This initial review generated a set of 298 comments. Each comment in the set was generated by Hypercritic, one or more of the panelists, or both.

In the second phase of the study, each panelist independently reviewed all 298 comments, without knowing whether it was generated by Hypercritic and/or by one or more of the panelists. Each panelist judged each comment to be either correct or

---

[5] The authors are grateful to Johan van der Lei and his colleagues for sharing the original data from their study.

incorrect. If Hypercritic generated a comment as it scanned the patient records during the initial review, it was assumed in the second phase of the study that Hypercritic considered the comment to be correct. (In other words, if Hypercritic were a "panelist" in the second phase, it would invariably consider its own comments to be correct.)

Table 8.5 illustrates partial results of the second phase of the study, including only 12 of the 298 comments. Each comment forms a row. The column in light grey indicates whether each comment was generated by Hypercritic in the first phase, and was thus considered by Hypercritic to be correct. The columns in darker grey, labeled A through H, portray the ratings of panelists. All ratings are coded as 1 for "correct" and 0 for "incorrect". Table 8.5 illustrates that Hypercritic and the panelists did not always agree, and that the panelists did not always agree among themselves. Comment 1 was generated by Hypercritic in the first phase—and perhaps by one or more panelists as well—and was subsequently rated as correct by Panelists A, B, and D. In the first phase, Comment 2 was not generated by Hypercritic but was generated by one or more of the panelists. It was endorsed by Panelists B, C, D, and G in the second phase.

Based on the structure of Table 8.5, the primary question directing a demonstration study of Hypercritic would be "To what extent do the panelists' judgements of correctness of comments align with those of Hypercritic?" According to the classical theory of measurement, the best estimate of the panelists' correctness score for each comment is the sum (or average) of the individual panelists' ratings, as shown in the rightmost column of Table 8.5. Hypercritic's assessments of its own correctness are in the second column of the table. For demonstration study purposes, the

Table 8.5 Results from a subset of 12 comments in the Hypercritic study

Objects: The Comments	Was comment generated by Hypercritic in the first phase? (Hypercritic's "correctness" score)[a]	*Observations*: Second phase ratings by each panelist[b]								Panelists' "correctness" score
		A	B	C	D	E	F	G	H	
1	1	1	1	0	1	0	0	0	0	3
2	0	0	1	1	1	0	0	1	0	4
3	0	1	1	1	1	0	0	1	0	5
4	1	1	1	0	1	0	0	1	1	5
5	0	1	1	1	1	1	0	0	1	6
6	0	0	1	1	1	0	0	1	1	5
7	0	1	1	1	1	0	0	1	1	6
8	0	1	1	1	1	0	0	1	1	6
9	1	1	1	0	1	1	1	1	1	7
10	1	1	1	1	1	1	0	1	1	7
11	0	1	1	1	1	1	1	0	1	7
12	1	1	1	0	1	1	1	1	1	7

[a]Values of "yes" are coded as "1" and values of "no" are coded as "0"
[b]Values of "correct" are coded at "1" and values of "incorrect" are coded as "0"

magnitude of this alignment can be computed as the correlation between these two columns of numbers.

From a measurement perspective, the variable "was the comment generated by Hypercritic?" (second column) has perfect reliability because Hypercritic could be expected to execute its own algorithms with perfect consistency. It is clear, however, from the variability in the panelists' ratings of each comment that the panelists' correctness score does not have perfect reliability. The reliability of these ratings can be computed from the objects-by-observations matrix data which is shown in darker grey shading on Table 8.5.

8.6.1 Reliability Estimate

The reliability of the panelists' ratings involves only the ratings shown in darker grey in Table 8.5. Framing this as a formal measurement process (see Sect. 6.2), the attribute is the correctness of each comment, the objects of measurement are the comments, and observations are the panelists' ratings. One rating by one panelist is an independent observation. From the complete data in this study with the full set of 298 comments instead of the 12 shown in the table, the reliability coefficient (Cronbach's α) is 0.65. The standard error of measurement of each correctness score is 1.2.[6]

8.6.2 Demonstration Study Correcting the Correlation for Attenuation

Knowing the reliability of the judges' correctness scores strengthens the demonstration study by enabling the correlation between Hypercritic's correctness scores and the panelists' correctness scores to be corrected for attenuation. Using the data in the second and rightmost columns in Table 8.5, the correlation coefficient between Hypercritic and the panelists, based on all 298 comments, is 0.50. This signals a moderate and statistically significant relationship. However, because of the measurement error in the judges' ratings, this observed correlation underestimates the "true" magnitude of the relationship between Hypercritic's output and the correctness scores. The formula to correct for attenuation (see Sect. 8.4.2 above) enables an estimation of the effect of measurement error on the observed correlation. Applying this formula, with reliability values of 1 for Hypercritic's correctness scores and 0.65 for the judges, yields a correlation coefficient of 0.62 after

---

[6] Readers familiar with theories of reliability might observe that coefficient alpha, which does not take into account panelists' stringency or leniency as a source of error, might overestimate the reliability in this example. In this case, however, use of an alternate reliability coefficient had a negligible effect.

correction for attenuation. Comparing the estimated corrected correlation of 0.62 with the observed correlation of 0.50, a failure to consider measurement error will lead, in a demonstration study, to underestimation of Hypercritic's accuracy by approximately 24%.

8.6.3 Additional Considerations from This Example

From a Measurement Perspective: The results of the measurement study described above allow a conclusion that the ratings by the panelists have sufficient reliability to be measuring something meaningful, but the result is below the desirable threshold of 0.7. A somewhat higher reliability could be obtained using the methods discussed earlier in this chapter: either the "brute force" method of adding equivalent panelists or the more elegant methods of identifying, using the method of part-whole correlations, the panelists whose ratings do not conform with the others.

A complete measurement study would also consider the validity of these ratings: the degree to which the ratings actually represent the attribute of correctness versus some other attribute(s) that might be "polluting" the measurement. Content or face validity of the ratings might be explored by examining the credentials of the judges, verifying that they had adequate qualifications for and experience in managing hypertension. A measure of criterion-related validity might be obtained by examining the incidence of hypertension-related complications, for cases where the corrective action recommended by the comment was *not* taken. A correlation between complication rates and correctness scores would be an indication of criterion-related validity in the predictive sense. Criterion-related validity of the correctness attribute, in the concurrent sense, might be assessed by comparing comments with published clinical guidelines.

From a Demonstration Perspective: The correlation coefficient is a useful way to portray the degree of relationship between two variables, but it says little about the nature of the disagreement between Hypercritic and the pooled ratings of the judges. Contingency table methods, discussed in detail in Chap. 12, might be used to look more deeply into the number and nature of the disagreements. In the actual published study the authors chose to view the panelists' correctness scores as an ordinal ("correct or incorrect") variable, viewing endorsement of a comment by 5 or more panelists as a threshold for considering a comment to be correct or not. With reference to Table 8.5, Comment 1 was considered correct by only three panelists and thus coded as "incorrect" by the study team. Comment 3, endorsed by five panelists, would be coded as "correct".

Table 8.6 Hypercritic demonstration study results in contingency table format, with endorsement by five or more judges as the criterion for "correctness"

| | Pooled rating by panelists | | |
	Comment rated as "correct" by 5 or more panelists	Comment rated as "correct" by 4 or fewer panelists	Total
Comment generated by Hypercritic	145	24	169
Comment not generated by Hypercritic	55	74	129
Total	200	98	298

With this ordinal representation of the panelists' correctness scores, the study results map into a contingency table as shown in Table 8.6.

The results portrayed in Table 8.6 reveal that two different kinds of disagreements occur in roughly equal proportion. Hypercritic failed to generate 55 of the 200 comments (28%) considered correct by five or more judges. Hypercritic generated 24 of the 98 comments (24%) considered incorrect by the judges. Keep in mind that the values in the cells of Table 8.6 depend critically on the choice of threshold.

Self-Test 8.4

Assume that the data in Table 8.5, based only on 12 comments, constitute a complete pilot study. The reliability of these data, based on 12 comments (objects) and 8 judges (observations) is 0.29. (Note that this illustrates the danger of conducting measurement studies with small samples of objects, as the reliability estimated from this small sample is different from that obtained with the full sample of 298 comments). For this pilot study:

1. What is the standard error of measurement of the "correctness of a comment" as determined by these eight judges?
2. Focusing on the ratings of panelists A-H. Use the method of corrected part–whole coefficients to determine who is the "best" panelist in terms of correlation with his or her colleagues. Who is the "worst?" What would happen to the reliability if the worst judge's ratings were removed from the set?
3. If there were four judges instead of eight, what would be the estimated reliability of the measurement? What if there were 10 judges?
4. Using the pooled ratings of all eight judges as the standard for accuracy, express the accuracy of Hypercritic's relevance judgments as a contingency table using endorsement by six or more judges as the threshold for assuming a comment to be correct.
5. For these data, the correlation between Hypercritic's judgments and the pooled ratings is 0.09. What effect does the correction for attenuation have on this correlation?

8.7   Conclusion

This chapter was an excursion into more practical aspects of measurement. When, for any of a variety of reasons, study teams find themselves in a position of having to conduct a measurement study, the first sections of the chapter provided a step-by-step process for carrying out the study and putting the study results to beneficial use. Realizing the benefits of a measurement study requires a more focused diagnosis of the causes of any measurement problems revealed by the study, and the method of part-whole correlation is a key element of that diagnostic process. Later parts of the chapter introduced the effects of measurement error on the results of demonstration studies, and through a detailed example described ways to estimate these effects.

This discussion and example provide a transition to Chap. 9 which follows. Chapter 9 introduces ways to design measurement processes to be high-performing from the outset.

Answers to Self-Tests

Self-Test 8.1

1. Stages 1–5 are explicitly represented in the example. Stage 6 is not represented because no validity study is described. This makes Stage 7 only partially complete and Stage 8 would be premature.
2. Ratings based on one judge, as prognosticated by the Prophesy Formula, have a reliability of .36, which is not acceptable.
3. Content validity could be explored by review of the credentials of the selected judges. Criterion validity could be explored by locating other cases similar to those used in the study, and then examining whether cases where the behavior recommended by the resource was taken (for whatever reason since they would not have had access to the resource) by clinicians caring for those patients. The system's advice would be valid to the extent that the cases where the recommended actions were taken exhibited better outcomes.

Self-Test 8.2

1. It would affect calibration, by altering the difference between the scores for Observation 5 and the other observations. The change reduces all values of Observation 5 by one scale point. This would not affect the correlation with the other observations.
2. More observations are needed to increase the reliability. The observations in the set are generally well behaved. There are not enough of them.
3. It appears that two attributes are being measured. Items 1–3 are measuring one attribute, and items 4–6 are measuring the other.

Self-Test 8.3

1. (a) Attribute is the acuity of each skin lesion.
Objects are the lesions, presumably images of them.
An independent observation is a diagnostic assessment of one lesion by one dermatologist.
(b) Because the machine learning approach uses a computer algorithm, as long as the code is stable and the hardware is functioning properly, the assessments made by machine would be expected to be completely consistent and reproducible.
(c) Using the attenuation formula, the "true" (unattenuated) correlation would be 0.54. (The two reliabilities are: 0.7 for the dermatologists' assessments and 1 for the machine learning assessments.)

2. The answer may be obtained by substituting $r_{corrected} \leq 1$ into the formula:

$$r_{corrected} = \frac{r_{observed}}{\sqrt{\rho_1 \rho_2}}$$

to obtain the inequality:

$$1 \geq \frac{r_{observed}}{\sqrt{\rho_1 \rho_2}}$$

Self-Test 8.4

1. SE_{meas} (eight judges) = 1.10.
2. Judge H displays the highest corrected part–whole correlation (0.55) and thus can be considered the "best" judge. Judge E is a close second with a part–whole correlation of 0.50. Judge C may be considered the worst judge, with a part–whole correlation of −0.27. Removing Judge C raises the reliability from 0.29 to 0.54 in this example. Such a large change in reliability is seen in part because the number of objects in this example is small. Judges B and D can in some sense be considered the worst, as they rendered the same result for every object and their part–whole correlations cannot be calculated.

3. Reliability (4 judges) = 0.17; reliability (10 judges) = 0.34.

4.

	Judges	
Hypercritic	Valid	Not valid
Generated	3	2
Not generated	4	3

5. Corrected correlation is 0.17.

References

Frey BB, editor. The Sage encyclopedia of educational research, measurement, and evaluation. Thousand Oaks, CA: Sage; 2018.

Mourougan S, Sethuraman K. Enhancing questionnaire design, development and testing through standardized approach. IOSR J Bus Manage. 2017;19:1–8.

Streiner DL, Norman GR, Cairney J. Health measurement scales: a practical guide to their development and use. 5th ed. Oxford: Oxford University Press; 2015.

Van der Lei J, Musen MA, van der Does E, Man in't Veld AJ, van Bemmel JH. Comparison of computer-aided and human review of general practitioners' management of hypertension. Lancet. 1991;338:1504–8.

Walsh CG, Sharman K, Hripcsak G. Beyond discrimination: a comparison of calibration methods and clinical usefulness of predictive models of readmission risk. J Biomed Inform. 2017;76:9–18.

Watson JC. Establishing evidence for internal structure using exploratory factor analysis. Measure Eval Counsel Develop. 2017;50:232–8.

Chapter 9
Designing Measurement Processes and Instruments

Learning Objectives

The text, examples, and self-tests in this chapter will enable you to:

1. Categorize a given measurement process according to whether the observation of primary interest is judges, tasks, or items.
2. For measurement problems where judges are the observation of primary interest:

    (a) identify the common sources of variation in measurement results;
    (b) describe the number of judges typically required for adequate measurement;
    (c) given a measurement process, describe ways to improve it.

3. For measurement processes where tasks are the observation of primary interest:

    (a) identify the common sources of variation in measurement results;
    (b) describe the number of tasks typically required for adequate measurement;
    (c) given a measurement process, identify the task domain and describe ways to improve the measurement process.

4. For measurement processes where items are the observation of primary interest:

    (a) identify the common sources of variation in measurement results;
    (b) describe the number of items typically required for adequate measurement;
    (c) given a measurement instrument, improve the items comprising the instrument.

© Springer Nature Switzerland AG 2022
C. P. Friedman et al., *Evaluation Methods in Biomedical and Health Informatics*, Health Informatics, https://doi.org/10.1007/978-3-030-86453-8_9

9.1 Introduction

This chapter addresses the pragmatics of quantitative measurement: how to design measurement processes that employ multiple observations and how to conduct the actual measurement process. This chapter concludes the topic of quantitative measurement, which began in Chap. 7, and forms a bridge to the topic of demonstration which begins in the following chapter.

This chapter organizes itself around three different categories of observations that arise frequently in quantitative measurement in informatics:

1. when the observations of primary interest are the opinions of human *judges*, also called "raters";
2. when the observations of primary interest are *tasks* completed by persons, information resources, or persons and resources working in combination;
3. when the observations of primary interest are *items* on attitude scales, questionnaires, surveys or other types of forms.

Figure 9.1 illustrates the specific form of the objects-by-observation matrices for each category of independent observation. Although the same general measurement concepts apply to all three categories, there are issues of implementation and technique specific to each.

For each category of observation, this chapter explores:

1. in actual studies, why the results for a given object vary from observation to observation and how much variation to expect;
2. in practice, how many independent observations are needed for reliable measurement;
3. techniques that can be useful improve this aspect of measurement.

For all categories of observations, the individual observation scores for each object are either summed or averaged to produce a total score which is the measurement result for that object. This is in accord with the classical theory of measurement which stipulates that the best estimate of an object's "true" score is the average of multiple independent observations. It is customary to assume that the objects' total scores have interval properties, as discussed in Sect. 6.4, even if the scores generated by each individual observation are ordinal.

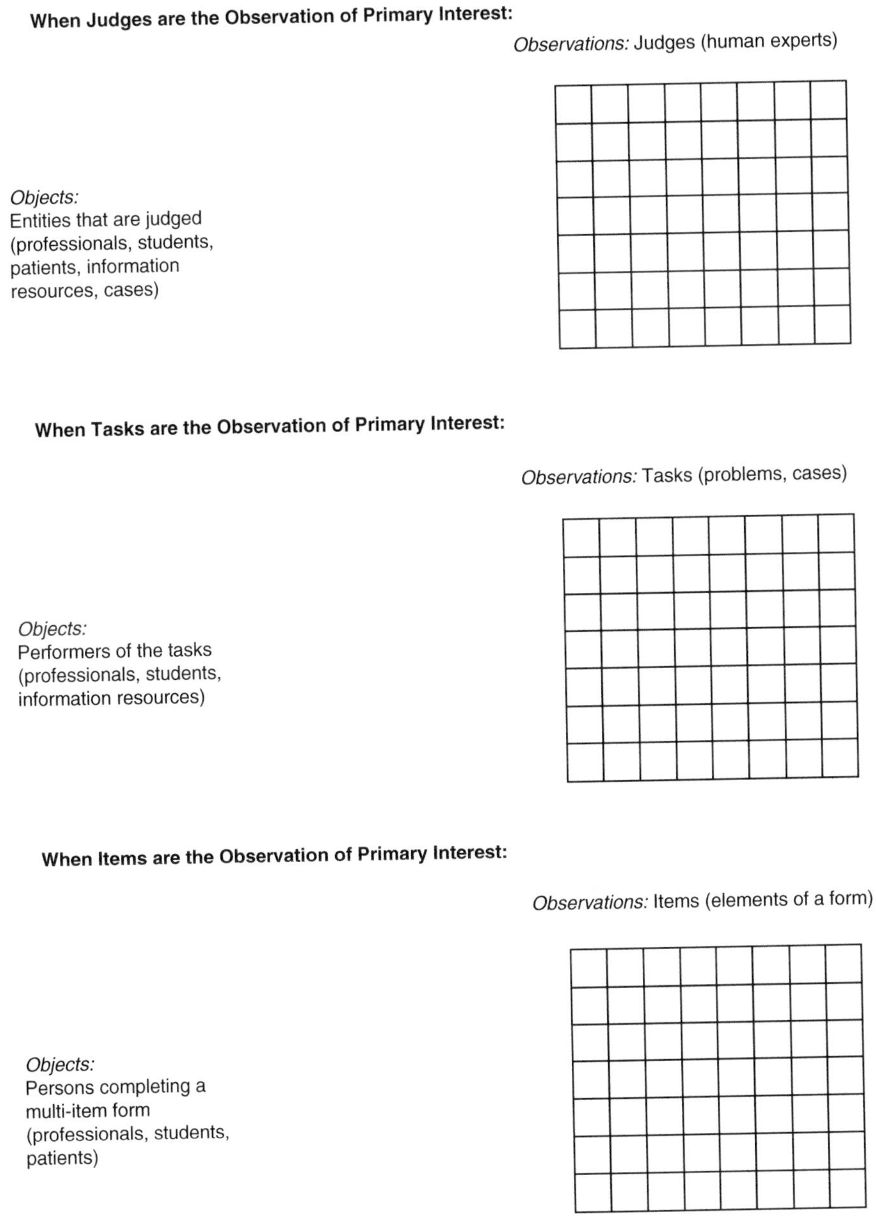

9.1   Introduction

When Judges are the Observation of Primary Interest:

Observations: Judges (human experts)

Objects:
Entities that are judged
(professionals, students,
patients, information
resources, cases)

When Tasks are the Observation of Primary Interest:

Observations: Tasks (problems, cases)

Objects:
Performers of the tasks
(professionals, students,
information resources)

When Items are the Observation of Primary Interest:

Observations: Items (elements of a form)

Objects:
Persons completing a
multi-item form
(professionals, students,
patients)

Fig. 9.1 Objects-by-observations matrices when judges, tasks, and items are the observations of primary interest

Self-Test 9.1

A team developing a new order entry system wants to measure clinicians' speed in entering orders, so they can evaluate the new system. They develop a measurement process where each clinician is asked to enter five order sets. They measure the time it takes each clinician to enter each one, with the intent of computing a clinician's "speed score" as the clinician's average speed over the five order sets. One clinician member of the team, develops the order sets, creating what that clinician believes is a diverse set based on personal experience.

1. Are the observations of primary interest in this process judges, tasks, or items? Explain.
2. A measurement study is conducted using 10 clinicians, generating the objects-by-observations matrix given in the table below. Each cell of the matrix is the time (in seconds) to enter each order set. Using the spreadsheet introduced in Sect. 7.6.2 or a statistical program, compute the reliability of this measurement process. Is this value sufficient for any measurement purpose?
3. What is the value of Clinician 1's "speed score"?

Data Table

Clinician	Order set				
	A	B	C	D	E
1	81	142	56	185	124
2	56	113	65	135	100
3	80	65	106	179	66
4	75	111	87	155	102
5	81	135	76	166	80
6	71	140	87	130	101
7	67	150	71	140	89
8	51	176	71	191	135
9	91	112	71	134	67
10	34	111	45	123	65

9.2 When Judges' Opinions Are the Primary Observations

Judges become central to measurement in informatics when the informed opinion is required to assess specific aspects of an activity or a product via observation or inspection. A study might employ experts to rate the quality of the interactions between patients and clinicians, as the clinicians enter patient data into an information resource during the interaction. In another example, judges may review the narrative information in patient charts to assess the presence or absence of specific clinical information. As with any measurement process, the primary concern is the concordance among the independent observations—in this situation, the judges—and the resulting number of judges required to obtain a reliable measurement. A set of "well-behaved" judges, whose opinions correlate with one another to an acceptable extent when rating a representative sample of objects, can be said to form a

Table 9.1 Objects-by-observations matrix with judges as the observation of primary interest

Objects being observed by the judges	Judges					Score for each object averaged across judges
	1	2	3	4	5	
A	4	5	3	5	5	4.4
B	3	5	5	3	4	4.0
C	4	4	4	4	5	4.2
D	3	2	3	2	3	2.6

scale. A substantial literature on performance assessment by judges, including studies specific to informatics, speaks in more detail to many of the issues addressed here (Gwet 2014; Hripcsak and Wilcox 2002; Hripcsak and Heitjan 2002; Hallgren 2012).

A generic objects-by-observations matrix for measurement involving judges' opinions is portrayed in Table 9.1. In most measurement involving judges, the judges record their opinions as numerical ratings. So in the table below, a score of "5" might correspond to a rating of excellent and a score of "1" might correspond to a rating of poor in relation to an attribute the judges are asked to score. Sometimes judges form a panel, all members of which observe an event as it is occurring; on other occasions, judges independently review a recording of the event which does not require simultaneous observation. To conform to the classical theory of measurement, the judges must carry out their actual ratings independently, with no discussion or other interaction during the rating process. The object scores in the rightmost column {4.4, 4.0, 4.2, 2.6} are typically assumed to have interval properties even though each individual judge's ratings are ordinal.

9.2.1 Sources of Variation Among Judges

Ideally, all judges of the same object, using the same criteria and forms to record their opinions, should render identical and thus perfectly reliable judgments—but this is rarely, if ever, seen in practice. Many of the factors that erode inter-judge agreement are well known and have been well documented over many years of research in the field of psychometrics (Guilford 1954):

1. *Interpretation or logical effects:* Judges may differ in their interpretations of the attribute(s) to be rated and the meanings of the items on the forms on which they record their judgments.
2. *Judge tendency effects:* Some judges are consistently overgenerous or lenient; others are consistently hypercritical or stringent. Others do not employ the full set of response options on a form, locating all of their ratings in a narrow region, which is usually at the middle of the range. This phenomenon is known as a "central tendency" effect.
3. *Insufficient exposure:* Sometimes the logistics of a study require that judges base their judgments on less exposure to the objects than is necessary to come to an informed conclusion. This may occur, for example, if study teams schedule

10 min of observation of end users working with a new information resource, but the users require 20 min to complete an assigned task.

4. *Inconsistent conditions:* Unless multiple judges make their observations simultaneously, some may do their ratings in the morning when they are alert and energetic, while others may do their ratings when fatigued in the evening.

9.2.2   Number of Judges Needed for Reliable Measurement

While specific steps described in the following section can reduce the effects of the factors listed above, it is not possible to completely eliminate differences among judges' ratings of the same performance. The greater the differences, the lower the reliability. The number of factors that influence judges' agreement and the varying ways study teams address them in the design of measurement processes make it impossible to predict in advance how reliable a measurement process engaging judges will be. When the judges' ratings are used primarily to rank-order individual objects, as in competitive ice skating, panels of 10 or more judges are required. The data from the study by van der Lei and colleagues, as described in Sect. 8.6, displayed a reliability of 0.65 with eight judges employed to rate comments on patient care as correct or incorrect. For attributes that are more straightforward to measure, such as the presence or absence of specific information in a medical record, two judges may be sufficient (Kern et al. 2013). Measurement based on one judge is never recommended unless justified by a previously conducted measurement study. Because so many factors can determine the reliability of measurement involving judges, measurement studies are essential to verify that "fit for purpose" reliability is obtained for any particular situation.

9.2.3   Improving Measurements That Use Judges as Observations

As with any approach to improve a measurement process, there are two possible strategies. The first is the "brute force" approach to increase the number of judges, which will improve reliability, and unlikely affect validity as long as the judges added are drawn from the same population as the original judges. For example, if a measurement process initially included a panel of academic cardiologists, any judges added should also be academic cardiologists. Any added judges should be trained for the task in the same way as their predecessors. Under these conditions, the Spearman-Brown prophecy formula estimates how much improvement will result from addition of any number of "equivalent" judges.

The second strategy is to improve the mechanics of the measurement process itself, which can affect both reliability and validity. Formal training or orientation of the judges can lead to significant improvement (Fowell et al. 2008). This can include a meeting where the judges discuss their personal interpretations of the attribute(s) to be rated. A more focused practice activity can be even more helpful. In a practice setting, judges first observe the phenomenon to be rated and make their ratings independently. Then the individual ratings are collected and summarized in a table, so all can see the aggregate performance of the group. This step is followed by a discussion among the judges, in which they share their reasons for rating as they did, and subsequently agree on clearer definitions or criteria for making judgments. This practice activity lays the groundwork for the "official" rating process, during which the judges must work independently.

Some simple logistical and practical steps, often overlooked, can also improve measurement using judges. Disqualify judges who, for a variety of reasons, are inappropriate participants in the study. An individual with a position of authority in the environment where the study is undertaken, such as the director of a lab or clinic, should not participate as a judge. Avoid using unwilling conscripts, who might have been "cordially required" by their supervisors to serve as judges. In some studies, under appropriate conditions, the activity to be judged can be recorded, and the judges can work asynchronously—as long as they do not informally discuss their ratings before all judgments have been rendered and as long as the recording captures all key elements of task performance in audio and video.

Self-Test 9.2

The TraumAID system (Clarke et al. 1994) was developed to provide minute-by-minute advice to trauma surgeons in the management of patients with penetrating wounds to the chest and abdomen. As part of a laboratory study of the accuracy of TraumAID's advice, Clarke and colleagues asked a panel of three judges—all experienced surgeons from the institution where TraumAID was developed—to rate the appropriateness of management for each of a series of cases that had been abstracted to paper descriptions. Ratings were on a scale of 1–4, where 4 indicated essentially flawless care and 1 indicated serious deficiencies. Each case appeared twice in the set: (1) as the patient was treated, and (2) as TraumAID would have treated the patient. The abstracts were carefully written to eliminate any cues as to whether the described care was computer generated or actually administered.

Overall, 111 cases were rated by three judges, with the following results:

Condition	Corrected part–whole correlations			Reliability of ratings
	Judge A	Judge B	Judge C	
Actual care	0.57	0.52	0.55	0.72
TraumAID	0.57	0.59	0.47	0.71

Condition	Mean ratings ± SD		
	Judge A	Judge B	Judge C
Actual care	2.35 ± 1.03	2.42 ± 0.80	2.25 ± 1.01
TraumAID	3.12 ± 1.12	2.71 ± 0.83	2.67 ± 0.95

1. What are the dimensions (number of rows and number of columns) of the two objects-by-observations matrices used to compute these results?

2. Is there any evidence of tendency errors (leniency, stringency, or central tendency) in these data?
3. Viewing this as a measurement study, what would you be inclined to conclude about the measurement process? Consider reliability and validity issues.
4. Viewing this as a demonstration study, what would you be inclined to conclude about the accuracy of TraumAID's advice?

9.3 When Tasks Are the Primary Observation

Many evaluation studies in informatics center on real-world tasks. The actors and/or agents undertaking these tasks are the objects of measurement. These may be persons (professionals, students, or patients) who are actual or potential users of an information resource, groups of these persons, information resources themselves, or interacting combinations of these entities. Depending on the nature of the tasks, these entities are asked to diagnose, intervene, compute, interpret, solve, analyze, predict, retrieve, propose, etc. The attributes of interest, what is actually measured, depend on the goals of the study. Examples of measurement with tasks as the primary observation include:

- usability studies where the tasks might be different functions that can be executed on a mobile app, the objects are people as users of the app, and the attribute is success at completing each task;
- technical studies where tasks are machine learning algorithms to execute, objects are different hardware configurations for executing these algorithms, and the attribute is execution speed;
- cognitive studies where the tasks are clinical cases requiring diagnosis, the objects are people and/or machines presented with these diagnoses, and the attribute is correctness of the diagnosis of each case.

In studies carried out in the "field", as discussed in Sect. 3.4.1, the tasks arise naturally within a practice environment or, in the case of personal health, in the activities of daily life. In studies conducted in the "lab", the tasks are invented, simulated, or abstracted. For many reasons, selection and design of tasks, for measurement as well as demonstration studies, are the most challenging aspects of quantitative study design.

Table 9.2 illustrates the general form of an objects-by-observations matrix when tasks are the observation of primary interest.

Table 9.2 Objects-by-observations matrix with tasks as the observation of primary interest

	Tasks					
Objects completing the tasks	1	2	3	4	5	Score for each object averaged across tasks
A	4	5	3	5	5	4.4
B	3	5	5	3	4	4.0
C	4	4	4	4	5	4.2
D	3	2	3	2	3	2.6

9.3.1   Sources of Variation Among Tasks

The task performance of persons, information resources, and the two in interaction, depends on many factors and is unpredictable. In the case of people executing tasks, how they perform will be highly dependent on what each individual already knows that is relevant to the specific task at hand, the amount of experience they have had with similar tasks, how recent these experiences were, and multiple other factors. Research over many decades has established the "case dependence" of health professionals as problem-solvers and task completers (Elstein et al. 1978).

Information resources are programmed to perform in response to a range of data inputs within a specified domain, with each set of inputs representing a task for the resource to complete. How well the resource will perform over a range of inputs depends on complex issues of resource design and construction. Inspection of the architecture, algorithms, and code that are incorporated into the system will rarely lead to clear conclusions about how well a resource will perform on any given task and how performance will vary across a range of tasks.

When the focus is on how well people will perform across tasks in interaction with information resources, the picture becomes even more complex. At each stage of an interaction, people will respond idiosyncratically to whatever information the resource generates in response to information they enter. These factors lead inexorably to a conclusion that, for informatics studies, a large number of factors contribute to variable performance across tasks, and that high task-to-task variability in performance should be expected, which in turn means that large numbers of tasks will be required for reliable measurement.

Although it is often challenging to determine what makes tasks similar to each other, it is reasonable to conclude that people and/or information resources will perform in similar ways on similar tasks. The greater the similarity between tasks, the greater the similarity in performance. Drawing on concepts of classical measurement theory and viewing tasks as observations as shown in Table 9.2, the greater the similarity of the tasks which are the independent observations, the smaller the number of tasks that will be required for reliable measurement.

Any study where tasks are the observation of interest must begin with identification of a *task domain* for purposes of that study. The task domain defines and limits the set of tasks that will be included in the measurement process of interest; for example, a usability study of a weight reduction app might be limited to just those app functions designated by the developers as "basic". The choice of task domain for any study is made by the study team. As illustrated in Fig. 9.2, the task domain selected by a study team must lie within the boundaries of the full-range of tasks that people could reasonably be expected to do, or that information resources were designed to do. *The tasks included in a study will then be sampled from that task domain, and the results of the study conducted with that sample of tasks will then apply only to that domain.* This creates a fundamental trade-off. If the selected domain is small, the tasks sampled from that domain for use in a study will be more similar to each other and, as a result, fewer tasks will be required for reliable

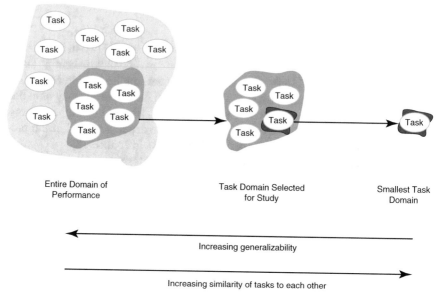

Fig. 9.2 The trade-offs in choosing a Task Domain

measurement. However, a choice of small domain will restrict the conclusions that can be drawn from the study. In the extreme limit where the chosen domain is so small that it includes only one task, the conclusions apply only to that single task. This dilemma for study teams may be rephrased as: "Do you want to know more about less, or less about more?"

As an example, consider a study to compare the ability of different information resources to answer patients' questions about their own health problems, spoken into the system in the patients' own words. In this case, the posed questions are the tasks presented to the resources which are the objects of measurement. The accuracy of each answer is the attribute of interest. As a first step in designing the study, the team must establish the task domain, which in this case would be a defined range of health topics. With reference to Fig. 9.2, this domain could be as large as the entirety of health care. Clearly this domain is too large and poorly defined to enable any meaningful study, so the team must then choose a smaller task domain, as shown in the middle of Fig. 9.2. This may, as one possibility, be questions related to the five most common chronic diseases as framed by patients with these diseases. Having chosen a "middle sized" task domain, the team could then conduct a measurement study to determine the number of questions required for adequately reliable measurement, following which the team could move on to the demonstration study. The results of the study would generalize only to questions falling within that chosen task domain.

It is also important, following a choice of task domain, to sample tasks in some systematic way from that domain. This is because it is usually not clear, simply by inspecting a task, how challenging it will be for people or information resources to complete it. Section 9.3.3 describes some specific approaches to task sampling and, more generally, how to approach measurement when tasks are the primary observations of interest.

9.3.2   Number of Tasks Needed for Reliable Measurement

When Tasks are Completed by People: Much research on problem solving in biomedical and health fields has clarified the number of tasks necessary for reliable measurement of human performance in those domains. The best research has been performed using standardized patient-actors (Roberts et al. 2006). To reach a reliability exceeding 0.70 in a test spanning a very broad domain of clinical practice, 30 cases (tasks) must be assigned to each person (object) (Newble and Swanson 1988). This requirement creates a challenge for study teams to include relatively large numbers of time-consuming tasks in their studies, but there are some ways, addressed in the following section, to place less burden on each human participant in the study. When presentations of diseases from a single organ system or scientific problems applying similar principles are tasks employed, relatively small numbers of these tasks may be required for reliable measurement. This strategy represents the "small domain" resolution of the trade-off described above. By including tasks relating only to management of diabetes mellitus in a clinical task set, a relatively small number of cases will be needed to achieve a target level of reliability, but we would only learn from a measurement study how well the measurement process works for diabetes cases. Generalization to other disease domains would be speculative.

When Tasks are Completed by Machines: For studies of information resource performance (where resources themselves are the objects of measurement), a mature line of research specifically related to the number of tasks necessary for reliable measurement has not developed to the same extent as it has for studies where tasks are completed by people (Hajian-Tilaki 2014). Thus it is difficult to give estimates, analogous to those cited above, for required numbers of tasks. Swets (1988), for example, pointed to the need for "large and representative samples" but did not quantify "large." Lacking guidelines for the number of tasks to use for studying information resource performance in biomedical domains, the most sensible approach is to conduct measurement studies, and estimate the reliability based on the number of tasks employed in the measurement study. If the results of an appropriately designed measurement study yield reliability that is unacceptable for the purpose of the intended demonstration study, the study team can use the methods described in Sects. 8.3 and 8.4 to improve the measurement process.

9.3.3 Improving Measurements That Use Tasks as Observations

When persons are the objects of measurement—when the tasks are completed by people--it is important to challenge these persons with a set of tasks that is large enough for adequate measurement, but no larger than necessary. The larger the task set, the greater the risk of fatigue, noncompliance, or half-hearted effort; data loss through failure to complete the task set; or study expense if the individuals are compensated for their work. The task-to-task variability in performance cannot be circumvented, but many other steps can be taken to ensure that every task included in a set is adding useful information to a study. The approaches to improve measurement in this domain are multiple: (1) consistent representation of tasks, (2) sampling systematically from a chosen task domain, (3) attention to how performance is scored, and (4) systematic assignment of tasks to study participants.

Consistency: Many informatics studies are conducted in laboratory settings where study participants work through a series of tasks under controlled conditions. These laboratory studies require representations of tasks that are as consistent as possible from task to task. In some instances, tasks are represented as abstracts of actual clinical cases containing information that is extracted from a medical record and summarized in a concise written document (Friedman et al. 1999). For consistency, the study team members who extract the information from larger records must follow a clearly stated set of guidelines specifying the rules for selecting findings from the full record for inclusion in the summary. These rules should be clearly stated even if the same person is doing all of the abstracting. Before any abstracting is done, these rules should also be carefully reviewed to ensure as much as possible that they are free of any evident bias, such as omission of information that is essential to successful completion of the task.

The need for consistency goes beyond the choice of which case findings to include. These findings must be represented in a consistent fashion, so they will have the same meaning within the set and across all who encounter the task. Using an example from clinical domains, an abstracter may decide to represent a laboratory test result as "normal" or "abnormal" instead of giving the quantitative value of the result. This is an effective strategy to counteract the tendency for persons who see only the quantitative value to come to different interpretations because of different standards for normalcy that exist across institutions. The abstracters must also decide how much interpretation of findings to provide and take steps to ensure that they do this consistently.

When tasks are "authored" by domain experts who are part of a study team, instead of abstracted from cases that occur in the natural world, the same requirement for consistency applies—but use of authored cases raises an additional challenge. While the information contained in abstracted tasks is guaranteed to be realistic, authored cases must be constructed in ways to ensure that they appear realistic. For example, a portrayal of a patient with a particular disease must not contain clinical findings that rarely if ever co-occur in that disease. Similarly, an authored portrayal of a scientific phenomenon must conform to known laws of physics, chemistry, and biology. Study teams should test the realism of authored cases, by asking other experts to review them and/or by testing them with small groups of resource users, before employing the cases in a study.

Sampling: Once a task domain is established, as described in Sect. 9.3.1, study teams must sample from that domain a set of tasks to be employed in their study. This must be done in a way that ensures that the task set is as representative of the domain as possible.

One way to do this is via feature-based sampling. In this approach, the study team first identifies features of a task that are important to be systematically represented in the sampled task set. The study team might decide that a feature should be represented in the task set with the same likelihood that it occurs in the task domain. For example, if the domain is a particular disease that is twice as likely to occur in women than men, the sampled task set will include women and men in that same proportion.

Another approach is based on natural occurrence, for example using consecutive admissions to a hospital, or consecutive calls to the help desk, as the criterion for including tasks in a sampled set. This approach is grounded in what happened in the real world, but the occurrence of specific task features in the sample is not under the study team's control. In studies of clinical decision support for rare but deadly problems or of biosurveillance to detect disease outbreaks, for example, cases that invoke the capabilities of these resources may not appear with sufficient frequency in a naturally occurring sequence. In this situation, tasks with desired features can be purposefully over-sampled, to make them more numerous, with statistical corrections applied later if necessary.

Whichever approach is followed for task selection, it is important for the study team to have a defensible selection plan that follows from the purposes of the study and in turn allows the study team to characterize the task sample. The implications of these strategies for demonstration study design will arise in several places in Chaps. 10–13.

Scoring: All studies using tasks require assignment of one or more scores to the result of task completion. From a measurement perspective, these scores occupy the cells of the objects-by-observations matrix. Some tasks employed in studies generate a result that can be scored manually by formula or computationally via algorithm, with no human judgment required to generate a score after the formula or algorithm is created. This is often the case when the task has a quantifiable performance metric or an unambiguous correct answer. For example, the accuracy of a resource performing protein structure prediction can be computed as the mean displacement of the atoms' predicted locations from their known actual locations as established experimentally. A task in clinical diagnosis may be scored in relation to the location of the correct diagnosis on the hypothesis list provided by clinicians.

In other circumstances, there may be no reference standard or correct answer, as in the earlier example of the resource that answers patients' questions about their own health problems. These tasks do not lend themselves to formulaic scoring, and

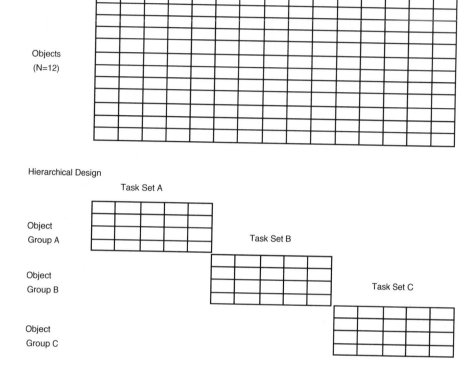

Fig. 9.3 Schemes for assigning tasks (cases) to objects

in these circumstances human judges must be employed to render an opinion or a verdict that becomes the performance score.[1]

Even when tasks can be scored formulaically or algorithmically, the development of scoring methods may not be straightforward and merits care. For example, the apparently simple assignment of a score to a clinician's diagnostic hypothesis list, for cases where the true diagnosis is known, can be undertaken in a variety of ways, each with advantages and disadvantages. When the tasks are scientific problems with correct answers, persons who work these problems may be assigned partial credit for taking the correct approach to a solution even if their provided answer is incorrect. The effects on reliability and validity of these alternate scoring schemes are not obvious and often must be explored through measurement studies.

---

[1] This example generates a more complex measurement problem, that includes both tasks and judges as observations of primary interest. These more complex problems require use of generalizability theory as discussed in Sect. 7.10.

Assignment: Many techniques can be used to assign tasks to objects in measurement studies, but whenever possible this assignment should be done by preordained design rather than chance encounter. As shown in Fig. 9.3, two common assignment modes include the "fully crossed" approach, where every object completes every task in the sample, and a "hierarchical" (or "nested") approach, where specific subsamples of objects are assigned to specific subsamples of tasks. The hierarchical approach is especially helpful when persons are completing the tasks; and the study team wishes to include a large number of tasks in the full study, but cannot burden any single person with the entire set. In informatics, the persons participating in studies are typically busy scientists, trainees, or care providers. Their time is scarce or expensive (or both). Using a hierarchical approach, 15 tasks can be randomly divided into three groups and the participating persons are similarly divided into three groups. Each group of people is then assigned to complete one group of tasks. In this scheme, each person works only five tasks, and the study team is not seriously limited in the conclusions that can be drawn from the study as long as the number of people in the study is reasonably large (and larger than illustrated in Fig. 9.3).

Self-Test 9.3
Refer back to the scenario in Self-test 9.1. Further assume that all of the tasks included in the measurement study pertain to order entry in critical care units.

1. How would you define the task domain for this measurement process?
2. Based on the literature on measurement described above, what advice would you give this team regarding the *number* of order sets to include in the measurement process?
3. Based on the literature on measurement described above, what advice would you give this team regarding the *selection* of the order sets to include in this measurement process?
4. If the task domain were narrowed to include entry of laboratory orders only, and five tasks were still used to sample that domain, what effect on the reliability of measurement would you expect?

9.4   When Items Are the Primary Observation

Items are the individual elements of an instrument to measure people's attributes such as opinions, attitudes, or perceptions of a situation. Survey questionnaires are a very common example of measurement where items are the key observations. Each item elicits one response and can be viewed as an independent observation. The instruments containing the items can be self-administered on paper or via computer, or can be read to a respondent in a highly structured interview format. Table 9.3 displays the objects-by-observations matrix corresponding to measurement when items are the observation of primary interest.

Table 9.3 Objects-by-observations matrix with items on a form as the observation of primary interest

Objects (persons) completing the form	Items					Score for each person averaged across items
	1	2	3	4	5	
A	4	5	3	5	5	4.4
B	3	5	5	3	4	4.0
C	4	4	4	4	5	4.2
D	3	2	3	2	3	2.6

For the same reason that a single judge or task cannot suffice for reliable assessment of performance, a single item on a form cannot serve to measure reliably a respondent's beliefs or test their degree of knowledge. However, the measurement strategy to obtain reliable measurement is exactly the same: use multiple independent observations (in this case, items) and average the results for each object (in this case, respondents) to obtain the best estimate of the value of the attribute for that object. If the items forming a set are shown to be "well behaved" in an appropriate measurement study, we can say that they comprise a scale.

A vast array of item types and formats is in common use (Streiner et al. 2015). In settings where items are used to elicit beliefs or attitudes, there is usually no correct answer to the items; however, in tests of knowledge, there exists a particular response that is identified as correct. In general, items consist of two parts. The first part is a stem, which elicits a response; the second provides a structured format for the individual completing the instrument to respond to the stem. Responses can be elicited using graphical or visual analog scales,[2] as shown in Fig. 9.4. Alternatively, responses can be elicited via a discrete set of options, as shown in the example in Table 9.4. The semantics of the response options themselves may be unipolar or bipolar. Unipolar response options are anchored at the extremes by "none" and "as much as is possible;" for example, "I never do this" vs. "I always do this." Bipolar response options are anchored at the extremes by semantic opposites, for example, "good" vs. "bad," "strongly agree" to "strongly disagree." In Table 9.4, the response options form a bipolar axis. In practice, questionnaire designers choose item formats largely based on the goodness of fit to the attribute being assessed, and the mechanics of the measurement process.

In the example shown in Table 9.4, each of the items addresses the perceived future effect of health information technology on a particular aspect of health and health care; but equally important, a person's response to the items *as a set* can be interpreted to reflect an overarching sense of impact of information technology on workflow. In this sense, each item on the form can be seen as a single observation of

---

[2] Note here the two ways in which the term scale can be used: (1) to mean a set of items addressing the same attribute, and (2) the structure of the response options for each item. It is usually possible from context to discern which meaning of "scale" is being used in a statement.

This patient was managed:

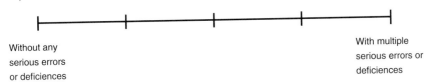

Without any
serious errors
or deficiences

With multiple
serious errors or
deficiences

Fig. 9.4 Rating item with a graphical response scale

Table 9.4 An eight-item scale assessing perceived effects of health information technology (HIT) on nurses' workflow (Dykes et al. 2007). The stem asked respondents to indicate their extent of agreement or disagreement with the following statements. Respondents were given six discrete response options along the bipolar axis of "strongly agree" to "strongly disagree". Reliability (α) of the scale was .89

1. The ways in which data/information are displayed using HIT improves access to data.
2. HIT depersonalizes care.
3. The HIT applications available at my site help me to process data and therefore improve access to information necessary to provide safe patient care.
4. The availability of electronic interdisciplinary documentation has improved the capacity of clinicians to work together.
5. HIT applications/tools support the nursing process.
6. The ways in which data/information are displayed using HIT reduces redundancy of care.
7. The ways in which data/information are displayed using HIT facilitates interdisciplinary care planning.
8. HIT applications/tools facilitate interdisciplinary treatment planning.

the attribute "HIT workflow impact". The items as a set will then comprise an "HIT workflow impact" scale. An individual who responds favorably to one item should have a tendency to respond favorably to the other items in the set, and a person's assessment of workflow impact is best estimated using the sum (or average) of the responses to the set of items comprising the scale.[3]

When a form comprised of multiple items is under development, the degree to which the items form a scale is a hypothesis rather than a certainty. Testing the hypothesis requires a measurement study that employs methods described in Sects. 8.3 and 8.4 to analyze the objects-by-observations matrix resulting from completion of the form by a representative sample of respondents. Items that are not well-behaved can then be revised or deleted.

Demonstration that the items form a reliable scale does not demonstrate that the scale is valid: that it actually assesses the intended attribute of HIT workflow impact. Additional studies of the validity of the scale, as discussed in Sect. 7.8, are required to make that determination. Complete and convincing validity studies require the collection of additional data.

---

[3] Summing and averaging items yield equivalent results as long as the respondent completes all the items composing the scale.

9.4.1   Sources of Variation Among Items

Forms measuring beliefs and knowledge through use of multiple items will inevitably display varying responses from item to item, for the apparent reason that the items themselves differ. It clearly makes no sense to build a form that contains identical items, and thus ask persons to respond repeatedly to the same question. By inspection of Table 9.4, no one would expect any thoughtful person to respond identically to all eight items that are included, but if the items form a proper scale the responses should reveal substantial part-whole correlations as described in Sect. 8.4.

The goal in constructing a form containing multiple items is to elicit the closest approximation to a true representation of the person's actual belief regarding what is being asked. In other words, if a person responds with "strongly agree" to the "HIT depersonalizes care" item in Table 9.4, does that person *really* believe that information technology will be highly detrimental to health care in the future? To the extent that responses are not eliciting actual beliefs, scores on the workflow scale, summed over all the items comprising the scale, will be less reliable and valid. Responses that are not reflective of actual beliefs can result from stems that are ambiguously phrased, items that have mismatched stems and response options, items with response options that do not accurately capture the full range of beliefs the respondents hold, and other factors. Specific ways to address these problems are described in Sect. 9.4.3 below.

Because those completing a form should exhibit beliefs that differ from item to item, results of a measurement study that reveal many respondents giving nearly identical responses to all items on the form is actually a source of concern. This is often due to the "halo effect", whereby respondents, in the process of completing a series of items addressing the same overarching attribute, may quickly develop an overall impression causing them to respond in an automatic way to each item in the set, rather than giving independent thought to each one. In the example in Table 9.4, respondents who recently had a bad experience with technology and then answer with "strongly disagree" for all eight items, will likely generate a very different score than would be the case if they thoughtfully considered each item. Halo effects can result in artificially inflated reliability, since all items are answered as if they were exactly the same, but this is accompanied by highly dubious validity. Ways to reduce halo effects are also discussed below.

9.4.2   Number of Items Needed for Reliable Measurement

This depends on the attribute being measured and the purposes of the measurement, and there is no *a priori* way to know precisely how many items will be necessary to achieve a desired level of reliability. Typically, a minimum of 8–10 items is needed to adequately measure a general belief or an attitude; however, if the targeted attribute is more concrete, high reliability with smaller numbers of items is possible.

Table 9.5 The reliability and number of items for selected published scales

Authors	Attribute	N of items	Reliability (α)
Yen et al. (2014)	Usefulness of Health IT	9	.94
Zhao et al. (2019)	mHealth App Ease of Use	5	.85
Roque et al. (2018)	Mobile Device Proficiency	16	.94
Dykes et al. (2007)	Health IT: Perceived Impact on Workflow	8	.89

Table 9.5 illustrates this point, listing the results of several measurement studies initially described in Sect. 6.7. The table indicates the attribute assessed by the scale, the number of items each scale contains, and the reliability achieved with that number of items. The more global attributes such as "usefulness" and "proficiency" require nine and 16 items respectively to achieve high reliability, whereas the scale focused on ease of use of a specific app achieved high reliability with five items.

In high-stakes standardized tests, where very high reliability is necessary to make decisions about each individual's competence, more than 100 knowledge questions (items) are routinely used within a knowledge domain. In this situation, large numbers of items are required both to attain the high reliability necessary to generate a small standard error of measurement and to sample adequately a broad domain of knowledge. For any particular measurement situation, a measurement study can determine how many items are necessary and which items should be deleted or modified to improve the performance of an item set hypothesized to comprise a scale.

In practice, a single form contains items that measure several distinct but related attributes. In these cases the term "sub-scale" is often used to designate the set of items that measures each of the attributes. For example, the scale measuring nurses' attitudes related to workflow, as shown in Table 9.4, is just one of the scales comprising a questionnaire more generally focused on nurses' attitudes toward information technology. In many cases, the items addressing each attribute will be clustered together on the form. In other cases, to combat halo effects, the items comprising each sub-scale are intermingled throughout the entire form.

Another practical consideration for measurement performed with items is the time required to complete a form and the desirability, therefore, to create a shorter but still adequate version of a scale (Webster and Feller 2016; Salsman et al. 2019). The effects of shortening a scale by reducing the number of items can be estimated by the Spearman-Brown formula introduced in Sect. 7.7.2. In the example of the nursing workflow scale (Table 9.4), the full eight item scale displayed a reliability of 0.89. The Spearman-Brown formula suggests that an five-item version of the same scale would still have a reliability of 0.80, which from a reliability perspective would still likely be adequate. However, the "prophecy" formula does not suggest which items to remove and removal of some of items could possibly alter the validity of the scale. When creating a shorter version of a scale, it is desirable to confirm that the validity of the scale has not been substantially altered by running a special kind of measurement study that compares responses to the full-length and shorter-length versions of the scale.

9.4.3 Improving Measurements That Use Items as Observations

This section offers several practical suggestions to maximize the reliability and validity of measurement through attention to item design. This discussion emphasizes elicitations of attitudes and beliefs because these arise frequently in informatics studies that are the focus of this book.

1. *Make items specific.* Perhaps the single most important way to improve items is to make them as specific as possible. The more information the respondents get from the item itself, about what exactly is being asked for and what the response options mean, the greater are the reliability and validity of the results. Consider a rather non-specific phrasing of an item that might be part of a form assessing users' satisfaction of an advice-giving system (Fig. 9.5a). As a first step toward specificity, the item should offer a definition or other elaboration of the stem

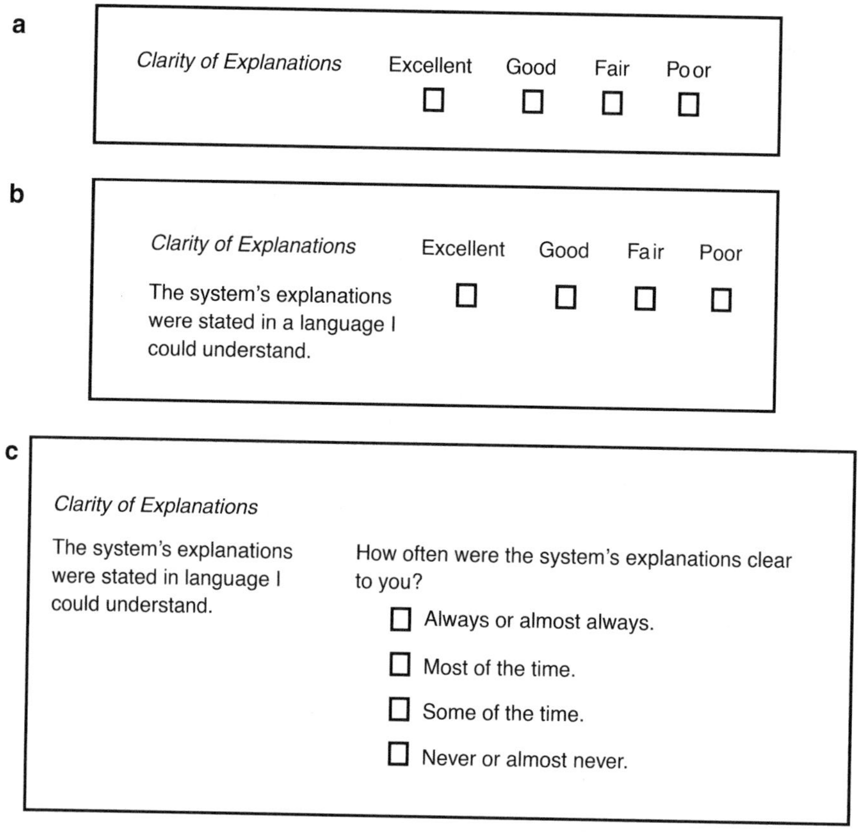

Fig. 9.5 (**a**) Basic rating item. (**b**) One improvement: define the attribute. (**c**) Second improvement: make the response categories correspond to what is directly observable

eliciting the response, as shown in Fig. 9.5b. The next step is to change the response categories from broad qualitative judgments to specific behaviors or events that the respondent might recall. As shown in Fig. 9.5c, this might entail changing the logic of the responses by specifically asking how frequently the explanations were actually clear.

2. *Match the logic of the response to that of the stem.* This step is vitally important. If the stem—the part of the item that elicits a response—requests an estimate of a quantity, the response formats must offer a range of reasonable quantities from which to choose. If the stem requests a strength of belief, the response formats must offer an appropriate way to express the strength of belief, such as the familiar "strongly agree" to "strongly disagree" format.

3. *Provide a range of semantically and logically distinct response options.* Be certain that the categories span the range of possible responses and do not overlap.

Bad Example 1
In your opinion, with what fraction of your clinic patients this month has the resource offered useful advice?

□ 0–25% □ 25–50% □ 50–75% □ 75–100%

Clearly it is necessary to begin the second option with 26%, the third with 51%, and the fourth with 76%.

Bad Example 2
Similarly, when response categories are stated verbally, the terms used should be carefully chosen so the categories are as equally spaced, in a semantic sense, as possible. Consider another mistake commonly made by novice item writers:
How satisfied are you with the new resource, overall?

□ Extremely □ Very □ Mostly □ Not at all

In this example, there is too much semantic space between "mostly" and "not at all." There are three response options that reflect positive views of the resource and only one option that is negative. To rectify this problem, a response option of "slightly" or "modestly" might be added to the existing set.

4. *Include an appropriate number of response options.* Although it may seem tempting to use a large number of response options to create at least the appearance of precise measurement, this may not achieve the intended purpose. In general, the number of response options should be limited to a maximum of seven.[4] For most purposes, four to six discrete options suffice. Using a five-option response format with a bipolar semantic axis allows a neutral response. A potential benefit of a neutral response option is that a respondent whose true belief is neutral has a response option reflective of that belief. In the opposing view, a neutral response option provides a way to respond that is safe and noncommittal, even though it may not be reflective of the respondent's actual belief.

5. *Invite a nonresponse.* Giving respondents permission to decline to respond to each item also contributes to successful measurement. When completing a form lacking non-response options, respondents may offer uninformed opinions because they feel they are expected to complete every item on the form. If an "unable to respond" category is explicitly available, respondents are more likely to omit items on which they do not feel confident or competent. An "unable to respond" category should be in a different typeface or otherwise set visually apart from the continuum of informed responses.

6. *Address halo effects.* There are two major ways to minimize halo effects through item design. The first is to include, within a set of items composing a scale, roughly equal numbers phrased positively and negatively. For example, the scale might include both of the following:

My ability to be productive in my job was enhanced by the new computer system.

☐ Strongly agree   ☐ Agree   ☐ Neither agree nor disagree
☐ Disagree   ☐ Strongly disagree

The new system slowed the rate at which I could complete routine tasks.

☐ Strongly agree   ☐ Agree   ☐ Neither agree nor disagree
☐ Disagree   ☐ Strongly disagree

In this example, the co-presence of some items that will be endorsed, along with others that will not be endorsed if the respondent feels positively about the system, forces the respondent to attend more closely to the content of the items themselves. This strategy increases the chance that the respondent will evaluate

---

[4] It is generally accepted that humans can process about seven (plus or minus two) items of disparate information at any one time (Miller 1956). The practical upper limit of seven response options may be attributable to this feature of human cognition.

each item on its own terms, rather than responding based on an overall impression. When analyzing the responses to such item sets, the negatively phrased items should be reverse coded before each respondent's results are averaged or summed to create a scale score.

7. *Request elaboration.* In some cases, asking respondents specifically for verbal elaborations or justifications of their responses can serve multiple purposes. It can stimulate respondents to be more thoughtful in their responses. Respondents may check off a specific option and then, when trying to elaborate on it, realize that their deeper beliefs differ from what an initial impression suggested. Elaborations are also a source of valuable data, particularly helpful when the items are part of a form that is in the early stages of development. Elaboration can also be informative as a source of data in a demonstration study. If the purpose of a study is to understand "why," in addition to "how much," these verbal elaborations may even be essential and will generate data that are particularly valuable in mixed methods studies discussed in later chapters.

9.4.4   The Ratings Paradox

When multi-item forms are employed to generate ratings, a major challenge is to identify the right level of specificity or granularity of the items themselves. The greater the specificity of the items, the less idiosyncratic judgment the raters exercise when offering their opinions, and this will usually generate higher reliability of measurement. However, rating forms that are highly specific can become almost mechanical. In the extreme, raters are merely observing the occurrence of atomic events—for example, "The end user entered a search term that was spelled correctly"—and their judgement and/or expertise are not being engaged at all.

As attributes rated by individual items become less specific and more global, agreement among raters is more difficult to achieve; as they become more atomic, the process becomes mechanical and possibly trivial. This can also be viewed as a trade-off between reliability and validity. The more global the ratings, the more "content valid" they are likely to be, in the sense that the world believes that the attributes being rated are important and indicative of what should be measured in a subsequent demonstration study. This may, however, come at a price of low (possibly unacceptably low) interrater agreement and thus low reliability. Study teams should seek to find a comfortable middle ground between these extremes.

Self-Test 9.4

1. Refer to Table 9.4 and the eight-item scale used to assess perceptions of health information technology "workflow impact".

   (a) If the scale were translated into Spanish and a measurement study with 100 Argentine nurses was carried out, how many rows and columns would the resulting objects-by-observations matrix have?

(b) If one nurse completing the form responded with a "4" to half of the eight items, and with a "5" to the other half, what would be the value of that nurse's "workflow impact" score?

(c) Inspect the scale from a content validity perspective. Which item seems, from that perspective, not to "belong" with the other items in this scale?

2. Using the guidelines offered in the previous section, find and fix the problems with each of the following items.

Item 1
Accuracy of system's advice

     □ Excellent   □ Good   □ Fair   □ Poor

Item 2
Indicate on a 1–10 scale your satisfaction with this system.

Item 3
The new system is easier to use than the one it replaced.

     □ Strongly agree   □ Agree   □ No opinion
     □ Disagree   □ Strongly disagree

Item 4
How often have you used the new laboratory system?

     □ Most of the time   □ Some of the time   □ Never

9.5 Conclusion

This chapter completes the discussion of quantitative measurement. This chapter and its two predecessors have sought to demonstrate the importance of measurement to sound quantitative study. This attention is motivated by the general lack of measurement instruments in biomedical informatics, resulting in the need for many study teams to develop and validate their own measurement approaches. This discussion closes almost where it began, by encouraging study teams who develop measurement methods to publish their measurement studies, so other study teams may benefit from their labors. The following chapters explore quantitative demonstration studies. Everything that follows in Chaps. 10–13 assumes that measurement issues have been resolved, in the sense that the reliability and validity of all measurement processes employed in a demonstration study have been established by the study team or were previously known from the work of other teams.

Answers to Self-Tests

Self-Test 9.1

1. Order entry is a task, so the observations of primary interest are tasks. Each order set to be entered is a separate task. In this situation, the objects of measurement are clinicians.
2. The reliability is .51. Insufficient for any measurement purpose.
3. Clinicians 1's speed score is the average time across the five order sets: 118 s.

Self-Test 9.2

1. Each (of two) objects-by-observations matrixes would have 111 rows (for cases as objects) and three columns (for judges as observations). One matrix would be generated for actual care cases and the other for TraumAID's recommendations.
2. There is no compelling evidence for rater tendency errors. The mean ratings of the judges are roughly equal and near the middle of the scale. Central tendency effects can be ruled out because the standard deviations of the ratings are substantial.
3. From a reliability standpoint, the ratings are more than adequate. However, the validity of the ratings must be questioned because the judges are from the institution where TraumAID was developed.
4. The data seem to suggest that TraumAID's advice is accurate, as the judges preferred how TraumAID would have treated the patients over how the patients were actually treated. However, the concern about validity of the ratings would cast some doubt on this conclusion.

Self-Test 9.3

1. The task domain is order entry in critical care.
2. Based on the research on measurement processes using tasks as observations, this low level of reliability is not surprising.
3. The major issue here is the lack of a systematic process to sample the tasks from the domain. One of the clinicians just "made them up" and this is not a recipe for success. The tasks should be created in some systematic way, so the sample set used in the study generalizes to some population of tasks. Any individual clinician's "speed" score should be based on a systematically selected set of tasks, which would leave the result less vulnerable to an accusation that a clinician's speed score is merely a function of the particular tasks there were included.
4. An increase of reliability would be expected since the five cases sampled from a smaller domain would be more similar to each other.

Self-Test 9.4

1. (a) There are eight items (observations) and 100 objects (nurses). The matrix would have 8 columns and 100 rows.
   (b) The object score is the average of the eight items, or in this case, 4.5.
   (c) The answer to this question is a matter of personal judgment, but a case can be made that Item 2 ("depersonalizes care") does not belong with others. All

other items refer to aspects of care process, whereas Item 2 addresses a global attitude that may invoke many other factors besides workflow.

2. *Item 1:* Accuracy should be defined. The response categories should be replaced by alternatives that are more behavioral or observable.
   Item 2: Ten response options are too many. The respondent needs to know whether 1 or 10 corresponds to a high level of satisfaction. The numerical response options have no verbal descriptors.
   Item 3: "No opinion" does not belong on the response continuum. Having no opinion is different from having an opinion that happens to be midway between "strongly agree" and "strongly disagree."
   Item 4: The logic of the response options does not match the stem. There are not enough response options, and they are not well spaced semantically.

References

Clarke JR, Webber BL, Gertner A, Rymon KJ. On-line decision support for emergency trauma management. Proc Symp Comput Applications Med Care. 1994;18:1028.

Dykes PC, Hurley A, Cashen M, Bakken S, Duffy ME. Development and psychometric evaluation of the impact of health information technology (I-HIT) scale. J AM Med Inform Assoc. 2007;14:507–14.

Elstein AS, Shulman LS, Sprafka SA. Medical problem solving. Cambridge, MA: Harvard University Press; 1978.

Fowell SL, Fewtrell R, McLaughlin PJ. Estimating the minimum number of judges required for test-centred standard setting on written assessments. Do discussion and iteration have an influence? Adv Health Sci Educ Theory Pract. 2008;13:11–24.

Friedman CP, Elstein AS, Wolf FM, Murphy GC, Franz TM, Heckerling PS, et al. Enhancement of clinicians' diagnostic reasoning by computer-based consultation: a multisite study of 2 systems. JAMA. 1999;282:1851–6.

Guilford JP. Psychometric methods. New York: McGraw-Hill; 1954.

Gwet KL. Handbook of inter-rater reliability: the definitive guide to measuring the extent of agreement among raters. 4th ed. Advanced Analytics, LLC: Gaithersberg, MD; 2014.

Hajian-Tilaki K. Sample size estimation in diagnostic test studies of biomedical informatics. J Biomed Inform. 2014;48:193–204.

Hallgren KA. Computing inter-rater reliability for observational data: an overview and tutorial. Tutor Quant Methods Pychol. 2012;8:23.

Hripcsak G, Heitjan DF. Measuring agreement in medical informatics reliability studies. J Biomed Inform. 2002;35:99–110.

Hripcsak G, Wilcox A. Reference standards, judges, and comparison subjects: roles for experts in evaluating system performance. J Am Med Inf Assoc. 2002;9:1–15.

Kern LM, Malhotra S, Barro'n Y, Quaresimo J, Dhopeshwarkar R, Pichardo M, et al. Accuracy of electronically reported "Meaningful Use" clinical quality measures. Ann Intern Med. 2013;158:77–83.

Miller GA. The magical number seven, plus or minus two: some limits on our capacity for processing information. Psychol Rev. 1956;63:46–9.

Newble DI, Swanson DB. Psychometric characteristics of the objective structured clinical examination. Med Educ. 1988;22:325–34.

Roberts C, Newble D, Jolly B, Reed M, Hampton K. Assuring the quality of high-stakes undergraduate assessments of clinical competence. Med Teach. 2006;28:535–43.

Roque NA, Boot WR. A new tool for assessing mobile device proficiency in older adults: the mobile device proficiency questionnaire. J Appl Gerontol. 2018;37:131–56.

Salsman JM, Schalet BD, Merluzzi TV, Park CL, Hahn EA, Snyder MA, Cella D. Calibration and initial validation of a general self-efficacy item bank and short form for the NIH PROMIS. Qual Life Res. 2019;28:2513–23.

Streiner DL, Norman GR, Cairney J. Health measurement scales: a practical guide to their development and use. 5th ed. Oxford: Oxford University Press; 2015.

Swets JA. Measuring the accuracy of diagnostic systems. Science. 1988;240:1285–93.

Webster KE, Feller JA. Comparison of the short form-12 (SF-12) health status questionnaire with the SF-36 in patients with knee osteoarthritis who have replacement surgery. Knee Surg Sports Traumatol Arthrosc. 2016;24:2620–6.

Yen PY, Sousa KH, Bakken S. Examining construct and predictive validity of the Health-IT Usability Evaluation Scale: confirmatory factor analysis and structural equation modeling results. J Am Med Inform Assoc. 2014;21:241–8.

Zhou L, Bao J, Setiawan IM, Saptono A, Parmanto B. The mHealth APP usability questionnaire (MAUQ): development and validation study. JMIR. 2019;7:e11500.

Chapter 10
Conducting Demonstration Studies

Learning Objectives

The text, examples, self-tests, and Food for Thought question in this chapter will enable the reader to:

1. Given a short description of a demonstration study, describe whether its design is primarily descriptive, interventional, or correlational.
2. For a given demonstration study design, identify the participants, the intervention (if any), the independent and dependent variables, the level of measurement of each variable, and for categorical variables (nominal, ordinal) the number of levels of the variable that can exist in the study.
3. Distinguish between issues of study design and interpretation that relate to internal validity from those that relate to external validity.
4. For each of the following types of biases that affect descriptive studies as well as other study types (assessment biases, Hawthorne Effect, data collection biases):

   (a) Explain why each type of bias can threaten the internal or external validity of descriptive studies;
   (b) Explain how, in descriptive study designs, each type of bias might call into question the credibility of the study and what might be done to improve the study's credibility.

5. Given a study description, identify which of the common threats to external validity of all kinds of demonstration studies (the volunteer effect; age, ethnic group and other biases; study setting bias; training set bias and overfitting) are present in the study.
6. Given examples of data generated by a descriptive study, describe how to analyze the data in the following cases:

   (a) Count data
   (b) Continuous data that is normally distributed
   (c) Continuous data that is not normally distributed

© Springer Nature Switzerland AG 2022
C. P. Friedman et al., *Evaluation Methods in Biomedical and Health Informatics*, Health Informatics, https://doi.org/10.1007/978-3-030-86453-8_10

10.1 Introduction

This is the first of four chapters describing how study teams can carry out and analyze the results of the three types of demonstration studies: descriptive, interventional and correlational studies. This chapter discusses some challenges to what are
called internal and external validity, common to all three kinds of studies, then
focuses on how to design and analyze the results of studies that are primarily
descriptive. However, it is not only study teams who need to understand study
design, validity, and how to maximize it: readers of articles describing demonstration studies also need to understand internal and external validity, so that they can
carry out critical appraisal of the article and decide whether they can rely on its
findings (see Chap. 5). So, an understanding of the threats to internal validity and
external validity in demonstration studies and how to address them can be useful
even if for someone who never plans to design a study.

Demonstration studies attempt to answer questions about an information
resource, exploring such issues as the resource's value to a certain professional
group or its impact on the structure, processes, or outcomes of health care, research,
or education (Donabedian 1966). Recall from Chap. 6 that measurement studies are
required to test, refine, and validate measurement processes before they can be used
to answer questions about a resource or its impact. Chapters 7–9 expanded these
ideas and explained how to conduct measurement studies in more detail. The material in this chapter assumes that measurement methods are available to the study
team which have been verified as reliable and valid—in the measurement sense—by
appropriate measurement studies.[1] To attempt to answer questions using a demonstration study, an appropriate study design must be formulated, the sample of participants and tasks defined, any threats to validity—in the demonstration
sense—identified and either eliminated or controlled for as much as possible, and
the resulting data analyzed. These issues are discussed in this chapter and the
next two.

Study teams carry out demonstration studies primarily to help others make decisions about the development, adoption, or implementation of the biomedical or
health information resources under study. For other people to trust the results of a
demonstration study to inform their decisions or actions, it is clearly important that
these results are worthy of their trust. Unfortunately, there are a wide range of challenges that can make demonstration studies less trustworthy. Some of these challenges can lead to results that, for a variety of reasons, are not credible. This is
called poor "internal validity". Other challenges lead to a situation where, even if
the results are correct, they cannot be confidently applied beyond the one setting
where the study was conducted. This is called poor "external validity". Unless these

---

[1] As will be seen later in this chapter, "validity" takes on very different meanings when discussed
in the contexts of measurement and demonstration. All the authors can do is apologize for this
confusion—we did not invent this terminology!

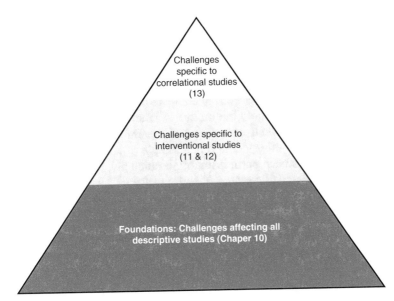

Fig. 10.1 Pyramid of challenges affecting the three types of demonstration studies

challenges are addressed, they will limit the trust that stakeholders have in the study results, and the relevance of the results to stakeholder decisions.

This chapter will first review the terminology used to describe all demonstration studies, review the three kinds of demonstration studies, and then describe challenges that reduce the usefulness of the simplest kind of demonstration study—descriptive studies—and methods to overcome these problems. As shown in Fig. 10.1, many of the challenges that affect descriptive studies, and the methods used to overcome them, also affect the other kinds of demonstration studies described in Chaps. 11 and 12.

10.2 Overview of Demonstration Studies

10.2.1 Types of Demonstration Studies

Recall from Chap. 6 that there are three kinds of demonstration study: descriptive studies, interventional studies, and correlational studies.

A *descriptive study* design seeks only to estimate the value of a variable or set of variables in a selected sample of participants. For example, to ascertain the usability of a nursing information resource, a group of nurses could use a rating form previously validated through a measurement study. The mean value of the "usability" variable, and its standard deviation, would be the main results of this demonstration study. Descriptive studies can be tied to the objectives-based approach to evaluation described in Sect. 2.7.1. When a study team seeks to determine if a resource has met a predetermined set of performance objectives, the logic and design of the resulting study are often descriptive.

In an *interventional* study, the study team typically creates or exploits a pre-existing contrasting set of conditions to compare the effects on a variable of one condition with another. Usually the motive is to attribute cause and effect, for example to evaluate to what extent use of the information resource *causes* improved performance. After identifying a sample of participants for the study, the study team assigns each participant, often randomly, to one or more conditions. This creates contrasting groups of participants, with each group experiencing the same condition. The conditions might perhaps be "no use of information resource" and "use of information resource." One or more variables of interest are then measured for each participant. The aggregated values of this variable, often the mean values, are then compared across the conditions. Interventional studies are aligned with the comparison-based approach to evaluation introduced in Sect. 2.7.1.

Correlational studies explore the relationships or association between a set of variables the study team measures, or inherits as already measured and stored in a data repository, but does not manipulate in any way. Correlational studies can be seen as primarily linked to the decision facilitation approach to evaluation discussed in Sect. 2.7.1, as they are linked to pragmatic questions and only in exceptional circumstances can they definitively settle issues of cause and effect. For example, a correlational study may reveal a statistical relationship between users' clinical performance at some task and the extent they use an information resource to accomplish those tasks, but it cannot definitively tell us which is the direction of causation, nor whether the association is due to a third, unidentified and thus unmeasured, factor. "Data mining" of previously collected data is also a form of correlational study.

10.2.2 Terminology for Demonstration Studies

This section introduces a terminology and several ideas necessary to understand and design demonstration studies of information resources. At this point, it may be useful to refer to Sect. 6.8, where terminology for measurement and demonstration studies was first introduced. For the purposes of the immediate discussion, the terms of most importance are as follows:

Participants: Participants in a study are the entities about which data are collected. A specific demonstration study employs one sample of participants, although this sample might be subdivided if, for example, participants are assigned to conditions in an interventional design. Although participants in demonstration studies are often people—resource users or individuals receiving health care—"participants" also may be information resources, groups of people, or even organizations.[2]

Variables: Variables are the specific characteristics of participants that are purposefully measured by the study team as a component of conducting the study, or that have been previously measured and are available for re-use. In the simplest descriptive study, there may be only one variable. In interventional and correlational studies, by definition there must be at least two variables, and there often are many more. "Big data" studies may include hundreds or even thousands of variables.

Levels of measurement: Each variable or data element in a study is associated with a level of measurement—nominal, ordinal, interval, and ratio—as described in Sect. 6.4. The level of measurement of each variable carries strong implications for how the study results are analyzed. Nominal and ordinal variables are often characterized together as "discrete" variables; interval and ratio variables are often characterized together as "continuous" variables.

Dependent variables: The dependent variables capture the outcomes of interest to the study team. For this reason, dependent variables are sometimes called "outcome variables" or just "outcomes." A study may have one or more dependent variables. Studies with one dependent variable are referred to as univariate, and studies with multiple dependent variables are referred to as multivariate.

Independent variables: The independent variables are those included in a study to explain the measured values of the dependent variables. A descriptive study has no independent variables, whereas interventional and correlational studies can have one or many independent variables.

Measurement challenges of the types discussed in Chaps. 7 through 9 almost always arise during assessment of the outcome or dependent variable for a study. Often, for example, the dependent variable is some type of performance measure such as "the quality of a medication plan" or the "precision of information retrieval" that invokes all of the concerns about reliability and validity of measurement described in those chapters. Depending on the study, the independent variables may also raise measurement challenges. When the independent variable is age, for example, the measurement process is relatively straightforward. However, if the independent variable is an attitude, level of experience, or extent of resource use, significant measurement challenges can arise.

By definition, the participants in a study are the entities on which the dependent variables are measured. This point is important in informatics because almost all professional practice—including health care, research, and education—is conducted in hierarchical settings with naturally occurring groups (e.g. a care provider's patients, care providers in a ward team, students in a class). This often raises

---

[2]Confusingly, participants in a demonstration study are sometimes called "subjects", and can be the equivalent of objects in a measurement study!

challenging questions about the level of aggregation at which to measure the dependent variables for a demonstration study. A branch of statistical analysis called multi-level modelling can be applied in these cases (Das-Munshi et al. 2010).

10.2.3 Demonstration Study Types Further Distinguished

This terminology makes it possible to sharpen the differences between descriptive, interventional, and correlational studies—as originally discussed in Sect. 6.9. Studies of all three types are, in a profound sense, designed by the study team. In all three, the study team chooses the participants, the variables, the measurement methods, and the logistics used to assign a value of each variable to each participant.

Table 10.1 summarizes how these demonstration study types differ in their goals, the kinds of variables used, and the timing of data collection.

Descriptive studies: In a descriptive study, however, there are no further decisions to be made. The defining characteristic of a descriptive study is the absence of independent variables. The state of a set of participants is described by measuring one or more dependent variables. Although a descriptive study may report the

Table 10.1 Summary of the main features of descriptive, interventional, and correlational studies

Study type	Study goal	Independent variables	Dependent variables	Timing of data collection	Analytical strategy
Descriptive study	To describe some aspects of a resource and/ or users as quantified variables	None	One or more	Prospective or retrospective	Descriptive statistics
Interventional study	To relate the study variables, often with primary intent of establishing a causal relationship	One or more, with at least one created by study team to describe participant allocation to groups	One or more as selected by the study team	Prospective	Estimation of magnitude of effects attributable to independent variables
Correlational study	To explore the relationships between study variables	One or more as selected by study team from an existing set	One or more as selected by the study team	Prospective or retrospective	Estimation of magnitude of the relationships among the variables

relationships among the dependent variables, there is no attempt to attribute variation in these variables to other factors. As discussed in Sect. 6.9.1, descriptive studies can be prospective or retrospective.

Interventional studies: The defining characteristic of an interventional study is the purposeful manipulation of independent variables to enable the study team to reliably infer cause and effect, for example, to estimate the effects uniquely attributable to an information resource. To this end, the study team creates at least one new independent variable, creating conditions which define the study groups. The value of this variable for each participant describes that participant's group membership; these values are often assigned descriptors, such as "intervention," "control," or "placebo." The study team assigns participants systematically or randomly to different study groups. If all sources of variation in the dependent variable, known and unknown, are controlled by random allocation, cause-and-effect relations between the independent and dependent variables can be rigorously attributed.

Correlational studies: In correlational studies, the study team hypothesizes a set of relationships among variables that are measured for a sample of participants in the study, but there is no intervention. The values of and variability in these measures are that which occurs naturally in the sample. The study team is at liberty to decide which variables will be the dependent variables and which should form the set of independent variables. Correlational studies can be retrospective, involving analyses of archival data, or prospective, involving data that will be accumulated and assembled in the future. Some statisticians believe that, in some cases, credible assertions of cause and effect can be derived from the pattern of statistical relations observed in correlational studies. This topic remains controversial and is discussed in Chap. 13 (Collins et al. 2020).

Self-Test 10.1

1. Classify each of the following demonstration studies as either descriptive, interventional, or correlational. Explain your answer briefly.

   (a) An enterprise information technology team maintains a dashboard of the number of requests to fix network outages during the current week.
   (b) The "h-indexes" (research productivity indexes) of faculty members at major universities are studied in relation to the student enrollment, endowment, and annual IT expenditure of each university.
   (c) An information technology group tests three telehealth platforms to determine which one patients find easiest to use. Ease of use is measured using a 10-item questionnaire.
   (d) A start-up company places a new consumer-facing app on the market, with a goal of having 10,000 active user accounts after 6 months. After 6 months, there are 12,000 active users.
   (e) In a geo-spatial analysis, rates of new asthma diagnoses are linked to the postal codes of the persons who were diagnosed.

2. For each part of Question 1 above:

   (a) Name the dependent and independent variables,
   (b) Classify each variable as either discrete (nominal, ordinal) or continuous (interval, ratio)
   (c) For categorical variables, indicate the number of possible values of the variable (the level of the variable).

10.2.4 Internal and External Validity of Demonstration Studies

Every study team wants its studies to be valid, and therefore credible.[3] There are two aspects to the validity of demonstration studies: internal and external. If a study is internally valid, decision makers or readers of the study report can be confident in the conclusions drawn from the specific circumstances of the study. These specific circumstances include: the population of participants actually studied, the measurements made, and any interventions provided. However, there are many potential threats to internal validity, as will be discussed in the next section of this chapter and in Chaps. 11–13. Even if all these threats to internal validity are overcome, study teams also want their study to have external validity. This means that the study conclusions can be generalized from the specific setting, participants, and intervention studied to the broader range of settings that others may encounter. Consider a sepsis risk assessment tool that implements the local organization's set of ICU admission criteria with the aim of improving the consistency of ICU admission rates in one health care organization. Even if a study team demonstrates convincingly that the tool reduces ICU admissions in their own health care organization, that may be of little interest to others unless the team can argue convincingly that the study design has features that enable the results to also apply to similar tools used in similar organizations with similar ICU admission criteria. The main threats to external validity are discussed in Sect. 10.5.

Recall from Table 10.1 that descriptive studies only include one class of variables, dependent variables, and there is no attempt to compare one set of dependent variables with another or to attribute cause and effect. A descriptive study simply describes the state of the world, usually at one moment in time. This makes descriptive studies the least complex of the three demonstration study types, and explains why they are considered first in this book. However, as suggested in Fig. 10.1, the problems affecting descriptive studies can, in fact, affect all kinds of demonstration studies, so they are foundational in a very significant sense.

Accordingly, this chapter focuses on the subset of internal validity challenges that descriptive study designers need to address, and that those appraising

---

[3] Note again the difference in the terminologies of measurement and demonstration studies. Validity of a demonstration study design, discussed here, is different from validity of a measurement method, discussed in Chap. 7.

descriptive studies need to know about. Chapters 11 and 12 describes the additional challenges that apply to interventional studies, and Chap. 13 describes additional challenges facing correlational studies. However, the same challenges to external validity apply to descriptive, interventional and correlational studies, so these are described in this chapter as well. So everything the study team needs to know about the internal and external validity of *descriptive studies* is included in this chapter.

10.3   The Process of Conducting Demonstration Studies

Figure 10.2, an elaboration of Fig. 2.3, shows how to design and conduct any kind of demonstration study.

The diagram shows how, after agreeing the study questions with stakeholders (as described in Chaps. 1–4) and finding or developing reliable and valid measurement methods (discussed in Chaps. 7–9), the study team needs to identify which type of demonstration study will be used to answer the study question (this chapter). The team then needs to consider and resolve the relevant biases and challenges to internal validity. While some of these challenges apply to all kinds of demonstration studies—described in this chapter—others only apply to interventional studies (described in Chaps. 11 and 12). A further category of challenges only applies to correlational studies, and especially when using the correlational design to attempt to infer causality or measure the effectiveness of an intervention using routine or "real world" data—see Chap. 13. Once the challenges to internal validity are addressed, the study team must turn its attention to external validity. Most of the major challenges to external validity in any kind of demonstration study are discussed in this chapter, but some additional challenges specific to interventional or

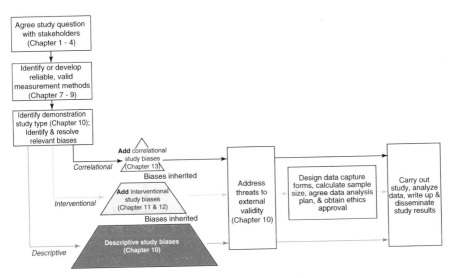

Fig. 10.2 The process of designing and conducting a demonstration study

correlational studies are mentioned in the applicable chapter. Finally, the study team can move on to the practicalities of carrying out the study: designing participant recruitment and data capture forms, developing a data analysis plan, calculating sample size (Chap. 11), discussing how recruitment will be monitored then seeking necessary regulatory approval and setting up a data monitoring committee for trials.

Once the study has completed, the data needs to be analyzed, and again there are differences in approach according to the type of study. For simple descriptive studies, this chapter describes the methods commonly used. Chapter 12 describes the analysis methods used for interventional studies, while the second half of Chap. 13 describes the methods used to analyze correlational studies. Finally, Chap. 20 includes a section on when and how to report the results of a study.

Self-Test 10.2

A usability study has identified a new EHR screen design that, based on the study data, reduces the time required to complete 10 common EHR tasks. Subjects in the study were nurses sampled randomly from a large academic medical center. For each of the following assertions that might be made about the conclusions of the study, indicate whether it primarily addresses internal validity or external validity.

(a) The results of the study can be expected to apply to any EHR system.
(b) The results of the study can be expected to apply to all nurses at the medical center where the study was performed.
(c) The results of the study can be expected to apply to physicians at the medical center where the study was performed.
(d) The improvement is not large enough in magnitude to justify an expenditure of funds to change the screen design.

10.4 Threats to Internal Validity That Affect All Demonstration Studies, and How to Avoid Them

10.4.1 Assessment Biases and the Hawthorne Effect

Assessment Bias: Suppose that judges who believe that an information resource had no value were asked to judge whether ICU admission recommendations generated by the resource were valid or not. This example is likely to lead to assessment bias, which occurs when anyone involved in a demonstration study can allow their own personal beliefs, whether positive or negative, to influence the results. This is a particular challenge if they participated in the development of the information resource, or if the criteria used for judging appropriateness of decisions or outcome are poorly formulated. Simply asking study participants to ignore their feelings about the resource so as to avoid biasing the study results is unrealistic: the study team must *ensure* that any prejudicial beliefs cannot affect the results, consciously or unconsciously. There are two main ways of doing this: obtaining an objective assessment of the outcome from a different source, or blinding the judges to the outcome. Sometimes it is possible to substitute a lab test result for a judge's opinion. An alternative is to ensure that any judges of the outcome are blinded to any factors that could bring their biases into play, such as whether a patient has used an app that is being evaluated. The people who could bias a study in this way include those designing it, those recruiting participants, those using the information resource, those collecting follow-up data, those who participate as judges in making measurements of dependent or independent variables, and those analyzing the data to create the study results.

Hawthorne Effect: This tendency for humans to improve their performance if they know it is being studied was discovered by psychologists measuring the effect of ambient lighting levels on worker productivity at the Hawthorne factory in Chicago (Roethligsburger and Dickson 1939). Productivity increased as the room illumination level was raised, but when the illumination level was accidentally reduced, productivity increased again, suggesting that it was the study itself, rather than changes in illumination, that caused the increase. During a study of a health or biomedical information resource, the Hawthorne Effect can lead to an improvement in the performance of all participants in all study groups, in reaction to their knowing they are being studied. This "global" Hawthorne effect is particularly likely to occur when performance can be increased relatively easily, for example by acquiring a small amount of knowledge or a simple insight. For example, it could affect a study of the patient-facing app to control diabetes since a patient could easily look up dietary recommendations, even if they were not using the tool.

In descriptive studies, the Hawthorne effect can inflate the values of the dependent variables, making the "state of affairs" that is being studied appear more positive or optimistic. In interventional studies, the Hawthorne Effect will improve performance in all groups, potentially causing the relative benefit from an information resource to be underestimated. A major advantage of correlational studies is the low risk of a Hawthorne Effect affecting the results since participants, even if they have consented to data about their performance being analyzed, will typically have a low level of awareness of being included in a study.

To quantify a Hawthorne effect requires a preliminary study of the performance of participants using a technique that does not alert them. This is conducted before the main study, probably using routine data.

10.4.2   Incomplete Response and Missing Data

In any demonstration study, it is rare to achieve 100% data completeness. For example, the response rate in many mail surveys is less than 50%, and for online surveys it can be less than 10% of those emailed a survey link (Sebo et al. 2017). However, it would be wrong to assume that missing data are missing at random. So, for example, if a study team produces a survey about the acceptability of AI and decision support in medicine (Petkus et al. 2020), the clinicians who respond are unlikely to be those with average opinions. Instead, unless the study team takes precautions to minimize this, respondents are more likely to be clinicians who are really keen on AI & decision support, or those who strongly oppose the idea because they may believe that AI systems threaten clinician autonomy. To take another example, a study team may wish to analyze 6 months' data from an app that logs the symptoms of patients with arthritis, to understand their burden of disease. If data is missing for a given patient for a 2 or 4 week period, what can the study team infer? Possible interpretations could be as extreme as that the patient was so unwell that they were admitted to the hospital—perhaps for a joint replacement—and did not use the app, or alternatively that they were so well that they went on vacation, forgetting to use the daily symptom-checking app. All anyone can say is that the data are missing, and it would be a serious mistake to assume that the patient's symptoms during the 2 or 4 week missing data period were the same as the average of the previous level—or the level that followed that period of missing data.

The approaches to the treatment of missing data are largely outside the scope of this book. The most straightforward approach is to delete in their entirety any cases with missing data, or to delete any variables with levels of missing data above a certain threshold, perhaps 5 or 10%. These "brute force" approaches result in more lost data than more sophisticated approaches such as multiple imputation, maximum likelihood, and Bayesian approaches (Hughes et al. 2019; Von Hippel 2021).

10.4.3 Insufficient Data to Allow a Clear Answer to the Study Question

Sometimes insufficient data undermines the results of a study so profoundly that no conclusions can safely be drawn. Imagine that a study team carries out a survey of user satisfaction for an ePrescribing system, but of the 100 users only 25 respond, a response rate of 25%. Even if the study team has good reason to believe that the responding sample is representative of the larger group, there will be uncertainty due to the small size of the sample. Say, for example, that 20 (80%) of the 25 responders claimed that they found this ePrescribing system very usable and would recommend it to others. That sounds good, but perhaps if there were 75 responders, the fraction reporting high usability might be only 60%, simply because of chance effects.

To address this source of doubt, the study team needs to examine the uncertainty around the 80% figure and check the likelihood that the "true" value might be 60% instead. The uncertainty around the mean of a continuous variable is based on the standard error of the mean (SE_{mean}), as described in Sect. 8.5, which enables calculation of the 95% confidence interval around the mean. In this case, because the 80% figure is based on 25 responses, the 95% confidence interval is plus or minus 16%, or a 95% CI of 64–96%. This shows that the "true" figure is very unlikely to include 60%. If the sample size had been 75, with the same 80% responding that they would recommend the resource, the 95% CI is narrower: 71–89%. One point to note here is that even though this second sample is three times larger than the first, the likely range (measured as 95% confidence interval) is not even halved. Every time a study team doubles the sample size, the confidence interval is reduced by 30%; to halve the confidence interval, a sample four times as large is required (Altman and Bland 2014).

When designing any demonstration study—even a simple survey—the study team should agree how precise the study results needs to be (i.e. the maximum confidence interval that would still enable them to answer the study question), then calculate the number of responses this will take, and design their recruitment strategy based on this figure. There is more detail on how this process applies to sample size calculation for hypothesis testing in interventional studies in Sect. 11.5.

10.4.4 Inadvertently Using Different Data Items or Codes

One specific problem that can arise in demonstration studies of information resources is that, due to history or the choice of application software, some organizations and users make specific clinical coding choices that differ from those used in other organizations or settings. For example, in UK primary care practices the definition of asthma—and thus the codes used to label patients with asthma— can vary from practice to practice. So, if a study team wishes to reliably identify all asthma cases they need to search for over 40 different asthma-related codes to find every asthmatic (Buchan and Kontopantelis 2009). This variability in coding practice in different settings can have serious consequences for the results of all types of demonstration studies. In the simplest case of a descriptive study, coding variability can invalidate counts of specific events. In a more complex example of determining the accuracy of a predictive algorithm, the measured accuracy of the algorithm could appear to differ between the two groups of practices, when in fact the accuracy is the same and the measured difference is due to differences in coding practices.

10.5 Threats to the External Validity and How to Overcome Them

This section describes the main threats to external validity for any kind of demonstration study, whether it is descriptive, interventional, or correlational.

10.5.1 The Volunteer Effect

A common bias in the sampling of participants for any kind of demonstration study is the use of volunteers. It has been established that people who volunteer as participants, whether to complete questionnaires, participate in psychology experiments, or test-drive new cars or other technologies, are atypical of the population at large (Myers et al. 1990). Volunteers tend to be better educated, more open to innovation, and extroverted. Although volunteers are often satisfactory as participants in measurement studies or pilot demonstration studies, use of volunteers should be avoided in definitive demonstration studies, as their participation considerably reduces the generalizability of findings. The best strategy is to randomly select a sample of all eligible participants, following up invitation letters or emails with telephone calls to achieve as near 100% recruitment of the selected sample as possible. (Note that this random selection of all participants is not the same as random allocation of participants to groups, as discussed in Chap. 11.)

The financial resources required for a study depend critically on the number of participants needed. The required number in turn depends on the precision of the answer required from the study and, in interventional studies, the risk that the study team is willing to take of failing to detect a significant effect, as discussed in Chap. 11. Statisticians can advise on this point and carry out sample size calculations to estimate the number of participants required. Sometimes, in order to recruit the required number of participants, some volunteer effect must be tolerated; often there is a trade-off between obtaining a sufficiently large sample and ensuring that the sample is representative.

10.5.2 Age, Ethnic Group, Gender and Other Biases

One problem of increasing concern is sampling biases that lead to underrepresentation in studies of participants from certain age groups, ethnic groups, genders or with what is known in many countries as other legally protected characteristics. One reason for this growing concern is that this problem can occur in the datasets used for machine learning, resulting in biased algorithms (O'Neill 2016). Careful development of more representative sampling strategies, and sometimes differential recruitment efforts designed to specifically target under-represented groups, also known as "oversampling", may help remedy this. To discover if such biases may exist, the study team can compare response rates in specific subpopulations with the overall response rate. If the response rate in a subpopulation differs markedly from the overall rate, this may indicate biased results that do not generalize to that subpopulation. Most usefully, a study team can compare the results of a key analysis (for example, the performance of an algorithm) in the overall study dataset and its performance on selected population subsets. As an example, the study by Knight et al. carefully assessed the potential population bias of a model to predict COVID mortality in specific age, ethnic and deprivation groups (Knight et al. 2020).

10.5.3 Study Setting Bias

One potent cause of bias is an inappropriate study setting or context. For example, if the study team wished to quantify the impact of an asthma self-management app on asthma control, they could mount a study in an emergency room or a tertiary referral center asthma clinic. However, both of these settings are likely to attract people with more severe disease, so the study could either underestimate or overestimate the benefit of the app in people with less severe disease. If the app is intended for anyone with asthma, it would be better in this case to recruit participants from a primary care or home setting. The general principle when selecting the study setting is that the study team must be clear about the study question, using the PICO formalism (Oxford Centre for Evidence-Based Medicine 2021) described in Sect. 3.2.4, which includes a clear statement of the participants and setting to which the study results will be generalized. Once this description of participants and setting is formalized, it needs to be adhered to.

10.5.4 Implementation Bias

This problem arises if, during a descriptive or other kind of demonstration study, extra effort is taken by the information resource supplier to support users, or even to adapt the resource to their specific needs or wishes, beyond what would be done in service to the typical user or organization that deploys the resource. This extra support might generate very promising results in a usability study or demonstrate striking impact in an interventional study that would be unlikely to be replicated elsewhere, once the resource is placed on the market and such extra support or tailoring is no longer available (Garg et al. 2005). The remedy, for all kinds of demonstration studies, is to ensure that the resource is implemented and supported during the study as it will be normally, once released on the market.

10.5.5 Training Set Bias and Overfitting

This problem primarily applies when estimating the performance of what is known as a classifier, such as a risk prediction algorithm based on machine learning. Typically such algorithms are developed using training data derived from a large dataset collected in one or more settings, often at some time in the past (and sometimes many years ago), or in a different setting from that in which the classifier will be used. The classifier performance is then estimated using a different part of the same large dataset—often selected randomly from the original dataset. While this approach of model development using a "training set" and then testing it using a separate "test set" overcomes random error and penalizes any "overfitting" of the algorithm to the training data as discussed below, it does not take account of other issues that may reduce model performance in a new setting (Wyatt and Spiegelhalter 1990; Wyatt and Altman 1995). These issues include:

- Overfitting: When a statistical model that powers an information resource is carefully adjusted to achieve maximal performance on training data, this adjustment may worsen its accuracy on a fresh set of data due to a phenomenon called overfitting (Moons et al. 2016). If the accuracy of the classifier is measured on a small number of hand-picked cases, it may appear spuriously excellent. This is especially likely if these cases are similar to, or even identical with, the training set of cases used to develop or tune the classifier before the evaluation is carried out. Thus, it is important to obtain a new, unseen set of cases and evaluate performance on this "test set".
- Changes in the definition or processes used to collect the input data for the algorithm. For example, differences in the use of disease or other codes between different healthcare organizations or settings, such as primary care practices versus secondary care, is likely to influence both the output of the algorithm for specific cases and also the reference standard for each case, so changing the overall performance of the algorithm in unpredictable ways.
- Changes in the definition, or processes to define, the outcome of interest. For example, the criteria used to diagnose many diseases changes over time, as people understand more about the underlying pathophysiology. This means that a classifier that worked well on data collected 5 years ago, or in a country that uses different disease definitions, may fail on new data collected once the disease definition changed.

The remedy for these problems is to use a relevant, contemporary, unseen test set, ideally collecting new data in a selection of centers that are representative of those in which the algorithm will be used. This and other precautions that should be taken when validating a prediction tool are included in the TRIPOD reporting guideline and checklist (Moons et al. 2016).

Sometimes developers omit cases from a test set if they do not fall within the scope of the classifier, for example, if the final diagnosis for a case is not represented in a diagnostic system's knowledge base. This practice violates the principle that a test set should be representative of all cases in which the information resource is likely to be used, and will overestimate its accuracy with unseen data. However, if the developer clearly indicates to users that the scope of the classifier does not include patients older than a certain threshold or with certain rare diseases, for example, then it would be valid to exclude these patients from the test set. So, it all depends on the claims that are being made by the developer. If the claim is that the classifier is 95% accurate across all diseases and patents, then the test set should include all diseases and patients.

Self-Test 10.3

A decision support system is developed to help ensure that children seen in Federally Qualified Health Centers get all appropriate care. When parents bring their children into the clinic, they are given a tablet computer which shows them, according to the electronic health record, a summary of each child's health record. The parents use the tablet to "flag" any data items that might have changed or that might be inaccurate, to be sure these are discussed between the parent and pediatrician. Before the tablet system is fully deployed, the developers invite and consent parents in three major metropolitan areas to be part of a study where parents first enter information into a tablet, following which the parent is interviewed by a nurse who elicits the same information sought by the tablet. Parents using the tablet can obtain assistance from members of the study team. A member of the team who developed the tablet application then compares the information entered into the tablet with the information derived from the interview. The study continues until an adequate number of visits, as determined by a statistician on the study team, have been included.

For each of the following "biases", indicate whether it may be a factor calling the credibility of the study into question. Explain each answer.

(a) Assessment bias

(b) Hawthorne effect

(c) Incomplete data

(d) Insufficient data

(e) Use of different codes

(f) Volunteer effect

(g) Age, ethnic group, gender biases

(h) Study setting bias

(i) Implementation bias

10.6   Analysis Methods for Descriptive Studies That Also Apply to All Study Types

This section summarizes some of the main techniques used to analyze the results of descriptive studies. These techniques are also relevant when analyzing data from other demonstration study types to obtain an overview of the results in descriptive terms. The techniques discussed here are:

- Graphical portrayal of the results: histogram or distribution plot;
- Indices of the central tendency: mean, median, or mode;
- Indices of variability: standard deviation SD, standard error SE, and confidence interval CI.

These techniques are described according to the level of measurement of data to be analyzed: discrete (nominal, or ordinal) and continuous (interval or ratio) variables. The discussion also includes continuous variables that are normally and non-normally distributed.

10.6.1 Graphical Portrayal of the Results

When the data to be analyzed are discrete counts, such as how many users of a resource responded positively to a yes/no question, it is usual to quote the rates as percentages. For each question in a survey, these percentage figures should be supplemented with the overall numbers of people surveyed and those responding to the question and those not expressing any opinion. When analyzing responses to more complex questions where there are more than two response options, the analysis can either use the median score or identify the most popular response, known as the mode.

For these more complex responses as well as for continuous data, a graph or histogram is often more informative than a table of figures. This is because, as well as identifying the central tendency and amount of variability, to be discussed below, a graph will indicate the shape of the distribution, providing a general impression of whether the statistical tests that apply to normally distributed data can be used, or whether so-called "non-parametric" statistical tests are required. For example, consider the following continuous data captured from a group of 18 participants in a study—Table 10.2. The results vary from 50 to 57, with some duplicates.

These continuous data can be graphed as a histogram by plotting the number of respondents with each value, as shown in Fig. 10.3. It can be seen from this graph (Fig. 10.3) that the data distribution is roughly symmetrical and the graphing tool has approximated the data to a normal distribution, the bell-shaped curve on the chart.

10.6.2 Indices of Central Tendency

When data follow a normal distribution, the mean is a valid index of the central tendency. However, for skewed or asymmetrical data, it is better to use the median as the mean can be biased by extreme values. For count data, the mode (most frequent value) can be helpful. In the histogram above, it can be seen that the mean (53.8), median (54), and mode (53) more or less coincide.

10.6.3 Indices of Variability

To estimate the variability of a distribution of values relative to an index of central tendency, it is necessary to calculate the confidence interval as discussed in Sect. 10.4.3. There are many online calculators to do this, such as (Creative Research Systems 2012).

The standard deviation (SD) describes the variability of a continuous variable such as the data in Table 10.2. In this case, the SD is 2.12.

Table 10.2 Example results from a study generating continuous data

Participant	Result
1	50
2	51
3	51
4	52
5	52
6	53
7	53
8	53
9	54
10	54
11	54
12	55
13	55
14	56
15	56
16	56
17	57
18	57

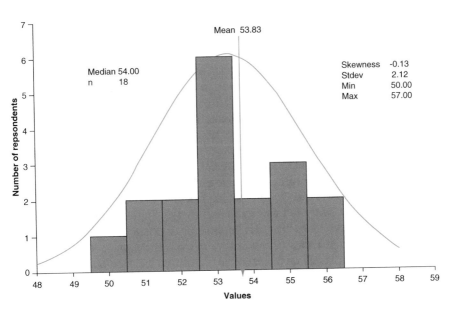

Fig. 10.3 Histogram plotting the example results, with superimposed normal distribution

The standard error of the mean (SE_{mean}) is the SD divided by the square root of the number of observations (N = 18), which computes to 0.50.

The 95% confidence interval is the mean plus or minus 1.96 times the SE_{mean}, which is 53.8 ± 1.0. This gives an estimate for the true mean as lying with a 95% probability between 52.8 and 54.8.

The appropriate measures of variability for non-normal data are the interquartile range (IQR, the value at 75th percentile minus the value at 25th percentile) and the range (the maximum value minus the minimum value). The standard deviation can also be used as a valid index of spread for non-normally distributed data. Sometimes a box and whisker plot (Wikipedia 2021) which shows the median, IQR and range on a chart, or an array of such plots, are useful ways to visualize non-normal data, especially if there are many data items to be displayed.

10.7 Conclusions

Study teams planning any kind of demonstration study need to assess potential challenges to both internal and external validity, then use the techniques discussed in this chapter to address in turn each of those challenges that could apply.

However, it is important to note here that there is often a tradeoff between achieving high internal validity and high external validity, making it hard to achieve both in the same study. For example, measuring the usability of a clinical information resource may entail rigorous usability techniques such as eye-tracking studies that can only take place in a usability lab, requiring the targeted health care professionals to spend some hours away from their place of work. While this may generate the most internally valid results, the results will be less externally valid as the only clinicians who are likely to take time off work to participate in such a study will be those who are somehow motivated to contribute, leaving the usability of the clinical resource in the hands of the average clinician unclear.

This again highlights the contrast between evaluation and research mentioned in Chap. 1. Teams carrying out evaluation studies are often less concerned about external validity than researchers, as their aims differ: evaluation often concerns questions such as whether a specific information resource worked here, while research specifically aims to generate transferable knowledge. So, for a research study, the study question would not be "Did *this* information resource work *here* or not?", but "Would information resources *like this* work in settings *similar to this*?"

The next chapter builds on this material to describe the additional challenges of designing interventional studies in which the aim is to compare outcomes at two or more different time points, or of multiple groups of people with different access to an information resource.

Self-Test 10.4

From this table of data:

Participant	Result
1	50
2	51
3	51
4	52
5	53
6	53
7	53
8	54
9	57
10	58
11	58
12	55
13	55
14	56
15	58
16	58
17	58
18	58

Q1: Plot this dataset using an appropriate method. What do you notice about the distribution of the data items?

Q2: How would you describe the central tendency and variability of this dataset?

Food for Thought

You find yourself in a discussion with a friend about the difference between research studies and evaluation studies. Your friend says that external validity is much less important in studies that are framed as evaluation than in studies framed as research. On the whole, do you agree? Explain your opinion.

Answers to Self-Tests

Self-Test 10.1

1. (a) Descriptive
   (b) Correlational
   (c) Interventional
   (d) Descriptive. (Note that there is no independent variable here because the company's goal was not measured in any way.)
   (e) Correlational

2. (a) The continuous dependent variable is the number of requests; no independent variable since the study is descriptive.

(b) Dependent variable is h-index (continuous); independent variables are student enrollment, endowment, and annual IT expenditure (all continuous).

(c) Dependent variable is ease of use (continuous due to 10 item survey with scores that will be summed across items, as discussed in Chap. 8); independent variable (discrete) is telehealth platform with three possible values.

(d) The continuous dependent variable is the number of user accounts; no independent variable since the study is descriptive.

(e) The continuous dependent variable is rate of new asthma diagnoses; the discrete independent variable is postal codes with the number of values equal to the number of distinct postal codes in the region being studied.

Self-Test 10.2

(a) External validity

(b) Internal validity (since the nurses were randomly sampled)

(c) External validity

(d) Internal validity

Self-Test 10.3

(a) Assessment bias is a factor. The staff member doing the comparison may be biased toward minimizing the difference because s/he participated in the tablet application's development.

(b) Hawthorne effect is a factor. Because parents are aware they are in the study, they may think "extra hard" about their child's medical history.

(c) Incomplete data is unlikely to be a factor because parents must use the tablet before their child will be seen by a clinician.

(d) Insufficient data is unlikely to be a factor because statisticians have predetermined the number of visits likely to be needed and the study will be extended until that number is reached.

(e) Use of different codes: it is likely that the resource uses internal codes for elements of children's medical records. This could be a factor unless codes are standardized across the three sites.

(f) The Volunteer Effect is unlikely to be a factor because it is unlikely that parents will refuse to consent to participate, so a high proportion of eligible parents will probably participate.

(g) Age, ethnic group or gender biases are a concern because the study is being run in a relatively small number of centers in metropolitan areas. The study sample will be skewed toward the population demographics of the selected metropolitan areas.

(h) Study setting bias is probably not a concern because the study is being conducted in the kinds of centers where larger scale deployment will take place. However, if metropolitan area centers are better funded, for example, this could be a concern.

(i) Implementation bias is likely to be a factor. Since this is a pre-implementation study, the level of assistance available to parents using the tablet is likely to be greater than that available during routine implementation.

Self-Test 10.4
The data plot looks like this:

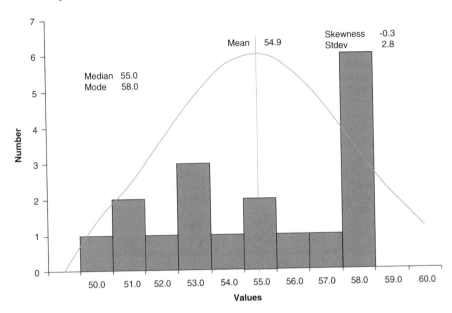

The distribution is almost bimodal, with a small peak around 53 and another higher peak at 58. The graphing tool has sketched a normal approximation, but in this case it is far from convincing.

It is a challenge to describe the central tendency here. Even though the mean and median as very similar, at 54.9 and 55 respectively, this is in many ways misleading. Equally, the spread is much wider than the SD of 2.8 suggests, as a third (6/18) of the observations lie at one extreme of the data values. The lesson here is that, especially with small numbers of observations, it can be extremely valuable to construct a simple plot of the data, even if the summary statistics seem to indicate a fairly normal data distribution. The only subtle warning sign here from the descriptive statistics is that the mode is 58, quite a distance from the mean and median.

References

Altman DG, Bland JM. Statistics notes: uncertainty beyond sampling error. BMJ. 2014;349:g7064.

Buchan IE, Kontopantelis E. Personal communication; 2009.

Collins R, Bowman L, Landray M, Peto R. The magic of randomization versus the myth of real-world evidence. N Engl J Med. 2020;382:674–8.

Creative Research Systems. The survey system: survey sample size calculator and survey confidence interval calculator. Sebastopol, CA: Creative Research Systems; 2012. https://www.surveysystem.com/sscalc.htm. Accessed 29 Jun 2021.

Das-Munshi J, Becares L, Dewey ME, Stansfeld SA, Prince MJ. Understanding the effect of ethnic density on mental health: multi-level investigation of survey data from England. BMJ. 2010;341:c5367.

Donabedian A. Evaluating the quality of medical care. Milbank Mem Fund Q. 1966;44(Suppl):166–206.

Garg AX, Adhikari NK, McDonald H, Rosas-Arellano MP, Devereaux PJ, Beyene J, et al. Effects of computerized clinical decision support systems on practitioner performance and patient outcomes: a systematic review. JAMA. 2005;293:1223–38.

Hughes RA, Heron J, Sterne JAC, Tilling K. Accounting for missing data in statistical analyses: multiple imputation is not always the answer. Int J Epidemiol. 2019;48:1294–304.

Knight SR, Ho A, Pius R, Buchan IE, Carson G, Drake TM, et al. Risk stratification of patients admitted to hospital in the United Kingdom with covid-19 using the ISARIC WHO Clinical Characterisation Protocol: development and validation of a multivariable prediction model for mortality. BMJ. 2020;370:m3339.

Moons KGM, Altman DG, Reitsma JB, Ioannidis JPA, Macaskill P, Steyerberg EW, et al. Transparent reporting of a multivariable prediction model for individual prognosis or diagnosis (TRIPOD): explanation and elaboration. Ann Intern Med. 2016;162:w1–w73.

Myers DH, Leahy A, Shoeb H, Ryder J. The patient's view of life in a psychiatric hospital: a questionnaire study and associated methodological considerations. Br J Psychiatry. 1990;156:853–60.

O'Neill C. Weapons of math destruction: how big data increases inequality and threatens democracy. New York: Crown; 2016.

Oxford Centre for Evidence-Based Medicine. Asking focused questions. Oxford: Oxford Centre for Evidence-Based Medicine, University of Oxford; 2021. https://www.cebm.ox.ac.uk/resources/ebm-tools/asking-focused-questions. Accessed 29 Jun 2021.

Petkus H, Hoogewerf J, Wyatt JC. What do senior physicians think about AI and clinical decision support systems: quantitative and qualitative analysis of data from specialty societies. Clin Med. 2020;20:324–8.

Roethligsburger FJ, Dickson WJ. Management and the worker. Cambridge, MA: Harvard University Press; 1939.

Sebo P, Maisonneuve H, Cerutti B, Fournier JP, Senn N, Haller DM. Rates, delays, and completeness of general practitioners' responses to a postal versus web-based survey: a randomized trial. J Med Internet Res. 2017;19:e83.

Von Hippel P. Missing data: software, advice, and research on handling data with missing values. Austin, TX: Missing Data; 2021. https://missingdata.org/. Accessed 29 Jun 2021.

Wikipedia. Box plot. Wikimedia Foundation; 2021. https://en.wikipedia.org/wiki/Box_plot. Accessed 29 Jun 2021.

Wyatt JC, Altman DG. Commentary: Prognostic models: clinically useful or quickly forgotten? BMJ. 1995;311:1539–41.

Wyatt JC, Spiegelhalter D. Evaluating medical expert systems: what to test and how? Med Inform (Lond) 1990;15:205-217. https://doi.org/10.3109/14639239009025268. Accessed 29 Jun 2021.

Chapter 11
Design of Interventional Studies

Learning Objectives

The text, examples, and self-tests in this chapter will enable the reader to:

1. Explain why control strategies are needed in interventional studies designed to measure the impact of an information resource on its users or on the problem it was intended to solve, and why association is not a reliable indicator of causation.
2. Given a description of an interventional study design, identify the control strategy employed.
3. Describe the various biases that can erode the internal validity of interventional studies of information resources.
4. Given a description of an interventional study, identify which of these biases are likely to be a challenge for that study, and recommend specific strategies to overcome them.
5. Distinguish between effect size and statistical significance; distinguish between Type I and Type II errors; and explain why statistical power depends upon sample size.

11.1 Introduction

The previous chapter described in some detail the structure and terminology of all quantitative demonstration studies; further distinguished descriptive, interventional, and correlational studies as three distinct types; and introduced biases that affect the internal and external validity of all kinds of demonstration studies. The chapter concluded with a discussion of descriptive studies, lacking an independent variable, as the simplest type. This chapter focuses on the more challenging interventional studies which are characterized by having at least one independent variable as well as a dependent variable (such as the rate of adverse drug events or research

© Springer Nature Switzerland AG 2022
C. P. Friedman et al., *Evaluation Methods in Biomedical and Health Informatics*, Health Informatics, https://doi.org/10.1007/978-3-030-86453-8_11

productivity) that the intervention is intended to affect in some way. In informatics this independent variable typically records the introduction or modification of an information resource into a functioning clinical, research, or educational environment. Interventional studies are prospective in the sense that they are designed prior to the measurement of the dependent variable(s). Interventional studies align with the "comparative" evaluation approach in House's typology, as introduced in Chap. 2. In informatics, as discussed in Chap. 3, they are used most frequently for field user-effect studies and problem impact studies.

The goal of an interventional study is for the study team to be able to confidently attribute to the intervention any change in the dependent variable(s). Because interventional studies are prospective, they are deliberately and thoughtfully designed by study teams who, in crafting these designs, must be attentive to a challenging set of potential biases that will affect the internal and external validity, and thus the credibility, of the study results. Accordingly, this chapter begins by describing the main challenges to assigning causation and estimating effectiveness in interventional studies of information resources (Sect. 11.2), followed by a discussion of several designs for interventional studies that can address these challenges and credibly assign causation (Sect. 11.3). The chapter progresses with a discussion of how to detect the desired level of effect or effect size (sample size and study power, Sect. 11.4) and concludes with a section on how to minimize the impact of several biases specific to interventional studies of information resources (Sect. 11.5). In Chap. 12, the discussion proceeds to focus on the analysis of demonstration study results.

This chapter will return frequently to the important point that an interventional study can be no better than the quality of the measurements of all variables. Thus, the success of the study rests heavily on the measurement processes used, as introduced in Chaps. 7, 8, and 9.

The available methods to design and analyze interventional studies are continuously improving. Those who wish to design and conduct state-of-the-art studies should consult statisticians for assistance in employing methods that are superior to the relatively basic approaches presented here. The limitations of the methods presented in this chapter and Chap. 12 will be apparent at many points. One goal of this chapter and the next is to facilitate communication between study team members who are primarily trained in health and biomedical informatics and their statistician colleagues.

11.2   Challenges When Designing Interventional Studies to Attribute Causation

One of the most challenging aspects of interventional study design is how to monitor all the other changes taking place in a working environment that are not attributable to the information resource, or, in more technical language, to obtain control. In health care, it is sometimes possible to predict patient outcome with good accuracy from a set of initial clinical findings, for example the survival of hospital patients with symptoms suggestive of COVID (Knight et al. 2020) or of patients admitted in intensive care (Knaus et al. 1991). In these unusual circumstances, where there exists a close approximation to an "evaluation machine" (see Sect. 1.3) that can foretell what would have happened to patients in the absence of any intervention, it is possible to compare what actually happens with what was predicted to draw tentative conclusions about the impact of the information resource. However, to trust such an approach, a predictive model is needed that is up-to-date, accurate and has excellent calibration (meaning that any probabilities it issues are themselves accurate). To work as an evaluation machine, a model also needs to be transferable, in the sense that it can be used confidently in a different site—or even country—from which it was developed. Unfortunately, as pointed out 25 years ago (Wyatt and Altman 1995), relatively few of the thousands of predictive models published every year across health and biomedicine meet all these criteria. This means that it is generally impossible to determine what students, clinicians or researchers would have done, or what the outcome of their work would have been, had no information resource been available. Instead, a control must be used. In an experiment a control is a strategy designed to isolate the effects of variables other than the independent variable—in this case, the information resource under study. This control can take the form of a group of observations (such as a distinct group of participants, tasks or measures on those tasks) that are unlikely to be influenced by the intervention being studied, or that are only influenced by a different intervention (in the case of studies comparing two or more information resources).

The following sections describe some of the many factors that require control in rigorous interventional studies.

11.2.1 Secular Trends

Secular trends are systematic changes, not directly related to the resource under study, that take place over time. These happen because practice improves over time, policies governing practice change, staff members move on, pandemics occur, and many other reasons. There are also more subtle changes, for example in the definitions of clinical findings or even in disease definitions. This can mean that, for example, risk scores that had performed well gradually become out of step with reality (Grant et al. 2012).

There are also shorter-term changes in many clinical findings due to the seasons, so that for example chest infections occur more frequently in the winter and skin rashes and sensitivity, as well as more serious skin cancers, are more likely to be diagnosed during the summer months.

For study teams, these secular and seasonal changes mean that it is difficult to reliably credit any changes in, for example, disease incidence or severity to changes related to information resources unless these are controlled for.

In the case of seasonal changes, one control solution would be to compare data collected on disease severity, for example, following the introduction of an app to support self-management in COPD, with the same data collected during the same months of previous years. However, even if this seemed to support the app, a sceptic could argue that a more effective drug had been introduced or that community physiotherapy resources had doubled, so the improvement was due to this, not to the app. The most reliable way to overcome such objections would be to conduct a controlled study, probably a randomized controlled trial of the app.

11.2.2   Regression to the Mean

A common health informatics project motivating an evaluation is to identify a problem (such as a high rate of adverse drug events—ADRs—in a certain hospital unit), develop a solution (such as an ePrescribing system), then implement the solution and determine its effects on the problem. Typically, the problem will reduce following implementation of the information resource, and everyone will be happy. However, there is a major challenge in using such a "before-after" study design to infer that the information resource itself solved the problem. Because the hospital unit was deliberately selected as having a higher than average ADR rate, a second measurement of its ADR rate is very likely to show improvement, in this case a lower figure, due to "regression to the mean".

Regression to the mean is a statistical tendency for any single measurement that lies far from the mean to "regress" or revert to a value that is closer to the mean. So, for example, the scores of outstanding (or weak) students in an exam tend to become closer to the mean (either lower for high scoring students, or higher for low scoring students) in a second exam on the same topic. This is not because the high scoring students "rest on their laurels" or the poor scoring ones "get a shock" from their first result, but because of random measurement error, as discussed in Chap. 7. Following an extreme low (or high) measurement, due to random error, the next measurement result is likely to be less extreme. So, in a selected sample of low-scoring students who subsequently receive an intervention designed to improve their learning, higher scores following the intervention may be due to regression to the mean affecting the entire sample, rather than the intervention itself (Bland and Altman 1994).

In the example of the ePrescribing system above, the way around regression to the mean would be to carry out the study in one or more units with a typical range of ADR rates, rather than only in a unit selected for its high ADR rate.

11.2.3   *Association Is Not Causation*

Chapter 13 discusses the challenges of uncontrolled correlational studies and offers several examples in which correlation did not signify causation. For example, assume that in a large dataset of weight loss app users, weight loss in overweight users is associated with the intensity of app use (or app engagement). It is obviously tempting for the app developers to assume that app use is responsible for the weight loss. However, an equally likely theory is that it is only people who really *want* to lose weight (and perhaps also have family and social support to do so) who will actually lose weight. It is equally likely that the intensity of app engagement is higher in those people who really want to lose weight. So, rather than being the cause of weight loss, intensive use of the app is a **marker** of the subgroup of overweight people who really want to lose weight and thus are more successful in achieving that goal.

In sum, secular trends, regression to the mean, and the difficulty of inferring causality from association highlight the need for well-designed controls in studies intended to measure the impact of an information resource in a user or problem impact study. The range of possible control strategies to overcome these challenges is discussed in the next section.

11.3   Control Strategies for Interventional Studies

The following sections present a series of specific control strategies, using an uncontrolled approach as an anchor point and moving to increasingly sophisticated approaches. An app provided by a hospital for use by clinicians to assess acutely ill patients for the presence and severity of sepsis is used as a running example to describe all strategies. In this example, the intervention is the availability of the app, and the participants are the clinicians. The dependent variables are the clinicians' rates of ordering antibiotics—an effect measure, in the parlance of Sect. 3.3—and the rate of intensive care admissions for sepsis averaged across the patients cared for by each clinician, which is an outcome measure. The independent variables in each example below are an inherent feature of the study design and derive from the specific control strategies employed.

11.3.1   *Uncontrolled Study*

In the simplest possible design, an uncontrolled study, measurements are taken after the sepsis app is deployed. There is no independent variable, so this is a purely descriptive study. Suppose it is found that the overall ICU sepsis admission rate is

5% and that physicians order prophylactic antibiotics in 60% of sepsis cases. Although there are two measured dependent variables, it is difficult to interpret these figures without any comparison. It is possible that there has been no change attributable to the information resource.

11.3.2 Historically Controlled or Before-After Studies

As a first improvement to the uncontrolled study, consider a historically controlled experiment, sometimes called a before–after study. The study team makes baseline measurements of antibiotic ordering and ICU sepsis admission rates before the app is made available and then makes the same measurements at some time after the app is in routine use. The independent variable is "time" and has two possible values: before and after app availability. Let us say that at baseline the ICU sepsis admission rate was 10% and clinicians ordered antibiotics in 40% of sepsis cases; the post-intervention figures are the same 5% and 60% as in the uncontrolled example above (Table 11.1).

After reviewing these data, the study team may claim that the information resource is responsible for halving the ICU sepsis admisison rate from 10% to 5%, especially because this was accompanied by a 50% relative increase (or 20% absolute increase) in clinicians' antibiotic prescribing. However, secular trends might have intervened in the interim to cause these results, especially if there was a long interval between the baseline and post-intervention measurements. New clinicians could have taken over the care of the patients, the case mix of patients on the unit could have altered, new antibiotics might have been introduced, or clinical audit meetings might have highlighted the sepsis problem causing greater awareness of infection. Simply assuming that the app alone caused the reduction in ICU admission rates is perilous.

Evidence for the bias associated with before–after studies comes from a systematic review that compared the results of many historically controlled studies of blood pressure drugs with the results of simultaneous randomized controlled trials carried out on the same drugs (Sacks et al. 1982). About 80% of the historically controlled studies showed that the new drugs evaluated were more effective, but this figure was confirmed in only 20% of better-controlled studies that evaluated the same drugs.

Table 11.1 Hypothetical results of a historically controlled study of a sepsis app

Time	Antibiotic prescribing rate (%)	ICU sepsis admission rate (%)
Baseline (before installation)	40%	10%
After installation	60%	5%
Relative difference	+50%	−100%

11.3.3   *Simultaneous External Controls*

To address some of the problems with historical controls, simultaneous controls might be used instead, making the same outcome measurements at the same times in clinicians and patients not influenced by the sepsis app (perhaps in a second unit in the same hospital but with less IT infrastructure), but who are still subject to the other changes taking place in the organization. If these measurements are made in both locations before and after the intervention in the target unit, it strengthens the design by providing an estimate of the differences in outcome rates due to other changes taking place during the study period. This second unit is called an "external control": it is a control as it is subject to the same non-specific changes taking place across the healthcare organization, but it is external as the staff there are not subject to the intervention being studied.

Table 11.2 gives some hypothetical results of such a study, focusing this time on ICU sepsis admission rate as the dependent variable. The independent variables are "time," as in the above example, and "group," which has the two values of intervention and control. There is the same dramatic improvement as in the previous example in the group of patients of clinicians to whom the app was available, but no improvement (indeed a slight deterioration) where no app was available. This design provides suggestive evidence of an improvement that is most likely to be due to the availability of the app.

Even though this example is better controlled, skeptics may still refute an argument that the app caused the reduction by claiming that there is some systematic unknown difference between the clinicians or patients in the app and control groups. For example, even if the two groups comprised the patients and clinicians in two adjacent units, the difference in the infection rates could be attributable to systematic differences between the units. Perhaps hospital staffing levels improved in some units but not others, or there was cross-infection by a multiply drug-resistant organism but only among patients in the control unit. To overcome such criticisms, a study team could expand the study to include all units in the hospital—or even other hospitals—but this requires many more measurements, which would clearly take more resources. The team could try to measure everything that happens to every patient in both units and build complete profiles of all staff to rule out systematic differences, but this approach is still vulnerable to the accusation that something

Table 11.2   Hypothetical results of a simultaneous controlled study of a sepsis app

Time	ICU sepsis admission rate (percent)	
	Patients of clinicians with the app	Patients of control clinicians
Baseline	10%	10%
After intervention	5%	11%
Relative difference	−100%	+10%

that was not measured—and the team did not even know about—explains the difference in results between the two units. This problem, of an unknown or unmeasured "confounder", is further discussed in Chap. 13, Sect. 13.7.3.

11.3.4 Internally and Externally Controlled Before-After Studies

An alternative approach to a before-after study with an external control is to add internal controls. An internal control is a control in the same unit as the information resource is deployed, but who for some reason would not be directly affected by the resource. Using internal controls, the study team adds to the study some new observations or dependent variables that are not expected to be affected by the intervention. The benefits of external controls were already discussed, comparing the rates of ICU admission and antibiotic prescribing with those in another unit where the app has not been implemented. However, the situation is strengthened further by measuring one or more appropriate variables in the same unit to check that nothing else in the clinical environment is changing during the period of a before-after study. If this works out, and there are no unexpected changes in the external site, one really can then begin to argue that any change in the measurement of interest must be due to the information resource (Wyatt and Wyatt 2003). However, the risk one takes in this kind of study design is that the results often turn out to be hard or impossible to interpret because of unforeseen changes in the dependent variable in the external or internal controls.

Pursuing our sepsis app example, for our internal controls, actions by the same clinicians and outcomes in the same patients need to be identified that are *not* affected by the app but *would* be affected by any of the confounding, nonspecific changes that might occur in the study unit. An internal clinical action that would reflect general changes in prescribing is the prescribing rate of antibiotics for chest infections, whereas an internal patient outcome that would reflect general care of sick patients is the rate of postoperative deep venous thromboses (DVTs).[1] This is because DVTs can usually be prevented by heparin therapy or other measures. So, any general improvements in clinical practice in the study unit should be revealed by changes in these measures. However, providing a sepsis app to clinicians should not affect either of these new measures, at least not directly. Table 11.3 shows the hypothetical results from such an internally controlled before–after study.

The absolute increase in prescribing for chest infections (5%) is much smaller than the increase in prescribing antibiotics for sepsis (20%), and the postoperative DVT rate increased, if anything. The evidence suggests that antibiotic prescribing in general has not changed much (using prescribing for chest infections as the internal

---

[1] DVTs are blood clots in the leg or pelvic veins that cause serious lung problems or even sudden death if they become detached.

Table 11.3 Hypothetical results of an internally controlled before–after study of a sepsis app

Time	Antibiotic prescribed (%)		ICU sepsis admissions	DVT rates
	For sepsis	For chest infections		
Baseline	40%	40%	10%	5%
After intervention	60%	45%	5%	6%

DVT deep venous thrombosis; *ICU* intensive care unit

control), and that care in general (using DVT rate as the internal control) is unchanged. Although less convincing than randomized controlled study, the results rule out major confounding changes in prescribing or postoperative care during the study period, so the observed improvement in the target measure, ICU sepsis admissions, can be cautiously attributed to introduction of the reminder system. This argument is strengthened by the 20% increase in prophylactic antibiotic prescribing observed. Unfortunately, interpretation of the results of internally controlled before–after studies that are performed in the real world is often more difficult than in this hypothetical example; adding in an external control as well may help (Wyatt and Wyatt 2003).

11.3.5   Simultaneous Randomized Controls

In the previous example, there may have been systematic, unmeasured differences between the patients and/or clinicians in the control group and the participants receiving the intervention. An effective way to remove systematic differences, due to both known and unknown factors, is to randomize the assignment of the clinical participants to the control or intervention group. Thus, half of the clinicians on both units could be randomly allocated to receive the app, and the remaining clinicians could work normally. ICU sepsis admission rate in patients managed by clinicians in the app and control groups would then be measured and compared. Providing that the clinicians never look after one another's patients or consult with one another, any statistically significant difference can reliably be attributed to the app, as the only way other differences could have emerged is by chance.

Table 11.4 shows the hypothesized results of such a randomized controlled study. The baseline ICU sepsis admission rates in the patients managed by the two groups of clinicians are similar, as would be expected because the clinicians were allocated to these groups by chance. After making the app available randomly to half of the clinicians, the sepsis admission rates for the app physicians are only 6%, a 25% reduction in the admission rates for the control physicians (8%). When comparing the figures for each group of patients with baseline, there is an even bigger difference: the admission rate in patients of app physicians shows a 36% fall while the rate in patients treated by control physicians shows a 20% fall. The only systematic difference between the two groups of patients is that some of the clinicians used the sepsis app. Provided that the sample size is large enough for these results not to be

Table 11.4 Results of a randomized controlled study of a sepsis app

Time	ICU sepsis admission rates (%)	
	Patients of physicians randomized to use the app	Patients of physicians randomized to not use the app
Baseline	11%	10%
After intervention	6%	8%
Relative difference pre-post	−36%	−20%

attributable to chance, the study team can conclude with some confidence that giving clinicians the app caused the reduction in ICU admission rates.

One lingering question is why there was also a small reduction, from baseline to post installation, in ICU admission rates in control cases. Four explanations are possible: chance (discussed in Sect. 11.4), the checklist effect or contamination (Sect. 11.5), and the Hawthorne Effect (discussed in Sect. 10.4.1).

11.3.6 Cluster Randomized and Step Wedge Designs

One variant on the randomized trial that is often used by study teams is the "cluster randomized" design. This design is required in field studies when the circumstances of real-world practice require randomization at the level of groups, even though the outcome of interest is measured at the level of individuals. For example, this can occur if the outcome of interest is patient infection rates but clinicians cannot reasonably expect to use an information resource to support the care of some patients and not others. In this case, the intervention must be applied to all patients of each clinician, and clinicians must be randomized to either employ the intervention or not. Similarly, in an educational setting, an intervention to improve student exam scores must be administered by teachers to entire classes, and the wisest strategy is to randomize at the level of teachers even though measurements will take place on individual students. This strategy is employed to overcome the problem of study contamination, discussed in Sect. 11.4.

A variation of the cluster randomized design that is increasingly used in health and biomedical informatics is the step-wedge design (Hemming et al. 2015; Sarrassat et al. 2021). In a step-wedge design, the study team randomly selects an entire unit to gain access to the resource first, then includes other units, one by one in random order, every few weeks or months until the resource is available in all parts of the organization—Fig. 11.1. The length of each phase of the study is carefully chosen to allow sufficient observations to be collected after a user orientation/ training period to allow calculation of the impact of the resource on the dependent variable. This design has become popular since it is difficult to randomize individual clinicians to use a pervasive information resource such as an ePrescribing or electronic patient record system in one organizational unit: either all the clinicians in the unit use it, or none. This step wedge design is useful to control bias as the

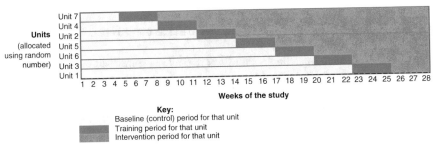

Fig. 11.1 Design of a sample step wedge study in which 7 units are allocated to receive the information resource in random order over a 28 week period. Baseline measures are made for at least 4 weeks in each unit, then a 4 week training period is allowed post resource implementation before further post-intervention measures are captured to maximize the staff benefit due to the resource

allocation of the resource to each unit is in random order. However, it is also useful ethically, as often the number of informatics staff is insufficient to allow a complex system to be implemented across a whole organization simultaneously, so use of a random number to decide who gets access to the information resource first is seen as fairer than an alternative approach.

When analyzing studies in which clinicians, teams or units are randomized but the measurements are made at the level of patients (so-called "randomization by group"), data analysis methods must be adjusted accordingly. In general, when randomizing at higher levels of aggregation, it is incorrect to analyze the results as if patients had been randomized. This is known as the unit of analysis error (Diwan et al. 1992; Cornfield 1978). Potential methods for addressing it are discussed in the section on hierarchical or nested study designs in Chap. 12.

11.3.7   Matched Controls as an Alternative to Randomization, and the Fallacy of Case-Control Studies

The principle of controls is that they should sensitively reflect all of the nonspecific influences and biases present in the study population, while being isolated in some way from the effects of the information resource. As argued earlier, it is only by random assignment that equivalence of the groups can be achieved for both known and unknown factors. Allocation of participants to control and intervention groups may be attempted by other methods, such as matching, when randomization is not feasible. Matching is achieved by including in the study a control participant who resembles a corresponding intervention participant in as many relevant features as possible. When this is done, participants and tasks in the control and intervention groups should be matched on the most important features known, or likely, to be relevant to the dependent variable. For example, assuming that participant age turns out to be an important predictor of participant use of an information resource, the participants for a study could be divided up into two groups, taking care that each older person in the group who is given access to the resource is matched by a similar age person in the control group.

The problem with "matched" controls is that all the important factors to use for matching are rarely known. Worse, some of these are hard or impossible to capture and match on, such as the factors that lead clinicians to prescribe a drug—or an app—for some patients and not for others. This problem is the focus of Sect. 13.7.4 which addresses the problem of "confounders".

11.3.8 Summary of Control Strategies

To summarize this section on controls and study designs, although study teams may be tempted to use either no controls or historical controls in demonstration studies, this discussion has illustrated, using a running example, why such studies are seldom convincing (Liu and Wyatt 2011). If the goal of an interventional study is to show cause and effect, simultaneous (preferably randomized) controls are required (Liu and Wyatt 2011). Using both internal and external controls within a before–after study design may be an alternative, but exposes the study team to the risk that their results will be impossible to interpret. The risk of other study designs—most clearly the case control design—is that there is no way of rebutting those who inevitably, and appropriately, point out that confounding factors, known or unknown, could account for all of the improvements the study team might wish to attribute to the information resource.

The motive for conducting demonstration studies is to provide reliable conclusions that are of interest to those making decisions. The primary interest is to inform decisions about the information resource in the context in which it was studied. To this end, study teams want their results to be free from threats to internal validity. In the sections that follow, some of the many potential sources of bias that jeopardize internal validity in studies of information resources are examined.

Self-Test 11.1

For each of the scenarios given below, (a) name the independent variables and the number of possible values for each, (b) identify the dependent variables and the method used to measure them, (c) identify the participants, and (d) indicate the control strategy employed by the study designers.

1. A new ePrescribing system is purchased by a major medical center. A study team administers a 30-item general attitude survey about information technology to staff members in selected departments 6 months before the resource is installed, 1 week before the resource is installed, and 1 and 6 months after it is installed.
2. An AI tool for interpreting images of skin lesions in an early stage of development is employed to offer advice on a set of test images. A definitive diagnosis for each test image had previously been established. The study team measures the accuracy of the resource as the proportion of time the computer-generated diagnoses agree with the previously established diagnosis.

3. At each of two metropolitan hospitals, 18 physicians are randomized to receive computer-generated advice on drug therapy. At each hospital, the first group receives advice automatically for all clinic patients, the second receives this advice only when the physicians request it, and the third receives no advice at all. Total charges related to drug therapy are measured by averaging across all relevant patients for each physician during the study period, where relevance is defined as patients whose conditions pertained to the domains covered by the resource's knowledge base.

4. A new discharge planning system is installed in a hospital that has 12 internal medicine services. During a 1-month period, care providers on six of the services, selected randomly, receive access to the system. On the other six services the output is generated by the system but not issued to the care providers. An audit of clinical care on all services is conducted to determine the extent to which the actions recommended by the discharge planning system were in fact taken.

5. A new computer-based educational tool is introduced in a medical school course. The tool covers pathophysiology of the cardiovascular (CV) and gastrointestinal (GI) systems. The class is divided randomly into two groups. The first group learns CV pathophysiology using the computer and GI pathophysiology by the usual lecture approach. The second group learns GI pathophysiology by the computer and CV pathophysiology by the lecture approach. Both groups are given a validated knowledge test covering both body systems after the course. (Example drawn from Lyon et al. (Lyon Jr et al. 1992)).

11.4 Biases and Challenges to the Internal Validity of Interventional Studies

The introduction to this chapter (Sect. 11.2) described some generic challenges to attributing causation, including secular trends and regression to the mean, that threaten any kind of demonstration study. Many of these generic challenges can be overcome by appropriate choice of controls, as discussed above. However, there are some additional challenges to attributing causation specific to controlled studies of information resources that also need to be considered before the study team can finalize the design for a controlled study. These challenges and methods to address them are discussed below.

11.4.1 The Placebo Effect

In some drug trials, simply giving patients an inactive tablet, or placebo, causes their conditions to improve. This placebo effect may be more powerful than the effect of the drug effect itself. Placebo effects can occur in health and biomedical informatics studies as well. For example, in a clinical information resource study, patients who watch their clinicians use impressive technology may believe they are receiving better or additional care. This can potentially overestimate the value of the information resource. But this can also go the other way: some patients may believe that a care provider who needs a workstation is less competent or resent the fact that the clinician is spending time with the computer instead of them. The placebo effect is most likely to arise when the attributes being measured are attitudes or beliefs, such the patients' satisfaction with their care or therapy, and when the technology is used in front of the patient. Blinding participants to whether they are in the intervention or control group is the standard way to address the placebo effect. In informatics studies, "blinding" could occur if all care providers left the patient for the same brief period, during which time the information resource would be used only for the intervention group.

The Hawthorne effect (discussed in Sect. 10.4.1) and the placebo effect can be difficult to distinguish. The Hawthorne effect is more likely to be in play when the study participants, those on whom measurements are made, are reacting to the perception of being studied. Placebo effects are more likely to be in play when the participants are clients, care recipients, or students who believe, and react to the belief, that they may be receiving exceptional or special treatment. Placebo effects are likely to greater when the study outcome measures are perceptual constructs such as perceived quality of care, level of pain, or quality of sleep—rather than more "concrete" metrics such as length of consultation, number of analgesic tablets consumed per day, or sleep quality measured by a wearable device.

11.4.2 Allocation and Recruitment Biases

Studies of information resources conducted early in the life cycle often take place in the environment in which the resource was developed and may arouse strong positive (or negative) feelings among study participants. In a study where patients are randomized and the clinicians have strong beliefs about the information resource, two biases may arise. In clinical studies, study teams may, subconsciously perhaps, but still systematically allocate easier (or more difficult) cases to the information resource group, which is **allocation bias**, or they may avoid recruiting easy (or difficult) cases to the study if they know in advance that the next patient will be allocated to the control group - **recruitment bias** (Schulz et al. 1995). These biases can either over- or underestimate the information resource's value.

To eliminate these potential biases, everyone involved in such judgments should be blinded to whether the information resource was used in each case. If follow-up data about a case are necessary to render a judgment, these data should ideally be obtained after an independent person removes any evidence of information resource use that may exist (Wyatt 1989). Finally, it is increasingly common for those who analyse the results of demonstration studies to be blinded to the identity of the groups until the analysis is complete ("triple blinding"). Only then does a third party reveal that in fact, "Group A" was the control or intervention group.

To address allocation biases, it is helpful to define carefully the population of participants eligible for the study, screen them strictly for eligibility, randomize them as late as possible before the information resource is used, and conceal the allocation of participants to intervention or control groups until they have firmly committed to participate in the study. This can be achieved by requiring everyone recruiting participants to a randomized study to use an online resource that registers the participant, checks their eligibility for the study, and allocates them to one or other group (Schulz et al. 1995).

11.4.3   Biases Associated with Data Collection

Data Completeness Effect   In some studies, the information resource itself may collect the data used to assess a dependent variable. Thus, more data are available in intervention cases than in controls. The data completeness effect may work in either direction in influencing study results. For example, consider a field study of an intensive care unit (ICU) information resource where the aim is to compare recovery rates from adverse events, such as transient hypotension between patients monitored by a new information resource with those allocated to use of earlier technology. Because the information resource records adverse episodes that may not be recorded by the older system, the recovery rate may apparently *fall* in this group of cases, because more adverse events are being detected. To detect this bias, the completeness and accuracy of data collected in the control and information resource groups can be compared against some third method of data collection, perhaps in a short pilot study. Alternatively, clinical events for patients in both groups should be logged by the newer system even though that information resource's output is available only for care of patients in the invention group.

"Second-Look" Bias   When conducting a laboratory study of the effects of an information resource on clinical decision making using case scenarios, a common procedure is to ask clinicians to read a problem or case scenario and state their initial decision. They are then allowed to use the information resource (e.g., an app or decision support system) and are asked again for their decision (Friedman et al. 1999). Any improvements in decision making might then be credited to the app or decision support system. However, there is a potential bias here. The participants are being given a second opportunity to review the same case scenario, which allows them more time and further opportunities for reflection, which can itself improve decision making.

The second-look bias can be reduced or eliminated by increasing the interval between the two exposures to the stimulus material to some weeks or months (Cartmill and Thornton 1992) or by providing a different set of case data, matched for difficulty with the first, for the second task (Suermondt and Cooper 1993). Alternatively, the size of the effect can be quantified by testing participants on a subset of the test data a second time without providing them access to the information resource. Another approach is to examine whether the information resource provided participants with any information of potential value and determine if the increase in performance was correlated with the utility of the information resource's advice.

11.4.4   Carryover Effect

The carryover effect or contamination results when a participant in a study of an information resource uses the resource in an early phase of the study and later, while still in the study, has the resource withheld. This can occur in crossover designs, not discussed in detail here, where a resource is initially "turned on" and later "turned off" to see if the performance of participants returns to baseline. Or it can occur when a resource is deployed in some practice environments (for example some floors of a hospital) and not others. If a participant rotates from a deployed environment to a non-deployed environment, the resource is effectively turned off for that participant. Now if the resource has an educational effect—participants learn in some persistent way from using the resource—then the effect of that learning will carry over to the environment where the resource is not in use. Such educational effects would be expected in any form of tutoring, critiquing, advisory or alerting resource. Working without the resource, study participants will often remember the advice they previously received from the resource, and the circumstances that generated it.

A carryover effect therefore reduces the measured difference in performance between information resource and control conditions. To eliminate the carryover effect, it is probably best to carry out a cluster trial (Sect. 11.3.6), randomizing at the level of the care provider instead of the patient (Tierney et al. 1994), the unit/department instead of the care provider (Adams et al. 1986), or the teacher instead of the student. This creates what is called a cluster or hierarchical study design.

11.4.5   Complex Interventions and Associated Biases

Often, the goal of an interventional study is to understand why, rather than simply ask if, something happens or not. Use of the phrase *information resource* does not define precisely what the resource or system includes. However, particularly when trying to answer questions such as "How much difference is the resource likely to make in a new setting?" it is important to isolate the effects due to the information resource itself from effects due to other activities surrounding its development and implementation.

For example, if a department were to implement a set of computer-based practice guidelines, a considerable amount of time might be spent on developing the guidelines before any implementation took place. Or a department that deploys a commercially purchased laboratory notebook developed for physics research might spend a great deal of time configuring it for biological use. Changes in practice following the implementation of the information resource might be partly or even largely due to the development or configuration process, not to the information resource itself (Grimshaw and Russell 1993). Transplanting an information resource to a new setting without repeating the development or configuration process at the new site might then yield inferior results to those seen at the development site.

When implementing new information resources it is usually necessary to offer training, feedback, and other support to resource users; these could also be considered part of the intervention. This insight—that many information resources are complex, in the sense that they consist of multiple components and activities that interact in different ways—has been recognized more widely as the field of "complex interventions", and guidance on the development and evaluation of these complex interventions is available in several publications (Medical Research Council 2019).

One approach to answering "why" questions is to split the information resource up into its components and evaluate each of these separately. A classic example from the literature is the Leeds Abdominal Pain Decision Support System, a diagnostic advisory system. Various components of this resource were tested for their effects on the diagnostic accuracy of 126 junior clinicians in a multicenter trial in 12 hospitals (Adams et al. 1986). Table 11.5 shows the average percentage improvement due to each component of the system. The data are extracted from a 2-year study of four groups of junior clinicians working for 6 months in 12 emergency rooms, each group being exposed to a different set of components of the full decision support intervention. The study team measured diagnostic accuracy after the clinicians had been in place for 1 and 6 months. The first row shows that there was minimal improvement in the clinicians' diagnostic accuracy over the 6-month period when no information resource was in place. The second row shows the sustained improvement in decision making when structured data collection forms were

Table 11.5 Impact of the different components of the Leeds abdominal pain system on diagnostic accuracy at two time points

Study group (Intervention components)	Improvement in diagnostic accuracy from baseline (%)	
	End of month 1	End of month 6
1: None	0	+1%
2: Data collection forms	+11%	+14%
3: Monthly feedback and data collection forms	+13%	+27%
4: Computer advice, feedback, and forms	+20%	+28%

Source: Data calculated from Adams et al. (Adams et al. 1986)

used solely to assist clinical data collection.[2] The third row shows the effects for clinicians who were given monthly feedback about their diagnostic performance as well as using the structured forms, and marked learning is seen. The fourth row reports results for those who received the full system intervention: diagnostic probabilities calculated by the Leeds computer-based advisory system at the time of decision making as well as monthly feedback and using data collection forms. From the last row in this table it can be seen that:

- Of the 20% improvement seen at the end of Month 1, only the difference between row 3 and row 4, which is 7%, is attributable to the computer's advice.
- In the first month, the computer advice is contributing less to improving diagnostic accuracy (net 7%) than the data collection forms (11%).
- If the information resource is defined as the computer advice alone, it is contributing only one third (7%) of the 20% improvement in diagnostic accuracy seen at Month 1
- Of the 28% improvement seen at month 6, only 1% appears due to the computer advice so the advice only contributed 1/30th of the total improvement at Month 6.

Thus, careful definition of the information resource and its components is necessary in demonstration studies to allow the study team to answer the explanatory question: "Which component of the information resource is responsible for the observed effects?". *It is critical that study teams define the constituents of the resource before the study begins and use that definition consistently through the entire evaluation effort.*

Complex interventions are associated with two types of biases:

Checklist Effect The checklist effect is the improvement observed in performance due to more complete and better-structured data collection about a case or problem when structured forms are used for this purpose (Arditi et al. 2012). Most information resources require that data be well structured and consistently represented. Perhaps it is the structuring of the data, rather than any computations that are performed, that generates performance improvement. As shown earlier (Table 11.5), the impact of a well-designed checklist on decision making can equal or exceed that of computer-generated advice.

To control for the checklist effect, the same data can be collected in the same way in control and information resource conditions, even though the information resource's output is only available for the latter group. To quantify the magnitude of this effect, a randomly selected "data collection only" group of patients can be recruited (Adams et al. 1986). Sometimes the checklist effect is ignored by defining the intervention to include both the revised data collection methods and the computation performed on the data after it is collected. While this approach may be scientifically unsatisfying, for purposes of evaluation it may be entirely satisfactory if the stakeholders have no interest in separating the issues.

---

[2] Due to what is called a checklist effect, described in more detail later in this section

Feedback Effect As mentioned in the earlier discussion, one interesting result of the study of the Leeds Abdominal Pain System (Adams et al. 1986) was that the diagnostic accuracy of the control clinicians failed to improve over the 6 month period, whereas the performance of the study clinicians in group 3 given both data collection forms and monthly feedback did improve, starting at 13% above control levels at month 1 and rising to 27% above control levels at month 6 (Table 11.5). Providing these clinicians with the opportunity to capture their diagnoses on a form and encouraging them to audit their performance monthly improved their performance, even though they did not receive any decision support per se. Many information resources provide a similar opportunity for easy audit and feedback of personal performance (Arditi et al. 2012). If a study team wishes to distinguish the effects of any decision support or advice from the effects of audit or feedback, control participants can be provided with the same audit and feedback as those in the intervention group. Alternatively, the study could include a third "audit and feedback only" group to quantify the size of the improvement caused by these factors alone, as was done in the Leeds study (Adams et al. 1986). As was the case with the checklist effect, study teams also have the option of ignoring it by considering audit and feedback to be components bundled in the overall intervention. If this is consistent with what stakeholders want to know, and the study team members are aware they are ignoring this effect, ignoring it can be a defensible strategy.

Self-Test 11.2
Consider the following table of data from an invented study of a chest pain decision support system in 4 groups of clinicians. The data are mean diagnostic accuracy across all patients managed by senior and junior clinicians during the same year.

Intervention components	Diagnostic accuracy for patients of clinicians in each group during the year (%)	
	Junior clinicians	Senior clinicians
None	45%	65%
Data collection forms	55%	75%
Monthly feedback and data collection forms	60%	80%
Computer advice, feedback, and forms	75%	85%

1. Describe how the various intervention components seem to influence the diagnostic accuracy of junior clinicians.
2. Describe how the various intervention components seem to influence the diagnostic accuracy of senior clinicians.
3. What do these results imply about the impact of the training program on clinicians diagnostic skills, and how could this training program be improved?

Self-Test 11.3

In each of the following short study scenarios, identify any of the potential biases or threats to validity, discussed in this chapter. Each scenario may contain more than one bias or threat to validity. Then consider how you might alter the resource implementation or evaluation plans to reduce or quantify the problem.

1. As part of its initiative to improve patient flow and teamwork, a family practice intends to install an electronic patient scheduling resource when it moves to new premises in 3 months' time. A study team proposes a 1-month baseline study of patient waiting times and phone calls between clinicians, starting at 2 months and conducted prior to the move, to be repeated immediately after starting to use the new resource in 3 months.

2. A bacteriology laboratory is being overwhelmed with requests for obscure tests with few relevant clinical data on the paper request forms. It asks the hospital information system director to arrange for electronic requesting and drafts a comprehensive three-screen list of questions clinicians must answer before submitting the request. The plan is to evaluate the effects of electronic test ordering on appropriateness of requests by randomizing patients to paper request forms or electronic requests for the next year. The staff members intend to present their work at a bacteriology conference.

3. A renowned chief cardiologist in a tertiary referral center on the west coast is concerned about the investigation of some types of congenital heart disease in patients in their unit. A medical informatics expert suggests that the cardiologist's expertise could be represented as reminders about test ordering for the junior staff looking after these patients. The cardiologist agrees, announces these plans at the next departmental meeting, and arranges system implementation and training. Each patient is managed by only one junior staff member; there are enough staff members to allow them to be randomized. After the trial, the appropriateness of test ordering for each patient is judged by the chief cardiologist from the entire medical record. It is markedly improved in patients managed by the clinicians who received reminders. Based on these results, the hospital chief executive agrees to fund a start-up company that could disseminate the reminder system to all U.S. cardiology units.

11.5 Statistical Inference: Considering the Effect of Chance

11.5.1 Effect Size and "Effectiveness"

When designing a demonstration study to measure the impact of an information resource on users or a health problem more generally, the underlying question is usually "Is the information resource effective?". However, the definition of "effective" needs careful thought. Is, for example, a 2% reduction in adverse drug events (ADEs) sufficient for an ePrescribing system to be judged "effective"? If the ADEs studied are those that are generally fatal, then most ePrescribing system developers and users would be thrilled with a 2% reduction. However, if the ADEs studied also include those that are merely minor biochemical disturbances and using the ePrescribing system adds 2 minutes to the time required to enter every prescription, users would probably require a much larger reduction in the number of ADEs to be evidence of effectiveness.

The measured magnitude of change in the key outcome variable(s) is the study's effect size. As discussed below, effect size is a very different concept than statistical significance. In an interventional study, deciding on the magnitude of change that is equated with effectiveness—the desired "effect size"—is very important. This decision has a crucial impact on the number of participants required to demonstrate with confidence that an effect of the desired size, and attributable to the intervention, has occurred.

11.5.2 Statistical Inference

When conducting an interventional study, there are four possible outcomes. In the context of a demonstration study exploring the effectiveness of an information resource, the four possible outcomes are:

1. The information resource was effective, and the study correctly shows this.
2. The information resource was ineffective, and the study correctly shows this.
3. The information resource was effective, but the study mistakenly fails to show this—a type II error.
4. The information resource was ineffective, but for some reason the study mistakenly suggests it was effective—a type I error.

Outcomes 1 and 2 are gratifying from a methodological viewpoint; the results of the study mirror reality. Outcome 3 is a false-negative result, or type II error. In the language of inferential statistics, the study team mistakenly accept the null hypothesis. Type II errors can arise when the size of the sample of participants included in the study is small relative to the size of the information resource's effect on the measure of interest (Freiman et al. 1978). Risks of type II errors relate to the concept of study power discussed in the following section. In Outcome 4 the study team has concluded that the resource is valuable when in reality it is not: a false-positive result, or type I error. In statistical terms, the team has mistakenly rejected the "null hypothesis". When the team accepts, for example, the value of $p < 0.05$ as a criterion for statistical significance, it is consciously accepting a 5% risk of making a Type I error as a consequence of using randomization as a mechanism of experimental control.

11.5.3 Study Power

Every interventional study has a probability of detecting a difference of particular size (the effect size) between the groups. This probability is known as the statistical power of the study design. All other things remaining equal, a larger number of participants is required in a demonstration study to detect a smaller effect size. So, by increasing the number of participants, the power can be increased, though this relationship is nonlinear. One challenge when designing studies is to decide how much of a difference to look for. Ideally, the study should be powered to look for a difference that would be just enough to lead to a change in practice—the "minimum worthwhile difference." For example, although looking for a 30% improvement in student knowledge scores after using a new resource would take very few students to demonstrate, the minimum worthwhile difference that might cause a university to adopt the resource could be only 15%. This suggests planning a larger study to detect this smaller, but still useful, effect.

Power is an important consideration in study design. A study with insufficient participants is unlikely to detect the minimum worthwhile difference between the groups, and will make poor use of the participant time and the study teams' resources (Freiman et al. 1978). While a detailed discussion of statistical power and sample size is beyond the scope of this volume, a clear and comprehensive discussion is found in the text by Blasey (Kraemer and Blasey 2016). If the calculation reveals that more participants are required than are likely to be available to the study team, a number of strategies are possible:

1. Extend the length of the study, to give the intended effect (e.g. disease complications or student knowledge) more time to develop.
2. Use a more easily changed measure of the effect, such as a laboratory test result instead of a self-reported patient outcome. However, a study that demonstrates even a large change in such a "surrogate outcome" may be criticized for being less relevant or important than demonstrating a small change in a real patient outcome. This means that such a study will be less likely to change patient or clinician behavior or the willingness of healthcare organizations to procure the information resource.
3. Use a more powerful study design, for example a crossover study (not discussed in this textbook). Any study design that takes repeated measures on the same participants will have greater power than a design that takes measurements on each participant only once
4. Contact potential collaborators and set up a multicenter study to engage more participants. As well as increasing the power of the study, this will also increase its external validity.

For simple two-group studies with equal numbers of participants allocated to each group, there are several reliable online calculators that can be used to compute the number of participants needed to achieve a desired level of study power (Schoenfeld 2021). However, when more certainty is needed, when the study is designed to demonstrate equivalence (as opposed to differences), when the intent is to measure the time until an event occurs, or when there are more than two groups or one outcome variable, study teams should consult a statistician for advice. Chapter 19, focused on writing of evaluation proposals, will stress again the importance of exploring statistical power.

11.5.4 Data Dredging

When analyzing the results from any kind of study it is always tempting to explore the data to identify "interesting" results, then quote these in the study report, ignoring the findings that turned out not to be "interesting" in the end. However, if the definition used for an "interesting" result is "p less than 0.05", then by definition one result in 20 will be "interesting" by chance alone. If carrying out dozens, or even hundreds, of such analyses, the study team will end up with several results that appear to be interesting just by chance. The internal validity threat here can be stated as retrospective outcome selection based on p values, and there is empirical evidence for this from a systematic review comparing the planned outcomes in protocols for randomized studies with the outcomes reported in the published article (Chan et al. 2004).

Several remedies can be taken by the study team before the study commences to avoid this problem. If multiple comparisons will be needed, the team should revise the criterion used for statistical significance to $p < 0.01$ or 0.001, to reduce the chances of a spurious finding of statistical significance. An alternative to using a standard figure such as 0.01 or 0.001 is to calculate the p threshold according to the number of comparisons that will be tested, using the so-called Bonferroni method (Wikipedia 2021). Also, the study team can pre-publish the study protocol including the agreed statistical analysis plan, which would then obligate them to adhere to this plan.

11.6 Conclusions

Designing rigorous, useful intervention studies makes an important contribution to our understanding of which health and biomedical information resources make a real difference to their users and to the problem they were designed to solve. However, as this chapter has demonstrated, study teams cannot simply rely on simple before-after comparisons to measure the size of an effect. The weakness of crediting all benefit to the information resource in such a study is highlighted by considering the likely response of the resource developers to a situation where performance of clinicians or outcome for patients *worsens* after installing the information resource. Most developers, particularly those directly involved in the creation of the resource, would search long and hard for other factors to explain the deterioration. However, there would be no such search if performance improved, even though the study design has the same faults. This chapter discussed a range of alternative control strategies to overcome this problem, and further biases that can undermine interventional studies, together with strategies to measure or eliminate them.

There is one challenge to intervention studies of information resources that has not yet been discussed, called the evaluation paradox, which in many ways is the opposite of the Hawthorne Effect. In an interventional study conducted in the context of real professional work, users may be understandably reluctant to employ and act on the output of an information resource until its value has been properly established. However, to establish the information resource's value, its output must be acted on, but its output is unlikely to be trusted if a study is in progress and the resource is clearly under evaluation. This "evaluation paradox" applies especially to so-called black box information resources, which provide no insight into the reasons for their output, for example a deep neural network that cannot "explain" its reasoning or a sequencing program that cannot reveal the underlying algorithm. This could cause professionals to ignore the information resource's output and therefore lead to its benefits being underestimated.

One ill-advised strategy to promote resource use and trust in the resource's output might be to deliberately exaggerate the benefits of using it during the study, but it is preferable to give users an honest account of the resource's scope and the performance observed in laboratory tests. This could include the differences between its computational method and that of other resources, or how the same tasks would be performed "by hand", and specific examples of cases where the resource was helpful and where it was not. This approach encourages users to treat the information resource as an aid, not as a black-box dictator. There is no justification for requiring users to always follow an information resource's output during a demonstration study, as this will certainly not be the case when the resource is made available following completion of evaluation studies.

In conclusion, once the study team understands the reasons for adopting a controlled study design and the potential biases that afflict these designs when applied to information resources, it is not hard to generate a plausible study design, which can then be tested and refined against the list of biases discussed here. Once the study has been carried out and the study dataset is available, a variety of analysis methods can be used to generate the results, and are the subject of the next chapter.

Self-Test 11.4

A study team plans to evaluate the impact of a diabetes app on blood glucose control, measured by the proportion of glucose readings recorded by a smart glucometer that are within the desired range. A preliminary study showed that this proportion is in app non-users 60% and the study team's ambition is to show that the proportion will increase to 80% in app users, though they accept that 70% is more likely. They consult a statistician to carry out a series of sample size calculations for their study, who provides these options[3]. Two sample size estimates are deliberately marked only as "X" and "Y".

---

[3] These numbers are estimated, not calculated.

Scenario	Control group	Projected Result for App group	Absolute difference	Desired Statistical significance	Desired Power	Total sample size needed
A	60%	70%	10%	0.05	80%	X
B	60%	70%	10%	0.05	90%	1400
C	60%	80%	20%	0.05	80%	350
D	60%	80%	20%	0.05	90%	Y
E	60%	75%	15%	0.05	80%	600

1. What is the effect size in each scenario?
2. Explain what is meant by "desired statistical significance"?
3. Will the value of "X" be lesser or greater than 350? Explain.
4. Will the value of "Y" be lesser or greater than 350? Explain.

Answers to Self-Tests

Self-Test 11.1

1. (a) Independent variables: time period before and after resource installation (four values).
   (b) Dependent variables and measurement method: attitude to information technology—30-item survey.
   (c) Participants: hospital staff members.
   (d) Control strategy employed: before–after.
2. (a) Independent variables: none; this study is descriptive.
   (b) Dependent variables and measurement method: accuracy of the resource—the proportion of time the computer-generated diagnoses agree with the previously established diagnosis, measured perhaps by an expert panel.
   (c) Participants: none.
   (d) Control strategy employed: none.
3. (a) Independent variables: hospital (two values), advice mode (three values).
   (b) Dependent variables and measurement method: total drug charges, averaged across all relevant patients for each physician during the study period.
   (c) Participants: physicians.
   (d) Control strategy employed: simultaneous randomized study.
4. (a) Independent variables: receipt of advice (two values).
   (b) Dependent variables and measurement method: extent to which the actions recommended by the reminders were in fact taken, measured by a case notes audit.
   (c) Participants: the 12 internal medicine services and the staff employed in them.
   (d) Control strategy employed: simultaneous randomized trial.

5. (a) Independent variables: body system (two values: cardiovascular or gastro-
        intestinal) and version of the resource accessed (two values).
   (b) Dependent variables and measurement method: knowledge scores for both
        disease areas, measured by a validated written test.
   (c) Participants: students.
   (d) Control strategy employed: randomized study or randomized cross-
        over design.

Self-Test 11.2

1. The baseline accuracy of junior clinicians is disappointingly low at 45%. This
   improves by 10% to 55% with forms, and a further 5% with monthly feedback.
   The computer advice improves diagnostic accuracy by a large 15% margin to 75%.
2. The baseline accuracy of senior clinicians is considerably higher at 65%. This
   again improves by 10% to 75% with the forms, and by a further 5% with monthly
   feedback. However, the computer advice only improves diagnostic accuracy by
   a modest 5% to 85%.
3. The training program the senior clinicians have experienced improves their baseline performance by 20% compared to the junior clinicians, but still allows their
   accuracy to improve by the same margin with forms and feedback. This suggests
   that training needs to focus more on the data items included in the checklist, and
   to use more feedback to reinforce these. The fact that the decision support only
   improves accuracy by 5% in the trained clinicians compared to 15% in the
   juniors implies that most of the content in the DSS is covered by the training, so
   it is of limited value to more senior clinicians.

Self-Test 11.3

1. Potential biases or threats to validity: Proposed study confounds impact of new
   premises on patient flow and teamwork with impact of new patient scheduling
   resource; fails to allow time for staff to train on new resource before making
   measures. Baseline measurement period ends on the day that move takes place,
   so last week or so may be disrupted by preparations for the move.
        Improvements to implementation/plan: Start 4-week baseline data collection
   at least 6 weeks before move. Postpone later data collection periods until at least
   4 weeks after move. Ideally, carry out a second data collection period before the
   new resource is implemented, to measure impact of new premises on patient
   flows and communication, followed by a third data collection period once staff
   are familiar with the new resource to quantify additional benefit of the resource
   on top of the move. Take care not to credit any improvements with the resource
   itself—other unknown changes may also have been associated with the move.
2. Potential biases or threats to validity: The trial may fail because the clinicians
   may refuse to fill out the three screens of data to request a test. It will be hard to
   compare the appropriateness of paper-based and detailed electronic requests
   relying on the supplied data alone, as there will be much more data once the
   electronic requesting resource is in place. There may be a carryover effect if

patients are randomized—it would be better to randomize clinicians. The study results may be specific to the tests, electronic requesting resource, and clinicians studied, so the generalizability of the study results to others attending the bacteriology conference may be limited.

Improvements to implementation/plan: Carry out a pilot study to ensure that the new electronic test request resource is usable and is likely to be used before the trial. Determine whether a request was appropriate or not by reference to case notes, not the data supplied. Randomize clinicians, not patients, to eliminate the carryover effect; analyze at the level of clinicians. Generalize from the study results to other settings with caution. Ideally, recruit other labs and conduct a multicenter study.

3. Potential biases or threats to validity: The generalizability of the findings from the study seem low, as this is a tertiary referral center handling particularly challenging cases, and attracting high-flying staff. The benefits therefore may not be replicated when the resource is rolled out to settings where most patients have simpler problems and the staff are less able to respond to the requests and interpret the resulting tests. The attention drawn to the study by an announcement at a departmental meeting could lead to a marked Hawthorne effect, thus reducing the apparent benefit from the reminders. The judgment of appropriate test ordering is carried out by a single cardiologist, whose views may not be shared by the community. The judge of appropriate ordering is the same person as the source of the rules, so the evaluation is circular: the testing is judged appropriate if it was done as she said it should be done.

Improvements to implementation/plan: Recruit a variety of hospitals to the study with a more typical case mix and staffing to enhance generalizability. Ignore the first 2 to 3 weeks of data during the trial to reduce the impact of Hawthorne effects. Carry out some kind of consensus process to develop broadly acceptable criteria of appropriate test ordering in cases such as these. Rather than the circular process above, measure a patient outcome, to see if more appropriate ordering actually helps the patients.

Self-Test 11.4

1. The effect size in the above table varies from an optimistic 20% improvement in the time spent by participants with their blood glucose in normal range to a more realistic and modest 10% improvement.
2. The desired level of statistical significance provided in the scenarios examined by the statistician is a constant 0.05 or 5%. This means that in any of the scenarios, the study team is willing to accept the chances of the result occurring by chance as only 1 in 20.
3. X will be greater than 350 because Scenario A seeks a smaller effect size (absolute difference of 10%) than Scenario C (absolute difference of 20%). All other things being equal, detection of smaller effect sizes requires larger samples.
4. Y will be greater than 350 because Scenario D seeks a high level of desired power than Scenario C. All other things being equal, higher power requires larger samples.

References

Adams ID, Chan M, Clifford PC, Cooke WM, Dallos V, De Dombal FT, et al. Computer aided diagnosis of acute abdominal pain: a multicentre study. BMJ. 1986;293:800–4.

Arditi C, Rège-Walther M, Wyatt JC, Durieux P, Burnand B. Computer-generated reminders delivered on paper to healthcare professionals: effects on professional practice and healthcare outcomes. Cochrane Database Syst Rev. 2012;12:CD001175.

Bland MJ, Altman DG. Statistics notes: regression to the mean. BMJ. 1994;308:1499.

Cartmill RSV, Thornton JG. Effect of presentation of partogram information on obstetric decision-making. Lancet. 1992;339:1520–2.

Chan AW, Hróbjartsson A, Haahr MT, Gøtzsche PC, Altman DG. Empirical evidence for selective reporting of outcomes in randomized trials: comparison of protocols to published articles. JAMA. 2004;291:2457–65.

Cornfield J. Randomization by group: a formal analysis. Am J Epidemiol. 1978;108:100–2.

Diwan VK, Eriksson B, Sterky G, Tomson G. Randomization by groups in studying the effect of drug information in primary care. Int J Epidemiol. 1992;21:124–30.

Freiman JA, Chalmers TC, Smith H, Kuebler RR. The importance of beta, the type II error and sample size in the design and interpretation of the randomised controlled trial. N Engl J Med. 1978;299:690–4.

Friedman CP, Elstein AS, Wolf FM, Murphy GC, Franz TM, Heckerling PS, et al. Enhancement of clinicians' diagnostic reasoning by computer-based consultation: a multisite study of 2 systems. JAMA. 1999;282:1851–6.

Grant SW, Hickey GL, Dimarakis I, Trivedi U, Bryan A, Treasure T, et al. How does EuroSCORE II perform in UK cardiac surgery; an analysis of 23 740 patients from the Society for Cardiothoracic Surgery in Great Britain and Ireland National Database. Heart. 2012;98:1568–72.

Grimshaw JM, Russell IT. Effect of clinical guidelines on medical practice: a systematic review of rigorous evaluations. Lancet. 1993;342:1317–22.

Hemming K, Lilford R, Girling AJ. Stepped-wedge cluster randomised controlled trials: a generic framework including parallel and multiple-level designs. Stat Med. 2015;34:181–96.

Knaus W, Wagner D, Lynn J. Short term mortality predictions for critically ill hospitalized patients: science and ethics. Science. 1991;254:389–94.

Knight SR, Ho A, Pius R, Buchan IE, Carson G, Drake TM, et al. Risk stratification of patients admitted to hospital in the United Kingdom with Covid-19 using the ISARIC WHO Clinical Characterisation Protocol: development and validation of a multivariable prediction model for mortality. BMJ. 2020;370:m3339.

Kraemer HC, Blasey CM. How many subjects?: Statistical power analysis in research. 2nd ed. Thousand Oaks, CA: Sage; 2016.

Liu JL, Wyatt JC. The case for randomized controlled trials to assess the impact of clinical information systems. J Am Med Inform Assoc. 2011;18:173–80.

Lyon HC Jr, Healy JC, Bell JR, et al. PlanAlyzer, an interactive computer-assisted program to teach clinical problem solving in diagnosing anemia and coronary artery disease. Acad Med. 1992;67:821–8.

Medical Research Council. Complex interventions guidance. Swindon, UK: Medical Research Council, UK Research and Innovation; 2019. Available from https://mrc.ukri.org/documents/pdf/complex-interventions-guidance/. Accessed 29 June 2021.

Sacks H, Chalmers TC, Smith H. Randomized vs. historical controls for clinical trials. Am J Med. 1982;72:233–40.

Sarrassat S, Lewis JJ, Some AS, Somda S, Cousens S, Blanchet K. An Integrated eDiagnosis Approach (IeDA) versus standard IMCI for assessing and managing childhood illness in Burkina Faso: a stepped-wedge cluster randomised trial. BMC Health Serv Res. 2021;21:1–19.

Schoenfeld DA. Sample size calculator. Boston: Hedwig, Massachusetts General Hospital, Harvard University; 2021. Available from http://hedwig.mgh.harvard.edu/sample_size/size.html. Accessed 29 June 2021.

Schulz KF, Chalmers I, Hayes RJ, Altman DG. Dimensions of methodological quality associated with estimates of treatment effects in controlled trials. JAMA. 1995;273:408–12.

Suermondt HJ, Cooper GF. An evaluation of explanations of probabilistic inference. Comput Biomed Res. 1993;26:242–54.

Tierney WM, Overhage JM, McDonald CJ. A plea for controlled trials in medical informatics. J Am Med Inform Assoc. 1994;1:353–5.

Wikipedia. Bonferroni Correction. Wikimedia Foundation; 2021. Available from https://en.wikipedia.org/wiki/Bonferroni_correction. Accessed 29 June 2021.

Wyatt JR. Lessons learned from the field trial of ACORN, an expert system to advise on chest pain. In: Barber B, Cao D, Qin D, editors. Proceedings of the sixth world conference on medical informatics, Singapore. Amsterdam: North Holland; 1989. p. 111–5.

Wyatt JC, Altman DG. Commentary: prognostic models: clinically useful or quickly forgotten? BMJ. 1995;311:1539–41.

Wyatt JC, Wyatt SM. When and how to evaluate health information systems? Int J Med Inform. 2003;69:251–9.

Chapter 12
Analyzing Interventional Study Results

Learning Objectives
The text, examples, and self-tests in this chapter will enable the reader to:

1. Select the appropriate analysis method for their study results based on the levels of measurement of the independent and dependent variables in the study.
2. Given a study result expressed as a 2 × 2 contingency table, compute and interpret the different indices of effect size.
3. Given the results of a study expressed as an ANOVA table, interpret the results.
4. Explain why an ROC curve is necessary to express the discriminatory power ("effect size") of a classifier.
5. Express the results of a study in terms of "number needed to treat".

12.1 Introduction

This chapter, the third of four focusing on quantitative demonstration studies, describes the wide range of methods that can be used to analyze the results of interventional studies, which will be the focus of this chapter. Nonetheless, many of these methods apply as well to correlational studies and even some aspects of measurement studies. Because of the many different kinds of analysis methods that are commonly used in biomedical informatics, the chapter starts by describing a strategy for selecting which methods to use based on the types of data to be analyzed, exposing four analysis method options. The chapter then discusses each of the analysis methods in detail, with a description of how to carry out the analysis, the main assumptions of each method, and how to interpret the analytic results. The chapter concludes with a discussion of several alternative methods to communicate effect size.

© Springer Nature Switzerland AG 2022
C. P. Friedman et al., *Evaluation Methods in Biomedical and Health Informatics*, Health Informatics, https://doi.org/10.1007/978-3-030-86453-8_12

Note that many of the calculations described in this chapter are best not performed by hand. The calculator supplied with most computers will perform arithmetic calculations, and many freely available Internet resources will perform more complex statistical calculations and tests, as will commercially available statistical packages. When choosing free resources from the Internet, preference should go to those available from universities, professional societies, and other likely trustworthy sources.

12.2 Grand Strategy for Analysis of Study Results

When the time comes to analyze the data collected during a demonstration study, the levels of measurement of the dependent and independent variables are the most important factors determining the approach taken. Even though there are four possible levels of measurement for any given variable, as introduced in Sect. 6.4, the discussion here requires us only to dichotomize levels of measurement as either discrete (nominal/ordinal) or continuous (interval/ratio). The structure of Table 12.1 reveals four types of analytic situations, depending on whether the dependent and independent variables are continuous or discrete. So the first step in study analysis is to determine into which cell of Table 12.1 the demonstration study falls.

This discussion is limited to studies with a single dependent variable, which are also known as univariate study designs. Having determined the relevant cell of Table 12.1 based on the type of dependent and independent variables in the demonstration study, the next step for the study team is to determine the appropriate index of the effect size. The effect size is a measure of the degree of association between the dependent variable and each of the independent variables. For example, consider a simple two-group demonstration study of a weight loss app with a discrete

Table 12.1 Suggested indices of effect size and statistical inference tests in relation to levels of measurement of the dependent and independent variables for demonstration studies

Dependent variable level of measurement	Independent variable(s) level of measurement	
	Discrete	Continuous
Discrete (binary or categorical)	*Index of Effect Size* Accuracy, sensitivity, kappa, other indices *Statistical Inference Test:* Chi-square	*Index of Effect Size* Area under the ROC curve *Statistical Inference Test:* Comparison of confidence intervals for ROC curve area
Continuous (interval or ratio)	*Index of Effect Size* Differences between group means or medians (for non-normal data); confidence intervals *Statistical Inference Test:* Analysis of variance (*t*-test); *t*-test of group means or Mann Whitney test for non-normal data	*Index of Effect Size* Magnitude of Correlation Regression Coefficients, R squared *Statistical Inference Test:* Tests of Significance of Coefficients

independent variable—use or non-use of the app—and continuous dependent variable such as the amount of weight lost over 3 months. This study fits in the bottom left cell of the table. So the appropriate index of effect size is the difference between the mean values of the dependent variable (weight loss) for each group.[1] At one extreme, if the means are the same in both groups, the effect size is zero. A range of indices of effect size are discussed in more detail with specific examples later in this chapter.

The next step in the grand strategy is to determine an appropriate test of statistical inference. Recall from Sect. 11.5.3 that tests of statistical inference allow estimation of the probability of making a Type I error, concluding that the dependent variable is related to the independent variable(s) when in truth it is not. When this probability is below a chosen threshold value (usually a probability lower than 0.05), the results of the demonstration study are said to be "statistically significant." All other things remaining equal, the larger the study effect size, the lower the probability of making a Type I error. However, the probability of making a Type I error is also related to the sample size (number of participants in the study) and other factors. It is vital to maintain the distinction, both conceptually and when analyzing study results, between effect sizes and results of statistical significance testing. The two concepts are often confused, especially because many analytical techniques simultaneously generate estimates of effect sizes and statistical significance. Nonetheless, there are many reasons to keep these concepts distinct. Among them is the fact that statistically significant results, particularly for studies with large numbers of participants, may have effect sizes that are so small that they have no clinical or practical significance.

The choice of methods to test statistical significance is primarily guided by the levels of measurement of the dependent and independent variables. Table 12.1 suggests some of the possible methods of testing statistical significance for these combinations. For example, analysis of variance (ANOVA) or *t*-tests of the sample means can be used when the independent variables of a study are discrete and the outcome variable is continuous.

When following this grand strategy, it is important to ensure that a study is placed into the cell of Table 12.1 where it naturally belongs, and not forced into a cell that allows analytical procedures with which the investigator is perhaps more familiar. For example, if the dependent variable (outcome measure) lends itself naturally to measurement at the interval or ratio level, there is usually no need to categorize it (or "discretize" it) artificially. Consider a study in which mean blood pressure, a continuous variable, is the outcome measure. To retain statistical power, the directly measured value can and should be used for purposes of statistical analysis. Categorizing measured blood pressure values as low, normal, and high neglects potentially useful differences between observations that otherwise would fall into the same category, and may make the results dependent on what might be arbitrary

---

[1] This assumes that weight loss is normally distributed; if not, a comparison of the medians may be more appropriate.

decisions regarding the choice of thresholds for these categories (Everson et al. 2020).

The discussion continues in the following sections by separately considering each cell of Table 12.1.

Self-Test 12.1

For each of the study scenarios below, indicate the cell of Table 12.1 where the study best fits.

1. An educational test development company conducts a study to discover if a computer administered version or a pencil-and-paper version of the test result in different failure rates.
2. Patients with hypertension are randomized to use three different versions of a smartphone app to determine which version leads to greater blood pressure control.
3. A study is undertaken to explore the degree to which health professionals' information literacy depends on years of education and years of professional experience.

12.3 Analyzing Studies with Discrete Independent and Discrete Dependent Variables: Basic Techniques

12.3.1 Contingency Tables

In health and biomedical informatics, many demonstration studies employ discrete dependent (outcome) variables and independent variables that are also discrete. The most common example from clinical domains compares the output of an information resource—for example, whether a patient has a given disease or not—with some type of accepted "gold standard" indicating whether the patient *really* has the disease or not.

The results of studies with both independent and dependent variables that are discrete (interval or ordinal) are best represented in terms of a contingency table. Contingency tables are matrices that include all combinations of the values of the independent and dependent variables. The dimensionality of the contingency table is equal to the total number of variables in the study design. For example, a study with two independent variables and one dependent variable could be represented as a three-dimensional contingency table. If the first independent variable had two possible values, the second independent variable three possible values, and the dependent variable had two possible values as well, the complete contingency table representing the results would have $2 \times 3 \times 2$ or 12 cells. The results of the demonstration study, for each participant, would fall uniquely into one of these 12 combinations, and the results of the study as a whole could be fully expressed as the total number of observations classified into each cell of the table.

While it is clear that the general case of this kind of study includes contingency tables with an arbitrary number of dimensions, and an arbitrary number of values for each dimension, a very common situation in informatics is the demonstration study that has two discrete variables, each with two possible values, generating a 2 × 2 contingency table to portray and analyze the study results. This important special case is discussed below.

12.3.2 Using Contingency (2 × 2) Tables: Indices of Effect Size

With 2 × 2 contingency tables, many indices of effect size, each with its own strengths and weaknesses, can be reported. Consider a study where the output of an information resource is dichotomized as the predicted presence or absence of some disease such as sepsis, and this output is being compared with assumed to be a proxy for the "truth" about the disease's presence or absence. The "truth" can be the opinions of judges or a more precise diagnosis that can only be made after the prediction. The results of such a study can be described in a 2 × 2 contingency table, as shown in Table 12.2 below, where the shaded cells of the table contain the counts of cases falling into each of the four possible outcomes. The values of the row and column totals are also known as the table's "marginals".

A key property of information resources such as this is its discrimination, or its ability to differentiate between cases that do or do not have the attribute of interest. One index of discrimination that can be reported is the percentage of agreements between the information resource and the presumed "truth". Here this is 75 + 18, or 93 out of 100, or 93%. Citing this crude accuracy alone can cause a number of problems. First, it gives the reader of the study report no idea of what accuracy could have been obtained by chance. For example, consider a diagnostic aid designed to detect sepsis, where the prevalence (prior probability) of sepsis in the test cases is 80%. If a decision support system always suggests sepsis, the measured accuracy over a large number of cases will be 80%. If the resource was slightly more subtle, still ignoring all input data but advising diagnoses solely according to their prevalence, it would still achieve an accuracy of around 64% by chance because it would diagnose sepsis on 80% of occasions, and on 80% of those occasions sepsis would be present.

Table 12.2 A 2 × 2 table showing the results of a hypothetical study of a sepsis diagnosis system

Information Resource's Prediction	"Gold standard"		Row Totals
	Sepsis present	No sepsis	
Sepsis present	75	2	77
No sepsis	5	18	23
Column Totals	80	20	100

Citing accuracy alone also ignores differences between types of errors. The two types of errors are interpreted very differently, and depending on the purpose of the resource, one type of error may be far more serious than the other. That said, a goal of information resource development is to diminish both types of errors. As shown in Table 12.3, an abstracted version of Table 12.2, errors can be classified as false positive (FP) or false negative (FN). The true positive (TP) and true negative (TN) counts represent both kinds of accurate predictions.

A number of metrics or indices of effect size can be calculated from the 2 × 2 table. These are summarized in Table 12.4 below:

Several items are important to know about these indices as they are used in practice:

- Sensitivity and specificity, related to the false negative and false positive rates respectively, are most commonly used. In a field study where an information resource is being used, care providers typically know the output and want to know how often it is correct, or they suspect a disease and want to know how often the information resource correctly detects it.
- In this situation, some care providers find the predictive value positive and the sensitivity, also known as the detection rate, intuitively more useful than the false-positive and false-negative rates.
- The positive predictive value has the disadvantage that it is highly dependent on disease prevalence, which may differ significantly between the test cases used in a study and the environment in which an information resource is deployed.
- Sensitivity and positive predictive value are particularly useful, however, with information resources that issue alarms, as the accuracy, specificity, and false-

Table 12.3 General structure of a 2 × 2 contingency table

Information Resource's Prediction	"Gold standard"		Totals
	Attribute present	Attribute absent	
Attribute present	TP	FP	TP + FP
Attribute absent	FN	TN	FN + TN
Total	TP + FN	FP + TN	N

Table 12.4 Metrics or indices of effect size in contingency tables

Effect size metric	Formula (see Table 12.3)	Value from counts in Table 12.2
Accuracy	(TP + TN)/N	93/100
False negative rate	FN/(TP + FN)	5/80
False-positive rate	FP/(FP + TN)	2/20
Positive predictive value:	TP/(TP + FP)	75/77
Negative predictive value	TN/(FN + TN)	18/77
Sensitivity (detection rate)	TP/(TP + FN)	75/80
Specificity	TN/(FP + TN)	18/20

positive rates may not be defined or obtainable. This is because, in an alarm system that continually monitors the value of one or more physiological parameters, there is no way to count discrete true negative events.

Self-Test 12.2
In a hypothetical study of a predictive model, the results generated by the model are compared with the true state of the patient or other object to generate the contingency table below.

Predictive model output:	Gold standard, i.e. What really happened		
	Patient had complications	No complications	Row totals
Patient will have complications	14	8	22
Patient will not have complications	6	72	78
Column totals	20	80	100

1. How many of the patients in the study developed complications?
2. Out of those patients who developed complications, how many did the model correctly predict and what index of effect size does this ratio correspond to?
3. What is the false negative rate, expressed as a ratio? Explain what this means in words.
4. What is the positive predictive value, expressed as a ratio? Explain what this means in words.
5. What is the false positive rate?
6. What is the overall accuracy of the model?

12.3.3 Chi-Square and Fisher's Exact Test for Statistical Significance

The basic test of statistical significance for 2 × 2 tables is performed by computing the chi-square statistic. Chi-square can tell us the probability of committing type I errors (i.e., incorrectly inferring a difference when there is none). Chi-square can be computed from the following formula:

$$\chi^2 = \sum_i \frac{(O_i - E_i)^2}{E_i}$$

where the summation is performed over all i cells of the table, O_i is the observed value of cell i and E_i is the value of cell i expected by chance alone. The expected values are computed by multiplying the relevant row and column totals for each cell and dividing this number by the total number of observations in the table. For example, Table 12.5 gives the results of a hypothetical laboratory study of an information resource based on 90 test cases. The columns give the gold standard verdict of a

Table 12.5 Hypothetical study results as a contingency table

System's prediction	Panel verdict, observed results (no.)		Total
	Disease	No disease	
Disease	**27** (19.6)	**14** (21.4)	41
No disease	**16** (23.4)	**33** (25.6)	49
Total	43	47	90

Numbers in parentheses are the expected results. Observed results are in boldface

panel as to whether each patient had the disease of interest, and the rows indicate whether the patient was predicted by the system to have the disease of interest. Observed results are in boldface type; expected frequencies for each cell, given these observed results, are in parentheses.[2]

The value of chi-square for Table 12.5 is 9.8. A 2×2 contingency table is associated with one so-called statistical degree of freedom. Intuitively, this can be appreciated from the fact that, once the row and column totals for the table are fixed, changing the value of one cell of the table determines the values of all the other cells. With reference, then, to a standard statistical table, the effect seen in the table is significant at about the .001 level, which means that the study team accepts a 1 in 1000 chance of making a type I error if they conclude that there is a relation between the system's predictions and the verdict of the panel. This of course is below the standard threshold of $p < 0.05$ for statistical significance, so most investigators would report this result as statistically significant.

As with any statistical test, there are cautions and limitations applying to its use. For example, with the small numbers in some cells, as in Table 12.2 above, Chi-square should not be used. Fisher's Exact Test is immune to this problem.

12.3.4 Cohen's Kappa: A Useful Effect Size Index

A very useful index of effect size is given by Cohen's kappa (κ), which compares the agreement between the variables in a contingency table against that which might be expected by chance (Cohen 1968). The formula for calculating κ is:

$$\kappa = \frac{O_{Ag} - E_{Ag}}{1 - E_{Ag}}$$

where O_{Ag} is the observed fraction of agreements (the sum of the diagonal cells divided by the total number of observations) and E_{Ag} is the expected fraction of agreements (the sum of the expected values of the diagonal cells, divided by the total number of observations).

---

[2] For example, the expected value for the disease–disease cell is obtained by multiplying the relevant row total (41) by the relevant column total (43) and dividing the product by the total number of participants (90).

Kappa can be thought of as the *chance-corrected proportional agreement*, (Fleiss 1975) and possible values range from +1 (perfect agreement) via 0 (no agreement above that expected by chance) to −1 (complete disagreement). Some authorities consider a κ above 0.4 as evidence of useful agreement (Fleiss 1975).

In our example in Table 12.4, $O_{Ag} = 0.67$ [(27 + 33)/90] and $E_{Ag} = 0.50$ [(19.6 + 25.6)/90], which makes the value of $\kappa = 0.33$. Note that even though the value of κ is corrected for chance, kappa is an effect size statistic and does not directly convey the result of a formal test of statistical inference.

A more sophisticated version of kappa, weighted kappa, is a similar statistic to Cohen's kappa but incorporates different weights for each kind of disagreement. Further discussion of the use of κ may be found in references (Altman 2018; Hilden and Habbema 1990).

12.4 Analyzing Studies with Discrete Independent and Discrete Dependent Variables: ROC Analysis

The previous section considered the basic techniques that are used when both independent and dependent variables are discrete. This section considers a more complex approach which uses a graphical approach to summarize a series of 2 × 2 tables obtained when an information resource or "classifier", as they are called in machine learning, has an internal threshold that can be varied to define when it issues advice. A classifier uses data about an individual to identify or assign that person into a specific category. The section starts with an explanation of why using a single cut point is usually mistaken, then goes on to describe the solution, a technique called ROC analysis.

12.4.1 Estimating the Performance of a Classifier Using a Single Cut Point

Consider a demonstration study with two variables: a dependent or outcome variable that is discrete and an independent or predictor variable that is continuous. In many circumstances, the purposes of the study are well served by treating the continuous variable as if it were a two-level discrete variable, allowing the results to be displayed in contingency table format. To make the continuous variable discrete, a threshold or cut-point must be selected.

For example, consider a classifier system that instead of simply classifying the patient as having sepsis or not predicts the probability of sepsis for a surgical patient and then sends an "alert" (or not) to clinicians that the patient is in danger of infection. The probability computed by the resource is a continuous variable. In a demonstration study, the investigator may want to relate the predictions of this computational resource to the "truth": whether patients develop an infection or not. The study could be done by treating the variables exactly as measured, with the patients' infectious state as a discrete outcome and the probability generated by the system as a continuous predictor. However, the reminder system, when deployed in the real world, will be programmed to either send an alert or not, so it may be more useful to see how this resource behaves over a range of choices of threshold probabilities for triggering an alert. (Should a computed probability of 0.5 trigger an alert, or should the threshold be higher, say 0.7?) To put it another way, if a suboptimal threshold is chosen for a study, the information resource's accuracy may appear lower than can actually be attained, and the resource will appear less useful than it could be.

12.4.2   ROC Analysis

Receiver operating characteristic (ROC) analysis is a technique commonly used in health and biomedical informatics to explore the relationship between the two variables in the study across a range of choices of threshold, and so becomes a useful tool to assess variation in the usefulness of the resource's advice as an internal threshold is adjusted (Hanley and McNeil 1982). The ROC curve is a plot of the true-positive rate against the false-positive rate for varying threshold levels. Each different choice of threshold level creates a unique 2 × 2 contingency table from which the true-positive and false-positive rates can be computed and subsequently plotted. The ROC curve is generated when these plotted points are connected. If an information resource provides random advice, its ROC curve lies on the diagonal, whereas an ideally performing information resource would have a "knee" close to the upper left hand corner. The area under the ROC curve provides an overall measure of the predictive or discriminatory power of the information resource (Swets 1988).

Let's consider a hypothetical example based on an algorithm was used to classify each case in a database of 600 cases as either having on not having a particular disease. In the database, 200 of the 600 patients were known to have the disease. The algorithm's computed output is a number from 1 to 9. A threshold from 1 to 9 can be used to decide if the algorithm's output for the case would be interpreted as "disease" or not. In an implementation of the algorithm, interpretation of the output as "disease" might then trigger an alert or an alarm.

Table 12.6 shows the results of the test of this algorithm using the 600 patient database. The table displays true positive and false positive rates for the algorithm's output across the range of threshold values. In Row 1, where the threshold is 1, the algorithm's output would be interpreted as no predictions of "disease" so the true and false positive rates are both zero.

Table 12.7 illustrates the more interesting example of the 2 × 2 table where the threshold is 2. The algorithm in this case would trigger very few alarms, 64 in fact, and 20 of these would be incorrect false positives.

Figure 12.1 illustrates the ROC curve for this algorithm, created by plotting the true and false positive rates for all 9 threshold values.

Table 12.6 Results for the example algorithm at different thresholds

Threshold	True positive rate	False positive rate
1	0	0
2	0.22	0.05
3	0.4	0.1
4	0.6	0.16
5	0.75	0.22
6	0.85	0.3
7	0.9	0.4
8	0.97	0.7
9	1	1

Table 12.7 2 × 2 table for example classifier with threshold of 2

	The known truth		
Algorithm prediction	Disease	No disease	Total
Disease	44	20	64
No disease	156	380	536
Total	200	400	600

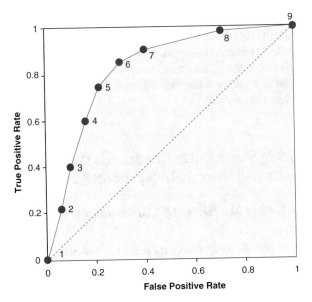

Fig. 12.1 Example receiver operating characteristic (ROC) curve. The threshold values are indicated next to each plotted point

The total area under the ROC curve or "C statistic" can be used as an index of the discrimination of a classifier, which can be viewed as an effect size for this type of study. The shaded region of Fig. 12.1 illustrates the area under the ROC curve, which is equal to .83.

The example described above represents the most common use of ROC curves in informatics. Other uses are possible. For example, ROC curves can be plotted from true positive and false positive results obtained as the number of input data items to an information resource is varied or the number of facts in a decision support system's knowledge base is changed from a small number to a larger number (O'Neil and Glowinski 1990).

Note also that there are occasions when the ROC curve, and the area under it, does not accurately reflect the true discrimination performance of classifier, for example when the actual occurrence of the disease is rare (around 5% or less). This can arise when developing a classifier to support a disease screening program, in which very low disease rates are expected—often 0.5% or less. In this situation, a precision-recall (P-R) curve is preferred, as this more accurately reflects differences in classifier performance at low event rates (Ozenne et al. 2015; Davis and Goadrich 2006). The P-R curve is constructed in a similar way to the ROC curve, but plots precision (i.e. positive predictive value) versus recall (i.e. sensitivity) at different threshold levels.

Self-Test 12.3

1. Table 12.7, drawing on the data in Table 12.6, provides the 2 × 2 table for the example classifier with a threshold of 2. Create the 2 × 2 table, analogous to Table 12.7, for the threshold value of 4.
2. With the threshold set at 9, the algorithm would predict "disease" for every case. Explain in words, why, in this case, the true positive and false positive rates would both be 1. First try doing this without drawing the table, then draw the table to confirm your answer.

12.5 Analyzing Studies with Continuous Dependent Variables and Discrete Independent Variables

12.5.1 Two Group Studies with Continuous Outcome Variables

The simplest case here is when the study team has recorded observations on two different groups of participants and needs to compare these. Assuming that these observations fall on a normal distribution—an approximate test for this is whether the SD is less than half the mean and the median is similar to the mean—the independent sample t-test applies. These generate a p value which can be compared with a selected threshold for statistical significance, typically $p < 0.05$, but often adjusted to a lower value if many such comparisons are being made. An alternative is to generate a 95% confidence interval (CI) for each of the two means, which can be compared[3]. The 95% confidence interval is computed as:

$$95\%CI = Mean \pm \left(1.96 \times SE_{mean}\right)$$

$$where\, SE_{mean} = \frac{SD}{\sqrt{N}}$$

If these CIs overlap, then there is no statistical difference between the two groups. Using confidence intervals also helps the study team make sense of the data and communicate the results to other stakeholders, to determine whether the results are significant in clinical, educational or other terms—irrespective of whether they are statistically significant.

There are two variations on this simplest-case approach. First, if the data are not normally distributed, an alternative method is needed to calculate the p value. The most commonly used is the Mann Whitney Test, but other "non parametric" statistical tests can also be used.

The other relatively common variation is when the two groups are not statistically "independent" of each other. This usually occurs when two sets of observations are made on the same individuals, for example measuring a person's weight before and 1 month after they start using a weight loss app, or when analyzing the results of a crossover trial. In that case, using the simple t-test is incorrect, even if the weights are normally distributed. The correct test to use here is the paired sample t-test.

In the case in which the data are both paired and not normally distributed, the correct test to use is the Wilcoxon signed rank test.

The next section introduces a more sophisticated approach to analyzing data from more complex study designs.

12.5.2 Alternative Methods for Expressing Effect Size

When analyzing the result of studies with two or more groups, each of which is measured as a continuous variable, there are a number of options to calculate and communicate the effect size. For example, imagine that the mean glucose level in diabetics using an app in a study is 7.8 mmol and for control patients it is 8.8 mmol. The difference of 1mMol in the mean figures is called the absolute difference. However, the difference can also be expressed as a percentage of the smaller value, which is 1/7.8 or 12.8%. This is called the relative difference.

A useful way of expressing effect sizes for this kind of study is Cohen's d, which, for any pair of cells of the design, is the difference between the mean values divided

---

[3]The confidence interval was first introduced in Sect. 10.4.3.

by the standard deviation of the observations. Use of Cohen's *d* allows standardized expression of effect sizes in "standard deviation units," which are comparable across studies. Traditionally, effect sizes of 0.8 standard deviations (or larger) are interpreted as "large" effects, 0.5 standard deviations as "medium" effects, and 0.2 standard deviations (or smaller) as "small" effects (Cohen 1988). If the SD here is 2, this gives a *d* of 0.5, which is conventionally considered a medium effect size.

12.5.3 Logic of Analysis of Variance (ANOVA)

Statistical methods using analysis of variance (ANOVA), discussed briefly here, exist specifically to analyze the results of studies with continuous dependent and discrete independent variables.

Recognizing that study design and ANOVA are the topics of entire textbooks, (Kleinbaum 1992) the basic principles are introduced here using the results of an actual study as an example. The example is based on the results of a biomedical information retrieval study conducted at the University of North Carolina (Wildemuth et al. 1998). The study explored whether a Boolean search tool or Hypertext access to a text database resulted in more effective retrieval of information to solve biomedical problems. The biomedical information available to participants was a "fact and text" database of bacteriology information and was identical across the two access modes. With the Boolean search tool, participants framed their queries as combinations of key words joined by logical *and* or *or* statements. With the Hypertext mode, participants could branch from one element of information to another via a large number of preconstructed links customary to the website.

In this study, medical students were randomized either to the Boolean search or Hypertext access mode. These participants were also randomized to one of two sets of clinical problems, each set comprising eight clinical infectious disease scenarios. Students were given two passes through their eight assigned problems. On the first pass they were asked to generate diagnostic hypotheses using only their personal knowledge. Immediately thereafter, on the second pass, they were asked to generate another set of diagnostic hypotheses for the same set of problems, but this time with aid from the text database—a within-subject design. On each occasion the student was scored for the appropriateness of their diagnostic hypotheses. The dependent variable is the improvement in the diagnostic hypotheses from the first pass to the second—the differences between the aided and unaided scores—averaged over the eight assigned cases. This variable was chosen because it estimates the effect attributable to the information retrieved from the text database, controlling for each participant's prior knowledge of bacteriology and infectious disease.

The study has two independent variables, each measured at the nominal level:

1. Access mode: Boolean or Hypertext.
2. The set of eight case problems to which students were assigned, arbitrarily labeled set A and set B.

Because each of the two independent variables has two possible values, the study design has four groups of subjects as shown in Table 12.8. Table 12.9 displays the result of the study.

The logic of analyzing data from such an experiment is to compare the mean values of the dependent variable across each of the groups. Table 12.9 displays the mean, standard deviations and confidence intervals of the improvement scores for each of the groups. For all participants, the mean improvement score is 16.4 with a standard deviation of 7.3 (95% CI: 14.2 to 18.6).

Take a minute to examine Table 12.9. It should be fairly clear that there are differences of potential interest between the groups. Across problem sets, the improvement scores are higher for the Hypertext access mode than the Boolean mode. Across access modes, the improvement scores are greater for problem set B than for problem set A.

In this table, the effect size d for the first column, the difference in improvement scores for hypertext versus Boolean access for problem set A, is: (15.7–10.2)/5.7, or 0.96—a large effect. For the second column, the difference in improvement scores for hypertext versus Boolean access for problem set B, d is: (21.8–17.5)/7.3 or 0.59—a medium effect.

12.5.4   Using ANOVA to Test Statistical Significance

The methods of ANOVA generate the probability that the effect sizes reflected in differences between group means, whatever the magnitude of these differences, arose due to chance alone. Differences in the mean of the dependent variable

Table 12.8 Structure of the information retrieval study example

Access mode	Assigned problem assigned set	
	Problem set A	Problem set B
Boolean	Group 1	Group 2
	($n = 11$)	($n = 11$)
Hypertext	Group 3	Group 4
	($n = 10$)	($n = 10$)

attributable to each of the independent variables are called *main effects*; differences attributable to the independent variables acting in combination are called *interactions*. The number of possible main effects is equal to the number of independent variables; the number of possible interactions increases geometrically with the number of independent variables. With two independent variables there is one interaction; with three independent variables there are four; with four independent variables there are 11.[4]

In the example with two independent variables, two main effects and one interaction need to be tested. Table 12.10 shows the results of ANOVA for these data. For purposes of this discussion, note the following:

1. The *sum-of-squares* is an estimate of the amount of variability in the dependent variable attributable to each main effect or interaction. All other things being equal, the greater the sum-of-squares, the more likely is the effect to be statistically significant.

2. A number of statistical *degrees of freedom* (*df*) is associated with each source of statistical variance. For each main effect, *df* is one less than the number of possible values of the relevant independent variable. Because each independent variable in our example has two possible values, *df* = 1 for both. For each interac-

Table 12.9 Results of the information retrieval study example

Access mode	Results: improvement in student score over baseline, by assigned problem set: mean ± SD, 95% confidence interval	
	Problem set A	Problem set B
Boolean	10.2 ± 5.7, CI 6.8 to 13.6	17.5 ± 7.3, CI 13.2 to 21.8
	(*n* = 11)	(*n* = 11)
Hypertext	15.7 ± 7.2, CI 11.2 to 20.2	21.8 ± 5.2, CI 18.6 to 25.0
	(*n* = 10)	(*n* = 10)

Table 12.10 Analysis of variance results for the information retrieval example

Source	Sum of squares	*df*	Mean square	*F* ratio	*p*
Main effects					
Problem set	400.935	1	400.935	9.716	.003
Access mode	210.005	1	210.005	5.089	.030
Interaction					
Problem set by access mode	0.078	1	0.078	0.002	.966
Error	1568.155	38	41.267		

---

[4]With three independent variables (A, B, C), there are three two-way interactions (AB, AC, BC) and one three-way interaction (ABC). With four independent variables (A, B, C, D), there are six two-way interactions (AB, AC, AD, BC, BD, CD), four three-way interactions (ABC, ABD, ACD, BCD), and one four-way interaction (ABCD).

tion, the *df* is the product of the *df*s for the interacting variables. In this example, *df* for the interaction is 1, as each interacting variable has a *df* of 1. Total *df* in a study is one less than the total number of participants.

3. The *mean square* is the sum of squares divided by the *df*.
4. The inferential statistic of interest is the *F* ratio, which is the ratio of the mean square of each main effect or interaction to the mean square for error. The mean square for error is the amount of variability that is unaccounted for statistically by the independent variables and the interactions among them. The *df* for error is the total *df* minus the *df* for all main effects and interactions.
5. Finally, with reference to standard statistical tables, a *p* value may be associated with each value of the *F* ratio and the values of *df* in the ANOVA table.

In the example, both main effects (mode of access and problem set) meet the conventional criterion for statistical significance (p < .05), but the interaction between the dependent variables does not. Note that the ANOVA summary (Table 12.10) does not reveal anything about the direction or substantive implications of these differences across the groups. Inspecting the mean values for the groups, as shown in Table 12.9, leads to a conclusion that the Hypertext access mode is associated with higher improvement scores and that the case problems in set B are more amenable to solution with aid from the database than are the problems in set A. Because there is no statistical interaction, this superiority of Hypertext access is consistent across problem sets.

A statistical interaction would be in evidence if, for example, the Hypertext group outperformed the Boolean group on set A, but the Boolean group outperformed the Hypertext group on set B. To see what a statistical interaction means, it is frequently useful to make a plot of the group means, as shown in Fig. 12.2, which depicts the study results represented in Table 12.9. Departure from parallelism of the lines connecting the plotted points is the indicator of a statistical interaction. In this case, the lines are nearly parallel.

12.5.5 Special Issues with ANOVA

In this section it was possible only to scratch the surface of analysis of study results using ANOVA methods. To close this section of the chapter, three special issues are mentioned:

1. In the special case where a study has one independent variable with two levels, this is the familiar two group study where the *t*-test (described earlier) applies. Applying ANOVA to this case yields the same results as the *t*-test, with $F = t^2$.
2. The analysis example discussed above pertains only to a completely randomized factorial design. The ANOVA methods employed for other designs, including nested and repeated measures designs, require special variants on this example.
3. Appropriate use of ANOVA requires that the measured values of the dependent variables are distributed roughly according to a "normal" distribution, and also meet other statistical requirements. If the measured dependent variables fail to

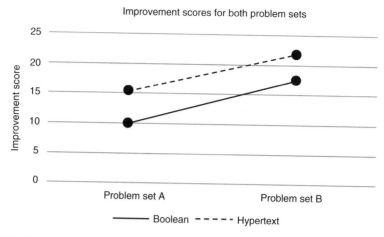

Fig. 12.2 Graphing results as a way to visualize statistical interactions

meet these assumptions, corrective actions such as log transformation of the data may be required, or ANOVA methods may not be applicable.

Self-Test 12.4

1. Given below are further data from the Hypercritic study discussed in Sect. 8.6. For these data, compute (a) Hypercritic's accuracy, sensitivity, and specificity; (b) the value of chi-square; and (c) the value of Cohen's κ.

Hypercritic	Pooled rating by judges		
	Comment valid (≥5 judges)	Comment not valid (<5 judges)	Total
Comment generated	145	24	169
Comment not generated	55	74	129
Total	200	98	298

2. At each of two metropolitan hospitals, 18 physicians are randomized to receive computer-generated advice on drug therapy. At each hospital, the first group receives advice automatically for all clinic patients, the second receives this advice only when the physicians request it, and the third receives no advice at all. Total hospital charges (in tens of dollars) related to drug therapy are measured by averaging across all relevant patients for each physician during the study period, where relevance is defined as patients whose conditions pertained to the domains covered by the resource's knowledge base. Hypothetical results of that study are summarized in the two tables below. The first table gives mean and standard deviation of the outcome measure, charges per patient, for each cell of the experiment. Note that $n = 6$ for each cell. The second table gives the ANOVA results.

Interpret these results in terms of main effects and interactions.

Means and standard deviations:

Hospital	Advice mode		No advice
	Advice always provided	Advice when requested	
A	58.8 ± 7.9	54.8 ± 5.8	67.3 ± 5.6
B	55.2 ± 7.4	56.0 ± 4.7	66.0 ± 7.5

ANOVA results:

Source	Sum-of-squares	df	Mean square	F ratio	p value
Main effects hospital	12.250	1	12.250	0.279	.601
Advice mode	901.056	2	450.528	10.266	<.001
Interaction					
Hospital by group	30.500	2	15.250	0.348	.709
Error	1316.500				

3. Consider an alternative outcome (shown below) of the information retrieval study described in Sect. 12.5.3. Make a plot of the results shown below that is analogous to those shown in Fig. 12.2. What would you conclude with regard to a possible statistical interaction?

Access mode	Mean ± SD, by assigned problem set	
	Problem set A	Problem set B
Boolean	19.3 ± 5.0	12.4 ± 4.9
	($n = 11$)	($n = 11$)
Hypertext	15.7 ± 5.7	21.8 ± 5.1
	($n = 10$)	($n = 10$)

12.6 Analyzing Studies with Continuous Independent and Dependent Variables

12.6.1 Studies with One Continuous Independent Variable and One Continuous Dependent Variable

In this most basic example, the study team makes measurements of the dependent or outcome measure and one independent variable, and both these variables are continuous in nature, such as user age or usage rate for an information resource. Many readers will be familiar with the use of simple regression analysis to analyze such data. The continuous dependent variable is plotted against a single continuous independent variable on an *x-y* graph and a regression line is fitted to the data points, typically using a least squares algorithm—see Fig. 12.3 for an example. Here, the

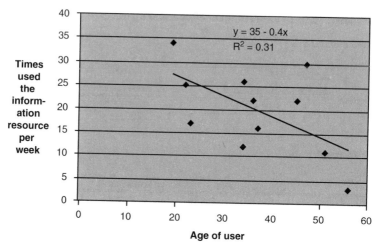

Fig. 12.3 Simple regression analysis of user age versus the number of times each user logged on to an information resource per week

number of times each of a set of 11 users logged into an information resource per week is plotted against the age of that user.

In addition, Fig. 12.3 shows the "regression line" that is a best fit to the data and displays the calculated algebraic expression of the line. However, it is clear from the graph that there is a lot of scatter in the data, so that although the general trend is for lower log-on rates with older users, some older users (e.g., one of 47 years) actually show higher usage rates than some younger users (e.g., one of 23 years).

In this two-variable example, the slope of the regression line is proportional to the statistical correlation coefficient (r) between the two variables. The square of this correlation, seen as R^2 in Fig. 12.3, is the proportion of the variation in the dependent variable (here, log-on rate) that is explained by a linear relationship with the independent variable (here, age). In this example, the R^2 is 0.31, meaning that 31% of the variance in usage rate is accounted for by variance in the age of the user, and the other 69% of the variance in usage rate is **not** accounted for. In this case, the value of R^2 is a measure of effect size. To obtain a test of statistical significance, standard statistical tables will reveal the probability that true value of R^2 differs from zero.

12.6.2 Studies with Multiple Independent Variables and a Continuous Dependent Variable

If the study team had access to additional data about these users, such as their number of years of computer experience or their scores on a computer attitude survey, they could try to more closely predict or explain their information resource usage rate by using multiple regression analysis. This might help to reduce the amount of variation in the dependent variable that is currently unexplained (69%). In multiple regression, values of the single continuous dependent variable are analyzed

Table 12.11 Formatting of an example data file for multiple regression (only three participants' data are shown)

	Dependent variable	Independent variables		
User ID	Usage rate per week	Age	Years of computer experience	Score on computer attitude scale
1	25	21	3	70%
2	30	47	25	65%
3	3	56	5	23%

simultaneously with those of two or more continuous independent variables to identify the unique contributions of each independent variable to these changes. The format of the data file required for such an analysis is shown in Table 12.11 below:

Multiple regression generates a "best fit" equation of the form:

$$y = b_1 x_1 + b_2 x_2 + b_3 x_3 + \text{constant}$$

where each x denotes a different independent variable and b_1, b_2, and b_3 are coefficients that can be computed from the study data.

In the example above, the study team might find that 55% of the variation in usage rate can now be explained by the following equation:

$$\text{Usage rate} = 45 - (0.4 \times \text{age}) + (0.2 \times \text{years of computer experience}) + (0.1 \times \text{computer attitude score})$$

This equation shows that user age remains a key factor (with older people generally using the resource less frequently), but that their number of years of computer experience is also an important independent factor, acting in the opposite direction. Users with more experience tend to use the resource more frequently. The user's score on a computer attitude scale is a third independent positive predictor. Users with a higher attitude score also tend to use the information resource more frequently.

The coefficients in these regression equations may be seen as indices of effect size. However, the magnitudes of the regression coefficients depend on the measurement units of each of the variables and thus cannot be compared directly to each other. Comparison of regression coefficients with each other requires use of standardized coefficients. This topic, and tests of statistical significance for regression coefficients are beyond the scope of this book but are available in standard references on regression analysis.

12.6.3 Relationships to Other Methods

From the above example it can be seen that multiple regression methods can be very useful for analyzing correlational studies with continuous variables as it helps to identify which independent variables matter, and in which direction. Such methods are further discussed in Chap. 13. However, the standard methods for multiple regression analysis often cannot be used when the independent variables include a mix of discrete and continuous variable types. This more general situation requires an expanded method called logistic regression that is beyond the scope of this volume and discussed in a range of texts (Kleinbaum 1992).

It is also the case that ANOVA and the regression methods discussed in this chapter belong to a general class of analytic methods known as general linear models. With appropriate transformations and mathematical representations of the independent variables, an ANOVA performed on demonstration study data can also be performed using multiple regression methods—with identical results. ANOVA and regression were introduced separately for different cells of Table 12.1 because ANOVA is better matched to the logic of interventional studies where the most important independent variables are discrete. Regression analysis is better matched to the logic of correlational studies, where most or all variables are continuous.

12.7   Choice of Effect Size Metrics: Absolute Change, Relative Change, Number Needed to Treat

This section offers a discussion of alternative ways of presenting effect sizes in demonstration studies. The goal of demonstration studies is to inform and enhance decisions about an information resource. Those carrying out such studies should not try to exaggerate the effects observed, any more than they would deliberately ignore known biases or threats to generality. Thus, it is important to describe the results of the study, particularly the effect sizes, in terms that effectively and accurately convey their meaning.

As an example, consider the study results in Table 12.12 below, showing the rates of postoperative infection in patients managed by clinicians before and after the introduction of antibiotic reminders. These results can be summarized in three main ways, each of which invites a slightly different interpretation:

1. *By citing the absolute difference (after intervention versus baseline) in infection rates.* The result can be portrayed as a "reduction in infection rates from 11% to 6%" which corresponds to an absolute difference of 5% as shown in the "abso-

Table 12.12 Hypothetical results of a simultaneous randomized controlled study of an antibiotic reminder system

Time	Postoperative Infection rate (%)	
	Reminder cases	Control cases
Baseline	11%	10%
After intervention	6%	8%
Absolute reduction in rate	5%	2%
Relative reduction in rate	−46%	−20%

lute difference" row of the table. But this is potentially misleading since the change in the control cases where reminders were not received was from 10% to 8%, presumably due to other factors. The absolute difference of 5% in the reminder group should thus be corrected by 2% to reflect the absolute difference seen in the control cases.

2. *By citing the relative difference in the percentage infection rates due to the reminders.* As shown in the "relative difference" row of the table, there is a 46% relative fall in the reminder cases, which should be compensated by the 20% fall in the control cases. So a more conservative estimate of the relative difference would be 26%: the 46% fall in the reminder group minus the 20% fall due to nonspecific other factors in the control cases.

3. *By citing the "number needed to treat" (NNT).* This figure suggests how many patients would need to be treated by the intervention to produce the result of interest in one patient, in this case prevention of an infection. The NNT is computed as:

$$NNT = \frac{1}{rate_1 - rate_2}$$

where $rate_1$ = absolute reduction in rate of the event in group 1 and $rate_2$ = absolute reduction of the rate of the event in group 2.

In this case, $rate_1$ (5% = .05) minus $rate_2$ (2% = .02) is 3% or 0.03, so the NNT is 33 (the reciprocal of 0.03) for these results. To put it another way, reminders would need to be issued for an average of 33 patients before one postoperative infection would be prevented.

Use of NNT to report effect sizes is highly recommended where appropriate. Several studies have shown that clinicians make much more sensible practice decisions when study results are cited as NNT rather than absolute or relative percentage differences (Bobbio et al. 1994). For studies in health and biomedical informatics, the NNT is often the most helpful way to represent the effects of implementing an information resource as determined by an interventional study.

12.8 Special Considerations when Studying Predictive Models

12.8.1 Use of Training and Testing Datasets

Several challenges arise when estimating the performance of a classifier, such as a risk prediction algorithm based on machine learning. Typically such algorithms are developed using training data derived from a large dataset collected in one or more settings, often at some time in the past (sometimes many years ago), or in a different country from that in which the classifier will be used. The classifier performance is then estimated using a different part of the same large dataset—often selected randomly from the original dataset. While this approach, of model development using a "training set" and then testing it using a separate "test set", overcomes random error and penalizes any overfitting of the algorithm to the training data, unfortunately it does not take account of some other issues that may impact model performance in a new setting (Peek and Abu-Hanna 2014). These issues include:

- Overfitting: When a statistical model that powers an information resource is carefully adjusted to achieve maximal performance on training data, this adjustment may worsen its accuracy on a fresh set of data due to a phenomenon called overfitting (Peek and Abu-Hanna 2014). If the accuracy of the classifier is measured on a small number of hand-picked cases, it may appear spuriously excellent. This is especially likely if these cases are similar to, or even identical with, the training set of cases used to develop or tune the classifier before the evaluation is carried out. Thus, it is important to obtain a new, unseen set of cases and evaluate performance on this "test set".
- Changes in the definition or processes used to collect the input data for the algorithm. For example, differences in the use of disease and other codes between different organizations such as primary care practices, already mentioned. In the retinopathy example discussed in more detail below (Beede et al. 2020), changes in the quality of test data collected in rural health settings led to most of the degradation in performance of the classifier.
- Changes in the definition, or processes to define, the outcome of interest. For example, the criteria used to diagnose many diseases changes over time, as people understand more about the underlying pathophysiology. This means that a classifier that worked well on data collected 5 years ago, or in a country that uses different disease definitions, may fail on new data collected once the disease definition changed.

The remedy for these problems is to use a relevant, contemporary, unseen test set, ideally collecting new data in a selection of centers that are representative of those in which the algorithm will be used.

Sometimes developers omit cases from a test set if they do not fall within the scope of the classifier, for example, if the final diagnosis for a case is not represented in a diagnostic system's knowledge base. This practice violates the principle that a test set should be representative of all cases in which the information resource will be used and will overestimate its accuracy with unseen data.

12.8.2 Some Cautionary Tales About Studying Predictive Models

Unfortunately, most published studies of medical applications of AI classifiers are retrospective, for example in a 2019 *Lancet* systematic review, 72 (88%) of 82 studies comparing AI with doctors were retrospective, and only 25 (36%) carried out their validation study on an external sample and were included in the meta-analysis, of the 69 studies that reported sufficient detail to enable results to be calculated (Liu et al. 2019).

One example is detecting malignant melanoma from skin photographs. A team of scientists in Stanford developed a very accurate algorithm for detecting melanoma that matched or exceeded the performance of a team of specialists (Estava et al. 2017). However, it later emerged that this algorithm was often basing its recommendations on the presence of a measurement ruler in images of melanoma, placed there by dermatologists following a biopsy (Narla et al. 2018). So, the algorithm was using data that would not be available at the stage in the diagnostic pathway when it would be most useful.

Another cautionary tale about studying predictive models comes from work by Google Health in Thailand (Beede et al. 2020), who developed a sophisticated deep learning model to detect retinopathy based on retinal images, then prospectively tested this on images captured from new patients in 11 rural health clinics. Unfortunately, their studies indicated that socio-environmental factors strongly impacted model performance, nursing workflows, and patient experience. They concluded that human-centered evaluative research needs to be conducted alongside prospective evaluations of model accuracy.

12.9 Conclusions

Given the wide range of possible interventional studies, it is important for study teams to have a strategy to guide them in choosing the appropriate analytical techniques. This chapter has focused on the nature of the dependent and independent variables as a way to decide which of the many analytical techniques make sense in different settings. However, as always before carrying out an analysis, it is wise to first review the study data to check that the assumptions that are made by the techniques selected are not being violated, and if necessary to consult with a statistician who can give advice and reassurance about which method is appropriate. The final section reviewed the range of methods that are available for communicating study results, to ensure that readers of the report are less likely to over-react to the findings.

Finally, even if they are not carrying out a study, this chapter can be used by those critically appraising study reports to check that the appropriate analysis methods have been used by the study team.

Self-Test 12.5

1. In a prospective survey of a carefully selected random sample of patients in a regional health care systems, respondents were asked the likelihood that they would use a patient portal. They responded on a visual analogue scale creating a continuous variable. Overall, 500 of the 600 patients surveyed returned a com-

pleted survey instrument. The survey analysts were asked to estimate the extent to which these "likelihood" responses were linearly related to respondents' age, their years of education completed, and how many digital devices they had in their households.

  (a) In what cell of Table 12.1 does this analysis lie?

  (b) Following the examples in Sect. 12.6.2, write the equation corresponding to the statistical model the regression analysis would fit to the survey data. Provide as much detail as possible.

  (c) Would you find the results of this analysis to be credible? Why or why not?

2. The table below is a variation on Table 12.12.

Time	Postoperative Infection rate (%)	
	Reminder cases	Control cases
Baseline	8%	9%
After intervention	2%	8%
Absolute reduction in rate	A	B
Relative reduction in rate	X	Y

  (a) Complete the missing cells of the table: the values of A, B, X, and Y.

  (b) Intuitively, would you expect the NNT for this case to be greater or lesser than that for the result in Table 12.12? To confirm your intuition, compute the "number needed to treat" for these results.

Answers to Self-Tests

Self-Test 12.1

1. Upper left cell: Both the independent and dependent variables are discrete.
2. Lower left cell: Independent variable is discrete and dependent variable (blood pressure) is continuous.
3. Lower right cell: Both the independent and dependent variables are continuous.

Self-Test 12.2

1. It can be seen that 20 of the 100 patients studied had complications, an overall complication rate of 20%.
2. Of these 20 people, the predictive model correctly predicted that 14 would have complications, so its sensitivity or true positive rate is 14/20 = 70%.
3. However, it missed 6 of the 20 who went on to have complications, a false negative rate of 30%.
4. The positive predictive value is 14/22. Of the 22 patients predicted to have complications, 14 in fact did.

5. Turning to the 80 patients who had no complications, the predictive model correctly predicted 72 of these, for a true negative rate of 90%, and labelled 8 as likely to have complications when they did not, a false positive rate of 10%.
6. Overall, of the 100 cases, the model correctly predicted the outcome in 14 + 72 = 86, giving an overall accuracy of 86%.

Self-Test 12.3

1. The 2 × 2 table for the example classifier and threshold value of 4 is:

Algorithm prediction	The known truth		Total
	Disease	No disease	
Disease	120	64	184
No disease	80	336	416
Total	200	400	600

2. If the algorithm predicted "disease" for every case, it would be correct for all 200 cases with the disease (TPR = 1) and incorrect for all 400 cases without the disease (FPR = 1). The corresponding 2 × 2 table is below:

Algorithm prediction	The known truth		Total
	Disease	No disease	
Disease	200	400	600
No disease	0	0	0
Total	200	400	600

Self-Test 12.4

1. Accuracy = (145 + 74)/298 = 0.73; sensitivity = 145/200 = 0.72; specificity = 74/98 = 0.75. (b) chi-square = 61.8 (highly significant with $df = 1$). (c) $\kappa = 0.44$.
2. By inspection of the ANOVA table, the only significant effect is the main effect for the advice mode. There is no interaction between hospital and group, and there is no difference, across groups in mean charges for the two hospitals. Examining the table of means and standard deviations, it is clear that the means are consistent across the two hospitals. The mean for all participants in hospital A is 60.3 and the mean for all participants in hospital B is 59.1. This small difference is indicative of the lack of a main effect for hospitals. Also note that, even though the means for groups vary, the pattern of this variation is the same across the two hospitals. The main effect for the groups is seen in the differences in the means for each group. It appears that the difference occurs between the "no advice" group and the other two groups. Although the F test used in ANOVA can tell us only if a global difference exists across the three groups, methods exist to test differences between levels of the dependent variables.
3. Non-parallelism of lines is clearly suggestive of an interaction. A test using ANOVA is required to confirm if the interaction is statistically significant.

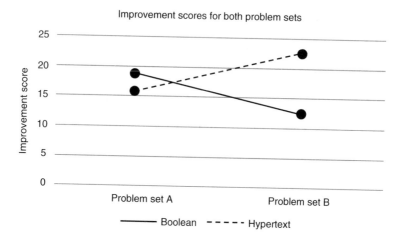

Improvement scores for both problem sets

Self-Test 12.5

1a. Lower right cell, as all variables are continuous
1b. (Likelihood of using the portal) =

$$b_1 \times (\text{age}) + b_2 \times (\text{years of education}) + b_3 \times (\text{number of devices}) + \text{constant}$$

The values of the regression coefficients b_i and the value of the constant would be the results of the analysis generating the "best fit" model to the data.

1c. The results of the study would appear to be highly credible since a large proportion of a randomly-selected sample responded. There might be some concern if the non-respondents were highly concentrated in one geographic area or shared particular//demographic characteristics.

2a. A = 6%, B = 1%, X = 75%, Y = 11%.
2b. Intuitively, the NNT should be lower since the effect size is greater.

To compute the NNT:

$$\text{rate}_1 = .06 \text{ and rate}_2 = .01, \text{ so NNT} = 1 / .05 = 20.$$

References

Altman DG. Practical statistics for medical research. London: Chapman & Hall/CRC; 2018.
Beede E, Baylor E, Hersch F, Iurchenko A, Wilcox L, Ruamviboonsuk P, et al. A human-centered evaluation of a deep learning system deployed in clinics for the detection of diabetic retinopathy. In: CHI'20: Proceedings of the 2020 CHI Conference on Human Factors in Computing Systems, Honolulu, Hawaii, USA. New York: Association for Computing Machinery, 2020;1–12. Available from https://doi.org/10.1145/3313831.3376718. Accessed 29 June 2021.

Bobbio M, Demichelis B, Ginstetto G. Completeness of reporting trial results: effect on physicians' willingness to prescribe. Lancet. 1994;343:1209–11.

Cohen J. Weighted kappa: nominal scale agreement with provision for scaled disagreement or partial credit. Psychol Bull. 1968;70:213–20.

Cohen J. Statistical power analysis for the social sciences. Hillsdale, NJ: Lawrence Erlbaum Associates; 1988.

Davis J, Goadrich M. The relationship between precision-recall and ROC curves. In: ICML'06: Proceedings of the 23rd international conference on machine learning. New York: Association for Computing Machinery, 2006;233–40. Available from https://doi.org/10.1145/1143844.1143874. Accessed 29 June 2021.

Estava A, Kuprel B, Novoa R, Ko J, Swetter SM, Blau HM, et al. Dermatologist level classification of skin cancer with deep neural networks. Nature. 2017;542:115–8.

Everson J, Rubin JC, Friedman CP. Reconsidering hospital EHR adoption at the dawn of HITECH: implications of the reported 9% adoption of a "basic" EHR. J Am Med Inform Assoc. 2020;27:11981205.

Fleiss JL. Measuring agreement between two judges on the presence or absence of a trait. Biometrics. 1975;31:357–70.

Hanley JA, McNeil BJ. The meaning and use of the area under a receiver operating characteristic (ROC) curve. Radiology. 1982;143:29–36.

Hilden J, Habbema DF. Evaluation of clinical decision-aids: more to think about. Med Inf (Lond). 1990;15:275–84.

Kleinbaum DG. Logistic regression. New York: Springer; 1992.

Liu X, Faes L, Kale AU, Wagner SK, Fu DJ, Bruynseels A, et al. A comparison of deep learning performance against health-care professionals in detecting diseases from medical imaging: a systematic review and meta-analysis. Lancet Digital Health. 2019;1:e271–97.

Narla A, Kuprel B, Sarin K, Novoa R, Ko J. Automated classification of skin lesions: from pixels to practice. J Investig Dermatol. 2018;138:2108–10.

O'Neil M, Glowinski A. Evaluating and validating very large knowledge-based systems. Med Inf. 1990;15:237–51.

Ozenne B, Subtil F, Maucort-Boulch D. The precision-recall curve overcame the optimism of the receiver operating characteristic curve in rare diseases. J Clin Epidemiol. 2015;68:855–9.

Peek N, Abu-Hanna A. Clinical prognostic methods: trends and developments. J Biomed Inform. 2014;48:1–4.

Swets JA. Measuring the accuracy of diagnostic systems. Science. 1988;240:1285–93.

Wildemuth BM, Friedman CP, Downs SM. Hypertext versus Boolean access to biomedical information: a comparison of effectiveness, efficiency, and user preferences. ACM Trans Computer-Human Interaction (TOCHI) 1998;5:156–83. Available from https://doi.org/10.1145/287675.287677. Accessed 29 June 2021.

Chapter 13
Designing and Carrying Out Correlational Studies Using Real-World Data

Learning Objectives

The text, examples, and self-tests in this chapter will enable the reader to:

1. Explain the advantages and challenges of correlational studies using routinely collected data.
2. Given a description of correlational study, offer an alternative explanation to a direct causal relation between variables.
3. Plan a complete correlational study on large scale linked datasets using a trusted research environment or safe haven.
4. Explain the differences between association and causation and the many biases that come into play when attempting to make causal inferences using routine data.
5. Given a description of a correlational study, identify the biases that could jeopardize the internal validity of the study.
6. Describe which methods to apply to estimate the size of or to overcome these biases.

13.1 Introduction

This chapter is the last of a set of four examining the design and analysis of quantitative demonstration studies. Chapter 10 focused on descriptive studies and also considered most of the external validity issues that impact all kinds of demonstration studies. The design and analysis of interventional studies were discussed in Chaps. 11 and 12. This chapter focuses on the design, conduct, and analysis of correlational studies and discusses how to maximize their internal validity.

© Springer Nature Switzerland AG 2022
C. P. Friedman et al., *Evaluation Methods in Biomedical and Health Informatics*, Health Informatics, https://doi.org/10.1007/978-3-030-86453-8_13

13.2   Reasons for Carrying Out Correlational Studies

Recall from Sects. 6.9.2 and 10.2 that the purpose of correlational studies is to explore the relationships between one or more independent variables and one or more dependent variables. These variables are recorded in one or more datasets, routinely generated as a by-product of ongoing life experience, health care, research, educational, or public health practice. Or these data may result from routinely conducted surveys by governmental agencies which then become publicly available. Correlational studies are increasingly called "real world evidence" studies and may also be called datamining or quality assessment studies. A common example would be the use of system log files to understand and compare user behavior at different times of day or by different grades of staff at the same time of day.

A second reason for carrying out a correlational study is to try to attribute cause and effect. For example, imagine a weight loss app, used by thousands of members of the public, that records both a person's engagement with the app (an index using a combination of the duration of each use and the overall number of uses over a given period) and their weight measured monthly as a component of app use. This app sends the data back, in this case to the developers, who can assemble it into a database with one row per user per month, as shown in Table 13.1.

Table 13.1 Simulated data (n = 1000) captured by a weight loss app. For simplicity, data from only 5 app users is shown here

Patient ID	Month	Weight (kg)	App engagement %
1	December	85	0
1	January	86	25
1	February	86	28
1	March	86	15
2	December	76	0
2	January	75	42
2	February	74	45
2	March	73	35
3	December	86	0
3	January	85	45
3	February	84	37
3	March	82	32
...	...	...	...
999	December	96	20
999	January	95	32
999	February	93	28
999	March	91	25
1000	December	88	0
1000	January	89	2
1000	February	87	4
1000	March	88	2

This table displays a snapshot of a simulated dataset of weights captured by the weight loss app at the end of each month for 1000 people, together with the mean app engagement over that month. Some observations can immediately be seen from the table. For example, some overweight people (e.g. participant 3) do lose weight over the 4-month period covered—as does one person (Participant 2) who may already be underweight. The table also shows that some people engage well with the app (Participant 2) while others appear to use it much less (Participant 1000).

One question of interest to the app developer and other stakeholders would be: Do people who are more engaged with the app lose more weight? Perhaps the most definitive way to answer this question is to perform a prospective interventional study as described in Chap. 11, randomizing half the potential users to access the app and comparing their weight over time with that of controls. However, such a study would require development of a placebo app that requires control participants to weigh themselves monthly but does not provide them with weight loss advice. By contrast, the data creating Table 13.1 are generated as a by-product of routine app use. So, one way to explore if weight loss is correlated with app engagement is to plot the relationship among these "real world data", as in Fig. 13.1.

Not surprisingly, the data exhibit a correlation, with people showing greater weight loss having higher figures for app engagement. This is confirmed by the positive slope of the line that best fits the data. However, just because this graph shows a relationship, this does not necessarily mean that increased app engagement *causes* the weight loss.

This example demonstrates both the main strength and the main weakness of correlational studies: they can help us understand whether one factor (app engagement) is associated with another (weight loss), but equating this association with causation is unjustified. Thanks to the greater availability of routine data, it is possible to discover correlations among many variables, but this brings the risk of assuming that the relationship among them is causal.

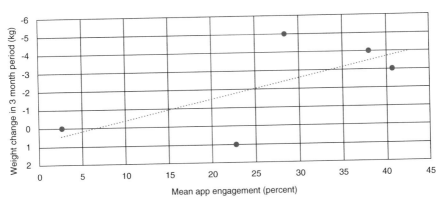

Fig. 13.1 Chart correlating weight change over 4 months versus percent engagement with the weight loss app (only 5 data points plotted for clarity)

The next sections consider the types of data used in correlational studies and the main advantages and disadvantages of such studies. As the chapter unfolds, the relationship between correlation and causation will be explored in greater detail.

13.3   Types of Data Used in Correlational Studies

This chapter is about carrying out studies using data that already exists. This can be data collected by organizations or individuals, devices or information resources, but the key distinction from an interventional study is that the data already exist, and "all" the study team needs to do is to legally and ethically negotiate access to it—which can often be a challenging task—then analyze it. Let us assume here that, following the procedures described in Sect. 13.5 below, the study team has obtained legitimate access to person-level anonymized data and has permission from the Institutional Review Board and/or other data controlling agencies to analyze it in a trusted virtual data enclave or safe haven.

Routine data is often obtained from person-level record systems such as electronic health records used to capture patient findings and actions taken by clinicians in care settings, or data generated by an app or wearable device used by the general public. Data from social media may also be analyzed to generate insights about the usage or impact of an information resource. Another rich source of data is information resource usage logs which can be analyzed to explore patterns of use by time of day, day of week, or type of user. Government agencies and other bodies collect huge quantities of data about people, climate, atmospheric pollution, geography, and travel. These data are often publicly available for analysis to address many different types of questions pertaining to individual, public, and population health. These data can often be linked with data about individual use of an app, for example, to answer questions such as:

- Are different levels or types of app use associated with user location?
- Is the level of pollution correlated with the severity of asthma, and can an app that warns about impending pollution help reduce this severity?

One development in recent years that has led to the conduct of many more correlational studies is the Learning Health System (LHS) (Friedman et al. 2010). LHS provides the data infrastructure, such as electronic health records and dashboards that summarize the data they contain, to enable multi-stakeholder interest groups to investigate, and then address, health problems of importance to these groups. These methods have been used to markedly improve health of individuals and populations; for example by increasing the remission rates of young adults with inflammatory bowel disease (Britto et al. 2018).

Finally, the data used in a correlational study may result from a previous research study, for example through long term follow-up of patients who were recruited into a randomized trial many years ago. An example of this is the use of electronic health and mortality records in 2005 by a team in Scotland to carry out long term follow up of thousands of people who were previously randomized to receive statin or placebo drugs in the 1990s (Ford et al. 2000). While this dataset started out being collected prospectively from participating subjects as part of a randomized trial, the investigators now have no contact with their former research participants and instead follow them up through routine health records and notifications of death or other events, so it has become a correlational study. When recruited, the original trial participants consented to the use of their data in this way.

The data in these real world evidence (RWE) studies do not always come packaged for analysis, but rather need to be transformed or manipulated in some way. For example, very often study teams will create indexes out of the raw data to express a more abstract but useful construct—such as usage index of the app discussed in Sect. 13.2. Doing this properly requires attention to the concepts of measurement discussed primarily in Chap. 8.

13.4 Advantages and Disadvantages of Correlational Studies Using Routine Data

13.4.1 Advantages

Potentially, as long as a study team can gain access to the relevant datasets, carrying out a study using routine data offer several advantages over experimental studies (Sherman et al. 2016; Deeny and Steventon 2015):

- The datasets can be 100–1000 times larger than those captured for interventional studies, making it possible, for example, to examine effects and side effects in relevant participant subgroups.
- Data is captured from routine activities and may be more representative of typical people having typical experiences in the world. In health care, real world evidence studies can include a valid cross-section of patients, rather than the kind of patients who attend the tertiary referral centers that usually conduct clinical research. Using the terminology introduced in Chap. 10, the results of studies based on routine data are likely to have a higher external validity than experimental studies.
- There is often a wider range of data items in a routine dataset versus an interventional study dataset, so it may be possible to address more questions. The fact that the data are recorded routinely in most or all cases can help to reduce what will later be called recall bias (Sect. 13.7.2) and allow the study team to explore further whether any correlations might in fact be causal relationships (Sect. 13.6).
- Since such studies use existing data, studies can begin sooner and offer a more economical way to address questions. However, this assumes that the infrastructure for generating and managing the data, such as very elaborate and expensive EHR systems, is already deployed in the study environment.

13.4.2 Disadvantages

Set against these advantages, there are several negative considerations (Deeny and Steventon 2015; Hemkens et al. 2016):

- While in theory the study can start as soon as the data are made available, there can be delays in linking, cleaning, and gaining access to the data for analysis. Many research network developments with routine mechanisms for rapidly allowing data access are changing this picture dramatically (Agiro et al. 2019).
- Even when access to the data is provided in a timely way, often the data turn out to be poorly documented and incomplete, or otherwise of poor quality. Using the jargon of the "data readiness level" (Lawrence 2017), datasets often fall into band C (poorly documented, of unclear origin and meaning), where data owners and study teams need them to be of at least band B and preferably band A to be useful.
- The data may lack key outcomes. For example, the widely-used UK Hospital Episode Statistics (United Kingdom National Health Service (NHS) Digital 2021) only includes diagnoses and procedures carried out, the length of stay, and place of admission and discharge. If the study team needs information about the outcome of the procedure carried out or the 30-day mortality, this dataset needs to be linked with other datasets such as UK Office for National Statistics data using the patient identifier. Because of confidentiality concerns, such a linkage is usually carried out by an "honest broker" or trusted third party.
- The data may lack key contextual variables such as risk factors necessary for understanding disease severity, other drug or non-drug interventions, or co-morbidities (Tsopra et al. 2019). Missing any of these can make it hard to interpret whether, for example, installation of an ePrescribing resource was responsible for a drop in drug-related adverse events.
- The ready availability of large datasets can lead analysts to become overconfident and carry out many informal, poorly planned analyzes. When one of these turns up "positive" results, in the sense of statistical significance $p < 0.05$, they may mistakenly seek to publish this result, forgetting that multiple tests of statistical significance carry the possibility that at least one will be significant by chance (see Sects. 10.4.3 and 11.5.4).
- When large amounts of person-level data are brought together in a data warehouse, and especially when different datasets are linked using a personal identifier, there is a significant risk of unmasking the identity of individuals, even if all the identifiers have been removed (Gymrek et al. 2013). This is because many datasets contain a sufficiently large number of items—especially if the data are longitudinal—that a person's record is unique, and can be matched with other sources of information about that individual (e.g. from social media accounts) to confirm their identity. Scenarios in which this threat might be realized include an attempt by a disgruntled data warehouse employee to shame their employer and make large sums of money by selling details a celebrity's medical record to the press, are offered large sums by an organized crime syndicate to reveal the

address of a witness who testified against them and has been given a new identity, or by a private detective employed by an individual investigating a spouse's alleged affairs and illegitimate children prior to divorce proceedings. This means that special measures must be taken to ensure that researchers, or anyone else with access to person-level data, cannot undertake unauthorized re-identification.

13.5   How to Carry Out Correlational Studies Using Person-Level Datasets

Given the benefits of correlational studies but also the privacy concerns described above, the process for carrying out studies analyzing person-level datasets has changed dramatically in the last few years. Not long ago, study teams were able to obtain copies of routine data from data controllers—the entities legally responsible for controlling access to and preservation of person-level data—and link these datasets on their own servers to form a personal data warehouse. This warehouse was maintained informally, the study teams were trusted to use it for any purposes they wished, and there were few if any physical or software barriers to others accessing the data. However, public, funder, ethical committee, and data controller attitudes have now changed, and study teams will find it increasingly difficult to hold their own data, or even to access useful data in a trusted research environment (safe haven) without following a defined process, illustrated in Fig. 13.2. This section is an overview of how to plan and carry out a correlational study in a professional way. The details of challenges such as confounding and biases that shape the statistical analysis plan are in Sects. 13.6 and 13.7.

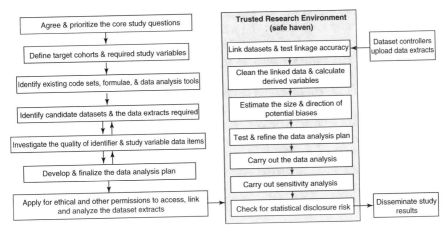

Fig. 13.2 The process for carrying out correlational studies with person-level datasets

13.5.1 Agreeing to the Study Questions, Cohorts, and Study Variables

The process starts by agreeing to the core study questions with stakeholders, through the negotiation process described in Sect. 3.2, and placing them in priority order. A sample correlational informatics question might be: *"What is the success rate of surgical robots in the obese compared to people of normal weight?"*. Each question will require data from one or more specific cohorts of patients, students, or others to answer it. Each cohort needs to be defined carefully in terms of the required characteristics, definitions of these, the number required, and the timeframe over which they should be sought in the datasets. An example cohort description might be *"Patients aged 50–70 admitted for robot-assisted total hip replacement between 2015 and 2020 with a body mass index (BMI) over 30"*. Each question will also require specific data about the members of the cohort. In our example, these might include the duration of the operation, any complications or difficulties encountered, the duration of post-operative recovery, total inpatient stay duration, and the success rate in terms of patient recorded outcome measures such as pain, walking distance, quality of life, and resumption of work or other responsibilities.

13.5.2 Identify the Datasets, Code Sets, and Data Analysis Routines

Once the required cohort and data items are clear, the next step is to find one or more datasets that contain these data on the included individuals, if necessary linking datasets using a unique identifier, if one is available. Often it is useful to copy the methods used in related studies, especially if they publish the list of clinical codes used to identify the cohort and relevant algorithms, such as the process for calculating body mass index from height and weight. It may also be useful at this stage to identify specific analysis code, such as SAS, STATA or R libraries to apply certain specialized analysis methods such as regression discontinuity design (see Sect. 13.7.3). To reduce the risk of unintended disclosure, most data controllers nowadays refuse to release their entire dataset to investigators, but will only supply the minimal data extract sufficient to answer the question of interest. Once the likely datasets are identified, it is useful to investigate their quality either by contacting the data controllers or reviewing published dataset descriptions or published studies using that dataset. Note that all relevant data items need to be investigated for quality, including the quality and completeness of identifier data used for record linkage, if used. Sometimes the results of this data quality review will require that the study team find an alternative dataset.

13.5.3   Develop and Finalize the Data Analysis Plan

To further clarify the thinking behind the study questions and the choice of cohort, study variables, data items, and datasets themselves before permissions are sought, it is very useful to develop a data analysis plan for the study. This should include the following topics, some of which have already been described above and others which will be described in later sections of this chapter:

- The methods and criteria used to define and select the study cohorts.
- What data items will be used from which datasets, measured at what times or intervals and combined in which ways, to form the key study variables (such as BMI, in our example).
- The specific method used for data linkage across multiple datasets, if needed. This should include the methods used to identify and merge duplicate records, if present.
- The approach to missing data. For example, will incomplete records be eliminated or will other methods such as multiple imputation be used (see Sect. 10.4.2)?
- How the study variables will be analyzed, and specifically, what descriptive and inferential statistical techniques will be used to answer the study question and to estimate the level of uncertainty in the answer.
- How the magnitude of specific biases that affect correlational studies (described later) will be estimated and, if possible, reduced or eliminated. This may include carrying out sensitivity analyzes, e.g. changing the assumptions about missing data (e.g. eliminating any records with missing data) and re-running the entire analysis to see if this changes the study conclusions.

If the study is to be published, at this stage it is useful to locate and refer to the relevant checklist from the EQUATOR website (Equator Network, UK EQUATOR Centre 2021) to confirm that all relevant checklist items have been covered in the study planning. The main reporting guidelines currently for correlational studies are STROBE (The Strengthening the Reporting of Observational Studies in Epidemiology Statement), and RECORD: REporting of studies Conducted using Observational Routinely-collected health Data (Benchimol et al. 2015), but this may change. Study teams are urged to consult the EQUATOR website to find the current relevant reporting guidelines. Note that this should be done before starting the study. The point of "reporting" guidelines is that they should improve the design and rigour of studies, so the study team should discuss and resolve all items on the checklist before the study is carried out, as after is too late.

13.5.4 Applying for Permissions

Once all this planning work is complete, the study team is then in a strong position to apply for ethical permission to link and analyze the data. They can also simultaneously apply to the various data controllers, supplying them with a copy of their study protocol including the data analysis plan and requesting them to prepare the relevant data extracts and place them into a trusted research environment once ethical permission is granted.

13.5.5 Virtual Data Enclave (Virtual Safe Haven)

It is rare nowadays—and some would say irresponsible—for data controllers to supply researchers with a copy of their entire database to be kept informally on a local server and used however the researchers wish. To enforce data protection regulations as described above, the default now is for data controllers to provide researchers with time-limited access to a restricted extract of the data, mounted in a monitored, secure virtual safe haven (virtual data enclave) to which researchers are granted temporary access to analyze the data using tools provided by the virtual safe haven. This virtual safe haven will prevent researchers copying or printing the data using a tool such as Citrix and may also provide a checking service to ensure that the outputs of the analysis do not contravene statistical disclosure rules (see below). It may also act as a trusted third party, linking data extracts that originate in two or more external organizations, neither of which trust each another enough to provide the data to the other.

13.5.6 Data Linkage

Once ethics permission is granted and the data controllers upload the relevant data extracts to the virtual safe haven, any linkage needs to be performed and the accuracy of this measured. Sometimes it is then possible to go back to the data controllers to request more data to improve the linkage accuracy. It is also useful at this point to check the completeness of key data required as study variables, as again it may be possible to request completion of these from the data controllers, if they have access to original patient record document data and transcription, natural language to clinical code software translation, or optical character recognition facilities.

13.5.7   Data Cleaning

Data cleaning is the process of removing incorrect, duplicate, or ambiguous data prior to carrying out the analysis. It is almost always required and may consume significant project effort before data analysis can commence. Cleaning is necessary to ensure that the data items to be analyzed are valid (conform to the documentation or metadata for that dataset, if available), accurate (close to the true values), complete (include all the data items needed to define the required cohort and calculate derived variables), and consistent (i.e. the same clinical concept is coded using the same code drawn from the same version of the clinical coding system; continuous variables all use the same units). The process of data cleaning includes a range of activities such as:

- Carrying out exploratory data analysis on key variables. If these are continuous, a histogram or box plot can indicate unexpected variable distributions (e.g. Some weights expressed as pounds while others are in kilograms), unexplained outliers or unauthorized missing data codes (e.g. users might have entered a patient's weight as 1 kg or height as 1 cm when in fact it was missing).
- Correcting invalid data values, e.g. updating outdated clinical codes from older records so that they are all compatible; substituting a missing data symbol for unauthorized values.
- Transforming data, e.g. carrying out a transformation on heavily skewed continuous data to make it amenable to standard statistical methods.
- Deleting duplicate records and merging partially complete records of the same person.
- Calculating derived variables such as body mass index, time interval between two events or percentage of app use in the earlier example.

The process of identifying and deleting duplicate records may sound trivial, but can take much time and thought, especially when data originates from multiple datasets that have been linked, or there are missing values, or the conventional identifiers have been removed. For example, consider the rows of data from a database of Type 1 diabetics shown in Table 13.2:

It appears at first sight that Records 1 and 2 are from the same individual, but this is not necessarily so, as there may be many diabetics with a diagnosis at age 12 who are aged 47 in 2021. If the dataset originated from only one city, or even from one neighborhood, the study team could be more confident that these records relate to

Table 13.2   Sample of de-identified data on Type 1 diabetics showing potentially duplicate records

Record ID	Year of diagnosis	Age at diagnosis	Gender	Age in 2021
1	1986	12	F	47
2		12	F	47
3	1986	12	M	47
4	1986		F	52
5	1986	17	F	53

the same person. Record 3 is obviously a different person because of the gender. Records 4 and 5 may be the same person, but again it depends on the disease frequency and typical age at diagnosis. The apparent difference in age in 2021 could be accounted for by different algorithms that calculate age, especially if the person's birthday is on January 1 or December 31. To resolve such problems, similar probabilistic techniques as are used for carrying out record linkage can be applied to the task of identifying duplicate records (Waruru et al. 2018; König et al. 2019).

13.5.8 Estimate the Size and Direction of Potential Biases

Correlational studies are subject to multiple biases and a phenomenon known confounding, as discussed in Sects. 13.6 and 13.7 to follow. Before the data are analyzed to answer the study questions, it is wise to check for the existence and likely magnitude of these phenomena that can profoundly affect the results. Would they increase or decrease the magnitude and/or direction of the study results, if present?

13.5.9 Test and Refine the Data Analysis Plan and Carry Out the Data Analysis

Often the data cleaning process will lead to minor changes to the data structure, so the data analysis plan needs to be modified to ensure that key variables are correctly analyzed. This means that the revised analysis methods need to be tested on a subset of the data to ensure they deliver the expected results. Once the study team is satisfied, the analysis can be run across the entire dataset and the results scrutinized by the team.

13.5.10 Carry Out Sensitivity Analyses

If the previous stage reveals that any of the biases might significantly influence the study results, it is wise to carry out a sensitivity analysis to explore this in more detail. At its simplest, this means re-running the analysis excluding the cases likely to be causing the bias—for example, with incomplete data or poor matching on record linkage. If the result differs markedly from the result including the entire cohort, the study team needs to decide whether further investigation is needed before reporting the results.

13.5.11   Check for Statistical Disclosure Risk

This process is an attempt to minimize the risk that someone reading the result of the study may be able to identify individuals whose data was analyzed. This may occur for individuals with rare diseases when the results include a cross tabulation of diagnosis with city of residence, for example. A range of measures can be taken to reduce this risk, such as using a symbol instead of a figure in table cells where the value is 5 or less, which are described in standard sources such as (O'Keefe and Rubin 2015; Hundepool et al. 2010; Matthews and Harel 2011).

13.5.12   Disseminate the Study Results

The final stage is to write up the study methods and results, using the relevant publication checklist, and to disseminate it to relevant audiences.

13.6   The Challenge of Inferring Causation from Association

It is widely known that "association does not equal causation", but unfortunately humans very readily attribute cause and effect (Kahneman 2011) when this is not justified. The following, somewhat extreme, example illustrated in Fig. 13.3 illustrates the natural human inclination to invent causal explanations for a strong association.

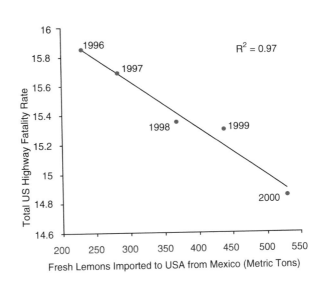

Fig. 13.3 Correlation between the volume of Mexican lemons imported to US over a 5 year period with highway fatality rates during the same period (from Baker L. (2018), with permission)

The graph in Fig. 13.3 shows a very strong inverse statistical association between US highway fatality rates and the mass of lemons imported from Mexico over a five-year period. The first observation is that the modest reduction of 7%, in highway fatality rates over the period, from 15.9 to 14.9 (per 100,000 inhabitants), is visually exaggerated in this graph by omitting a large part of the vertical axis from the value of zero up to the value of 14.6. This misleading representation is known as a suppressed zero graph (Tufte 1992). The second observation is that this 7% reduction in mortality is associated with a doubling in the mass of lemons imported from Mexico, from less than 250 tons per annum to over 500 tons. It would seem, based on common sense, that there is no possible causal link between these two variables, but on presenting this graph to generations of students they have suggested several possible reasons why increased lemon importation could cause, directly or indirectly, decreased highway fatalities, including:

1. *A direct causal effect:* Perhaps Mexican lemon truck drivers drive more slowly on unfamiliar US highways, slowing down all traffic and reducing serious accident rates.
2. *Reverse causation (discussed in more detail in* Sect. 13.7.3*):* There may be a direct effect in the opposite direction. Perhaps the lower US highway fatality rates over the years in question have increased Mexican truck drivers' willingness to negotiate the extra distances required to deliver their lemons north of the border.
3. *A common cause for both observations*: Perhaps climate change has led to fewer icy winter days in the US, lowering highway fatality rates but also increasing the winter survival of insects that reduce the yield of the US lemon crop, causing higher prices that now make Mexican imports more affordable.

The most likely explanation is a variant of Explanation 3 above. A third unmeasured variable is at play here, which both reduces highway fatality rates and increases the demand for Mexican lemons. This variable is likely to be the gradual increase in wealth of society, which leads both to safer vehicles (through both better technology and better maintenance, enforced via regulation) and to higher import rates of better quality, organic fruit. This effect is known as spurious correlation.

13.6.1   Spurious Correlation

A biomedical example further demonstrates the risks of spurious correlations. Agniel et al. (2018) sought to identify predictors of which hospital inpatients would die over the following 3-year period, and to this end conducted a correlational study examining the results of 272 common lab tests and 3 year mortality in a very large patient cohort (n = 669,452 at two hospitals from 2001–2006). Surprisingly, the mere *presence* of a test order, regardless of the result of the test, was significantly associated (P < 0.001) with survival in 233 (86%) of the 272 lab tests they studied. In addition, the *time of ordering* for certain tests (i.e. hour of day, day of week or interval between tests) such as white blood cell (WBC) count was more accurate at predicting 3-year survival than the result of the test in 118 (68%) of 174 tests. Survival was highest for patients in whom all WBC tests were ordered during the normal hospital working day (5 am to 4 pm) and lowest for patients with any tests ordered between midnight and 2 am.

The most likely explanation for this strong association is not hard to discern. Laboratories do not work at full staff capacity over the 24-hour period so will only analyze specimens out of hours if there is a strong case to do this. The usual case for requesting a test at 1 or 2 am is that the patient is seriously ill, so clinicians need the test result to guide treatment. Inpatients who are seriously ill at 1 or 2 am have increased 3-year mortality compared to inpatients only requiring routine investigation during working hours.

In both of the examples above, that was a strong correlation between the independent and the dependent variable, but the independent variable was not the cause: there was a third, unmeasured variable that was responsible for the spurious correlation. This challenge is further explored in Sect. 13.7.3 below, but meanwhile it is useful to review how epidemiologists (who study the association between people's exposure to potential toxins at home or at work and the development of disease) examine the problem of association versus causation, and some criteria for causation proposed 50 years ago that are still in common use.

13.6.2   Bradford Hill's Criteria to Support Causality in Correlational Studies

Bradford Hill, a British statistician, developed several criteria to help epidemiologists assess the likelihood that a correlation between exposure to an environmental toxin and development of disease was likely to be causal in nature (Hill 1965). Some of these criteria are specific to biological phenomena or exposure to toxins (Grimes and Schulz 2002) and are not further discussed here, but most are relevant to correlational studies in health and biomedical informatics. The relevant criteria are:

- Temporality: the temporal order between exposure (or resource use) and outcome must be correct. For example, any trend in weight loss should be faster following use of a weight reduction app than before the person started using the app.
- Association strength: stronger associations are more likely to be due to an intervention (such as an app) and less likely to be due to bias or confounding. While this may be the case, debate continues on what threshold to use. One frequent suggestion is a risk ratio above 3, so for example, the number of people losing weight with an app would need to be three times the number losing weight without to make bias unlikely. However, this criterion excludes many effective drugs, procedures, and other interventions with useful benefits. It does, however, explain

why parachutes and some drugs such as insulin in diabetes and penicillin for bronchial pneumonia have not been exposed to testing in a randomized trial: they are so effective that no RCT is needed.

- Consistency: consistency of results across a variety of study settings and methods suggests that the cause-effect relationship is real, rather than due to some bias or confounding. One way to assess the consistency of results is by conducting a meta-analysis of relevant studies and checking for statistical heterogeneity. If significant heterogeneity is present, this should be investigated using meta regression methods.
- Dose-response relationship: in drug studies, the size of the benefit should be greater with a larger dose than a smaller dose. In informatics studies, the equivalent is that a larger effect should be seen with more intensive engagement with an information resource, such as an app (measured as the number of minutes of use per day, for example), or a longer period of app use than in people with a lower intensity or duration of use.
- Specificity of effect: in informatics, this means that, for example, a weight reduction app should reduce weight and perhaps also the rates of diseases related to weight (such as diabetes, blood pressure, heart attacks, and arthritis or back pain), but not change the rate of diseases unrelated to weight, such as upper respiratory or urinary tract infections.

13.7   Factors Leading to False Inference of Causation

Bradford Hill's criteria are useful to consider when interpreting or critically appraising the results of a published correlational study, but may not help when designing such a study. When designing a study, and especially when deciding whose data to include in the study cohort, how to derive the outcome measure using the available data and how to analyze the data, the study team needs to understand and take account of many possible biases that can interfere with correlational studies. While hundreds of biases in such studies have been described—enough to fill an entire book (Andersen 1990)—the next section describes the nine most relevant biases in informatics correlational studies, summarized for convenience here in Table 13.3.

In any correlational study, especially those using large datasets which are increasingly available, it is highly likely that a study team will uncover associations between one variable and another (Grimes and Schulz 2002), but what do these associations mean, and do they provide reliable evidence of causation? Therefore, it is important to understand the circumstances under which such misleading conclusions can be drawn about causal association using routine data. Spurious correlation has already been discussed (Sect. 13.6). The additional factors, as Table 13.3 shows, fall into three broad types: selection bias, information bias and confounding (Grimes and Schulz 2002). The next 3 sub-sections examine each of these factors in turn, discussing how each can impact correlational studies in health and biomedical informatics and what to do about them.

Table 13.3 The main biases that are relevant to the design of correlational studies in informatics, and how to overcome each

Broad category	Bias name	Example in this chapter	How to overcome it
Spurious correlation (Sect. 13.6.1; see also 13.7.3)	Common cause	Correlation between Mexican lemon imports and road accidents, or timing of lab tests and mortality	Seek the common cause (e.g. for lemons: increased wealth; for lab tests: case severity)
Selection bias (Sect. 13.7.1)	Membership bias	When comparing resource users versus non-users (e.g. Rochester library study), resource usage may be a **marker** of the outcome, not the cause. The outcome may even predict resource usage ("reverse causation")	Randomize a group of users to access the resource or not and measure the outcomes in both groups.
	Non-response bias	Using the smoking rate people report in a smoking cessation app may overestimate app effectiveness	Seek external data sources for outcome variables
	Unmasking bias	A pre-diabetes app does not cause diabetes, it just unmasks the undiagnosed disease state	Convert from retrospective to prospective cohort study
Information bias (Sect. 13.7.2)	Measurement noise or missing data	Relying on clinician-recorded data in an EMR rather than patient-recorded outcomes may reduce effect size	Seek external data sources
	Assessment bias	Relying on outcomes assessed by clinicians who know if the resource was used or not may exaggerate the effect size	Carry out separate outcome assessment by judges blind to whether resource was used or not
	Differential recall bias	People with sight loss caused by diabetes may recall use of a diabetes app more often than normally sighted diabetics	Check app usage data from external source; convert from retrospective to prospective cohort study
	Immortal time bias	Benefit of suicide reduction website is overestimated if period before first use of website included in analysis	Ignore outcome-free period before first use of resource in the resource user group
Confounding (Sect. 13.7.3; see also 13.6.1)	Simpson's paradox, confounding by indication	Misleading raised heart attack rates in patients prescribed statins vs. not prescribed; diabetic control better in diabetics recommended to use app than not recommended app	Multiple regression using all available data, including confounders. Add a propensity score to the regression, use propensity matching, case-crossover, or instrumental variable design.

13.7.1 Selection Bias

Selection bias occurs when the two or more groups of participants being compared in a correlational study differ in ways other than their use of an information resource. For example, people who engage more with a weight loss app may be driven to do so by strong family or social pressures to lose weight; whereas those who engage little with the app have no such pressures or support to lose weight. It is known that family and social support are an effective means to change behavior and lose weight, which is why organizations that promote weight loss have emphasized the importance of social support. So, simply comparing the amount of weight lost by people who engage well with the app versus those who do not is a biased comparison, as the high engagers probably also have higher levels of social and family support. Matching up app users with control non-users with a similar level of motivation and family support to lose weight may help, but unfortunately these may not be the only factors that predict success in weight loss—see Sect. 11.3.8.

A study with selection bias is an example of a "case-control" study. In a case-control study, investigators try to infer whether a dependent variable is associated with one or more independent variables by analyzing a set of data that has already been collected, so it is a retrospective study design. For example, investigators could measure attitudes to computers in participants who happened in the past to use an information resource ("cases") and compare them to attitudes of participants who, in the past, did not ("controls"). This is an invalid comparison, as the fact that certain participants chose to use the resource is a clear marker of different levels of skill, attitude, experience, uncertainty, etc., compared to those who ignored it. Thus, any differences in outcome between participants in the two groups are much more likely to follow from fundamental differences between the participants involved than from use of the information resource.

One published example is the Rochester Library study that tried to attribute reduced length of stay in hospital inpatients to use by their physician of medical library services (Klein et al. 1994). In the study, patient lengths of stay were compared in two groups of patients: those for whose physicians a literature search on a patient problem had been conducted by a librarian, and a control group consisting of patients with the same disease and severity of illness but for whom no review had been carried out. The length of stay was lower in those patients for whom a literature search had been conducted, suggesting that this caused a lower length of stay. However, an alternative explanation is that those clinicians who request literature searches may also be more efficient in their use of hospital resources, and so discharge their patients earlier. This would imply that the literature search was a *marker* that the patient was being managed by a good doctor, rather than a *cause* of the patient spending less time in hospital. A cynic might even argue that the study shows that clinicians who order literature searches want to spend more time reading the literature and less time with their patients, and thus tend to discharge their patients earlier! All of these alternative explanations are consistent with the data, showing the dangers of such a case-control study.

Grimes et al. (Grimes and Schulz 2002) divide selection bias into several sub-types, considered below.

- *Membership bias:* This is similar to the scenario described above, in which those who use an information resource "belong" to a different category of people from non-users. They might be different in age, gender, computer or health literacy, level of education, or something else that influences the dependent variable and thus makes it hard to ignore when comparing resource users versus nonusers. Since it is often unclear if there is a membership bias, and the effect can be quite powerful, the best advice is to avoid comparing users versus nonusers in correlational studies. "User-non user" studies are analogous to case control studies in epidemiology.
- *Non-respondent bias:* This bias can result from differential response to data collection between subgroups, making any comparison misleading. For example, it is well known that cigarette smokers are less likely to return survey forms about their smoking history, especially if they are continuing to smoke (Seltzer et al. 1974). So, it would be a serious mistake to evaluate the impact of an app to reduce smoking by correlating app usage rates with self-reported smoking data collected by the app, especially if the study team assumes that no report is equivalent to no longer smoking.
- *Unmasking bias:* Sometimes the routine use of a new test or an information resource, such as an app designed to support people with symptoms of pre-diabetes, can unmask an existing disease such as diabetes. This might cause some app users to be investigated earlier, apparently increasing diabetes rates in app users. What is really happening is that it is accelerating the diagnosis of diabetes, rather than causing diabetes to occur.

13.7.2   Information Bias

Information or measurement bias concerns problems with determining either the outcome or the "exposure". In epidemiology, researchers are interested in exposures such as occupational or environmental toxins or infectious agents, but in the informatics context the "exposure" of interest is whether a person for whom the study team has data used the information resource or not. Given access to the usage logs of an information resource and a unique identifier, it should be easy to determine "exposure" in informatics studies. However, as with any measurement process (as discussed in detail in Sect. 8.5), determination of each participant's outcome (meaning whether they developed the disease, complication, or other feature of interest) can be unreliable, adding random noise and reducing the measured effect size. An example would be incomplete capture by busy clinicians of patient recorded outcomes (PROMs) into an EHR, reducing the apparent benefit of a patient-controlled analgesia device. The remedy here might be to give each patient a paper optical mark reading (OMR) form that can be scanned or an app to self-report their PROMs, which are then uploaded with greater fidelity and frequency into the EHR. When measurement error is related to validity, the effect may be seen more in one group than the other, this can lead to misleading results.

There are three broad kinds of information bias in correlational studies: assessment bias, recall bias, and immortal time bias.

Assessment bias has previously been covered in Sect. 10.4.1. It might affect the correlational study of the prediabetes app described above using an existing database if those who decided whether the patient really had diabetes or not and recorded this in the EMR knew in advance if the patient had used the prediabetes app or not, and had a strong wish for the app to succeed (perhaps because they were involved in its development). A better approach is to seek an independent dataset listing all diabetics, or to carry out a new outcome assessment using judges blinded to whether the information resource was used or not.[1]

Recall bias (or differential recall) occurs when people are asked to recollect an event (e.g. an exposure to a toxic chemical or a traumatic event such as a car crash) after they developed a disease. Because of the human tendency to try to make sense of life events such as developing a serious illness, seriously ill people tend to recall such exposure more often than people who had the same exposure but did not develop the illness. Consequently, it may be seriously misleading to rely on human recall of a prior event to determine if this event is causally related to a later one. An informatics example is asking if clinicians ask people who develop a disease or disease complication (such as blindness following poorly controlled diabetes) if they have ever used a certain app to monitor their diabetes, and record this in the EMR. In such a study, investigators might falsely conclude that using the app caused the illness or complication (perhaps by falsely reassuring the patient, causing them to delay seeking medical treatment), whereas the study team are simply picking up differential recall. The remedy is to use existing records such as an EMR or a dataset maintained by the app to check if the people who developed the illness had used the app.

Immortal time bias occurs in correlational studies that use mortality as an end point if study teams mistakenly include the pre-exposure period (i.e. the time period before people started using the information resource) in the time period analyzed (Suissa and Ernst 2020).

This is best understood by example. As illustrated in Fig. 13.4, an informatics example would be an attempt to measure the effectiveness of a suicide prevention website and online support group using routine data collected on a cohort of people considered to be at suicide risk—and doing this by comparing suicide rates per month for users and non-users of the intervention. If the study team mistakenly includes the period of time from the identification of a person being at high risk of suicide to the first time that person used the website in the website user group (the dashed line in the top row of the figure, from month 1 to month 15), this will over-count the number of months free of suicide since the website could not have

---

[1] Assessment bias is also known as ascertainment bias.

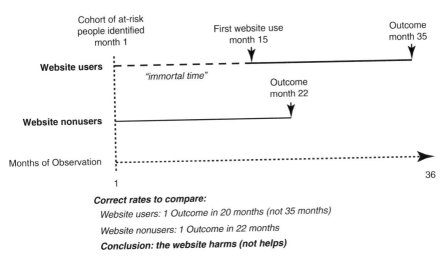

Fig. 13.4 An illustration of immortal time bias for a hypothetical study of website users

prevented suicide during this "immortal time" period prior to using the intervention. This will overestimate the intervention's effectiveness. Sometimes this bias is so large that it can overwhelm all other biases, so it needs to be eliminated by only analyzing the time period after actual exposure.

In this example, the website appears to be very effective if the immortal time period is included (1 suicide in 35 months in website users compared to 1 in 22 months in nonusers), a 59% drop in suicide rates in website users (1/22 divided by 1/35). However, when the immortal time is excluded, the period free of suicide is in fact longer in nonusers, at 22 months, compared to 20 months in the users, giving a 9% *increase* in suicide rates in website users (1/20 divided by 1/22). Including the immortal time here would lead to completely the wrong conclusions about website effectiveness.

13.7.3   Confounding

Confounding occurs when an additional factor, as well as exposure to the information resource, may be responsible for an observed outcome (Grimes and Schulz 2002). In many ways confounding is similar to the spurious correlation due to common cause described in Sect. 13.6.1, but since confounding is such a problem in real world data analysis we discuss this problem in greater detail here. When this occurs, the additional factor is labelled a "confounder". The previous chapter described how interventional studies can sometimes generate misleading results due to confounders such as the placebo or checklist effect. Similar problems exist for correlational studies and are described later in this section. However, while selection bias or measurement bias in a correlational study are very hard or impossible to overcome (see Sects. 13.7.1 and 13.7.2), it is sometimes possible to overcome confounding. This section starts with an example of confounding then discusses Simpson's Paradox and two major causes of confounding before discussing the four main ways to overcome confounding: restriction, matching, stratification, and multivariate methods.

A classic example of confounding in epidemiological studies is the apparent increased rate of heart attacks in women taking the oral contraceptive pill (OCP) (Ory 1977). While biologically plausible, on further analysis it turned out that more of the women using OCPs smoked than women who were not using OCPs. Smoking is a potent cause of heart attacks, so this was enough to explain the apparent increase associated with use of OCPs. To give an informatics example, consider a weight loss app that happens to be used more frequently by people with type 2 diabetes than people without. If the analyst simply compared outcomes between app users versus nonusers and ignored the diabetes, they might falsely conclude that app use somehow caused peripheral vascular disease or loss of eyesight (complications of diabetes that occur more rarely in the general population). This error might occur because the analyst was unaware that diabetes is a potential confounder and ignored it in their analysis, or because the dataset they were analyzing did not include the diabetes diagnosis. A final possibility is that the data analysis might take place sometime before it is known that diabetes causes those complications, or that the app is used more frequently by diabetics. These would be examples of "unmeasured confounders"—factors that can influence the outcome but are not yet known about or are otherwise not included in the analysis.

The presence of unmeasured confounders can lead to a phenomenon called Simpson's Paradox (Simpson 1951). An example of this comes from an analysis of death rates in the Poole Diabetes cohort (Julious and Mullee 1994). Strangely, the death rate of 40% from Type 2 diabetes (a less severe variant of diabetes) appeared to be higher in this cohort than the 29% death rate from Type 1 diabetes, the more severe form (top row of Table 13.4). Knowing that Type 2 diabetes mainly affects older people unlike Type 1 diabetes, and that older people are more likely to die, the investigators then analyzed death rates broken down by age group—see the second and third rows of Table 13.4.

Simpson's Paradox emerges when the mortality rates are broken down by age group. In direct contrast to the figures in Row one, the death rate in Rows 2 and 3 for both age groups (up to 40 and over 40 years), is higher in Type 1 than in Type 2 diabetes, even though the overall mortality is higher in the type 2 diabetics!

Table 13.4 Mortality rates in the Poole diabetes cohort broken down by age (Julious and Mullee 1994)

	Type 1 diabetes	Type 2 diabetes
Overall mortality	29%	40%
Up to 40 years old	1%	0%
Over 40 years old	46%	41%

Table 13.5 Mortality rates and actual figures in the Poole diabetes cohort broken down by age (Julious and Mullee 1994)

	Type 1 diabetes	Type 2 diabetes
Overall mortality	29% (105/358)	40% (218/544)
Up to 40 years old	1% (1/130)	0% (0/15)
Over 40 years old	46% (104/228)	41% (218/529)

Table 13.5 examines this result in more detail, showing the actual figures rather than percentages.

The reason for the mystery is now revealed in Table 13.5. Looking at the actual numbers of diabetic patients (the denominator figures) in Rows 2 and 4 it is now clear that a far higher proportion, 529 (97%) of the 544 type 2 diabetics are over 40 years old compared to the 228 (64%) of 358 type 1 diabetics. So, even though Type 1 diabetes causes more severe complications and higher age-specific mortality than Type 2 (seen in both Rows 3 and 4), it affects younger people, so the overall mortality rate in the Type 1 diabetics was lower.

The important lesson here is that the study team could have falsely concluded that Type 2 diabetes is a more severe illness than Type 1 if they did not have access to this age data. So, age is an important confounder in this example. This is why failure to identify an important confounder is a recurrent nightmare for epidemiologists, pharmaco-epidemiologists, and real-world data analysts, and is probably the most common cause of incorrect inference of cause and effect.

An informatics example arises when comparing an outcome—disease severity, for example—between people who did and did not use an information resource—a disease self-management app, for example. Assume that disease severity is less in people who use the app compared to non-users, with the conclusion that it must be the app that caused this difference. However, there are many reasons why disease severity might be different between two groups of people. One group might have a greater level of health literacy, for example, so are better able to read about their long-term condition and discuss with their clinicians how to control it. Alternatively, they might simply be more highly motivated to self-manage—perhaps because they need to keep their disease under good control to perform a demanding job. It is also known that people who have high disease literacy and high motivation to self-manage are also the kind of people who are more likely to use a self-management app. So, far from the app being the *cause* of better self-management and less severe illness, app use could simply be a *marker* of the kind of person who is better at self-management!

An even more extreme example of this fallacy is that in some diseases, people may become less able to use information resources such as self-management apps through worsening eyesight, pain, or manual dexterity, for example. This can lead to what is called reverse causation: app use is associated with less severe illness not because the app is *causing* people to develop less severe illness, but because it is *only* people with less severe illness who are able to use the app. Measuring the level of health literacy or disability in the study population would help to disentangle the direction of causality here, but another approach that might help, called propensity scores, is described after an introduction to confounding by indication.

Confounding by indication: A common and somewhat intractable example of confounding in healthcare is confounding by indication. This occurs in studies where treatments or other interventions were *ordered* by clinicians. A classic example is the apparent effects of therapy with the class of drugs known as statins to reduce blood lipids and thus reduce risk of cardiovascular disease such as heart attacks. The data for this study derived from the comprehensive national linked record system in Denmark and showed that the death rate from heart attacks in people receiving statins appeared to be 25% higher than average (Collins et al. 2020). However, it is now known from many randomized trials and even systematic reviews of randomized trials that this association is not causal: statins do not cause heart attacks; rather, they do the opposite by lowering the risk. The problem of interpreting results of this simple correlational study was that physicians prescribe statins to patients who they know are already at high risk of heart attacks, so receiving a statin was a *marker*, not a *cause*, of high cardiac risk.

An informatics example would be if study teams find that patients who were recommended to use an app by their clinician demonstrate better diabetes control than patients who were not recommended to use the app. While it is easiest to conclude that the app *caused* the superior diabetes control, this may well not be the case. For example, the clinicians may have formed a judgement that some patients are more motivated or have higher health or digital literacy, so are better able to use the app, and only recommended the app to those individuals. This means that the patient groups who were recommended to use the app are not comparable with those who were not recommended it, and this difference alone may explain the observed differences in diabetes outcome.

The propensity score as a remedy for confounding by indication. One approach to confounding by indication is to acknowledge the difference in app usage rates and try to take it into account in the analyzes. This can be done using a technique called *propensity analysis* (Rosenbaum and Rubin 1983). This is a two-stage process. First, the study team develops a statistical model to predict which patients are most likely to be recommended to use the app. This generates a propensity score for app use. Then this propensity score is either incorporated into the model used to predict which patients show improvements in diabetes control, or used to match cases and controls (*propensity matching*). In theory, this approach *should* reduce the effects of confounding by indication when estimating treatment effect by simply comparing outcomes among people that received the treatment (used the app) versus those who did not, but this does not always work (Zhou et al. 2020). There is now a reporting guideline for studies that make use of propensity scores (Yao et al. 2017). Possible alternative approaches to propensity scores are discussed in later sections of this chapter. These include the case crossover design (Sect. 13.8.1) and instrumental variable methods such as regression discontinuity design, discussed in Sect. 13.8.2.

13.7.4   Some General Methods to Detect or Control for Confounding

The good thing about confounding is that it can *sometimes* be controlled for in the analysis, as long as the study team is aware of it and has access to the necessary data. While there is insufficient space here to go into details, in summary the main techniques used to detect or control for confounding are (Grimes and Schulz 2002):

- *Restriction or exclusion of participants* with the confounding variable. So, for example, if it is known that smoking is a confounder for heart attacks in a correlational study of the effects of oral contraceptives, analysts can simply exclude smokers from the analysis in their cohort selection process.
- *Matching on the confounding variable.* The smoking example here would be to match women who both take the OCP and smoke with women who smoke but do not take the OCP, and then compare heart attack rates in these more comparable groups. Propensity matching has already been mentioned in Sect. 13.7.3 as an alternative method to reduce confounding by indication.
- *Stratification* by the confounding variable. Here, the analyst compares outcome rates in all relevant groups, including those with confounders. For the OCP example, this means comparing heart attack rates in 4 groups: non-smokers who take the OCP, non-smokers who do not take the OCP, smokers who take the OCP, and smokers who do not take the OCP.
- *Use multivariate modelling*, including known confounders in the model. This approach includes all the patients and all the data items known or suspected to correlate with the dependent variable in a single model—see Sect. 12.6.2. An example is calculation of the propensity score for smoking (Sect. 13.7.3) among those who do or do not use the OCP, then including this score in a multivariate model to predict heart attacks. This approach presumes that the data required for this analysis are available, which is certainly not always the case.

Self-Test 13.1
A company marketing an app to promote weight loss entered into an agreement with a large health care delivery network to study the effect of the app for its intended purpose. After the app has been available to patients for 1 year, the average weight of app users (n = 3500) is found to have decreased by 3 pounds while the average weight of all other adult patients in the network (n = 247,000) has increased by 2 pounds. Similarly, the average weight of app users with confirmed diagnoses of diabetes or hypertension (n = 2700) is found to have decreased by 4 pounds while the average weight of all other patients (n = 65,000) with these diagnoses has decreased by 1 pound. Patients in the network consent, when they enroll for care, to use of their data anonymously for research purposes, so this study was approved by all research oversight committees in the network.

Based on these findings with large numbers of patients, the company wants to promote its app as "effective".

You object to this conclusion and argue that this correlational study is subject to several biases. With reference to Table 13.3, describe how this study might be subject to:

1. Spurious correlation
2. Selection bias
3. Information bias
4. Confounding

13.8   Advanced Analysis Methods for Correlational Studies

This section introduces three additional techniques to address confounding: the case crossover design, instrumental variable analysis, and regression discontinuity design.

13.8.1   The Case-Crossover Design

The case crossover design (Maclure 1991) or self-controlled case series is a method for determining whether an event (such as a patient's adverse drug event) was triggered by something unusual that happened just before the event, such as their clinician using a decision support system or a website. In simple terms, the analyst would search a database of adverse drug events (ADE) to check whether the clinician used the CDSS or website in the preceding short period of time—perhaps dividing up the preceding day into six 4-h intervals. The analyst also checks over a longer period—the previous 3 months, say—how often the same clinicians used the information resource and no adverse event followed within 24 h, and estimates the monthly rate of adverse events in all patents managed by the same clinicians during the period under study. The analyst uses conditional logistic regression to compare the rate of CDSS use with no consequences against the overall rate of ADEs and the number of times an ADE followed use of the CDSS (within a clinically plausible timeframe) to estimate the probability that using the CDSS will trigger an ADE. A published informatics example is a study to measure the importance of advice from a member of a person's social network in prompting a person to make an unscheduled primary care visit (Eriksson et al. 2004). The study authors compared the rates of advice from family and friends to consult a GP in the 24 h preceding an unscheduled visit with rates of the same advice several days before the day of consultation, and showed that advice from a person's social network made it five times more likely that they would consult (Eriksson et al. 2004).

13.8.2   Instrumental Variable Analysis

This approach originates in econometrics and has only recently been applied in health and health care, where it is sometimes called Mendelian randomization (Riaz et al. 2018). The technique relies on the identification of an instrumental variable (IV) which is a factor that determines whether a participant is exposed to an intervention or not, *but is not related to the outcome of interest*—either by causing that directly, or by being caused by it. If the study can identify and measure the values of an IV, then demonstrate a correlation between the IV and the outcome, this is an indirect way of demonstrating that the intervention causes the outcome and estimating the effect size.

In the standard correlational approach, the analyst estimates the effectiveness of the intervention by calculating the probability of the outcome following it using the direct approaches described above. In the instrumental variable approach, the analyst seeks to identify an instrument that determines whether the intervention is administered or not but is not directly correlated with the outcome. Once such an instrument is found, the analyst estimates the effectiveness of the intervention indirectly, by comparing the probability that the outcome occurs when the instrument is present versus when it is absent.

The challenge is usually in locating a suitable instrument in the dataset and demonstrating that this instrument is only correlated with the outcome via the intervention, not directly or through some other mechanism. One example of an excellent IV design that demonstrates all these properties is use of a random number to allocate participants to an intervention. When this is done, the result is a randomized prospective trial. However, when randomization is not possible, some alternative potential instruments that may be documented in a dataset include:

- Allocation by clinician, time period, day of week, distance from a hospital, district of residence, etc., any of which may be used by healthcare organizations to decide who is eligible for therapy.
- Another factor that makes treatment possible, e.g. an HLA-matched sibling allowing bone marrow transplant in children with acute leukemia (Davey Smith 2006).
- Allocation of a patient's treatment or other procedure by clinicians according to a threshold in a continuous data item, such as by a test result or risk score. This practice, often based on a guideline recommendation, leads to a specific type of IV study called a regression discontinuity design and is described in more detail below.

13.8.3 Regression Discontinuity Design

In a regression discontinuity study design (RDD) (Hilton Boon et al. 2021), a study team can compare outcomes in a group of participants lying just below a threshold with those of a group lying just above the threshold in a continuous variable that has been used to allocate treatment as part of routine clinical care. An example would be statin prescriptions: according to NICE guidance, statins should be prescribed in people with a cholesterol level above 5 mM, and not in those with levels below 5 mM. Knowing that the drug will (mainly) be given to members of first of these groups and that the exact value of the allocation variable, serum cholesterol, is subject to random measurement error, this means that the two groups of patients are likely to be very similar (almost as close as if they had been randomized, in fact). This means that any difference in outcome between the two groups can be reliably credited to the drug.

The main assumptions for a regression discontinuity design to be valid are that:

- The therapy is assigned (deterministically or probabilistically) according to a known cut off point in a continuous assignment variable (the instrumental variable).
- The assignment variable is measured before treatment and is not changed by treatment.
- There must be continuity in the outcomes at the threshold (i.e. no measured or unmeasured confounders). This is typically due to random measurement error.
- The values of any other important variables that are correlated with the outcome of interest must be similar above and below threshold.
- A sufficiently large number of participants lie just below and just above the threshold to allow the comparison of outcomes in the two groups to show reasonable statistical power.

Often, however, the intervention is not allocated strictly according to the allocation variable, diluting the measured intervention effect—a bit like reduced adherence in a randomized trial. An extreme example of this was our attempted RDD study of the impact of chemotherapy on survival in women with breast cancer in Scotland using routine data—see Fig. 13.5 and (Gray et al. 2019). The reason for attempting to use RDD in this case was that oncologists treating women with breast cancer should follow guidelines stating that chemotherapy should not be given if the predicted chance of benefit using the NHS Predict score is less than 3%, and to administer chemo if the predicted chance of benefit is above 5%. This should make it possible to carry out an RDD study to estimate the effectiveness of chemo in older women who have been excluded from RCTs, by comparing outcomes in women with predicted risks below 3% and above 5%. However, when the authors attempted this they found that oncologists appeared unaware of these thresholds and many women received chemo even when their predicted chance of benefit was only 1% (see Fig. 13.5). Instead, they identified some alternative instruments to carry out this estimation, see (Gray et al. 2019).

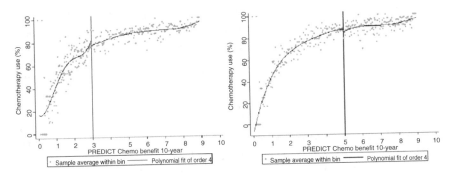

Fig. 13.5 The cumulative proportion of 17,000 patients with breast cancer receiving chemotherapy (y axis) in the SATURNE study (Gray et al. 2019) related to their predicted chance of benefit (percent, x axis). Guidelines suggest that few patients should receive chemotherapy at less than 3% predicted benefit, and most should receive it above 5%. The lack of clinical adherence to guidelines shown here made it impossible to carry out the planned regression discontinuity study (This is an unpublished figure from the study)

13.9   Conclusions

This chapter began with a step-by-step approach to carrying out correlational studies using person-level data, to reduce concerns about the erosion of privacy and enhance the reliability of study conclusions by pre-specifying the main study questions and analytical methods. Increasingly, funding organizations, ethical review boards, and data controllers are insisting that study teams adopt such an approach before they will release funding or data. In addition, research organizations are building the safe haven, training, and support infrastructure to facilitate this approach to conducting studies using what has increasingly become known as "big" health data.

It should be clear from this chapter that correlational studies have many attractions, but also that there are many challenges in using them, if the goal is to rigorously assign causality. However, for many study purposes in informatics, rigorous establishment of causality may not be required to address the study questions to the satisfaction of the stakeholders and decision makers. Application of Bradford Hill's criteria, introduced in Sect. 13.6.2, can strengthen an argument that associations identified in correlational studies are "likely enough" to be causal, and because causation is "likely enough" these associations are qualified to influence decisions related to the development and use of information resources. As such, correlational studies can be used to understand usage patterns of an information resource from log files and explore the factors that may influence its uptake, to develop predictive models or risk scores, or to compare clinical practice from one site to another or with evidence based standards.

The increasing availability of routine data from electronic records, together with the attraction of being able to study the use of drugs and other innovations in routine care, has led to much interest in evidence from real world data. However, as pointed out 40 years ago, databases will never replace randomized trials (Byar 1980). These issues have been summarized more recently in a 2020 New England Medicine article by Collins et al. (Collins et al. 2020). They point out that, while routine data will help in generating hypotheses and answering some questions, if the question is about estimating the effectiveness or impact of an intervention (and especially if the size of the effect is likely to be small, i.e. an effect size less than 0.2), then the analysis of routine data will rarely yield reliable results. (See Sect. 12.5.2).

One conclusion from this description of the many biases affecting correlational studies is that study teams should use the best possible method to analyze their data, given the resources available. One justification for this conclusion is a systematic review of research publications in major medical journals that were cited more than 1000 times. This review found that 9 of the 10 most cited studies whose results were later disproved were correlational in design (Ioannidis 2005). For highly cited randomized controlled trials, 23% were disproved. This suggests that using a correlational study design to assign causality and estimate effectiveness is risky, though even using a randomized design does not guarantee enduring results.

Overall, there are many reasons for carrying out correlational studies using real world evidence, and much current methodological research on how to use such a study design to understand causality is likely to improve study teams' abilities to draw convincing conclusions from this approach (Cooper et al. 2015). Over the next few years it is likely that novel methods such as using instrumental variables or Bayesian models (Cooper et al. 2015) will become more common in informatics, and that increasing numbers of more sophisticated correlational studies will realize the potential of this line of investigation.

Self-Test 13.2

A university offers a one-year preparatory program to boost students' performance in basic science subjects prior to their entering formal medical training. After running this program for 5 years, they conduct a 5 year review of the impact of online learning methods. From automatically maintained records, they were able to estimate the percent of student learning time that was based on online instruction each year. They studied this in relationship to the changes in the mean student "added value score" (the difference between the students' scores on entering and leaving the program) for three subjects: anatomy, physiology, and pathology. The results are shown in the table below.

Year	2018	2019	2020	2021	2022
Percent online time	25%	27%	56%	45%	35%
Anatomy value added	12%	13%	9%	10%	14%
Physiology value added	17%	16%	12%	13%	16%
Pathology value added	22%	21%	17%	17%	19%

1. Create a single histogram depicting the percent online time and added value data for each subject, by year. What trends do you see, across these years, in the online time and the value added scores in each subject?
2. For each year, average the added value scores across the three subjects. Then plot these averages (on the Y axis) against the percent online time on the X axis. Using a statistical package or online resource, add a linear regression line and determine the value of R squared. What is your interpretation of this finding?
3. Compare the histogram you created for Question 1 to the plot you created in Question 2. What general conclusions would you draw from this comparison?
4. What further data might you collect to better understand these relationships?

Answers to Self-Tests

Self-Test 13.1

1. Spurious correlation

    The key observation of concern here is that the 3500 app users form only 1.4% of the overall population (3500/247,000) eligible to use the app. So, a study team needs to ask themselves what factors *caused* those relatively few individuals to use the app, and might these factors also have led to the observed weight loss ? Some possible factors could include:

    - Pressure from family and friends (and support) to lose weight
    - A forthcoming significant future event, e.g. a wedding or graduation
    - A health scare, e.g. a heart attack, suffered by the patients themselves or a close family member
    - Developing a new relationship that caused them to "turn over a new leaf" in their lives

        Any of these factors are plausible explanations for why weight loss app users might start and continue to use the app, but are also reasons why that person might lose weight even without the app. So, use of the app might be a *marker* of the small proportion of people in the population who really intend to lose weight, and are being supported by others close to them to lose weight—rather than the cause. To investigate this possibility, a random subset of app users and non-users could be surveyed, asking them about any of the above factors. A qualitative investigation (using interviews or focus groups) of factors associated with success at weight loss (both promoters and inhibitors) would be useful to carry out before finalizing the survey questions to ensure that factors identified as important by responders are included.

2. Selection bias

    Generally, consistent app users are younger than non-users and more likely to own and use a smartphone, which puts them in a income group than non-app users or people who do not own a smart phone. So, what we could be seeing here is bigger weight loss in the subset of people who own and use a smart phone compared to those who do not. This selection bias could be investigated by comparing data on the educational level and household income of app users and non-users, and adding these variable into a regression model to predict weight loss based on a number of factors that are known to be associated with success at losing weight. The person's use or not of the smart phone app could then be added to this model to investigate if this new model explains more of the observed variation in weight loss, and if the coefficient for this variable is significant.

3. Information bias

If the person's weight at the start and the end of the study period is recorded by the app in app users but is recorded informally (in a notebook, diary or vague remembrance of weight a year ago) in the non-users, this could lead to an information bias. Assume that the weight loss based on data recorded in the app is correct. However, it is possible that the weight loss for non-users is underestimated because the non-app users have forgotten or misplaced their weight recording from a year ago, and innocently substitute a figure that is closer to their current weight than it was in reality. If this is the case, the lack of an objective record in non-app users of their weight a year ago would tend to reduce their apparent weights loss, and thus make it appear less than the measured weight loss in app users. To eliminate this bias, we would need an objective record of the baseline weight for the entire population.

4. Confounding

It is possible that the app was provided to a subset of the population specifically selected by the health care delivery network as part of a risk management exercise to reduce their chances of developing a weight-related illness (such as diabetes, stroke, heart attacks). In addition, other actions may also have formed part of this program, for example regular contacts with a dietician, feedback on their weight and exercise regime compared to that of others in the program, or incentives such as a lower renewal fee if they were successful at reaching their weight loss goal at year end. All of these would be confounders in any study that claimed to measure the impact of the app on weight loss. A searching question directed back to the health care delivery network about potential reasons for app use would help the study team uncover whether the app was targeted to a specific population, and if the users also received other weight loss support.

Self-Test 13.2

1. The histogram shows how the percent online time increased dramatically in 2020, then reduced slightly in 2021 and 2022, but still remained higher than for the earlier years. Two of the added value scores (physiology and pathology) remained roughly the same, but the anatomy added value score appeared to drop in 2020 and the following years.

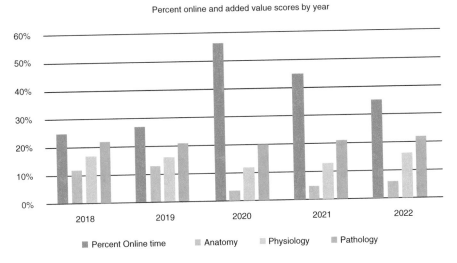

Percent online and added value scores by year

2. The X-Y plot of mean added value score against percent online time shows a negative correlation, with a slight drop in the mean score with increasing online time. This is confirmed by the regression line, showing all data points close to the line, and R squared of 0.97. From this analysis, one would be inclined to conclude that online learning does not promote student learning.

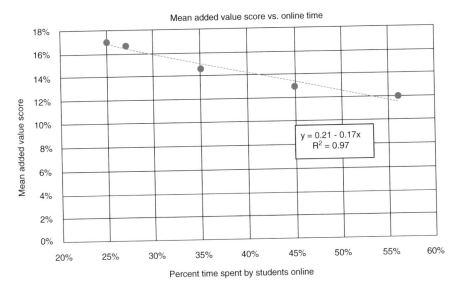

Mean added value score vs. online time

3. The histogram clearly indicates that the effects differ across the subjects, and the averaging across the subjects conceals this effect. It appears that the increased time spent online had little effect on the added value scores for two subjects, but

was associated with a large drop in the anatomy program added value scores. This might be because anatomy is better taught face to face, or it might be another factor, such as a change in program admission criteria, in teaching staff or the materials used. For example, an increase in the baseline anatomy skills of the students in 2020 and onwards could explain the decrease in added value. Equally, the explanation for reduced anatomy added value scores may be an effect of the COVID pandemic which started in 2020, but this would be expected to affect all three subjects.

To investigate if the "online effect" is larger for anatomy than for the two other subjects, the slopes of the regression lines for anatomy alone versus the mean of all three subjects could be compared. The final graph below shows that the reduction in added value with online teaching is greater for anatomy with a regression coefficient of 0.29, almost double the coefficient of 0.17 for the mean across all three subjects.

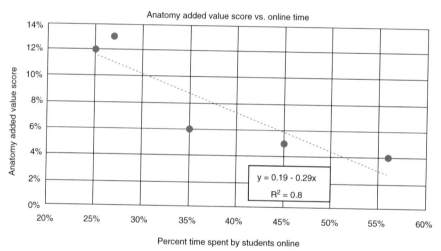

4. Further data that could be sought to understand the reason for the observed changes includes:

- The numbers of students in the program each year, as some of these changes could simply be due to small numbers in certain years.
- The actual entering scores of students for each subject and year, as interpretation of the added value score varies depending on the starting point.
- Data about teaching staff numbers and skills and student participation rates and satisfaction year by year and across the three subjects, to help understand why online learning may have led to lower added value for anatomy, if this is true.

- Further information about the types (and quality) of online learning experience offered across the three subjects. It is possible, for example, that the online offerings for anatomy differed from those in physiology and pathology.

References

Agiro A, Chen X, Eshete B, Sutphen R, Bourquardez Clark E, Burroughs CM, et al. Data linkages between patient-powered research networks and health plans: a foundation for collaborative research. J Am Med Inform Assoc 2019;26:594–602. Available from https://doi.org/10.1093/jamia/ocz012. Accessed 2 July 2021.

Agniel D, Kohane IS, Weber GM. Biases in electronic health record data due to processes within the healthcare system: retrospective observational study. BMJ. 2018;361:k1479.

Andersen B. Methodological errors in medical research. Oxford, UK: Blackwell Scientific Publications; 1990.

Baker L. Hilarious graphs (and pirates) prove that correlation is not causation. Data Science Central. Issaquah, WA: Data Science Central, TechTarget; 2018. Available from https://www.datasciencecentral.com/profiles/blogs/hilarious-graphs-and-pirates-prove-that-correlation-is-not. Accessed 29 June 2021.

Benchimol EI, Smeeth L, Guttmann A, Harron K, Moher D, Petersen I, et al. RECORD Working Committee. The REporting of studies conducted using observational routinely-collected health data (RECORD) statement. PLoS Med 2015;12:e1001885. Available from https://doi.org/10.1371/journal.pmed.1001885. Accessed 29 June 2021.

Britto MT, Fuller SC, Kaplan HC, Kotagal U, Lannon C, Margolis PA, et al. Using a network organisational architecture to support the development of learning healthcare systems. BMJ Qual Safety. 2018;27:937–46.

Byar DP. Why data bases should not replace randomized clinical trials. Biometrics. 1980;36:337–42.

Collins R, Bowman L, Landray M, Peto R. The magic of randomization versus the myth of real-world evidence. N Engl J Med. 2020;382:674–8.

Cooper GF, Bahar I, Becich MJ, Benos PV, Berg J, Espino JU, et al. Center for Causal Discovery Team. The center for causal discovery of biomedical knowledge from big data. J Am Med Inform Assoc 2015;22:1132–6. Available from https://doi.org/10.1093/jamia/ocv059. Accessed 29 June 2021.

Davey Smith G. Capitalising on Mendelian randomization to assess the effects of treatments. James Lind Library Bull (JLL Bull) 2006;1:1. Available from https://www.jameslindlibrary.org/articles/capitalising-on-mendelian-randomization-to-assess-the-effects-of-treatments/. Accessed 29 June 2021.

Deeny SR, Steventon A. Making sense of the shadows: priorities for creating a learning healthcare system based on routinely collected data. BMJ Qual Safety. 2015;24:505–15.

Equator Network, UK EQUATOR Centre. Results of Search for Reporting Guidelines: Observational Studies. Oxford, UK: Equator Network, UK EQUATOR Centre, Centre for Statistics in Medicine (CSM), NDORMS, University of Oxford; 2021. Available from https://www.equator-network.org/?post_type=eq_guidelines&eq_guidelines_study_design=observational-studies&eq_guidelines_clinical_specialty=0&eq_guidelines_report_section=0&s=+&eq_guidelines_study_design_sub_cat=0. Accessed 29 June 2021.

Eriksson T, Maclure M, Kragstrup J. Consultation with the general practitioner triggered by advice from social network members. Scand J Prim Health Care. 2004;22:54–9.

Ford I, Murray H, Packard CJ, Shepherd J, Macfarlane PW, Cobbe SM. Long-term follow-up of the West of Scotland Coronary Prevention Study. N Eng J Med. 2000;357:1477–86.

Friedman CP, Wong AK, Blumenthal D. Achieving a nationwide learning health system. Sci Transl Med 2010;2:57cm29. Available from https://doi.org/10.1126/scitranslmed.3001456. Accessed 29 June 2021.

Gray E, Marti J, Wyatt JC, Brewster DH, Hall PS, SATURNE Advisory Group. Chemotherapy effectiveness in trial-underrepresented groups with early breast cancer: a retrospective cohort study. PLoS Med. 2019;16:e1003006.

Grimes DA, Schulz KF. Bias and causal associations in observational research. Lancet. 2002;359:248–52.

Gymrek M, McGuire AL, Golan D, Halperin E, Erlich Y. Identifying personal genomes by surname inference. Science. 2013;339:321–4.

Hemkens LG, Contopoulos-Ioannidis DG, Ioannidis JP. Routinely collected data and comparative effectiveness evidence: promises and limitations. CMAJ 2016;188:158–164. Available from https://doi.org/10.1503/cmaj.150653. Accessed 29 Jun 2021.

Hill AB. The environment and disease: association or causation? Proc R Soc Med. 1965;58:295–300.

Hilton Boon M, Craig P, Thomson H, Campbell M, Moore L. Regression discontinuity designs in health: a systematic review. Epidemiology. 2021;32:87–93.

Hundepool A, Domingo-Ferrer J, Franconi L, Giessing S, Lenz R, Longhurst J, et al. A CENtre of EXcellence for Statistical Disclosure Control Handbook on Statistical Disclosure Control Version 1.2. Brussels, Belgium: CENtre of EXcellence for Statistical Disclosure Control (CENEX SDC), Portal on Collaboration in Research and Methodology for Official Statistics (CROS), ESSnet, European Commission; 2010.

Ioannidis JPA. Contradicted and initially stronger effects in highly cited clinical research. JAMA. 2005;294:218–28.

Julious SA, Mullee MA. Confounding and Simpson's paradox. BMJ. 1994;309:1480–1.

Kahneman D. Thinking, fast and slow. New York: Macmillan; 2011.

Klein MS, Ross FV, Adams DL, Gilbert CM. Effect of online literature searching on length of stay and patient care costs. Acad Med. 1994;69(6):489–95.

König K, Pechmann A, Thiele S, Walter MC, Schorling D, Tassoni A, et al. De-duplicating patient records from three independent data sources reveals the incidence of rare neuromuscular disorders in Germany. Orphanet J Rare Dis. 2019;14:152.

Lawrence ND. Data readiness levels: turning data from palid to vivid. In: Lawrence ND, editors. inverseprobability.com: Neil Lawrence's Homepage. Cambridge, UK: Neil D. Lawrence; 2017. Available from https://inverseprobability.com/2017/01/12/data-readiness-levels. Accessed 29 June 2021.

Maclure M. The case-crossover design: a method for studying transient effects on the risk of acute events. Am J Epidemiol. 1991;133:144–53.

Matthews GJ, Harel O. Data confidentiality: a review of methods for statistical disclosure limitation and methods for assessing privacy. Stat Surv. 2011;5:1–29.

O'Keefe CM, Rubin DB. Individual privacy versus public good: protecting confidentiality in health research. Stat Med. 2015;34:3081–103.

Ory HW. Association between oral contraceptives and myocardial infarction. A review. JAMA. 1977;237:2619–22.

Riaz H, Khan MS, Siddiqi TJ, Usman MS, Shah N, Goyal A, et al. Association between obesity and cardiovascular outcomes: a systematic review and meta-analysis of Mendelian randomization studies. JAMA Netw Open. 2018;1:e183788.

Rosenbaum PR, Rubin DB. The central role of the propensity score in observational studies for causal effects. Biometrika. 1983;70:41–55.

Seltzer CC, Bosse R, Garvey AJ. Mail survey response by smoking status. Am J Epidemiol. 1974;100:453–7.

Sherman RE, Anderson SA, Dal Pan GJ, Gray GW, Gross T, Hunter NL, et al. Real-world evidence - what is it and what can it tell us? N Engl J Med. 2016;375:2293–7.

Simpson EH. The interpretation of interaction in contingency tables. J R Stat Soc Ser B (Methodol). 1951;13:238–41.

Suissa S, Ernst P. Avoiding immortal time bias in observational studies. Eur Respir J. 2020;55:2000138.

Tsopra R, Wyatt JC, Beirne P, Rodger K, Callister M, Ghosh D, et al. Level of accuracy of diagnoses recorded in discharge summaries: a cohort study in three respiratory wards. J Eval Clin Pract. 2019;25:36–43.

Tufte E. The visual display of quantitative information. New York: Graphics Press; 1992.

United Kingdom National Health Service (NHS) Digital. Hospital Episode Statistics for Admitted Patient Care and Outpatient Data: Official Statistics Database. Leeds, UK: United Kingdom National Health Service (NHS) Digital, United Kingdom National Health Service (NHS); 2021. Available from https://digital.nhs.uk/data-and-information/publications/statistical/hospital-episode-statistics-for-admitted-patient-care-outpatient-and-accident-and-emergency-data. Accessed 29 June 2021.

Waruru A, Natukunda A, Nyagah LM, Kellogg TA, Zielinski-Gutierrez E, Waruiru W, et al. Where no universal health care identifier exists: comparison and determination of the utility of score-based persons matching algorithms using demographic data. JMIR Pub Health Surveill. 2018;4:e10436.

Yao XI, Wang X, Speicher PJ, Hwang ES, Cheng P, Harpole DH, et al. Reporting and guidelines in propensity score analysis: a systematic review of cancer and cancer surgical studies. J Natl Cancer Inst 2017;109:djw323. Available from https://doi.org/10.1093/jnci/djw323. Accessed 29 June 2021.

Zhou Y, Matsouaka RA, Thomas L. Propensity score weighting under limited overlap and model misspecification. Stat Methods Med Res. 2020;29:3721–56.

Part III
Qualitative Studies

Chapter 14
An Introduction to Qualitative Evaluation Approaches

Learning Objectives

The text, examples, Food for Thought questions, and the self-test in this chapter will enable the reader to:

1. Differentiate between inductive and deductive thinking.
2. Outline the types of questions that can best be answered using qualitative approaches.
3. Describe why qualitative approaches are especially useful in informatics evaluations.
4. Explain why qualitative studies are "emergent" in design.
5. Describe how qualitative methods can establish cause and effect.
6. Explain the five main steps in the natural history of a qualitative study.

14.1 Introduction

This chapter is the first of three devoted to describing qualitative methods for evaluating informatics interventions. It is a turning point, since these methods represent a different way of thinking about evaluation. According to House's typology (House 1980), described in Chap. 2, the illuminative/responsive study type, which will be described in detail in the following chapters, is open to and is in fact embracing of novel concepts. The quantitative approaches described in prior chapters are deductive in that a theory is being tested and, to have a theory in mind, team members must already know a good deal about the topic. Qualitative methods are designed to develop theory, so they are inductive and open to discovery. They are ideal for investigation of topics about which little is known. The major goals of this chapter are to describe when qualitative methods are appropriate and to offer a general framework for understanding how studies using these methods are conducted. Chapters 15 and 16 provide much more detailed tours through the methods of qualitative evaluation.

© Springer Nature Switzerland AG 2022
C. P. Friedman et al., *Evaluation Methods in Biomedical and Health Informatics*, Health Informatics, https://doi.org/10.1007/978-3-030-86453-8_14

Consider the following quote, often falsely attributed to Albert Einstein:

"Not everything that can be counted counts, and not everything that counts can be counted." (date and origin unknown) (Toye 2015, p. 7)

Quantitative approaches depend on numbers, on what can be "counted." Sometimes numbers are available, but they may not be important or useful. The quote is applicable here because it illustrates that quantitative approaches are useful for answering some, but by no means all, of the interesting and important questions that challenge study teams in any discipline. Quantitative approaches are ideal for answering questions about "everything that can be counted." The qualitative approaches, introduced here and expanded upon in Chaps. 15 and 16, address the problem of evaluation from a more nuanced point of view. They are best at answering questions when "not everything that counts can be counted." (Toye 2015). Qualitative approaches derive from philosophical views that may be less familiar and perhaps even discomforting to some readers. They challenge some traditional beliefs about the scientific method and the validity of an understanding of the world that develops from quantitative investigation. They argue that, particularly within the realm of evaluation of information resources, the kind of "knowing" that develops from qualitative studies may be more useful than that which derives from quantitative studies. In the past, some study teams in health care have dismissed qualitative methods as informal, imprecise, or "subjective," (Crabtree and Miller 1999). When carried out well, however, these studies are none of the above. They are equally rigorous, but in a different way. Professionals in informatics, even those who choose not to conduct qualitative studies, have come to appreciate the validity and value of this approach. As a testament to this increasing appreciation, a special issue of the Journal of the American Medical Informatics Association has been published in honor of Diana Forsythe, who was a strong proponent for the use of these methods in informatics (J Am Med Inform Assoc 2021).

14.2 Definition of the Qualitative Approach

The methods discussed in this chapter have many different labels. They are sometimes called "interpretive" or "inductive" to connote that they seek to go beyond simple documentation and description of phenomena. They are often called "qualitative" to connote the predominantly non-numerical nature of the data that are collected and the non-statistical methods of data analysis. Some call them "constructivist" to acknowledge the social construction of reality. They are at times called "naturalistic" to connote that studies are performed without purposeful manipulation of the environment under study (Crabtree and Miller 1999). Since no single term does full justice to these methods, we use multiple terms throughout this chapter, and we consider the terms *qualitative, subjectivist, constructivist, inductive, responsive, naturalistic,* and *interpretive* as practically synonymous.

Strauss and Corbin, in their seminal work, defined qualitative research (and, by extension, evaluation) as research that produces findings not using statistical procedures or other means of quantification (Strauss and Corbin 1990). Qualitative studies generally gather data in an open-ended way through interviews or observation, and the data are then progressively interpreted by the study team. These data are usually words and pictures, not numbers. Qualitative studies are generally conducted at the site where the work of interest is actually being done; typically, these are sites where an information resource is or will be in actual use. The studies themselves are often called field studies. Novel methods were also developed during the COVID pandemic so that data from the field could be gathered in a virtual manner and, though not always as rich as when data are collected on site, the results can be almost as useful (University of Technology Sydney 2020).

The authors of this chapter define qualitative methods as those that seek to provide an understanding of "the meaning individuals or groups ascribe to a social or human problem." (Cresswell, 2019, p. 232) The approach is inductive in that the study team collects information from the field with few preconceived notions, learning about the culture from those within it, and draws conclusions and some limited generalizations based on those data. The qualitative approach to evaluation is designed to address the deeper questions: the detailed "whys" and "according to whoms" in addition to the aggregate "whethers" and "whats." This approach seeks to represent the viewpoints of those who are users of the resource or otherwise significant participants in the context where the resource operates. The goal is "illumination" rather than judgment. The study team members seek to build an argument that promotes deeper understanding of the information resource and/or environment of which it is a part. The methods used derive largely from ethnography, a branch of anthropology that studies culture (Crabtree and Miller 1999). As such, the study team members immerse themselves physically or virtually in the situation where the information resource is or will be operational and collect data primarily through observations, interviews, and reviews of documents, naturally occurring data, or other text-producing media.

The designs—the data collection plans—of these studies are not rigidly predetermined and do not unfold in a fixed sequence. They develop dynamically and nonlinearly as the study team's experience accumulates. The study team begins with a minimal set of orienting questions; the deeper questions that receive more thorough study evolve from initial investigation. Team members keep records of all data collected and the methods used to collect and analyze them. Reports of qualitative studies tend to be written narratives. Such studies can be conducted before, during, or after the introduction of an information resource. Informatics professionals attribute the early use of ethnographic methods and the subsequent acceptance of their use to the late Diana Forsythe, who wrote convincingly about the need to consider context when studying health information technology implementations (Forsythe 1992, 1996, 1998).

Qualitative studies are designed so that the topic of interest is considered within a larger context. For example, a clinical researcher's success in using a new protocol management system depends in part on whether the system is consistent with other features of the local research environment such as institutional human subject procedures and perceptions of the system by colleagues. Qualitative methods might be used to evaluate the system within such a context, and the data collected can be the source of both rich descriptions and explanations. Qualitative methods can illuminate the evolution of important phenomena over time if the study has an historical component; they can indicate what led up to certain consequences; they can lead to new questions and insights; and their results can be presented in particularly vivid ways by using quotes to illustrate points. As discussed in the previous chapter, well-executed qualitative studies are credible, dependable, and replicable.

14.3   Motivation for Qualitative Studies: What People Really Want to Know

Chapter 2 presented some prototypical evaluation questions:

- Is the information resource working as intended?
- Can it be improved?
- Does it make any difference?
- Are the differences it makes beneficial?
- Do the observed effects match those envisioned by the developers?

A study team could also append "Why is it working well or not working well?" to each of the questions listed above. The reader should take a moment to examine these questions carefully and begin to think about how to go about answering them. When subjected to such deeper scrutiny, the questions quickly become more ornate and intricate. Instead of inviting yes/no answers, the questions might be reworded (see italics) to elicit more nuanced answers:

- How well is the resource working as intended?
  As who intended? Were the intentions set realistically? Did these intentions shift over time? What is it really like to use this resource as part of everyday professional activity?
- How can it be improved?
  How does one distinguish important from idiosyncratic suggestions for improvement? Which suggestions should be addressed?
- How much difference does it make?
  Was it needed in the first place? What features are making the difference?
- How much are the differences it makes beneficial?
  To whom? From whose point of view? Are all the pertinent views represented?
- How well do the observed effects match those envisioned by the developers?
  How do you detect what you do not anticipate?

These more specific, more explanatory, and more probing questions, shown in italics, are often what those who commission evaluation studies—and others with interest in an information resource—want to know. Some of these deeper, complex, and more nuanced questions are difficult to answer using quantitative approaches to evaluation. It may be that these questions are never discussed, or are deferred as interesting but "subjective" issues during discussions of what should be the foci of an evaluation study. These questions may never be asked in a formal or official sense because of a perception that the methods either do not exist to answer them in a credible way or are too difficult and time consuming to use.

Many of these deeper questions derive their importance from life in a pluralistic world. As discussed throughout this book, information resources are typically introduced into complex organizations where there exist competing value systems: different beliefs about what is "good" and what is "right," which translate into different beliefs about whether specific changes induced by information resources are beneficial or detrimental. These beliefs are real to the people who hold them and difficult to change. Indeed, there are many actors playing many roles in any real-world setting where an information resource is introduced. Each actor, as an individual and a member of multiple groups, brings a unique viewpoint to questions about inextricably fuzzy constructs such as need, quality, and benefit. If these constructs are explored in an evaluation study, perhaps the actors should not be expected to agree about what these constructs mean and how to measure them. Perhaps need, quality, and benefit do not inhere in an information resource. Perhaps they are dependent on the observer as well as the observed. Perhaps evaluation studies should be conducted in ways that document how these various individuals and groups "see" the resource, and not in ways that assume there is a consensus when there is no reason to believe one exists. Perhaps there are many "truths" about an information resource, not just one. For example, an individual user of an EHR may see it as both beneficial and detrimental: it is a benefit in that all members of the health care team can view a patient record no matter where the team member is located, but it is a detriment in that the user must take the time to enter the information into structured fields in a highly prescribed way (Winograd and Flores 1987).

14.4 Support for Qualitative Approaches

As suggested by the quotation at the beginning of this chapter, the results of a study, when reduced to tables and tests of statistical significance, may be elegant but useless if they assess an issue that does not matter. By contrast, the results of a rigorous qualitative study addressing an issue that is difficult to measure but extremely important can be of great use.

Chapter 7, introducing quantitative measurement, employed the analogy of the archer attempting to hit a stationary target with an unknown bulls-eye. The target is passive, so the archer's equipment, method and expertise determine how closely the arrow gets to the bullseye.

However, what happens if the target is an animated human and technical system moving unpredictably? The information resource under study usually cannot remain static, so it is like a moving target. In fact, it might be unethical to hold an electronic system rigid since it must constantly change and improve. In this case, the study team requires strategies to evaluate the context and a large area surrounding the target and, in fact, the team should aim for multiple targets. In qualitative studies, sometimes targets are outliers, perhaps out of sight, but they may be more important than those within sight. Their importance does not become evident until other targets have been explored. The targets are giving feedback and actually teaching the archer which way to shoot.

Although qualitative approaches may run counter to many notions of how one conducts empirical investigations, these methods and their conceptual underpinnings are not at all foreign to the worlds of information and computer science. The pluralistic, nonlinear thinking that underlies qualitative investigation shares many features with modern conceptualizations of the information resource design process. Consider the following statements from two highly regarded classical works addressing issues central to resource design. Winograd and Flores (Winograd 1987) argued as follows:

> In designing computer-based devices, we are not in the position of creating a formal "system" that covers the functioning of the organization and the people within it. When this is attempted, the resulting system (and the space of potential action for people within it) is inflexible and unable to cope with new breakdowns or potentials. Instead, we design additions and changes to the network of equipment (some of it computer based) within which people work. The computer is like a tool, in that it is brought up for use by people engaged in some domain of action. The use of the tool shapes the potential for what those actions are and how they are conducted... Its power does not lie in having a single purpose... but in its connection to the larger network of communication (electronic, telephone, paper-based) in which organizations operate (Winograd 1987, p. 170).

Norman (Norman 1996) in another seminal work, noted:

> Tools affect more than the ease with which we do things; they can dramatically affect our view of ourselves, society, and the world (Norman 1996, p. 209).

These thoughts from the work of system designers alert us to the multiple forces that shape the "effects" of introducing an information resource, the unpredictable character of these forces, and the many viewpoints on these effects that exist. These sentiments are highly consonant with the premises underlying the qualitative evaluation approaches.

Another connection between information/computer science and qualitative evaluation approaches is to the methodology of formal systems analysis, generally accepted as an essential component of information resource development (Davis 1994). Systems analysis uses many methods that resemble closely the qualitative methods for evaluation that are introduced here. It is recognized that systems analysis requires a process of information gathering about the present system before a design for an improved future system can be inferred. For example, the first scenario described in Chap. 4 outlines the needs assessment process. Systems analysis requires a process of information gathering, heavily reliant on interviews with those who use the existing system in various ways. Information gathering for systems analysis is typically portrayed as a cyclical, iterative process rather than a linear process (Kaplan 1987). In the literature of systems analysis we find admonitions, analogous to those made by proponents of qualitative evaluation, about an approach that is too highly structured. An overly structured approach can mis-portray the capabilities of workers in the system's environment, incompletely interpret or ignore the role of informal communication in the work accomplished, underestimate the prevalence of exceptions, and fail to account for political forces within every organization that shape much of what happens. Within the field of systems analysis, then, there has developed an appreciation of some of the shortcomings of quantitative methods and the potential value of qualitative methods drawn from ethnography that is discussed here (Davis 1994; Kaplan 1987).

14.5   Why Are Qualitative Studies Especially Useful in Informatics?

Because informatics applications are like moving targets, evaluation of them must include the entire context surrounding them. Variables cannot be held constant in an attempt to just target one or two. These applications are developed and implemented longitudinally, so the iterative and sometimes continuing nature of inductive studies is especially suitable.

These strategies can be used throughout the life cycle. Certainly, inductive approaches are applicable at all stages of development of an information resource, but they are most clearly applicable at two points in this continuum. First, as part of the design process, a qualitative study can document the need for the resource and clarify its potential niche within a given work environment. Indeed, it is possible for system developers to misread or misinterpret the needs and beliefs of potential users of an information resource in ways that could lead to failure of an entire project. Qualitative methods, if applied appropriately, can clarify these issues and direct resource development toward a more valid understanding of user needs. There is already a sense of general support for use of qualitative methods at the design stage of a resource (Davis 1994). At this point, the relation between qualitative evaluation and the methods of formal systems analysis is most evident.

Second, after an information resource is mature and has been tested in laboratory usability studies, further evaluation using qualitative approaches can describe the impact of the resource on the work environments in which it is deployed. At this developmental stage, the insights that can derive from both deductive and inductive studies are different and potentially complementary (Davis 1994). Deductive methods, and specifically the comparison-based approach, have dominated the literature on the impact of information resources. The randomized clinical trial has been put forward as the standard against which such studies should be measured. Although the randomized trial can estimate the magnitude of an effect of interest for an information resource, this method cannot elucidate the meaning of this effect for users of the resource and other interested parties, and typically sheds little light on whether the effect of interest to the evaluation as conducted was the effect of most importance. A survey could be conducted to answer questions about users' perceptions, but the survey designers might be imposing their limited knowledge and bias on the questions. They might not know enough to ask the right questions. Whether the impact of a resource is better established by deductive methods or by inductive methods derived from the ethnographic tradition, is and should be a matter of ongoing discussion with those who are commissioning the study. It is likely that mixed methods studies making simultaneous use of both approaches, as described in Chap. 17, is ideal, though costly.

Self-Test 14.1

1. At which two points in the life cycle of a project are qualitative approaches most useful and why?
2. In what ways are procedures for systems analysis similar to qualitative methods?
3. What does it mean when people refer to qualitative methods as being inductive and interpretive?

14.6 When Are Qualitative Studies Appropriate?

A study design needs to be appropriate for the investigative questions and for the setting. For evaluation of informatics resources, knowledge of the context or environment within which the implementation takes place is critical. As noted earlier, Diana Forsythe, an anthropologist who was influential in promoting the use of qualitative studies for evaluation of biomedical and health information systems, offered strong arguments in favor of studying these resources as they are used in context (Forsythe 1996). Kaplan, in another classic paper, outlined four areas for which qualitative evaluations are most useful: when the focus is on communication, care, control, and context or, as she calls them, the 4 C's (Kaplan and Duchon 1988). More specific to assuring the safety of information resources during and after implementation, the socio-technical 8-dimensional model describes eight areas that must be included in any thorough context-driven assessment (Sittig and Singh 2010).

If study team members seek to evaluate human interactions, the effects on personal health or the delivery of care, the political aspects of a system, or the importance of the practice setting on the success of an implementation, all of which involve context and the social environment, the team should consider using qualitative methods. Qualitative techniques are especially useful, therefore, when studying informatics applications at the organizational level, such as enterprise-level implementation of systems designed for use by health care providers or biomedical researchers.

14.7   Rigorous, but Different, Methodology

Qualitative studies have become an accepted alternative or complement to quantitative studies in the health care and informatics literature because they make it possible to address certain research questions that cannot be investigated any other way. They are ideal for evaluating information resource implementation processes, for example. Like their quantitative counterparts, qualitative studies can be done either well or poorly. Examples of both exist in the literature. It is therefore critical that individuals knowledgeable in informatics be able to distinguish the useful studies from the flawed ones, as well as be able to do high-quality evaluation studies themselves.

Qualitative approaches to evaluation, like their quantitative counterparts, are empirical methods. Although it is easy to focus only on their differences, quantitative and qualitative approaches to evaluation share many general features. In all empirical studies, for example, evidence is collected with great care. The study team members are always aware of the plan for what they are doing and why. The evidence is compiled, interpreted, and ultimately reported. Study team members keep records of their procedures, and these records are open to subsequent audit by the team members themselves or by individuals outside the study team. The principal investigator or evaluation team leader is under an almost sacred scientific obligation to report use of methods in detail, ideally in enough detail to enable another study team to replicate the study. Failure to be able to do so invalidates any study.

The two approaches also share a dependence on theories that guide study teams toward explanations of the phenomena they observe, and share a dependence on the pertinent empirical literature: published studies that address similar phenomena or similar settings. Within quantitative and qualitative approaches, there are rules of good practice that are generally accepted. It is therefore possible to distinguish a good study from a bad one. Finally, a neophyte can learn either or both types of approaches, initially by reading textbooks and other methodological literature, by formal learning, and ultimately by conducting studies under the guidance of experienced mentors.

There are, at the same time, many fundamental differences between quantitative and qualitative approaches. First and foremost, qualitative studies are "emergent" in design. Quantitative studies typically begin with a set of hypotheses or specific questions and a plan for addressing each member of this set. There is also an assumption by the study team members that, barring major unforeseen developments, the plan will be followed exactly. (When quantitative study teams deviate from their plans, they do so apologetically and view their having done it as a limitation of their study.) Not following the plan is seen as a source of bias, because the study team member who sees negative results emerging from the exploration of a particular question or use of a particular measurement instrument might change strategies in the hope of obtaining more positive findings. By contrast, qualitative studies typically begin with some general orienting issues that stimulate the early stages of investigation. Through these initial investigations, the important questions for further study begin to emerge. The qualitative study team member is willing, at virtually any point, to adjust future aspects of the study in light of the most recent information obtained. Qualitative team members are incrementalists; they live from day to day and have a high tolerance for ambiguity and uncertainty. (In this respect, they are again much like good software developers.) Also, like software developers, skilled qualitative team members must develop the ability to recognize when a project is finished, when it has reached the point of saturation—when further benefit can be obtained only at great cost in time and effort.

A second distinguishing feature of qualitative studies is a "naturalistic" orientation—a reluctance to manipulate the setting of the study, which in most cases is the work environment into which the information resource is introduced. Because qualitative studies avoid altering the environment in order to study it, these studies can be done in an appealing "ecological context" (Crabtree and Miller 1999, p. 10) There is no question that the results apply to the exact setting, work process, and culture within which the information resource under study is deployed. The extent to which the results can be safely generalized from that specific setting to other similar settings depends very much on local circumstances. In qualitative investigations, however, the aim is rarely to generalize to other settings, and more usually to gain better insight and understanding into the specific setting under scrutiny. Control groups, placebos, purposefully altering information resources to create contrasting interventions, and other techniques central to the construction of quantitative studies are typically not used in qualitative work. Qualitative studies do employ quantitative data for descriptive demographic purposes and may additionally offer quantitative comparisons when the study setting offers up a natural experiment where such comparisons can be made without altering how work is organized or performed in that environment. Qualitative study team members are opportunists where pertinent information is concerned; they use what they see as the best information available to illuminate a question under investigation.

A third important feature of qualitative studies is seen in their product or "deliverable;" qualitative studies result in reports written in narrative prose. Although these reports can be lengthy and may require a more significant time investment on the part of the reader, no technical understanding of quantitative methods or statistics is required to comprehend them fully. Results of qualitative studies are therefore accessible to a broad community—and often entertaining—in a way that results of many quantitative studies are not. Reports of qualitative studies seek to engage their audience, often by the use of pithy quotes from subjects.

14.8   Qualitative Approaches and Their Philosophical Premises

Qualitative studies do not seek to prove or demonstrate. They strive for insightful description—what has been called "thick description" (Berg and Lune 2012)— leading to deeper understanding of the phenomena under study. They offer an argument; they seek to persuade rather than demonstrate. They progress from data gathering in the field so the study team can learn about the culture to analysis that seeks to interpret those data and discover the meaning of that culture. Hence, they are interpretive in approach.

It has been emphasized that the purpose of evaluation is to be useful to various "stakeholders": those with a need to know. These needs vary from study to study, and within a given study the needs vary across the different stakeholder groups over time. A major feature of qualitative approaches is their responsiveness to these needs. The foci of a study are formulated through a process of negotiation, to ensure their relevance from the outset. These foci can be changed in light of accumulating evidence to guarantee their continuing relevance. As with quantitative methods, qualitative methods are therefore concordant with the basic tenets of evaluation as a process that, in order to be successful, must be useful in addition to truthful.

As the discussion of qualitative methods unfolds in this and the following two chapters, it should become clear that there are numerous features working to ensure that well-executed studies meet the dual criteria of utility (usefulness) and credibility (believability). At this point, the reader might ask whether a method that is so open-ended and responsive can also generate confidence in the veracity of the findings. In so doing, the discussion must turn to the general issue of what makes evidence believable. Quantitative studies rely on methods of quantitative measurement, on *quantitative objectivity*, described in great detail earlier in this book, which in turn are based on the principle of intersubjectivity. Simply stated, this principle holds that the more independent observers who agree with an observation, the more likely it is to be correct. (Chapter 7 developed specific methods for implementing this principle.) Indeed, within the quantitative mindset, unless the study team can show that several observers agree to an acceptable extent, their observations are *prima facie* not believable. One observer is not to be trusted. By contrast, the principle of *qualitative objectivity* is central to qualitative work. It holds that an experienced, unbiased observer is capable of making fundamentally truthful observations that may, in fact, be superior to those of a panel of observers who agree but are all wrong because of some bias they share. In this light, qualitative approaches can be seen to be as objective (i.e., truthful) as quantitative studies. They rely, however, on a different concept of objectivity.

Finally, the two approaches address issues of cause and effect differently. How can cause-and-effect relationships be established without the experimental control customary to randomized trials? In qualitative investigation, a case for cause and effect can be made in much the same way that a detective determines the perpetrator of a crime or a forensic pathologist infers cause of death. Through detailed examination of evidence, the investigator recreates the pertinent story, often depicting in great detail a number of critical events or incidents. Via this portrayal, the detective, pathologist, or study team member crafts a logical, compelling case for cause and effect. In the end, such a portrayal can be as compelling as the result of a controlled experiment that is subject to the manifold biases described in Chap. 10.

14.9   Natural History of a Qualitative Study

As a first step in introducing the techniques of qualitative evaluation, Fig. 14.1 illustrates the stages or natural history of a study. These stages comprise a general sequence, but, as mentioned earlier, the qualitative study team members must always be prepared to revise their thinking and possibly return to earlier stages in light of new evidence. Backtracking is a legitimate aspect of this model.

14.9.1   Negotiation of the "Ground Rules" of the Study

First, during any empirical research, and particularly for evaluation studies, it is important to negotiate an understanding between the study team and those commissioning the study. This understanding should embrace the general aims of the study, the kinds of methods to be used, access to various sources of information including health care providers, patients, social media, and documents, and the format for interim and final reports. The aims of the study might be formulated in a set of initial "orienting questions." Ideally, this understanding is expressed in a memorandum of understanding, analogous to a contract, signed by all interested parties. By analogy to a contract, these ground rules can be changed during a study with the consent of all parties. (Although essential to a qualitative study, a memo of understanding or evaluation contract is recommended for all studies, irrespective of methods employed.) If needed, approval from a human subjects committee should be sought at this point.

14.9.2   Immersion into the Environment and Initial Data Collection

At this stage the study team members should begin spending time in the fieldwork environment. The activities range from formal introductions to informal conversations and the silent presence of the study team members at meetings and other

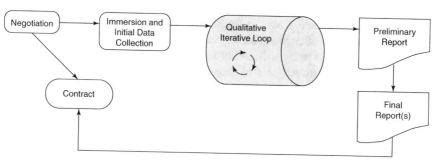

Fig. 14.1 Natural history of a qualitative study

events. Qualitative methodologists use the generic term *field* to refer to the setting, which may be multiple physical locations, where the work under study is carried out. The field may also be entered virtually when in person visits are not possible. Trust and openness between the study team and those in the field are essential elements of qualitative studies. If a qualitative study is in fact to generate insights with minimal alteration of the environment under study, those who live and work in the field (clinicians, patients, researchers, students, and others) must feel sufficiently comfortable with the presence of the study team to go about their work in the customary way. Time invested by the study team in building such relationships pays compound interest in the future.

Initial data collection aims to focus the questions. Even as immersion is taking place, the study team is already collecting data to sharpen the initial questions or issues guiding the study. The early discussions with those in the field and other activities primarily targeted toward immersion inevitably begin to shape the study team's views. Almost from the outset, the study team is typically addressing several aspects of the study simultaneously.

14.9.3   Iterative Qualitative Loop

At this point, the procedural structure of the study becomes akin to an iterative qualitative loop (see Fig. 14.2) as the study team engages in cycles of data collection, analysis, reflection, interpretation, and reorganization. Data collection involves interviewing, observation, document analysis, and other methods. Data are collected on planned occasions as well as serendipitously or spontaneously. The data are carefully recorded and interpreted in the context of what is already known. Reflection entails the contemplation and interpretation of the new findings during each cycle of the loop. Reorganization results in a revised agenda for data collection in the next cycle of the loop.

Although each cycle within the iterative loop is depicted as linear or unidirectional, even this portrayal is somewhat misleading. The net progress through the loop is clockwise, as shown in Fig. 14.2, but backward steps within each cycle are both natural and inevitable. They are not reflective of mistakes or errors. A study

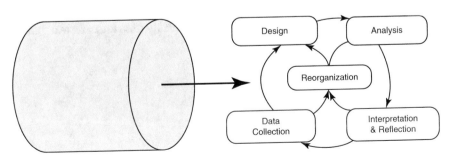

Fig. 14.2 Iterative qualitative loop

team may, after conducting a series of interviews and studying what participants have said, decide to speak again with one or two participants to clarify their positions on a particular issue.

An important element of the iterative loop, which can be considered part of the reflection and interpretive process, is sharing of the study team's own thoughts and beliefs with the participants themselves. (This step is called "member checking" and is discussed in more detail in Chap. 15.) Within the quantitative tradition, member checking might "unblind" the study and introduce bias. In the qualitative tradition, the views of informed participants on the study team's evolving conclusions are considered a key resource.

Food for Thought

Return to Self-test 2.1. After rereading the case study presented there, consider the following questions:

1. What specific evaluation questions in this case are better addressed by quantitative methods and which are better addressed by qualitative methods?
2. If you were conducting a qualitative study of the project, consider the various data collection modalities you might employ. Whom would you interview? What events would you observe? What artifacts would you examine?
3. How would you immerse yourself into the environment of this project?

14.9.4   Preliminary Report

The first version of the final report should itself be viewed as a form of investigative instrument. By sharing this draft report with a variety of individuals, a major check on the validity of the findings can be obtained. Typically, reactions to the preliminary report generate useful clarification and a general sharpening of the study findings. Sometimes (but rarely if previous stages of the study have been carried out with care), reactions to the preliminary report generate needs for further data to be collected. Because the qualitative study report is usually a narrative, it is vitally important that it be relatively concise and well written, in language understood by all intended audiences. Circulation of the report in draft can ensure that the final document communicates as intended. Liberal use of anonymous quotations from interviews and documents, distinguished typographically from the main text, makes a report highly vivid and meaningful to readers.

14.9.5 Final Report

The final report, once completed, should be distributed as negotiated in the original memorandum of understanding. In qualitative evaluation studies, distribution of the report is often accompanied by "meet the study team" sessions that allow interested persons to explore the study findings interactively and in greater depth.

Qualitative study requires a frame of mind on the part of the study team different from that in quantitative study. The agenda is never closed. The study team members must always be alert to new information that may require a systemic reorganization of everything they have done so far. For these reasons, many heuristic strategies and safeguards are built into the process. Just as there are well-documented procedures for collecting data while conducting qualitative studies, there is also a set of strategies used by study teams to validate results and insights. These are discussed in more detail in Chap. 15.

Food for Thought

Consider the following about qualitative studies:

1. How can they be believable in the world of informatics?
2. What personal attributes must a qualitative study team member have?
3. Which of these attributes do you personally have?
4. Should informatics professionals themselves perform these studies, or should they be "farmed out" to anthropologists or other social scientists?

14.10 Summary: The Value of Different Approaches

When all is said and done, how does the study team know that the findings of a qualitative study are "correct"? How does the team know if the findings carry any truth? What makes a study of this type more than one person's opinion, or the opinion of a study team that may share a certain preexisting perspective on the resource under study? To explore this question fairly, both qualitative and quantitative studies should be seen as belonging to a more general family of methods for empirical investigation. Neither approach should be placed on the defensive and required to prove itself against a set of standards produced by proponents of the other. When seen in this light, the credibility of both quantitative and qualitative approaches derives from five sources:

- Belief in the philosophical basis of the approach
- Existence of rules of good practice

- The study team's adherence to these rules
- Accessibility of the data to others if necessary
- Value of the resulting studies to their respective audiences

Previous chapters addressed how these factors apply to quantitative studies. How they apply to qualitative studies is covered to some extent in this chapter. A more thorough discussion is found in Chaps. 15 and 16. Mixed methods studies, incorporating both, will be described in Chap. 17.

Ultimately, each reader of this volume must make a personal judgment about the credibility of any of the evaluation approaches presented. None of the approaches is beyond challenge. The authors specifically caution the reader against establishing a quantitative approach as a standard, and then assessing qualitative approaches using the specific characteristics of this standard. This would inequitably frame the competition using the logic, definitions, and assumptions unique to one of the competitors. For example, consider the question of whether qualitative approaches can establish causality as well as their quantitative counterparts. If cause and effect are defined as proponents of quantitative methods see the world, of course the answer is no. (Quantitative work establishes cause and effect through randomization and experimental control. Since qualitative work does not employ randomization and control, it cannot therefore establish cause and effect.) The argument changes if cause and effect are defined more generically, however. If both sides accept that one can establish cause and effect by building a logical, believable case, they will conclude that both quantitative and qualitative studies can approach such issues. They will just do it differently or, in the case of mixed methods studies, together.

It is also human nature to compare anything relatively new to an idealization of what is familiar. Because quantitative studies may be more familiar, it is tempting to compare qualitative methods against the perfect quantitative study, which is never realized in practice. Every quantitative study has limitations that are usually articulated at the end of a study report. Many such reports end with a lengthy list of limitations and cautions and a statement that further research is needed. For these reasons, rarely has any one study, quantitative or qualitative, ended a controversy over an issue of scientific or social importance.

14.11 Two Example Summaries

To convey both the substance and some of the style of qualitative work in informatics, we include below summaries of two published studies. The first is Forsythe's classic 1996 work, which advocated for gathering input from real users at the design stage and also had a groundbreaking impact on the field of informatics, convincing readers of the value of ethnographic approaches (Forsythe 1996). The second is a paper by Goedhart et al., published nearly three decades later, making similar arguments based on more recent research (Goedhart et al. 2021).

Example Summaries
Summary of Forsythe's 1996 paper
The problem of user acceptance of knowledge-based systems has always been a concern in informatics. User acceptance should increase when system-builders understand both the needs of potential users and the context in which a system will be used. Ethnography is one source of such understanding. This paper, a classic dating back to 1996, describes the contribution of ethnography (and an anthropological perspective) during the first year of a 3-year interdisciplinary project to build a patient education system on migraines. Systematic fieldwork produced extensive data on the information needs of migraineurs. These data called into question some of the assumptions on which the entire development project was based. Although it was not easy for the research team members to rethink their assumptions and the implications for design, using ethnography enabled them to undertake this process relatively early in the project at a time when redesign costs were low. It greatly improved the system so that it met the need of real users, thus avoiding the troublesome problem of user acceptance (Forsythe 1996).
Summary of Goedhart et al. 2021 paper
This paper reflects the spirit of Diana Forsythe, reporting results of a study focused on designing a patient portal. The title describes it well: "Persistent Inequitable Design and Implementation of Patient Portals for Users at the Margins." Interviews with developers, implementers, and policy makers indicated that potential users of the portal who were included in the initial design process were health care professionals rather than patients. While potential use of the portal by vulnerable patient populations seems like it should have been of paramount interest to the designers, these populations were not being involved (Goedhart et al. 2021). This is a situation that would have disappointed the late Diana Forsythe, who was always an advocate for those without a voice in society, but also would have pleased her in that the evaluation did stimulate future involvement of patients.

Answers to Self-Tests

Self-Test 14.1
1. Qualitative methods can be especially useful at the beginning of a project because they can yield information in the form of a needs assessment. Later, after implementation, they can be extremely valuable if used to provide feedback for improving the information resource.
2. Systems analysis processes attempt to be sensitive to the context into which the system will be placed and they are also iterative like qualitative processes.
3. These methods are inductive and interpretive because they require the study team to be open minded, to learn from participants, and to offer explanations and describe the meaning of cultural attributes.

References

Berg BL, Lune H. Qualitative research methods for social scientists. 8th ed. Pearson: Boston, MA; 2012.

Crabtree BF, Miller WL. Doing qualitative research. 2nd ed. Thousand Oaks, CA: Sage; 1999.

Cresswell JW. Research design: qualitative, quantitative, and mixed methods approaches. Thousand Oaks, CA: Sage; 2019.

Davis WS. Business systems design and analysis. Belmont, CA: Wadsworth; 1994.

Forsythe DE. Using ethnography to build a working system: rethinking basic design assumptions. Symp Comput Applications Med Care. 1992;16:505–9.

Forsythe DE. New bottles, old wine: hidden cultural assumptions in a computerized explanation system for migraine sufferers. Med Anthropol Q. 1996;10:551–74.

Forsythe DE. Using ethnography to investigate life scientists' information needs. Bull Med Libr Assoc. 1998;86(3):402–9.

Goedhart NS, Zuiderent-Jerak T, Woudstra J, Broerse JEW, Betten AW, Dedding C. Persistent inequitable design and implementation of patient portals for users at the margins. J Am Med Inform Assoc. 2021;28(2):276–83.

House ER. Evaluating with validity. Beverly Hills: Sage; 1980.

J Am Med Inform Assoc 2021;28(2):197–423. (special issue).

Kaplan B. Initial impact of a clinical laboratory computer system. J Med Syst. 1987;11:137–47.

Kaplan B, Duchon D. Combining qualitative and quantitative methods in information systems research: a case study. MIS Q. 1988;12:571–86.

Keikhosrokiani P. Perspectives in the development of mobile medical information systems: life cycle, management, methodological approach and application. Amsterdam: Elsevier; 2019.

Norman DA. The design of everyday things. New York: Basic Books; 1996.

Sittig DF, Singh H. A new sociotechnical model for studying health information technology in complex adaptive healthcare systems. Qual Saf Heath Care. 2010;19(suppl 3):i68–74.

Strauss A, Corbin J. Basics of qualitative research: grounded theory procedures and techniques. Newbury Park, CA: Sage; 1990.

Toye F. Not everything that can be counted counts, and not everything that counts can be counted. Br J Pain. 2015;9:7.

University of Technology Sydney. Adapting research methods in the COVID-19 pandemic: resources for researchers, 2nd ed. December, 2020.

Winograd T, Flores F. Understanding computers and cognition: a new foundation for design. Reading, MA: Addison-Wesley; 1987.

Chapter 15
Qualitative Study Design and Data Collection

Learning Objectives

The text, examples, Food for Thought questions, and self-tests in this chapter will enable the reader to:

1. Explain why the qualitative study data collection process is described as an iterative loop.
2. Describe five mechanisms for assuring that a qualitative study is rigorous and, given a study description, identify the mechanisms employed in that study.
3. Describe the major steps in the qualitative data gathering process and, given a study description, identify what constitutes each step employed in that study.
4. Distinguish between interviews at different levels of structure: unstructured, semi-structured, fully structured.
5. Describe the processes of qualitative data collection for observing, interviewing, focus groups, and naturally occurring data. Given a study description, identify the processes employed in that study.
6. Explain why sometimes it is best to use a combination of qualitative strategies for data gathering.

15.1 Introduction

While the prior chapter set the stage for an understanding of the nature of qualitative evaluation, this chapter will offer strategies for planning a study and making decisions about how to gather data. The process is depicted as an iterative looping through steps beginning with idea generation to dissemination of results. It is critical that strategies for rigor be incorporated throughout the process. This chapter outlines methods for data collection using interviews, observation, focus groups,

© Springer Nature Switzerland AG 2022
C. P. Friedman et al., *Evaluation Methods in Biomedical and Health Informatics*, Health Informatics, https://doi.org/10.1007/978-3-030-86453-8_15

and naturally occurring data, and also describes combinations often used together, which constitute toolkits of complementary techniques.

This chapter proceeds in the context of an example study setting, a variation on the "YourView" patient portal example introduced in Chap. 1. The example is presented here first, and then revisited to illustrate different aspects of qualitative methods arising throughout this discussion.

Nouveau Community Hospital and Health System Case Example
Eighteen months ago, the leadership of Nouveau Community Hospital and Health System decided to develop locally and then deploy a patient portal named YourView, a comprehensive system allowing patients to make appointments, to communicate via secure email with staff and care providers, and to view their medical record information that is stored on the hospital's EHR. Immediately following a brief pilot, YourView went live for all clinic patients. Anticipating some problems with patient access, usability, and privacy concerns based on the pilot, the leaders asked a consultant, Dr. Yu, to evaluate the new system and the implementation process over a 1-year period. The leaders expect the evaluation to provide the basis for smoothing out the rough spots in the portal and to guide further development and ongoing support. Their working assumption is that the patients will grumble at first, but will love the portal after a short period of time.

To begin considering what this consultant might do, it is possible to imagine a wide range of alternatives in terms of the focus (workflow and efficiency issues, quality of data entered by patients and patient safety, attitudes of personnel, and others), the calendar for the evaluation (a short project to help develop the training to be provided or a longer study to find and address currently unknown issues), and the boundaries of the problem (a narrow look into the new system, or a wider look at the other issues that are inevitably attached to it). Selection among these alternatives would be required at the beginning of the study, particularly in view of the management expectation, whether justified or not, that the patients will accept the portal after a short transition period. It is also critical to realize that the initial decisions might need to change as new facts are clarified and new questions identified. More fundamentally, it would be essential to clarify the character of the evaluation, and then maintain contact with the client group at Nouveau Community Hospital and Health System to ensure that the outcomes are understandable and present no major surprises when submitted in a final report. These points, and many others, all imply the need for serious discussion early in the life of the study.

15.2   The Iterative Loop of Qualitative Processes

Figure 14.2 in Chap. 14 depicted what the authors view as the iterative loop of qualitative processes, shown in greater detail here in Fig. 15.1. This shows how fluid the process is, but it is always aimed at answering the evaluation questions. At any point in the process, a decision can be made to loop back to revisit steps and even to start over with new or revised research questions, however.

   Within a qualitative study, the processes of gathering and interpreting data happen continuously in a cyclical iterative looping fashion. There are periods of data collection in which the deliberate effort is to explore—to find data that suggest new questions as well as answers. This requires an openness to and curiosity about events and details. There are times when the purpose is to seek data that either confirm or conflict with an emerging explanation. Confirmatory work includes the effort to ensure that the findings from the most recent experience "in the field" are cross-checked against data from other times and places. It also includes a search for other kinds of data that should be evident or absent if current thinking is accurate. Work that explores conflicting data is also necessary so that many perspectives can be understood and the study can be as thorough as possible (Pope and Mays 2020).

   This iterative loop process entails recognizing, finding, collecting, and recording data according to the emergent design of the study. (Specific data gathering methods will be discussed in detail later in this chapter.) During data gathering, the mode of thinking required of the study team members is best described as "open minded." This thought process is inevitably shaped by personal and formal theories and assumptions that evolve with the study. The data gathering process generally begins with an effort to gain an overview of the issues and context and moves toward more focus later. It might entail initial strategies such as visiting the site, gathering

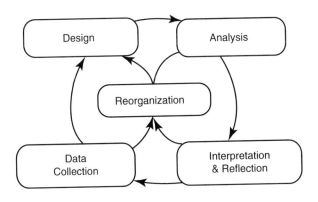

Fig. 15.1   The Iterative Loop of Qualitative Evaluation

information about the organization from its web site, partnering with insiders, talking to colleagues who may know something about the organization, having a virtual demonstration of the information resource, and/or learning more about the information resource that is to be evaluated from vendors and others. Open-mindedness in qualitative studies is critical. In contrast with quantitative work, where data are gathered according to a predetermined plan and deviations are to be minimized, qualitative investigators are actively seeking opportunities to change the plan (Tolley et al. 2016).

Qualitative study teams can only plan in general terms, however, because the feedback loops in the process (see Fig. 15.1) often instigate revisiting prior activities, taking detours based on what is being learned, or halting data gathering early when saturation is reached. Qualitative inquiry is distinguished by the organization and sequence of investigative actions within a study. Within quantitative studies the familiar sequence progresses along a linear path from the research problem, through a literature review, the development of a research design, the collection and analysis of data, and finally arrives at a statement of conclusions. Each of these activities is also included in qualitative work, but the sequence is neither linear nor quite so predictable. As discussed in Chap. 14, every element of a qualitative study—the questions, relevant literature, data collection, and interpretation of results—is continually examined and refined. Work proceeds continuously through an iterative loop with frequent adjustments as the data and theory are gradually combined into explanation. The team can define a starting place for a study, where a decision is made to invest resources into answering a set of questions. The team can also define a stopping place, where a decision is made that an argument has been assembled that adequately captures the significant data and answers the initially posited questions. The path between beginning and end is flexible by design and intention.

15.3 Theories, Frameworks, and Models that Can Guide the Process

Qualitative inquiry both draws upon and builds theory. Many kinds of theory come into play at different stages of a study. Personal theory stems from an investigator's own professional training and background. "Hunches" based on personal experience fall into this category. For example, a clinician informatician might notice that an alert is no longer firing and has a hunch that a recent system upgrade may be responsible. This hunch might lead to a broad evaluation of clinical decision support problems. A second type of theory, somewhat more formal in nature, is based on prior research by others who do work closely related to that of the investigator. Theories of why information resources succeed or fail, derived from the informatics or general information technology literatures, fall into this category. As described in Chap. 5, seeking theories used in past studies can be helpful when designing new studies.

The third kind of theory is formal theory found in the literature of the organizational, social, behavioral, and other related sciences. Such theories, for example those relating to the communication and adoption of innovations, like the classic Technology Acceptance Model (TAM), (Davis et al. 1989) have applicability to many domains, and certainly can be supportive of studies within informatics. These theories can be used to guide planning a study. For example, a study team might use elements from TAM to help guide development of an interview guide. Since qualitative methods are also useful in developing new theories, a review of established theories later in the study might reveal revisions that could be made to the original theory. For example, TAM primarily applies to acceptance of a new technology when the technology use is optional. A qualitative study of implementation of an informatics tool that clinicians are required to use, that is not optional, could lead to a revision of TAM that includes the concept of required use. RE-AIM is another frequently used framework, especially in health care, with five evaluation components: Reach, Effectiveness, Adoption, Implementation, and Maintenance. Using this framework, evaluation teams can study the impact of innovations at both the individual (i.e., end-user) and organizational (i.e., delivery agent) levels (Gaglio et al. 2013).

The role of theory in qualitative studies can be confusing because many guidelines for the proper conduct of such studies emphasize that theory should be articulated early in the study so that it can help in formulating the design process. In this case, it is probably best to think of the use of theory at this stage as providing a framework for the study. In fact, the terms framework, model, and theory are all applicable here. A confusing and almost contradictory aspect of the use of theory early in the study is the option of using a "theory of no theory," of choosing to be "atheoretic". The choice of being atheoretic is itself also a theory. This atheoretic approach is often called grounded theory, which seeks to develop new theory entirely from the newly collected data (Glaser and Strauss 1967).

The grounded theory approach in its purest form is the most inductive approach possible. Glaser and Strauss (1967) and then Strauss and Corbin (1990) first described and then expanded on the notion of grounded theory and this approach has become widely used, though in a modified manner, in informatics research and evaluation. It is an approach that is firmly based or grounded in the data. This means that few preconceived notions are allowed to structure the outcomes of the study. For example, when Crabtree did his early work studying what it is like becoming a physician, he gathered observational and interview data, and did his coding by selecting the residents' own words as a beginning coding scheme. He saw patterns as codes were grouped together, but the larger theme became clear in what he calls a "thunderbolt" moment (Crabtree and Miller 1999, p. 155). The central organizing theme describing the lives of residents, he realized, was "surviving." He then revisited his data and became more and more confident that this was the central issue. His subjects did not talk directly about survival, so it was not until he reached the highest level of interpretation that in the study team's view, full understanding emerged. The insight emerged directly from the data and is an excellent example of the outcome of a grounded theory approach (Crabtree and Miller 1999).

Using the principles of evidence-based health informatics (EBHI), described in Chap. 5, the study team should query the evidence base throughout the process. The study team should turn to the literature during a study, as the team members' theories begin to be confirmed through fieldwork and they seek a broader theoretical framework to gain a better understanding. The dominant mode of thought in this process is akin to traditional scholarship. Sometimes the analysis and the report require the investigation to venture into unfamiliar literature and theory. When a potential answer lies wholly or partially outside the familiar intellectual ground of the study team member, there is a new challenge in finding useful literature and theory. Assistance from colleagues who have different backgrounds may be fruitful at these points. For example, if the study team has a hunch that TAM might be applicable to the project at hand, the team may want to ask a colleague in engineering or marketing for advice; if RE-AIM seems applicable, the team may need to approach a friend with a background in public health. A multidisciplinary approach that takes advantage of literature and expertise outside of informatics can expose the team to relevant theories that might otherwise go undiscovered. The following is an example of development of a new theory, which, unlike the fictitious Nouveau Community Hospital running example, actually did take place.

Example of Development of a New Theory

As results become more finalized, the non-theory based atheoretical study may wondrously produce a new exciting theory. There is nothing more thrilling than the moment when a team experiences a theoretical breakthrough. Years ago, after analysis of data from a site visit to Kaiser Permanente Northwest, the study team led by one of the authors sat in a conference room surrounded by flip charts summarizing findings. For the first time, the team had seen an information system in use for nearly every activity within clinics. A team member said it was as if an umbilical cord connected everyone to the computer system. The team built on that idea, developing a theory they called "the hub," (Ash et al. 2005) predicting that as more healthcare organizations implemented more systems, the computer would become the hub of all activities. The theory has been verified over time: when systems go down, when this hub disappears, organizations fall into disarray. Even if a spoke in the hub breaks, the rest of it is weakened. The new concept assisted the study team in interpreting the results of the study, but it also generated a new theory to be tested in the future.

Again, an established framework—or a theory that emerges early in a study--can be used both at the beginning and end of a study. One example is the use of the multiple perspectives framework (Linstone 1984). Using this framework, the study team identifies the foci, or information resources, of interest and the stakeholder perspectives that need to be explored. A multiple perspectives framework may be developed for studying any information resource, gaining perspectives from any stakeholder groups. This framework assures that the study team deliberately talks with and about those groups. At the end of the study, themes may revolve around what has been learned about each perspective and the perspectives provide a framework for describing the results (Linstone 1984). The following is an example of use of the multiple perspectives framework, which, unlike the fictitious Nouveau Community Hospital running example, actually occurred.

Example of Use of a Framework

When proposing to a funding agency an evaluation study of clinical decision support available in EHR's, a multidisciplinary team of informaticians, clinicians, and social scientists discussed theories, frameworks, and models that might help them design the study. A team member who had been assigned to seek evidence about prior evaluation efforts reported that at that time very few similar evaluations had been done, so the team decided to use a very open-ended grounded theory approach. However, the team members felt some structure was needed. They decided to use a multiple perspectives framework, which is an application of general systems theory, as a guide. This framework requires identification of the most important perspectives, which in this case included clinicians, information technology and informatician experts, individuals in leadership and management positions, and both EHR and CDS vendor personnel. The framework also requires a clear focus of the study, the information resource itself, in this case elements of CDS (Ash et al. 2015).

15.4   Strategies for Study Rigor

There are a number of important considerations when designing and conducting rigorous qualitative studies. The following sections describe methods for safeguarding the integrity of the work. Safeguards imply a deliberate rigor to achieve a descriptive and explanatory argument that can be used by others with confidence. There are five strategies that should be considered part of every qualitative study: (1) reflexivity, (2) triangulation, (3) member checking, (4) saturation in the field, and (5) an audit trail.

15.4.1   Reflexivity

This concept, also known as "self-reflection," should operate in the mind of each study team member during every phase of the study. It is natural and perfectly acceptable that the study team members enter any study with their own biases.

Reflexivity is the conscious recognition of these biases and the equally conscious design of the study to address them (Crabtree and Miller 1999, p. 14). Recognizing bias from the beginning is essential because additional study team members who do not have the same biases may need to be called upon to view the data as well.[1] In a qualitative study, the study team member is the primary tool for data gathering, and must also be aware of personal interests and predilections. For example, it is often difficult for persons with clinical backgrounds to observe patient care without being judgmental. Their tendency is to focus on the validity of diagnostic and management decisions when perhaps, in the interest of the evaluation study, they should be watching work processes and technology usage. Once aware of this bias, however, they can put their attention where it should be, while still exploiting their clinical knowledge by making acute observations that would elude most non-clinicians.

Nouveau Community Hospital and Health System Case Example
At Nouveau, Dr. Yu made an initial visit to get a sense of its physical layout. Dr. Yu was surprised when noticing a patient smile when the nurse reminded the patient to sign up for the patient portal and set a password for it. On reflection, Dr. Yu recognized that Dr. Yu had expected the patient to express annoyance and perhaps some guilt about not having done it earlier. When later asked about this situation during a qualitative interview, the patient responded that the nurse had always been kind to the patient in the past, so the reminder was a welcome piece of information from someone the patient trusted. Dr. Yu moved from an original sense of surprise to a new curiosity about how the non-provider staff in the clinic could help champion use of the portal. Without such self-awareness when recognizing this sense of surprise, Dr. Yu would have missed out on gaining an important new insight.

15.4.2 Triangulation

Triangulation is a strong check on the credibility of study findings (Patton 1980, pp. 31–32). The qualitative study team member looks across different types of information (observation of work events, interviews of individuals from a variety of roles, analysis of documents) to determine if a consistent picture emerges for any given theme in the results. In some ways this is the qualitative analogy to the quantitative strategy of using multiple, independent measures to estimate the quantitative error in a measurement process.

Triangulation is a term borrowed from surveying and navigation, and represents a process by which the location of a third point is deduced from the locations of two

---

[1] This is of course a major point of departure between qualitative methods and their quantitative counterparts. In quantitative work, investigators rarely acknowledge bias, and if they do, they may be disqualified from participating in the study.

known points a fixed distance apart. Triangulation was originally used in the social sciences to describe multiple data collection techniques employed to measure a single concept. The definition of the term has expanded to include the use of multiple theories, multiple researchers, and multiple methods, or combinations of these, to obtain a "fix" on a specific issue. In qualitative studies, triangulation means the weaving together of different data gathering techniques, data elements, or study team member views to help ensure that the resulting descriptions and interpretations are as useful as possible. Comparing and contrasting data from these varying sources verifies and strengthens the results that emerge. External verification, which also constitutes triangulation, can be sought by asking an experienced investigator not associated with this particular study to review for logical consistency the data and the derived conclusions. What is sought here is not necessarily agreement with the conclusions themselves but, rather, an affirmation that the conclusions were reached in a scientifically competent and responsible manner, and that the conclusions are consistent with the data on which they are based. Members of a study team routinely audit each other, but the addition of external reviewers reduces the possibility that some perspective affecting the entire team will skew the results.

The following is an example about development of a triangulation metaphor, which, unlike the fictitious illustrative Nouveau Community Hospital running example, actually did take place.

Example of Triangulation Metaphor Development
Years ago, one of the authors of this book and a colleague were struggling to find a metaphor that would describe the concept of triangulation and emphasize its importance. They both happened to see an hour-long TV documentary about making mole in Mexico and the next day agreed they had found the perfect analogy. Mole is a sauce, often served over meat such as chicken. It has great complexity and depth. Many diners are surprised that it contains chocolate, since it is primarily quite spicy. The film had scenes of cooks selecting their ingredients at open markets, each cook having different preferences. The ingredients were for making three different sauces: a corn sauce, a pepper sauce, and a chocolate sauce. While all cooks selected peppers at the market, each found favorites and made the pepper sauce slightly differently. The same went for finding the right kind of corn for the corn sauce. For the chocolate sauce, the cooks selected from among gorgeous piles of cocoa beans and roasted them themselves for the special recipe each had for the sauce. After all of this hard work shopping and making the sauces, they blended them into different versions of their finished mole sauce.

The attention to detail, the blending, the individual recipes, the richness of the product are all similar to the notion of triangulation in qualitative studies.

15.4.3 Member Checking

Member checking employs the actual subjects in the study, at various points, to explore whether the study team's evolving findings are reasonable (Patton 1980, p. 329). Member checking can take the form of brief, individual contacts ("Did I get the main point from the meeting yesterday?") or of repeated contact with an intact group within the organization that will serve this function progressively throughout the study. The team members ask if their notes and results are logical and to the point. To the extent that the informants, or "members," confirm that the results make sense, this constitutes an important kind of support for the results. Member checking can be used both to validate and sharpen preliminary hunches at the early stages of a study, and to confirm almost-finalized results toward the end. Such verification by individuals external to the study and by participants themselves is another important check on the veracity of the findings. When people familiar with the setting of a study read a report or a preliminary document, the message should be meaningful or insightful to them. They should say, perhaps with enthusiasm, "Yes, that's right. You've portrayed it correctly."

15.4.4 Data Saturation

A useful strategy when trying to ensure that all issues of high relevance to a study have been identified and explored is to seek data saturation, which means closure or convergence, at which point study team members are seeing and hearing about what they already know (Crabtree and Miller 1999, p. 42). In very general terms, these three concepts suggest that if the investigators remain properly open-ended throughout their approaches to participants, and no longer hear anything substantially new, it is likely that they have identified the full range of issues as well as the full range of views about each one.

At some point in the life of the study, the next cycle in the field seems like it will be repetitive. During the analysis process, little that is new emerges. The search for confirmation and contradiction only supports the interpretation that has already been done. During interviews, answers given are similar to those offered by prior interviewees. During periods of observation in the field, observers find themselves taking fewer notes because they are seeing nothing new. It is hard to predict how much time a study will take before this point of data saturation is reached. Often the constraints of the resources available and the study team members' other roles limit the study and bring it to an end sooner than might otherwise come through a natural sense of saturation. However, if saturation is reached first, the fieldwork can stop at that point.

15.4.5 Audit Trail

The audit trail (Miles and Huberman 1994, pp. 277–278) is a record of the study, one sufficiently detailed to allow someone else to follow the study's history and determine if the investigation and the resulting data provided an adequate basis for the argument and conclusions. Such a record would also enable someone external to the study to determine if there were flaws in the investigative process. Most records composing the audit trail emerge as natural by-products of properly executed fieldwork. If team members have maintained files of field notes, a chronology of coding, notes on efforts to find contradicting data, and a log of experiences, the audit trail is essentially complete.

The audit trail becomes more complex in a study conducted by a larger team. There is a greater amount of data collected, and the audit trail must also capture the interactions among the team members. A team audit trail might include minutes of analysis and planning meetings involving all or a subset of the team members. Overall, the clarity of the trail and the depth of its detail are indicators of the care with which a study was conducted.

Self-Test 15.1

Which of the strategies to ensure study rigor is primarily employed in the qualitative study scenarios below?

1. Data from interviews about the usability of a resource are analyzed thematically. The evaluation study team looks to see if and how similar themes have arisen in earlier meetings of the team.
2. A member of the study team, who has recently participated in another study of a similar kind of resource, becomes concerned that that person's views about the current study are being shaped by that previous experience. That person sits with another member of the study team to share that person's concerns and put them in perspective.
3. At a "town hall" meeting called to present the results of a qualitative study, the sponsor of the study raises deep and serious questions about the validity of the findings. The study team returns to notes from their team meetings to review how and based on what data they came to this conclusion.
4. During an evaluation project team meeting, one of the study team members finds themselves deeply repelled by off-color comments made by one of the project staff. The team member makes a note of this personal response as part of their field notes.
5. After interviewing 10 patients participating in a study, a study team member perceives that they are hearing the same points raised by all interviewees. The team member requests a study team meeting to consider reducing the total number of interviews from 20, as previously planned, to 12.
6. A study team member "corners" a participant in a system development effort following a meeting and asks for that person's impressions on what transpired in the meeting.

Nouveau Community Hospital and Health System Case Example
Dr. Yu, the evaluation consultant for Nouveau, has been told that the consultant's job is to evaluate the success of the system over the next year. After Dr. Yu's initial visit to get the lay of the land, Dr. Yu knows the team will need to interview and observe patients using the portal, but team members will also need to observe how messages from patients sent through the portal are received and handled in the office.

Nouveau will be undergoing major change as the portal is implemented, and Dr. Yu decides to try to identify an "insider" to play a key role in the study by becoming a member of the study team. It turns out that one of the nurses has expressed interest in the evaluation. This nurse agrees to collaborate and assist, especially in making introductions and in gaining patient involvement.

Dr. Yu, an astute practitioner of reflexivity, knows that Dr. Yu has some bias about the assignment from past experience: Dr. Yu already sympathizes with the users of the portal. Fortunately, Dr. Yu knows a number of graduate students at a nearby university who are seeking practical experience in evaluating informatics projects, and have taken a qualitative methods course. Dr. Yu recruits several of them to be part of the study team. The team subsequently meets and outlines its strategy. Each member comes from a different background, so the members discuss their personal biases and how they will provide checks on one another. To facilitate triangulation, they decide that all members of the team will participate in observing workflow, performing interviews, and reviewing portal records. They outline a timeline bounded by the 12 months specified by the client, so they will gather data throughout the implementation period and beyond or until they sense they have reached data saturation. They will keep a careful audit trail. To this end, one team member volunteers to search for evidence in the literature, another offers to gather, disseminate, and archive files as field notes and transcripts become available. Another, who has project management experience, agrees to keep records of analysis meetings, enter into the qualitative data analysis software the themes identified in team discussion, and track the project. They outline a schedule for delivering reports to the clinic, and once each report is in draft form, they will do a member check by meeting with as many of the staff as they can to go over the preliminary results.

The team must make some additional early decisions about how to design the evaluation. It meets with the clinic manager and nurse insider and together they formulate the overarching question, which will be "What are the factors that affect the success of this portal implementation?" Since communication, care, control, and context are all involved, the team members know they will be using qualitative methods. They especially want to focus on access, usability, and privacy concerns. This stage can be problematic, since many study teams are comfortable in either the quantitative or qualitative school of thought, and will see the question from the beginning in relation to their preferred strategies. It is important to be able to recognize the full range of methods that may be appropriate. The team identifies this as an appropriate situation for a qualitative study. First, the group of insiders and study teams agrees that they want and need to do a high-quality, rigorous study since Nouveau has invested heavily in the new system. The team then needs to make decisions about how to gather the data and analyze them.

After searching the evidence, a team member reports that a trusted health professional recommending use of the portal increases acceptance among patients, so the team wants to find out if and how well this is occurring. The team members might observe at the computer stations to see what kinds of patient messages are arriving, at the nurses' stations for the same reason, in the hallways and exam rooms while shadowing patients interacting with providers and other clinicians or in offices where patients talk with providers and providers receive portal messages. The team members might shadow the staff members who make use of the EHR to answer patient queries. They might talk with some providers and staff members casually and then ask a selection of them to give more detailed information through formal interviews. The team might also ask to review minutes of staff and committee meeting as well as records of portal activity.

During site visits, team members may also interview or schedule interviews with patients with the assistance of the insider nurse. They may plan to conduct some interviews in the clinic and others in patients' homes or wherever they access the portal. For example, some patients may regularly use their public library's computers, so data gathering through interviews and patients' demonstrating their use would be useful.

15.5 Specifics of Conducting Qualitative Studies

The preceding sections have provided strategic background information on qualitative approaches to evaluation. This section provides an overview of some of the specific decisions about techniques and procedures that need to be made while planning and executing a qualitative study.

While what follows may seem overly prescriptive and formulaic in light of the numerous options available within the qualitative framework, this section seeks to remove some of the mystery that is often attached to qualitative study. The references provide more detail and other methodological options.

The basic steps are as follows: Once the study questions have been articulated and it is clear that qualitative methods are appropriate for answering them, it is necessary to select the sites for fieldwork if these are not constrained, select the study team members and the individuals to be interviewed or observed, determine the techniques for gathering the data, outline a time frame for data collection, and decide what resources and budget are needed.

15.5.1   Site and Informant Selection

In many evaluations, the sites for field investigation are predetermined, but this is not always the case. When there is flexibility in site selection, the choices are invariably driven by the study goals or research questions. For example, if the goal is to understand how house staff (interns and residents) view a new clinical decision support intervention, teaching hospitals should be the focus. On the other hand, to compare views of clinicians at teaching and nonteaching hospitals, it would be necessary to do fieldwork in at least one site of each type. Even if the evaluation site is predetermined, the investigators still need to decide, for example, which unit(s) of the organization might be the focal points for fieldwork.

Whether selecting a study site or selecting the individuals to interview or observe, the logic of the selection procedure is similar. The individuals are called subjects, informants, or participants by different qualitative methods experts and here we use the terms interchangeably for individuals being studied. Qualitative methods usually entail a "purposive" selection, meaning that deliberate selection is based on the purpose of the study, and the selection strategy can evolve as the progress of the study reveals initially unanticipated needs for subjects who bring potentially novel viewpoints (Leedy and Ormrod 2016, p. 165). Study team members use their judgment to select appropriate sites and subjects. They may begin selecting individual study subjects by finding the most knowledgeable people, those who know something about the focus of the evaluation (clinic staff impacted by the portal, for example) and who also know other people in the organization. These important contacts can then identify types of patients who may be productively included in the study. The goal is to have a purpose for each person to be included, usually because that person is knowledgeable about the topic. Random sampling is generally not used in qualitative studies because there is little value in gathering data from people who are unable to answer the study team's questions.

Another approach might be convenience sampling, (Leedy and Ormrod 2016, p. 165) which is less deliberate, and based purely on availability and happenstance. For example, a study team focused on emergency department use of dictating software might talk with any providers scheduled to work a particular day. However, the study team has already purposively selected the hospital, the emergency department unit, and providers as a group, so purposive sampling has dominated the strategy.

Careful purposive selection of informants is almost always the preferred method for qualitative investigation. This is especially the case when choosing subjects for interviews that are very expensive to conduct when the costs of transcription and analysis are added to the very real costs of the time of the interviewer and interviewees. Informants should be selected based on the information or expertise that they can share. As a general rule, both expert and nonexpert users of an information resource should be included. It is useful to seek out the outlier and the skeptic in addition to those who are known to be heavy system users. Generally, the study team will not have a complete list of participants at first; the list will be constantly expanded and adjusted. Names of skeptics and other new informants with special viewpoints are often suggested by previously identified interviewees, for example. In the case of the portal, it would make sense to interview some portal users, some former users, and some skeptical non-users.

When considering the numbers of informants to interview or observe over an extended period, Crabtree and Miller suggest five to eight individuals if most people are like-minded, or 12 to 20 if there is a good deal of variation (Crabtree and Miller 1999, p. 42). Obviously, these numbers may increase as additional issues emerge, bringing with them requirements for additional points of view. For observation, the number of people will depend on the focus of the study and the context. If one were evaluating a large-scale information resource implementation in a large, complex environment, such as a medical school and hospital setting with 1500 faculty members, a larger number of informants would be needed than would be the case for a study set in a small rural clinic.

15.5.2 The Team

Informatics evaluation studies are akin to applied research projects and in applied research ethnographers are normally members of multidisciplinary teams (Erickson and Stull 1998). A team approach to qualitative study is almost always desirable. Whatever the focus of the evaluation, it should be viewed through different lenses so that the most complete picture has the opportunity to emerge. Teamwork can sometimes be frustrating and seem slow; nonetheless, it provides a higher level of credibility to the study. A team study needs an overall leader. Leaders must recognize that they will need to spend considerable time managing the team and performing administrative duties such as writing interim reports for the client(s) and tracking expenses. If a project manager is hired, however, some of the managerial burden can be assumed by the person in that role.

Team members need to be effective collaborators in every sense of the word, but above all in producing what they have promised to do on time. Effective teams often include individuals with different backgrounds and roles in the study. For example, clinicians and non-clinicians can be deliberately paired during observations in the field so that they can offer different perspectives on the same activity. As noted earlier, if resources permit, it is helpful to have an experienced qualitative researcher employed to scan all of the data and provide judgment about the reasonableness of the interpretations made by team members.

Team planning includes extended periods of time for meetings to develop strategies and timelines. Early in the life of the project, the team might gather for a half-day or full day in a retreat setting, using flip charts or project management software to plan tasks, timelines, and resource use. During regular team meetings once a study is underway, it is important to set time aside to track progress and do further planning. Dividing the meeting agenda into two major segments, separating study planning from data interpretation and assigning fixed amounts of time to each, helps assure that all tasks will be achieved.

15.6   Techniques for Data Collection

For qualitative studies largely conducted in the field, the selection of data gathering techniques depends on the purpose of the evaluation, the questions being asked, and the resources available. As described below, plans for entering the field must be carefully formulated and executed.

This section discusses in some detail procedures that help generate data useful to building a cohesive argument by the end of the study, beginning with the process of entering the field. This section then emphasizes interviews, focus groups, observation, and gathering naturally occurring data--because they are the techniques most commonly used in informatics studies--and describes "mix and match" toolkits of complementary techniques.

15.6.1   Entering the Field

Entering the field can be a daunting experience, even for team members who are clinicians, teachers, or researchers to whom the health care, educational, or research environments are familiar. Simply learning about the physical layout of the study settings can be challenging. Gathering some background information about the site—by looking at routine organizational publications, Web sites, or perhaps vendor information about the information resource under study—is invariably helpful. One central contact person within the organization who can act as a key informant and sponsor, someone who knows well the people involved and the resources being studied, can provide names of other possible informants and assist with gaining human subjects approvals. It is a good idea to do a "lay of the land" visit first so the study team members not only know their way around the physical space, but also feel comfortable when the work of actual data collection starts. Each participant with whom the study team will spend considerable time should be given a fact sheet outlining the purpose of the study, and a consent form if it is required by the local human subjects committee. While both necessary and important, the consent process can make for awkward introductions, so it is helpful to have an informative verbal statement ready that outlines the purposes of the study, the data collection methods to be employed, safeguards of confidentiality, and perhaps something a bit personal about members of the study team to build rapport.

Nouveau Community Hospital and Health System Case Example
The evaluation team, which has named itself the Evaluation Team (ET), now includes the original consultant, graduate students, and an "insider" who is an employee of the organization. The consultant discusses the project with an additional person, a colleague who has qualitative analysis experience and asks if the person would serve as the objective outsider who would provide an

overview of the data and audit trail later in the project to make sure bias has not crept into the study. This person agrees and the clinic approves the additional expense.

The nurse insider gives the ET a tour and makes brief introductions. The insider recommends that the ET members provide a pizza lunch and several coffee breaks and be in the staff room to chat with everyone so that their faces are known and they can distribute fact sheets and explain about informed consent. They do this and find that a large number of staff members, including busy clinicians, agree to participate in these activities. The staff members also encourage the ET members to conduct their short patient interviews in the clinic when patients are checking out and agree the checkout clerk will introduce team members to patients at that point. During those short interviews, some patients will be asked if they would volunteer to be interviewed further in whatever place they use when accessing the portal.

15.6.2   Interviews

Qualitative studies rely heavily on interviews. Formal interviews are occasions where both the study team member and interviewee are aware that the answers to questions are being recorded for direct contribution to the evaluation study. Formal interviews vary in their degree of structure. At one extreme is the fully unstructured interview where there are no or few predetermined questions. Between the extremes is the semi-structured interview where the study team members specify in advance a set of topics they would like to address but are flexible as to the order in which these topics are addressed and open to discussion of topics not on the prespecified list. Both unstructured and semi-structured interviews should normally be recorded so that a complete and accurate depiction of the interview is available for analysis. At the other extreme in the degree of interview structure is the fully structured interview with a schedule of questions that are always presented in the same words and in the same order. They are actually survey questions delivered and answered orally and most often they need not be recorded. In general, the unstructured and semi-structured interviews are preferred for qualitative studies (Berg and Lune 2012). A fully structured interview that includes only Likert scale or yes/no answers cannot be considered qualitative since it is basically a survey. Informal interviews, spontaneous discussions between the study team and persons in the field that occur during routine observation, are also part of the data collection process and are a source of important data. If the setting allows, these spontaneous events should be recorded as well, though handwritten notes may need to suffice in hurried situations.

The semi-structured interviews preferred for qualitative evaluation require an interview guide that provides focus while allowing flexibility in the topics addressed. The ability of the interviewer to offer gentle guidance, without interrupting the continuity of the interviewee's thoughts, is the art of interviewing. There are numerous variations of qualitative semi-structured interviews, including: cultural interviews to find out what shared meanings group members hold; oral histories, which gather reminiscences about a focus or theme; life histories, which gather memories about one's life, which can be oral histories or written documents; and so-called evaluation interviews, which discover whether new efforts are meeting expectations. With evaluation interviews, "the researcher learns in depth and detail how those involved view the successes and failures of a program or project" (Rubin and Rubin 1995, p. 6).

Self-Test 15.2

Label each of the following interview scenarios, conducted as part of a qualitative study, as representing the fully structured, semi-structured or unstructured approach.

1. A study team member "corners" a participant in a system development project following a meeting and asks for that person's impressions on what transpired in the meeting.
2. A study team member schedules time with a patient who is using an information resource to acquire specific information about the patient's medical history.
3. A study team member works with partners on the study team to develop a set of questions to be asked of all interviewees. Each question is to be followed up with the question: "Why do you think this is the case?". At the end of the interview, subjects will be asked: "What else would you like to tell us to shed light on these matters?"
4. An interview begins with the statement: "In general, what has been your experience using this EHR?" The remaining questions depend on how the interviewee answers this opening question.
5. A set of specific questions are read verbatim from an interview guide. No other questions are asked. The interviewees' responses are recorded.

The following list offers practical suggestions for conducting successful semi-structured interviews.

Interviewing Details

- *Preparation*

  - *Prepare an interview guide prior to each interview. This should outline five or six main areas you would like to explore and should also list more detailed sub-questions that you can ask if the interviewee does not spontaneously cover them. These will guide your probing questions.*
  - *Organize what you need to have with you: the interview guide, background information on the interviewee, a notebook for writing or a recorder and backup recorder, a consent form, an information sheet outlining the project, and a token thank you gift (if appropriate).*

- *Getting Started*

  - *Review the information sheet and consent form and briefly outline verbally what is on the interview guide. Make it clear that although you have a set of issues you wish to address, you are interested in hearing everything the interviewee thinks is important to the subject. If the study plan calls for sending a summary back to the interviewee for approval or if you will be making a copy of the consent form to give them later, state that at the beginning. Give appropriate reassurance about anonymity and confidentiality.*

  - *Tell the interviewee that you will make every effort to end the interview on time, where "on time" means whatever time has been negotiated. That gives the interviewee a chance to say that it is all right to run overtime—or not.*

  - *Begin the interview with an open-ended, easy question about the person's background and what led up to their involvement in the project being evaluated. Then ask about their role in the project and move on to more specific questions. It is easier to go from more general to more specific questions and, if asked early on, specific questions may constrain the person's thinking too much.*

  - *Glance down at your interview guide as necessary and jot a note or two to remind you to follow up on something later, but keep eye contact most of the time. Recording is recommended so that eye contact is not broken by your having to write notes.*

  - *Having an assistant interviewer in the room is highly recommended. That person can take notes and after the main interviewer has finished, can ask follow-up questions the first interviewer neglected to ask. If the second interviewer has a different background, perhaps a more technical background than the first interviewer, the follow-up questions might target more technical areas, for example. The interviewer can minimize the awkwardness of having a third person involved in the interview by introducing them and by explaining that person's role.*

- *Topic Flow and Question Format*

  - *You have provided some structure to the interview by this point, so you can now allow and invite the interviewee to dictate the flow of topics. However, if time is running out and you still have not covered some important topics, you need to take more control of the agenda. Do it with an apology and note that you are watching the clock so that you will not inconvenience your interviewee; often the person is enjoying talking so much that they agree to go over the time limit.*

  - *Before making any major shift in subject, ask the interviewee if they have anything else to add on this issue.*

- *Always avoid leading questions, questions that tell the interviewee the answer you want, questions like "You don't approve of the way this is being done, do you?"*
- *Do not be drawn into a conversation with the interviewee. You are there to listen, not to talk. You can offer encouragement as the interviewee talks, but not a value judgment.*
- *Ask for clarification if you do not understand what the interviewee is saying or if you would like them to expand on a point.*

• *Ending the Interview*

- *When time is up, say so and make it clear that the interview can end now. Many interviewees want to keep on talking and some save their most insightful remarks until the end. For this reason, do not schedule back-to-back interviews and be flexible with your time.*
- *Thank the interviewee, ask if they have any questions, present a token gift if appropriate, and explain the next steps.*
- *Always send a formal thank you note afterward.*

When studying informatics resources, it can be useful to incorporate some oral history techniques into the interview. An oral history technique can be used with great success to explore recollections, for example, of different phases of informatics system implementations. Oral history is a technique "for obtaining first person accounts of how modern society has been shaped by causative factors of historical significance" (Brunet et al. 1991, p. 251). Oral history questions might be designed to elicit memories about the history of an information resource implementation. (Often, one major goal of a qualitative study is to establish "what really happened.") The goal of semi-structured interviews is to generate perspectives on a specific topic by asking questions that open the door to informants' beliefs. Oral history questions that evoke memories and stories are ideal for generating these narratives.

People enjoy talking about themselves. Any interview can begin with a few oral history/life-history questions about their education, work, and roles, because it is always helpful to understand people's background and learn the roots of their involvement in the project. Life history questions are also good rapport builders. One then builds questions into the interview that can elicit descriptions of organizational culture, such as "From your viewpoint, what group or groups were influential in shaping the plans to implement machine learning predictive models in critical care units?"

Although it is not always possible or practical, interviews should ideally be recorded so that the interviewer does not need to take detailed notes and eye contact can be retained. Human subjects committees often have special requirements for interviews that are to be recorded. For example, these committees will usually require that the protocol for the study specify where the recording will be stored and for how long. As part of further preparation, the interviewer should gather information about the interviewee ahead of time and develop an interview guide listing several main questions and many sub questions that can probe for answers if the interviewee does not spontaneously respond. Some questions may be asked of everyone who is interviewed, while other questions will be reserved for specific types of individuals. It is absolutely critical that the interviewer be a good listener and avoid making any value judgments during the interview. Judicious use of silence can often stimulate responses. Interrupting is to be avoided (Patton 1980, pp. 195–263). Many investigators prefer to allocate a full hour to conduct an interview of this type; however, shorter interviews can be useful if they have a well-defined focus or if the informant is being contacted on multiple occasions. After the interview, the recordings should be transcribed by someone with experience in capturing nuances such as laughter and sighs; it can take 3 h or longer to transcribe 1 h of recording, depending on the quality of the recording, clarity of the speech, and availability of software assistance. This can be expensive, and not all studies require this level of detail in transcripts. Voice recognition software, often used as a starting point by transcriptionists who then amplify what the software produces, can also be used instead of transcriptionists if rough transcripts will suffice. Another option for reducing the expense of transcription is to use voice recognition transcription first and to then select especially rich portions of the rough transcripts to fully transcribe.

15.6.3 Focus Groups

Focus groups differ fundamentally from the one-on-one interviews discussed above. They should not be viewed as a way to gather more interview information in less time. Focus groups are not easy to do well and, if several participants speak at the same time, audio recordings of focus groups can be hard to transcribe. When managed properly, however, focus groups have distinct advantages. In particular, useful synergy among the participants can develop. When this happens, participants build on the thoughts of one another to generate new insights or more accurate recollections of past events. (It is interesting to listen to members of a focus group correct one another's personal recollections of a past event, until a more accurate consensus develops.) Focus groups can employ many of the same types of questions as those used in one-to-one interviews. Gathering a group of up to 10 informants over pizza at lunchtime can generate lively narratives. The moderator needs an interview guide and must set some fairly strict ground rules to discourage participants from interrupting or monopolizing the discussion. An assistant moderator can take general notes that specify who is speaking (which will help the transcriptionist), and can also manage the audio recording (Morgan and Krueger 1998). It is ideal if focus groups are held in person, but with video conferencing platforms becoming increasingly sophisticated, they are now a reasonable and practical substitute.

Consensus or expert panels can also be treated like expanded focus groups. The advantage here is that informants may be together for extended periods of time, perhaps for days, so rapport and synergy can build over time and settled opinions of the full group can be generated. Transcripts of these discussions can be formally analyzed as qualitative data sources. Often the conversation leading up to an agreement is filled with vivid stories and examples, and worthy of capture and analysis (Mohan et al. 2016).

Nouveau Community Hospital and Health System Case Example
Back at Nouveau, the nurse insider has urged the team to interview at least 12 of the staff members who receive patient portal messages regularly, clinicians who encourage patients to use the portal, and technical experts who developed and assisted with training. The team decided also that 12 interviews with a variety of patient types would be reasonable. The team then developed two different interview guides, one for staff and one for patients, but each will be tailored for individuals. For example, if a patient has been purposely selected because that person was using the portal but stopped at some point, a question would be added to the interview guide asking why the person no longer used the portal.

15.6.4   Observations

Ethnographic observation, in which study team members are immersed in the daily life and culture of a group for extended periods of time, is an excellent way to confirm what is discovered through other data-gathering methods such as interviews, and to generate new hypotheses and explanations (Berg and Lune 2012). There are times when participants will describe their own behavior in interviews, but will act differently when observed. This occurs not because informants are disingenuous, but rather because they may not be conscious of the differences between what they state and what they actually do. This phenomenon is common when asking participants about their workflow. They usually describe the ideal workflow rather than the actual messy, interruptive workflow normally experienced. Only by watching can evaluation team members capture the "lived experience" of persons in the field (Berg and Lune 2012).

Data collection by observation grew out of anthropological research methods employed initially to understand unfamiliar cultures. For the purposes of evaluation within biomedical informatics, those original methods have been significantly modified. In this respect, Berg (Berg and Lune 2012) offers a useful distinction between *macroethnography*, which strives to describe a way of life in general terms, such as how cancer researchers around the world collaborate, and *microethnography*, which focuses on specific activities within a culture, such as how cancer researchers use bioinformatics tools in their research. The original work of anthropologists falls largely into the first category, and informatics evaluations fall into the second.

Macroethnography requires that an ethnographer observe a social unit over a long period of time so the person can be immersed in the culture. This level of immersion consumes considerable resources and is often not practical in informatics, nor does it provide triangulation of data gathered by multiple team members. When evaluating biomedical information resources, a micoethnographic approach characterized by short but intense periods of observation, ideally conducted by a multidisciplinary team, can be of greater value (Berg and Lune 2012).

There is a spectrum of roles for observers, ranging from passive following of informants to full participant observation, during which the study team member, if qualified, contributes to the work being done in the study setting. For informatics studies, participant observation can be problematic in active clinical and biomedical research settings, because the primary focus of the observation needs to be the information resource under study. Writing field notes absorbs an observer full time; practicing medicine or conducting research cannot be done at the same time. Full participant observation, for this reason, is not often used in informatics studies. More typically, members of the team who are practicing clinicians or researchers function in a more passive observer role, following the practitioners and quietly watching while taking notes in real time. The length of time needed to observe a participant depends on the repetitive nature of the workflow. It is sometimes more efficient to observe more people for shorter periods of time. For example, 4 h is usually sufficient for shadowing a clinician if the work is repetitive, such as in primary care in the outpatient setting or in the emergency department when each provider tends to follow a repetitive process for seeing each patient. Each provider may have a somewhat different workflow due to personal preferences, so shadowing more individuals for half days rather than fewer individuals for longer can gain more complete data about workflow variations. To follow intact groups of workers, such as research lab teams or clinic teams in hospital settings, study team members should allocate longer periods of time because the workflow is less repetitive. Using the emergency department as an example again, patients are admitted with a wide variety of problems, so watching the activities of a group of clinicians in a large workroom with many desktop computers, for example, may need a longer observation period to gather sufficient data. With a larger number of team members in the field, more data can be gathered and the time needed will be lessened before saturation is reached. There may be opportunities, during a break in the work routine, to perform brief informal interviews of one or more informants to ask questions about what has been observed. This can be particularly useful if an important event in the work of the group (e.g., analysis of clinical trial data that reveals a positive result) has just occurred. Focused follow-up interviews can capture informants' immediate reactions to the event before these are forgotten. For this reason, observers should routinely carry recording devices approved by the IRB: some require encryption. The possible occurrence of spontaneous interviews should be reflected in the general consent form, so informants do not have to be "reconsented" before such an interview can occur.

15.6.5   *Field Notes*

Field notes are usually handwritten notes, sometimes called "jottings," about the setting or physical layout of the facility, activities, and events under observation. Field notes can be taken unobtrusively in a health care setting since many people are routinely taking notes. Pocket-sized notebooks are recommended for these jottings. Clipboards should be avoided because, in professional cultures, these evoke images of surveillance by accrediting agencies or managers.[2] Some team members may prefer writing on tablet computers that will automatically transcribe the handwritten notes into electronic form. However, these jottings will not be complete, so time will still need to be set aside to correct and complete these electronic notes. Team members benefit from translating their own jottings into full field notes because the typing process encourages their cognitive processes, enhancing their memories.

Jottings or initial notes taken in the field are meant to capture key ideas, for later expansion into full field notes. Just as with interviews, there is a spectrum of structure to field notes. If the notes are completely structured, akin to checklists, they are not actually qualitative field notes because they do not allow recording of interpretations and unanticipated events. Some level of structure—for example, 5–10 foci identified from prior investigation—can help organize initial field notes and facilitate recording of new observations. Completely unstructured initial field notes are compatible with, and indeed recommended for, forays into the field that occur at the beginning stages of a study.

Field notes from an episode of observation often begin with the observer's assumptions, which are preconceived notions about what the observer expects. Writing such pieces ahead of time is a form of reflexivity. These notes can be followed by a diagram of the observed physical space, and can move on to a description of the events actually observed. Team members often record thoughts that come to mind regarding theory and future plans for investigation. It is best to write these down immediately, for otherwise they are frequently forgotten. Personal notes, thoughts about the team member's own feelings about the events as they are being observed, should also be jotted down to enhance reflexivity. It is extremely important that what is actually observed or discussed during observation is separated in the field notes from the description of the observer's interpretation of the activity. During the analysis process, memos based on these interpretations while in the field and on ongoing interpretations are produced to keep track of team members' thoughts (Crabtree and Miller 1999, p. 65).

Details about the general structure of observations are described in the following list.

---

[2] For the same reasons, the observers should not dress too formally. They should dress as comparably as possible to the workers being observed in the field. Always ask ahead of time about dress codes.

Observation Details

- *Preparation*

    - *Negotiate entry into the site by making contacts with people within the organization who can grant you access, give you information, and make arrangements such as getting identification badges, space for your personal items, access to schedules, etc.*
    - *Put aside the time for observation; there should be no other demands on your time. Also put aside time for writing field notes after the observation is complete.*
    - *Have some initial research questions in mind so that you can focus your attention in a general way.*
    - *Prepare an "elevator speech" so all team members will know what to say if they are queried about their presence. This should include a very short explanation of the project in general, the name of the local contact person, and mention that the study is IRB approved.*
    - *Dress so that you blend into your surroundings, and dress comfortably, since you will probably be doing a lot of walking. Follow the local dress code.*

- *Getting Started*

    - *If you have an appointment to observe someone, have your contact person introduce you, deliver your elevator speech, have information sheets available in case the informant wants to know more, and go over and sign the consent form if it is needed.*
    - *If you are doing a "lay-of-the-land" visit, pick a place in the flow of activity and simply watch for a while and pay particular attention during transitions in the activities, such as shift changes.*

- *Data Recording*

    - *Take notes briefly and occasionally and step out of the situation often to work on your jottings in more detail.*
    - *If you want to dictate notes into a recording device, do this outside the situation; if you do a short informal interview with the recorder, step away so no one is disturbed.*
    - *Include jottings about personal, theory, observational, and methods issues in your notes.*

- *Ending the Session*

    - *Express your thanks and describe the next steps; offer a token gift if appropriate.*
    - *Review your jotting notes soon after leaving the site. Keep adding to them.*
    - *Put ample time aside for transforming jottings into full field notes.*

As soon as possible after a session in the field, the observer should type out or dictate more complete "full" field notes. At the beginning of a study, if a very open approach is being taken, it could take 3 or 4 h to think through and produce full field notes based on jottings from 1 h of observation (Berg and Lune 2012, p. 234), so time should be set aside for this task. This time can be decreased later in the study when observations become more focused and if the jottings are written in a somewhat structured format, because the researcher does not need to labor over how to organize the full field notes. Debriefings of team members, during which each shares impressions of the field experience and a description of accomplishments, should take place often.

Video recording can be an enhancement to both interviews and observation, but the use of video presents special challenges related to confidentiality and analysis. It is fairly easy to de-identify transcripts and audiotapes by deleting headers or introductory information. Deidentification can be done with videotapes, but confidentiality is a much greater problem with this medium. Anyone acquainted with an informant can identify that person from a segment of video. Team members need to be sensitive to these considerations. Also, in order to take full advantage of the medium, analysis of video requires explicit attention to the nonverbal aspects of the subjects' behavior (Crabtree and Miller 1999, pp. 336–337). While this can provide especially useful and rich data, the analysis process is time intensive and expensive. When on site visiting is not practical, an option is to ask subjects to video their own experiences. With human subjects permissions in place, providers could video patient visits, for example, or patients could video their own use of patient portals.

Nouveau Community Hospital and Health System Case Example
The ET members meet several times to map out strategies for this study. They have already decided to use observation and interviews, but they need to outline interview questions and foci for observations, along with a format for field notes. Although the consultant has agreed to interview a core set of informants to maximize consistency, and although the consultant will vary the questions for each interviewee, the consultant spends considerable time with the team outlining six main questions to ask each person. The team agrees that field notes will include personal, theory, observational, and methods sections when they are in their final form. The team also agrees that the focus will be on describing the attitudes and actions of clinic staff and patients as they use computers in general, and the portal in particular.

15.6.6 Gathering Naturally Occurring Data

Organized human activity produces a trail of paper or electronic documents. In biomedical informatics, these include electronic patient charts, blogs and social media posts produced by patients, the original researchers' lab notes, various versions of computer programs and their documentation, memos prepared by the project team, minutes of decision-making committees, and others. These, once produced, do not change. They can be examined retrospectively and referred to repeatedly as necessary over the course of a study. Records accrued as part of the routine use of an information resource, such as automatically generated user log files, are key naturally occurring data sources for biomedical informatics projects. Data from these records are often quantifiable, and are frequently analyzed quantitatively even within the framework of a qualitative study. Most important, because these sources of data are produced naturally, without study team members guiding their generation, they are not subject to any researcher bias in their generation (Kiyimba et al. 2019).

Naturally occurring data are the normal products of human activity. Organization charts of a complex health system, e-mail correspondence of a team of researchers about their research, notes taken in class by students, and paper hospital forms are all examples of entities that can be used as data sources. An e-mail message from an informant in response to a question posed by a member of the study team produces "data" but not naturally occurring data. Informants will often offer naturally occurring data sources to the investigators. For example, even if a hospital has an electronic health record, some of the care process may still be documented on paper forms. A nurse may reach into a drawer while team members are informally interviewing the nurse and hand a team member a blank form as an example. If the observers see something they would like to keep or copy, they should, of course, ask permission. Evaluation team members need to be careful about naturally occurring data sources that are offered by well-meaning informants but in fact represent a breach of confidentiality—for example, a printout of a completed form with patient data already written on it.

A naturally occurring document can stimulate useful questions and answers. If a subject produces such a document, the study team can ask about the original motivation for the creation of the document. During interviews, team members can refer to the document and ask questions about it: How was the text recorded in the document collected? Who collected it? How was it generated? Why was it generated? Was the artifact creation mandated or spontaneous? Answers to such questions will assist the study team in interpreting the history and meaning of a document or other data source.

Food for Thought

With reference to the Nouveau patient portal example described at the beginning of this chapter, the study team decided it wanted to visit patient homes to learn about the context of using the portal in that setting.

1. If the team could not visit homes because of pandemic guidelines, how could they involve patients in sending them information?
2. What kinds of information in what format should they request?

15.7   Toolkits of Complementary Techniques

As qualitative methods have become more common in informatics, study teams have experimented with combinations of techniques that work especially well together. This has led to the creation of what the authors of this volume are labeling "toolkits" that can guide decision making when planning studies. Some of the most-used toolkits are those for case studies, action research, rapid assessments, and virtual site visits.

15.7.1   Case Studies

Case studies are detailed descriptions of real situations in which the context and decision-making processes are explored (Yin 2003). Case studies are an ideal way of summarizing the results of an evaluation study by combining what has been learned using different methods. They can include one or more sites and can be either quantitative or qualitative, but are generally a blending of both. They can be used for evaluation purposes or as teaching aids, but would be written differently for those different purposes. Evaluation or research case studies must be accurate and rigorous; teaching cases need not be, as long as they stimulate dialogue. We will limit our discussion to evaluation cases.

Methods for data gathering must be selected as appropriate for the study. For example, a case study in the industrial sector, of a company that provides medication decision support, will not benefit from observations if staff members sit in cubicles all day, and the company might not welcome observers. In this case, interviews and focus groups, plus written materials, might need to suffice.

Case studies can be descriptive or explanatory and, like the qualitative methods described above, they can help to answer research or evaluation questions that ask why or how something happened. Also as outlined above, there must be a clear strategy for selecting sites and informants and specific foci or measures appropriate for the research questions. Cases can be either representative of the norm or unique. Evidence for the study can come from any of the qualitative methods described and from any documents, including those presenting facts and figures from quantitative data. The goal is to gather enough information in different ways so that the multiple sources of evidence help the study team paint an accurate picture. Analysis consists of gaining a broad view of the data as well as finding patterns and themes. Again, it is recommended that multiple researchers partake in analysis in order to have more perspectives on interpretation. The final report might be prepared in different versions, depending on the audience. Usually, a case is written in traditional narrative form that describes and explains the situation. If multiple cases are to be described, they can be reported separately and followed by a cross-case analysis, or the cross-case analysis can stand alone. Another possible format is to organize the report by issue and describe the individual cases to illustrate each issue. Either the cross-case analysis or issues-based description can help keep the study sites anonymous, if that is a goal. Above all, the case study report must be well written and engaging (Yin 2003).

15.7.2 Action Research

Another set of complementary techniques, action research or participatory action research (PAR), is being used in healthcare though it emanated from the discipline of organizational behavior. An Institute of Medicine workshop explored the use of action research for including the public in the design, review, and setting of the clinical research agenda to increase health equity (Weinstein and Caciu 2017). Action research involves a partnership between people who will be impacted by whatever is being studied and those conducting the research with the purpose of taking action. Participatory action research, participatory inquiry, and action research are all terms that refer to the cooperative actions of researchers and those who work inside an organization to improve the situation. Qualitative evaluation studies that involve the stakeholders in the design, execution, analysis, and determination of what will be done with the results fit this definition. The Rapid Assessment Process described below is an excellent example, one that is being increasingly used in informatics.

15.7.3 Rapid Assessment Process

The rapid assessment process (RAP), or quick or rapid ethnography, is a toolkit or package of methods developed in the public health arena (Beebe 2001). Investigators needed to enter the field and quickly learn about a local culture so that training programs could be tailored for specific contexts. The idea is similar for teams entering health care settings. A requirement is having an insider on the team to help with cultural understanding. RAP also requires a multidisciplinary team to expedite data gathering. As much work as possible is done ahead of a visit. During a site visit, multiple researchers conduct observations, interviews, and other appropriate activities such as short interview surveys. A fieldwork manual includes a schedule, fact sheet to be distributed to subjects, interview guides for subjects in different roles, information about information systems at the site, an observation guide, and contact information for the team and insiders (McMullen et al. 2011). The list below outlines what a fieldwork manual might include.

Example of RAP Fieldwork Manual Contents
- *Results of a structured survey sent to an insider contact person with technical questions about the system under study.*
- *Notes summarizing a virtual demonstration given by the insider to the study team prior to the visit.*
- *Copies of interview guides for different roles within the organization.*
- *Schedules of who is interviewing or observing whom for the site visit.*
- *A cell phone contact list for all team members and insider sponsors.*
- *A copy of a template for writing field notes, listing a few foci/things to look for.*
- *A copy of any field survey being conducted as a short interview survey during the site visit to augment and complement interviews and observations.*
- *A map of the study site.*
- *A general agenda for nightly debriefings of the team during the site visit.*

15.7.4   Virtual Ethnography Techniques

Covid 19 affected many qualitative researchers around the world when the pandemic halted their on-site work in the spring of 2020. While virtual (off-site) data gathering techniques have been previously used, largely for projects with low budgets, study teams generally believed that these techniques yielded inferior data. It is hard to capture the spirit of an organizational culture when you conduct a video-conference interview of someone in their office. On the positive side, no travel is needed and health concerns during a pandemic are not an issue. Researchers have invented creative new ways of gathering data to substitute for not being there in person. These methods can be both innovative and rigorous (University of Technology Sydney 2020).

Interviews can be conducted by video and saved in either video or audio form depending on privacy considerations. Having video available can be better than just audio for an in-person interview because the entire team can view it afterwards. Substitutions for in-person observation are harder. An insider, a participant-observer who works actively in the field setting, can do a walk-around and video-record activities if allowed; interviewees can be asked to give demonstrations or take pictures; meetings can be attended by video.

Other innovative methods have become increasingly common. For example, subjects may be asked to write in a journal, with the text later analyzed. Especially when a study is of a patient centered consumer resource, virtual techniques can be superior to in person visits. Patients can use video to show researchers the layout of their homes, for example, while explaining what they are recording.

15.8 The Future of Qualitative Data Gathering Methods

A scoping review of qualitative research papers published in JAMIA since its first issue in 1994 to 2019 found that under 5% of total papers were qualitative in nature and that nearly all had focused on healthcare personnel rather than patients (Hussain et al. 2020). Most included interviews, observations, or both. The authors of the scoping review encouraged more funding and publishing of qualitative work, especially involving health care consumers as participants. Because of changes brought about by the COVID-19 pandemic, virtual methods will undoubtedly continue to be used and their use expanded. New technology for recording both voice and video is not only becoming more sophisticated, but is also becoming more familiar to subjects. Study participants who have access to video conferencing applications will feel more at ease during interviews using these applications than they might have in the past, for example. However, the inequities between those with access to the technology and those who do not will increase (Weinstein and Caciu 2017; Goedhart et al. 2021). Another equity issue that needs addressing is language translation, including the process for gaining consent (NIH Office of Behavioral and Social Science Research 2001, p. 12). Too often studies can only select English speaking interviewees because evaluation team members are not bilingual. Observations of non-English speakers by these team members in the field are possible, especially if a translator is present during a clinical encounter that is being observed, but some information will be incomplete. Another trend is that the team method for data gathering will become even more broadly accepted as the rigor of triangulation becomes standard procedure, especially in informatics where information technology and clinical knowledge merge. This is because informaticians have different disciplinary backgrounds, so by working together on study teams, more complete and accurate data can be gathered. Finally, more study team members with qualitative data gathering experience will become available: informaticians-in-training usually enjoy interviewing and observing to such an extent that they are anxious to do more.

Answers to Self-Tests

Self-Test 15.1
Which of the strategies to ensure study rigor is primarily employed in the qualitative study scenarios below:

1. Data from interviews about the usability of a resource are analyzed thematically. The evaluation study team looks to see if and how similar themes have arisen in earlier meetings of the team.
   Audit trail

2. A member of the study team, who has recently participated in another study of a similar kind of resource, becomes concerned that that person's views about the current study are being shaped by that previous experience. That person sits with another member of the study team to share that person's concerns and put them in perspective.
   Reflexivity

3. At a "town hall" meeting called to present the results of a qualitative study, the sponsor of the study raises deep and serious questions about the validity of the findings. The study team returns to notes from their team meetings to review how and based on what data they came to this conclusion.
   Member checking

4. During an evaluation project team meeting, one of the study team members finds themselves deeply repelled by off-color comments made by one of the project staff. The team member makes a note of this personal response as part of field notes.
   Reflexivity

5. After interviewing 10 patients participating in a study, a study team member perceives that they are hearing the same points raised by all interviewees. The team member requests a study team meeting to consider reducing the total number of interviews from 20, as previously planned, to 12.
   Data saturation

6. A study team member "corners" a participant in a system development effort following a meeting and asks for the participant's impressions on what transpired in the meeting.
   Member checking

Self-Test 15.2
Label each of the following interview scenarios, conducted as part of a qualitative study, as representing the fully structured, semi-structured or unstructured approach.

1. A study team member "corners" a participant in a system development project following a meeting and asks for that person's impressions on what transpired in the meeting.
   Open-ended

2. A study team member schedules time with a patient who is using an information resource to acquire specific information about the patient's medical history.
   Likely fully structured, though it could generate discussion, in which case it could veer towards semi-structured.

3. A study team member works with partners on the study team to develop a set of questions to be asked to all interviewees. Each question is to be followed up with the question: "Why do you think this is the case?". At the end of the interview,

subjects will be asked: "What else would you like to tell us to shed light on these matters?"
Semi-structured

4. An interview begins with the statement: "In general, what has been your experience using this EHR?" The remaining questions depend on how the interviewee answers this opening question.
Unstructured

5. A set of specific questions are read verbatim from an interview guide. No other questions are asked. The interviewees' responses are recorded.
Fully structured

References

Ash JS, Chin HL, Sittig DF, Dykstra R. Ambulatory computerized physician order entry implementation. Proc Am Med Inform Assoc. 2005;2005:11–5.

Ash JS, Sittig DF, McMullen CK, Wright A, Bunce A, Mohan V, Cohen DJ, Middleton B. Multiple perspectives on clinical decision support: a qualitative study of fifteen clinical and vendor organizations. BMC Med Inform Decision Making. 2015 Apr 24;15:35.

Beebe J. Rapid assessment process: an introduction. Lanham, PA: AltaMira Press; 2001.

Berg BL, Lune H. Qualitative research methods for the social sciences. 8th ed. Boston: Pearson; 2012.

Brunet LW, Morrissey CY, Gorry GA. Oral history and information technology: human voices of assessment. J Org Comput. 1991;1:251–74.

Crabtree BF, Miller WL. Doing qualitative research. 2nd ed. Thousand Oaks, CA: Sage; 1999.

Davis FD, Bagozzi RP, Warshaw PR. User acceptance of computer technology: a comparison of two theoretical models. Manag Sci. 1989;35:982–1003.

Erickson K, Stull D. Doing team ethnography: warnings and advice. Thousand Oaks, CA: Sage; 1998.

Gaglio B, Shoup JA, Glasgow RE. The RE-AIM framework: a systematic review of use over time. Am J Public Health. 2013;103:e38–46.

Glaser BG, Strauss A. Discovery of grounded theory. Strategies for qualitative research. Mill Valley, CA: Sociology Press; 1967.

Goedhart NS, Zuiderent-Jerak T, Woudstra J, Broerse JEW, Betten AW, Dedding C. Persistent inequitable design and implementation of patient portals for users at the margins. J Am Med Inform Assoc. 2021;28:276–83.

Hussain MI, Figuerredo MC, Tran BD, Su Z, Molldrem S, Eikey EV, Chen Y. A scoping review of qualitative research in JAMIA: past contributions and opportunities for future work. J Am Med Inform Assoc. 2021;28:402–13.

Kiyimba N, Lester JN, O'Reilly M. Using naturally occurring data in qualitative Health Research: a practical guide. Amsterdam: Springer; 2019.

Leedy PD, Ormrod JE. Practical research: planning and design. 11th ed. Pearson: Boston, MA; 2016.

Linstone H. Multiple perspectives for decision making: bridging the gap between analysis and action. North-Holland Elsevier: Amsterdam, NE; 1984.

McMullen CK, Ash JS, Sittig DF, Bunce A, Guappone K, Dykstra R, et al. Rapid assessment of clinical information systems in the healthcare setting: an efficient method for time-pressed evaluation. Methods Inform Med. 2011;50:299–307.

Miles MB, Huberman AM. Qualitative data analysis. 2nd ed. Thousand Oaks, CA: Sage; 1994.

Mohan V, Woodcock D, McGrath K, Scholl G, Pransat R, Doberne JW, et al. Using simulations to improve electronic health record use, clinician training and patient safety: recommendations from a consensus conference. AMIA Ann Symp Proc. 2016;2016:904–13.

Morgan DL, Krueger RA. The focus group kit. Thousand Oaks, CA: Sage; 1998.

NIH Office of Behavioral and Social Science Research. Qualitative methods in health research: opportunities and considerations in application and review. NIH Publication No. 02-!5046, December 2001.

Patton MQ. Qualitative evaluation methods. Thousand Oaks, CA: Sage; 1980.

Pope C, Mays N. Qualitative research in health care. 4th ed. Hoboken, NJ: Wiley; 2020.

Rubin HJ, Rubin IS. Qualitative interviewing: the art of hearing data. Thousand Oaks, CA: Sage; 1995.

Strauss A, Corbin J. Basics of qualitative research: grounded theory procedures and techniques. Newbury Park, CA: Sage; 1990.

Tolley EE, Ulin PR, Mack N, Robinson ET, Succop SM. Qualitative methods in public health: a field guide for applied research. Hoboken NJ: Wiley; 2016.

University of Technology Sydney. Adapting research methods in the COVID-19 pandemic: resources for researchers, 2nd ed. UTS and University of Washington, December, 2020.

Weinstein JN, Caciu A, editors. Communities in action: pathways to health equity. New York: National Academies of Sciences, Engineering, and Medicine, National Academies Press; 2017.

Yin RK. Case study research: design and methods. 3rd ed. Thousand Oaks, CA: Sage; 2003.

Chapter 16
Qualitative Data Analysis and Presentation of Analysis Results

Learning Objectives

The text, examples, Food for Thought questions, and self-tests in this chapter will enable the reader to:

1. Explain why the qualitative data analysis process needs to be iterative.
2. Outline the different approaches to coding qualitative data and the rationale for each.
3. Given samples of qualitative data, perform the coding process and compare the samples.
4. Describe how qualitative data analysis software can assist with the analysis process.
5. Describe the unique challenges of performing data analysis procedures on social media and other naturally occurring data.
6. Describe why qualitative evaluation team members usually use graphics to assist in their interpretation of data and reporting of the results of their analysis.

16.1 Introduction

Qualitative inquiry, when successfully undertaken, answers research and evaluation questions at two levels: descriptive and explanatory. At the descriptive level, the conclusions that emerge offer a "thick description," a term initially used by Geertz (1973) in application to ethnography. Thick description portrays "particular events, rituals, and customs" in detail (Denzin and Lincoln 2000, p. 15). At the explanatory level, qualitative inquiry also offers interpretation of the events that are described. If the conclusions of a study are to be useful, they must go beyond the telling of a story, no matter how interesting that story may be. To offer a useful indication of the value inherent in an information resource or service, the study must do more than

© Springer Nature Switzerland AG 2022
C. P. Friedman et al., *Evaluation Methods in Biomedical and Health Informatics*, Health Informatics, https://doi.org/10.1007/978-3-030-86453-8_16

assemble the apparent facts. Qualitative inquiry, like all other serious inquiry, must combine description with explanation. How the explanation part is accomplished is the major focus of this chapter.

16.2 What Is Qualitative Analysis?

Qualitative data analysis is so different from quantitative analysis that it can seem rather mysterious and dauntingly difficult to those who have never attempted it. However, as is true of quantitative studies, the quality of qualitative data analysis depends on the quality of the data that have been gathered. If triangulation, member checking, data saturation, the audit trail, and reflexivity—as discussed in Chap. 15—have guided a rigorous data collection effort, and if analysis is conducted with rigor as well, the results of the study will be credible. If the research team starts the analysis process at the same time it starts the data collection process, analysis is far less daunting. The two processes should be done in parallel because otherwise, corrections and course changes in the data gathering process might come too late. Although informal analysis will likely occur as soon as the team discusses the first few hours of data collection, formal line by line analysis of data generally starts as soon as the first interview transcripts are available or when study team members type their field notes. Unlike quantitative data analysis, which is conducted when data gathering is complete, qualitative data analysis should normally be concurrent with data gathering.

Qualitative data analysis aims to make sense of what is sometimes a vast amount of text. It is the process of interpreting the data and understanding meaning through analytical acuity, creativity, and intuition as the data and personal/formal theory are progressively brought together, from the hunches that give the argument initial shape to the interpretation in its final form. Data are first organized, described, interpreted and then corroborated. Interpretation is the central intellectual work of the study. Over time, interpretation progresses from small, tentative fragmented possibilities to a coherent, confident answer.

16.3 Approaches to Analysis

There exists a spectrum of types of approaches to qualitative data analysis, just as there is for data collection. Also, just as in data collection, different approaches are chosen and interwoven by investigators in producing their results. At one extreme of the analytical spectrum is a quasi-statistical style, such as word counting, to find how often a particular term occurs in the data. Some scholars of qualitative methods use the term *content analysis* to refer to word counting, but others broaden the idea of content analysis to include both counting/quantitative analysis and interpretation/qualitative analysis (Berg and Lune 2012). At the other extreme is completely unstructured data analysis, in which the study team members read and consider the data and reach conclusions without formal intermediate steps.

Crabtree and Miller (1999) describe three points on the data analysis spectrum. They categorize the chief organizing styles used in qualitative data analysis as template, editing, and immersion/crystallization. The template style is the most structured. It uses a *preexisting* list of terms, called codes in the language of qualitative analysis. In some projects, the study team might develop a code list at the outset of the data analysis process, and use it to count occurrences of a particular word or phrase representing a concept, and to index text. For example, a list could be developed based on a framework like the multiple perspectives model described in Chap. 15. If the focus of the study was CDS, and interviews and observations were about CDS, it would make sense to have "CDS" as a predetermined term on the list. If the selected perspectives based on this model were those of users, information technology specialists, and administrators, terms representing those groups would be on that list. These predetermined codes are "structural codes" (Pope and Mays 2020) and the development of the list is similar to the common practice of constructing an index to a book.

Another option for using the template style is to apply a preexisting list developed and previously used. For example, Nielsen's heuristics for evaluating usability, which will be described in more detail in Chap. 17, include: visibility of system status; match between system and the real world; user control and freedom; consistency and standards; error prevention; recognition rather than recall; flexibility and efficiency of use; aesthetic and minimalist design; help users recognize, diagnose, and recover from errors; and help and documentation (Nielsen and Mack 1994). A qualitative evaluation team might decide to include Nielsen's list in its list of predetermined codes if the evaluation question related to usability. Note that the template style could be used after all the data are collected since the codes will not change. Although the template style permits the creation of new codes throughout the analysis process, this is considered exceptional.

The editing style (Crabtree and Miller 1999) generates codes in a different way. Investigators develop codes as they review the data, making notes as they read and reread the various texts they have assembled. These are "emergent codes" (Pope and Mays 2020) that emanate directly from the data. The resulting code list is continuously modified as new data are collected and reviewed. The editing style requires frequent recoding of textual items, as the codes themselves evolve, and for this process, qualitative data analysis software can be very helpful. It is the information that goes beyond the predetermined terms, the nuanced issues surrounding them, the deeper meanings, that can be sought out with a newly developed coding scheme built from the bottom up. The aim is to go beyond the expected to elicit newer and deeper insight. Most qualitative studies in informatics use the editing style because it offers the ability to update the coding scheme as more data are collected. This style is especially effective when data collection and data analysis are done at the same time.

The immersion/crystallization style (Crabtree and Miller 1999) is the least structured, with study team members spending extended periods of time reading and interpreting the text and gaining an intuitive sense of the data prior to writing a description of their interpretation. The coding process using this style is much less formal and systematic. It is not often used in informatics, largely because it is hard to convince an audience, whether stakeholders, grant reviewers, or reviewers for publications, of the credibility of the results.

The template and editing approaches together contribute in different ways to the attributes of qualitative rigor. Figure 16.1 illustrates the top-down approach of the template style and the more bottom-up approach of the editing style. The template and editing styles work well to improve the credibility of a study. These methods impart a natural internal consistency to the analysis, and they allow analysis to be conducted collaboratively. The different approaches can also be used in combination over the lifecycle of a study. In the beginning stages of fieldwork, the editing approach can be used in an open-ended way to produce codes, and later the more developed codes can be used in a template style as a more stable list when new data enter the study. In some cases, the use of both approaches in analyzing the same data set can increase understanding of the topic with great depth. This form of triangulation during the analysis process can be envisioned as approaching the data in two complementary ways to maximize what can be learned from the same data set. The template style, as noted above, can be used at the end of data collection to re-code the same data that were coded using the editing style, but with a different list, such as Nielsen's (Nielsen and Mack 1994).

Fig. 16.1 Top-down and bottom-up approaches

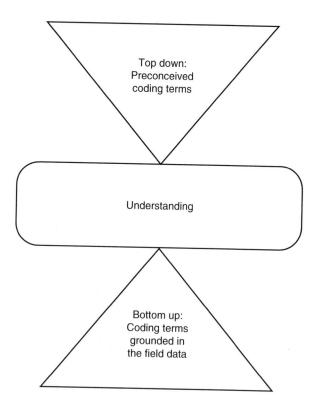

Top down:
Preconceived
coding terms

Understanding

Bottom up:
Coding terms
grounded in
the field data

The analysis process is fluid, with analytical goals shifting as the study becomes more mature. At an early stage the goal is primarily to identify and then focus the questions that are to be the targets of further data elicitation. At the later stages of study, the primary goal is to collate data that address these questions. The study team must recognize that the data often raise new questions in addition to answering preexisting ones. Sometimes new data do not alter the basic conclusions of a study but reveal to the study team how a significant reorganization of the results will lend greater clarity to the exposition. (This situation is analogous to a linear transformation in mathematics; the same information is contained in the result but is expressed relative to a different and more revealing set of axes.)

16.4   Qualitative Data Management

Qualitative study team members must keep careful records of their procedures and be extremely diligent in their handling of qualitative data. As noted in Chap. 14, a detailed audit trail is necessary for assuring the rigor of the study, but it is also necessary for the convenience of the team members. At the outset, study teams should develop a method for naming files and organizing them in a manner that is easily accessible to team members. If the study is conducted under oversight from an Institutional Review Board (IRB), the IRB may require this level of detail regarding data recording and management before any field work begins. The files must be stored securely for a designated period of time and then destroyed at a time required by the approving IRB. Audio files should be screened to eliminate private information such as personal health information or completely extraneous information and then be transcribed into text files so team members do not need to listen to audio unless listening is part of the protocol. Audio from video recordings is usually treated the same way, but the video can be analyzed separately using a different list of terms/codes to identify nonverbal details.

Data must be put into formats that can be analyzed by team members. Recordings should be transcribed as soon as possible. There are several different ways that transcription can be done, ranging from completely manual transcription to completely automated transcription using voice recognition software. There are advantages and disadvantages to both and selection of a mechanism depends on requirements of the study. The more a human is involved, the more tailored the transcription will be to the needs of team members: a transcriptionist can heavily edit an automatically produced transcript according to the study team's instructions, but that will cost more than accepting an unedited automatically produced transcript. A perfect transcript will take some time to produce and an automated one will be available quickly. Voice recognition can be used the same way for speech in video, and AI-assisted sentiment analysis can help evaluate tone, intent, and emotions. Analysis of visual recording can be greatly aided by special software developed for that purpose.

Handwritten field note jottings should be transformed by the study team members into full field notes into digital files within 2–3 days to avoid forgetting the significance of the moment that led to their creation. Team members should clearly mark verbatim quotes from subjects with quote marks and distinguish their thoughts or immediate interpretations in brackets. The field notes should be written in a manner that allows teammates who will be analyzing the notes to feel that they were there. Field notes that are not well written, especially if they do not distinguish between what the team member heard or saw directly and their interpretation of what they heard or saw, might bias results, so they should be analyzed selectively or not at all.

16.5 Qualitative Analysis Software

Qualitative data analysis software (QDA) (Tolley et al. 2016) allows the study team to share data, store, organize, count, search and retrieve, code, group codes into themes, and visualize patterns across codes, themes, and documents. Some software packages provide voice recognition and AI assisted transcription and sentiment analysis. Though perhaps one day a study team member will be able to load a group of transcripts into the program, select an analytic method from a drop-down menu, and immediately see lists of suggested patterns and themes in the text, this is not possible at the present time. Study team members must read and code every line of text and record the coding manually, though the software provides a good deal of assistance. There is benefit to this extensive coding process since the team becomes immersed in the data. The software analysis program assists in organizing the data and thus promoting ongoing evolution of results, in providing graphics capabilities to help team members literally see connections, and to index and provide reports of what the team has captured.

There are many software packages available. Some with limited capabilities are free and others that have more functions are available for purchase. Those that are downloadable and free are usually sufficiently powerful enough for a small study. The more expensive software like NVivo (QSR International n.d.) and Atlas.ti (Atlasti n.d.) have attributes for researcher sharing, managing large numbers of documents, sophisticated graphics, and management of multiple types of media. Some provide AI-enhanced transcription and sentiment analysis, aids for analyzing video, automatic coding of synonyms, assistance for surveys and other more quantitative data, and numerous visualization tools. Often even vendors of the more expensive software will allow potential buyers to download a program for a limited time at no charge so team members can try using it. Dedoose (Dedoose n.d.) is a package that is quite sophisticated with a charging structure on a per use basis. Informaticians tend to learn how to select and use the software quickly. Online tutorials and YouTube videos are excellent for the more complex software, international user groups provide support in addition to vendor support, and the packages are rapidly becoming more sophisticated (Pope and Mays 2020).

Self-Test 16.1
A study team has been asked to evaluate the implementation of a patient and family digital educational program in one hospital of five in a system before the program is rolled out systemwide. The program offers a tablet to each patient or a family member with educational content about clinical procedures, hospital layout and information, and how to get answers to questions. The study team finds no evidence of similar programs elsewhere, so team members are excited that the evaluation will be novel because so little is known already.

1. Should they use a template style of analysis or an editing style and why?
2. What should they look for when selecting a qualitative data analysis program and why?

16.6   How to Code

Briefly, the coding process unfolds as follows. First, each person conducting the analysis obtains an electronic version of a document. The team member reads through the document to gain an overall impression. Then the person reads each line and assigns labels to meaningful words or phrases. Often, more than one label is applicable to the same phrase. As the team member reads on, the same label may be used many times, indicating there is a pattern. As the study team members read and code more documents, new codes and patterns will be discovered and at some point the patterns will become clearer and themes will emerge. It is also possible that new patterns will continue to emerge as more analysis takes place.

If the reader is unfamiliar with qualitative data analysis coding, Self-test 16.2 will offer a useful introduction. The exercise includes two short segments of transcripts of oral history interviews, among many freely available through the U.S. National Library of Medicine (National Institutes of Health, National Library of Medicine, Oral History Division n.d.) conducted by one of the authors. Two interviewees were asked to offer advice to future informaticians. Their answers are vivid and thoughtful, with both similarities and differences. Although it is most likely that manual coding will not be used by a qualitative study team, the steps in the exercise are identical to those used in any software program.

Self-Test 16.2
Qualitative Data Analysis Exercise
    Instructions: Code the following two transcript excerpts. To do this,

1. Write down on a sheet of paper a list of terms you might write in the right margin of the transcript next to important phrases.
2. After you have two lists of codes, one for each transcript, make a list of concepts that are alike across the two transcripts and concepts that differ.

Interview with A by JA

- **JA**: *And if you have some advice to offer one of these young new fellows about being an informatician, what would you say?*
- **A**: *Oh god! Well, a couple of things is to basically to have fun while you're doing it because your fellowship years are some of the best years you are gonna have (laugh); be willing to take some degree of risk and work hard in doing it, that is very hard to quite know what's the right thing to do [as you go]; certainly to get to know all the fellows, all the peers, to get to know the people in the field and to listen at meetings you've gotta be open to a whole variety of things; to not get discouraged too easily because there will be bad times and you hope that [sometimes] you're lucky and that basically things work out right; it is very much … circumstances very much … time … I mean certainly I would, I could, never have planned my career on [a sort of] rational basis …. ok, here is what I am gonna do step 1, step 2, step 3.*
- **JA**: *So serendipity obviously played a great role in your life. Do you think future informaticians will benefit from serendipity too?*
- **A**: *You can't demand a hope for serendipities … you can't plan for them so I'm saying I think there are people who I suppose basically need to be aware that over there is an opportunity … And be willing to take chances on something that might work out because [they] are very hard to predict.*
- **JA**: *And these hard times that you just mentioned that all informaticians go through, what did you mean by that?*
- **A**: *Oh, there are things that don't work out. I mean, the chief of medicine one time said to me, I really don't know what you do and that caused me a little bit of concern … but he turned out to be one of my great fans after 3 or 4 years. Oh, you get grants turned down, you will get a project that doesn't work, you've got papers that [don't] get accepted, it's not all a bed of roses (laugh) that's for sure.*

Interview with B by JA

- **JA**: *Let me then ask you my last question. As I told you, I'd ask you about—in your wisdom, what advice would you give to those coming into informatics today?*
- **B**: *Well, first of all, I think an appreciation of the importance of informatics. And I think informatics is an increasingly important—and it ought to become a lot more dominating in influencing what happens. The second thing is don't have preconceptions of what is possible and what is not possible. And even though there's a law that says you can't do that, then look to see what the ideal solution is, and that's where you set the bar. And it may take you a career to get there, but if you set the bar here (low), you'll never get to here (high). And so just simply declare where you really want to be, and say, "This is what we need," and then expend the energy to get there. Look at history, and look at it from the perspective of what was done here that becomes usable by me in solving the problems that I face now in today's world. I look to see what's the lesson. When I read a paper, and when my students write papers, what I really want them to know is write for*

me three bullets of what you want the person who reads this paper to take with them that they can use. And a lot of papers I can do is, I say, "Gee, they've got a great system, but I don't have anything I can take away to make my system great." So it's that sharing and recognizing that the sum of the parts has the potential always to be greater than the whole. And we're lucky. I mean, this is one of the few professions in which my best friends are my professional colleagues. And I just think that's great. I really envy, in some ways, the youth coming in today, because they've really got—. It's an open area, and there's so many things that's available to learn. I've never been restricted and said, "Gee, that's another area for me." I mean, genetics, business, technology, all of those kinds of things are things that—. Gene Stead, in the tape that he made, said, "If you can get B interested in something, something would happen. And it's not difficult to get B interested in anything." And that really is the true part of this.

I think the thing about standards is that everybody pretty much knows that if they put me into a group, something would happen. And that's what I would really say, is do something! Don't be passive—be active. And there's nothing you can't do. Every problem is solvable. I may not know how to do it my entire lifetime, but I firmly believe—. I believe knowledge is perfect, and every problem is solvable. And that's the way I approach everything.

When using the editing style of analysis, the study team must develop the list of codes and maintain them in a "code book" (Crabtree and Miller 1999). At the outset of this code book development process, all team members will be generally familiar with the data since they will have discussed it during the data gathering process, during briefing sessions among themselves, and with stakeholders as a form of member checking. Once enough transcripts are available, team members can break into dyads, usually a clinician matched with a non-clinician to promote diversity. Each dyad is assigned two transcripts for coding and given a deadline, at which point the dyad reports back to the full team the codes they have generated from those two transcripts. What the dyad reports back is the product of the two team members' discussions about codes based on their reaching consensus. Some qualitative experts, especially those using content analysis, report kappa scores denoting how much agreement there was between dyad members. Most, however, believe that there is little meaning in a kappa score because the goal is consensus and the score will always become perfect when consensus is reached (Tolley et al. 2016). The whole team, usually composed of three or four dyads, generates an initial code book as comparisons are made across dyads. The code book lists main topics, sometimes called parent codes, and sub-topics, or child codes. Each is defined carefully so that future coding is simplified. The definitions should include not only descriptions of what should be labeled with that code, but also what should not be assigned that code, but assigned another code. Armed with this code book, team members can recode the initial transcripts, this time using the software, and code new transcripts. The code book may need modification over time, but usually changes are not major. The software eases this entire process.

As noted earlier, in many cases, the same text can be coded under several different code terms. This is a concept that quantitative researchers find hard to understand. However, it is necessary for complete analysis of the data. For example, one interviewee once said of an EHR: "It's timely, legible, accurate, comprehensive information then, so I think it beats the heck out of the paper chart, and the one big drawback then and still to this day is that it takes longer for the doctors I think to go back and type it." This one sentence could be coded "benefits," "drawbacks," and "time issues." If the study team later wanted to find all text that related any of these terms, and used the software to search on any of them, this quote would be found. In addition, by coding text with multiple terms, overlaps and relationships between and among concepts can be identified, especially with the aid of QDA software.

During the coding process, terms can be grouped into higher level patterns of terms and examined for relationships to build themes and develop theory. See Chap. 14 for discussion of theory. The software packages that allow sharing of codes among different researchers are most desirable for team research. Within this class of software, graphics capabilities for building conceptual networks are especially helpful. For example, the ability to identify overlaps in codes can lead to making connections: if interviewees often speak of two concepts together, team members can be made aware of this common pairing of concepts through graphics produced by the software.

As coding progresses, a process of writing "memos" (Crabtree and Miller 1999) is done by study teams as well. These take the form of "notes to self" about how methods are working, perceptions of possible bias, potential interpretations, thoughts about theory, and anything else a team member wants to revisit or share with the larger team. The software has the capability of organizing these memos, which should be revisited periodically as further analysis takes place.

During coding, the team members will read each segment and assign terms just as they would do manually, but the software keeps track of the assignments so that later the segments assigned to each code can be retrieved using searches by code words. This feature of generating reports of text associated with codes is especially helpful when writing study results. By reviewing all of the text pieces associated with a particular code, it is easier to find relevant material to quote, for example.

The process outlined above suits the discipline of informatics because it has been designed to take advantage of the multidisciplinary nature of the field. Because team members have varied backgrounds, and because even the more senior members of the team should take part in the analysis so everyone becomes steeped in the data, team members should be offered options to minimize the more clerical tasks. For example, some team members may prefer either hand writing codes in the margin, as in Self-test 16.2, or typing the codes as comments in a word processor to using the QDA software. These coded documents can then be given to another team member to enter into the software program. This process works well to encourage busy senior evaluation team members to take part in the coding process since it is faster and easier than using the QDA software, and it also gives a second team member an opportunity to double check the coding before final coding steps are accomplished.

Nouveau Community Hospital and Health System Case Example

The ET members decide to analyze data using the "editing style" described by Crabtree and Miller (Crabtree and Miller 1999), and to manage the large amount of data to be collected by multiple researchers, they select an appropriate software package. They decide to have two team members code each document and reach consensus on coding before that coding is presented to the entire team. During team meetings, each dyad of coders will suggest changes to the coding scheme based on the dyad's experience coding the assigned transcripts. The team holds regular analysis meetings to review the coding that each dyad has done and then they agree on "team" codes to be included in their code book.

16.7   How to Interpret

Interpretation basically means making sense of the data. It allows an evaluation effort to go beyond description and provide explanations. Interpretation of qualitative data has been called a "dance" (Crabtree and Miller 1999, p. 127) because it is iterative, responsive, has rhythm, is creative, and may have a variety of audiences.

Interpretation is needed to gain a deeper understanding of the cultural phenomena under study. It is the meaning making process. It is why teams enter the field to gather data from those within the culture. Often the meaning is latent, in that it is not obvious. For example, during a study of CDS, team members noticed during the interpretation process that providers had complained that ICT professionals were telling them what to do, by which they meant that when CDS was implemented the providers were losing their autonomy. The informaticians and ICT staff were proud of their ability to implement CDS. The evaluation team identified an underlying latent theme of "power shifts" in the data. Subjects did not use this term, but as it turned out, it helped explain much of the discomfort and frustration providers were feeling (Ash et al. 2006).

At naturally occurring checkpoints, such as the end of coding of data from a particular site, the team should meet to discuss interpretation. Basically, team members answer the questions: Where are there patterns or themes across the data and what do the data say about the research question? What have we learned that is new? Is there a new theory emerging?

The data may end up being coded under 50 codes or more, many of them actually sub-codes of parent or major codes. The team can group them into patterns and themes through discussion. This exercise can be aided by low level technology: sticky notes and flip chart paper or cards that can be sorted into piles. The result is aggregating codes into themes. These themes could be reported directly, but a talented study team can add value by interpreting the themes.

During the interpretation process, revisiting theory can help with explanations and understanding. If a theoretical framework was used to guide study design, revisiting it can help guarantee that all of the areas outlined in the framework have been covered. Theory can also be sought out to explain and interpret results. For example, diffusion of innovations theory (Rogers 2003) has outlined the roles of opinion leaders and champions, with an opinion leader being a respected expert in the field and a champion being someone especially forward thinking and innovative. The roles are quite different: if, during a field study of providers, an evaluation team concluded that these roles existed and were played by different people, the team was once again validating the applicability of that theory.

The interpretation process should include safeguards against going too far beyond the data into the realm of speculation. This is a point in the qualitative process that is especially vulnerable to drawing the wrong conclusions. The evaluation team members might be biased in favor of or against the information resource, for example. Study teams should not abuse the power they hold by imposing unwarranted meaning on the results. Usually, results of a complex qualitative study point out both the positive and negative nature of the information resource and these should be reported without bias.

Nouveau Community Hospital and Health Center Case Example
The ET members meet to review coding, interpret the data, and generate themes. They have used a multiple perspectives framework for gathering data, so they make sure they have enough data from individuals in all the targeted roles and have reached saturation. They make sure they have learned about usability and the other important aspects of the patient portal. They know they will be reporting on these, but they also found many novel ideas within the data and were surprised at how many suggestions patients had about improving the portal. They developed a theory that patient portal use can not only empower patients, but can also benefit the clinics by getting more accurate structured data entered by the patients themselves rather than having medical assistants enter it. The team wants to test this theory as a next step.

16.8   Other Analysis Approaches

16.8.1   Narrative Analysis

Narrative analysis frames informants' observations as "stories" and addresses both the story content and structure. For example, the structure of informants' stories can be analyzed for plot development. Stories generally follow a standard plot development format: the stage is set; something happens; there are consequences. The stories of different informants, once projected onto a common structure, can then be compared with one another. For example, one study (Stavri and Ash 2003) found that stories about successful implementations of computerized physician order entry usually began with descriptions of prior failures. This finding was important because it raised a new theme: how might failure of previous CPOE systems breed future success? It also provided insight into the thinking of the storytellers (their memories focus on the failures first, perhaps because those are more vivid than memories of success). Narrative analysis has been especially useful for studying provider-patient relationships during clinical encounters as well (Crabtree and Miller 1999).

16.8.2 Analysis of Naturally Occurring Text or Visual Data

Evaluation experts are rapidly developing creative methods for analyzing data from equally rapidly developing sources available online. These sources and data have advantages in that they can be gathered unobtrusively, at no cost, and they represent up-to-date current cultural attributes (Rogers 2016). Each source of data also represents a different challenge when planning analysis of data, however. The sources include online discussion groups, Facebook groups, YouTube, podcasts, and even digital mapping and geospatial technologies (Kiyimba et al. 2019).

Analysis of Web based sources offers unique challenges. Identifying what data to analyze from among very large data sets is especially difficult since the data may be synchronous or asynchronous, public or private, and in text, audio, or video formats. On the positive side, software is available to help team members download selected data, which is an enormous benefit, and the more sophisticated qualitative data analysis programs can assist as well (Lupton 2020).

While the basics of coding and theme development are the same as those described above for more traditional qualitative data, there are additional issues to be addressed in analysis of Web based data. Transcripts from semi-structured interviews generally have a well-organized structure because an interviewer is guiding the conversational journey. Field notes, once in electronic form, usually follow some structured format. Structure makes these data sources easier to analyze. Social media sources may not have this level of structure. As the importance of social media in our society continues to increase, however, it will provide a rich source of data and researchers will continue to improve analysis methods.

Food for Thought Questions
1. Why are sometimes even the small sounds that are not words, like laughs or the sound of pounding on a table, important when analyzing interview data?
2. What are some examples of areas where narrative analysis might be useful?
3. If you were asked to use analysis of Web based data to discover what recently diagnosed breast cancer patients are concerned about, what sources might you go to?

16.9   Using Graphics in the Analysis and Reporting Processes

Writing reports for qualitative studies has some challenges that do not exist for quantitative studies. When reporting results of statistical studies, there exist conventions for what kinds of tables to include. No such conventions exist for qualitative studies, though normally it is wise to include at least a table explaining the sources of data. Excellent guidance in the literature offers a multitude of possibilities from which to select (Miles 2013). Information about sites or informants can be illustrated in a matrix which may list locations across the top and relevant attributes along the left side. The cells of the matrix could include numbers of people in each category who were interviewed or observed. Table 16.1 is an example of a matrix summarizing data sources during five hypothetical site visits.

If the evaluation study has used a theory, model, or framework as an organizing structure, the conceptual framework can be artfully described with the help of a diagram. Figure 16.2 illustrates a hypothetical study that could be conducted at Nouveau Community Hospital and Health System. This is a simple diagram to

Table 16.1 Example of matrix of sites and subjects

Site Visit Study						
	Site A	Site B	Site C	Site D	Site E	Total
Geographic location	Northwest	Southwest	Southeast	Midwest	Northeast	5 sites
Setting	Teaching Hospital and Clinics	Community Health System	Community Hospital	Teaching Hospital	Urgent Care Clinics	N/A
Dates of site visits	Oct-17 to Jan-18	Jan-18 to Feb-18	Aug-20	Oct-20	Dec-18 to Jan-19	Oct-17-to Jan-19
EHR used	X	X	Y	Z	Z	3
Total interview time	12 h	7 h	11 h	12 h	5 h	47 h
Number of interviews	13 total (14 people: 4 providers; 4 staff; 6 admin)	15 total (18 people: 6 providers; 5 staff; 7 admin)	18 total (18 people: 8 providers; 6 staff; 4 admin)	19 total (19 people: 6 providers; 7 staff; 6 admin)	11 total (12 people: 6 providers; 5 staff; 1 admin)	76 total (81 people: 30 providers; 27 staff; 24 admin)
Number of clinics observed	2 clinics	3 clinics	1 clinic	3 clinics	3 clinics	12 clinics
Number of people observed	5 total	12 total	8 total	16 total	11 total	52 total
Total observation time	17 h	20 h	6 h	25 hs	12 h	80 h

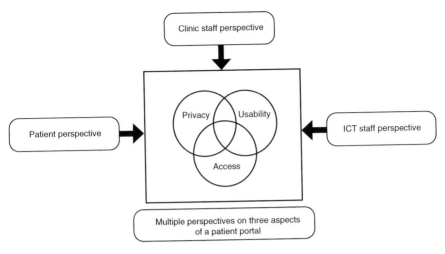

Fig. 16.2   Example of conceptual diagram of framework

illustrate how a multiple perspectives framework could be employed to analyze the views of clinical staff, ICT staff, and patients about the new patient portal. The foci of the study, in the center, are privacy, usability, and access, which are depicted as overlapping circles because one impacts the other. Such a diagram can quickly engage and educate the reader.

Often reports of qualitative evaluations include tables of quotes. If the report discusses themes and sub-themes, the table might include especially pithy quotes about each. Photographs can be extremely effective as well, though of course permissions must be obtained. For example, patients might be willing to share pictures of the setting in which they use the patient portal. Diagrams of various kinds, flowcharts, and timelines are all useful for illustrating important aspects of some studies (Miles et al. 2013). The format for these representations cannot be prescribed, and is specific to the study methods and results being reported.

A final study report or manuscript can be organized into standard sections: an executive summary, introduction, background, methods, findings or results, discussion, and conclusions. Chapter 19 describes this format in more detail. This is a conservative approach that almost always communicates effectively as long as each section itself is well written. Most often, the results section of a qualitative study is organized by theme or concept, but such reports can also be arranged in more creative ways, depending on the purpose or the requirements of the report. For example, May and Ellis (2001) present a study of a telemedicine project as a narrative story written in a dramatic tone, complete with a plot and actors. These authors have also used a chronological approach in the results section of another paper (May et al. 2001), tracing informants' viewpoints at different points in the evaluation process through representative quotes. Patton (1980) suggests that a formal qualitative evaluation report might include sections on the purpose of the evaluation, the methods, presentation of the findings (describe the information resource, describe the findings, and offer interpretations); validation and verification of the findings; and conclusions and recommendations. While a team approach to data collection and analysis is almost always fruitful, preparation of the actual report usually falls to one or a small number of study team members. Nonetheless, a team discussion is ideal for reaching consensus about how to best communicate in a succinct, aesthetically pleasing, and easily grasped manner.

Nouveau Community Hospital and Health System Case Example
After member checking by sending a summary of results to participants for feedback and after writing several drafts, the ET members prepare a final report for the health system following Patton's outline. However, they also ask for and receive permission from their sponsor to prepare several papers for publication. They mask the identity of Nouveau and, of course, the informants. They write a case study paper for publication in an informatics journal and a paper about their methods for a social science and medicine journal. The ET story is a success story: the ET members, led by Dr. Yu, have helped smooth the transition to the implementation of a patient portal by providing ongoing feedback and a summary report, the students have gained valuable fieldwork experience, and the team eventually has both of its papers published in prestigious peer reviewed journals.

16.10   Evaluating the Quality of Qualitative Analysis

The end product of qualitative analysis is a set of conclusions—the interpretation—built on the data. Study teams seek trustworthiness, confirmability, credibility, and transferability. While quantitative studies also strive for analogous characteristics, they are conceptualized, approached, and attained in qualitative work in distinctive ways that are not dependent on measurement and statistical constructs such as confidence intervals and inference. In qualitative studies, trustworthiness implies total authenticity of findings. Confirmability denotes objectivity or freedom from bias. Credibility is analogous to internal validity or gaining a true and believable picture. Transferability is an analog to generalizability—the degree to which the results are applicable to other contexts. Dependability, like reliability, is the extent to which the process of the study has been undertaken with consistency and care (Tolley 2016). Table 16.2 provides a checklist that can be used for evaluating any proposal or report of qualitative work.

Table 16.2 Checklist for proposals and reports of qualitative work

Checklist for Qualitative Proposals and Reports
Description of the evaluation study team
Does it include an appropriate mix of methods, ICT, and health expertise?
Does it include both insiders and outsiders?
Rationale for the match/fit between study questions and methods
Are the methods appropriate for answering the questions?
Details of the sampling strategy
Are subjects selected in a purposive manner that is well explained?
Description of data collection and analysis rigor
Are the five essential strategies for rigor utilized?
Delineation of description vs. interpretation
Is it clear what emanates from the field and what springs from the evaluation team's interpretive process?
Veracity of Results
Do the results ring true?

16.11 The Future of Qualitative Analysis

As noted earlier, software for qualitative data analysis is fortunately keeping pace with the needs of those who conduct qualitative studies. This became evident when, early in the COVID-19 pandemic, many evaluation teams needed to change methods from in person to digital and a number of researchers turned to naturally occurring data as a substitute for fieldwork. Also helpful are the increasing capabilities of software to help collect data for analysis of Web based data by organizing and downloading it. There is a shift towards more team analysis and the software is becoming more adept with this as well. Finally, informaticians will be pleased with progress being made in the development of tools to assist qualitative analysis that take advantage of existing AI technology for voice recognition and transcription.

Answers to Self-Tests

Self-Test 16.1

1. They should use an editing style. Using a template style, they would be imposing a preconceived list of terms upon the data, as if they were indexing the data. However, this project enters new territory and little is as yet known about this information resource, so it would be difficult developing a list of applicable codes at this early point. The editing style would let them develop a code book of terms that arise from the data.

2. The team's selection of software depends on how big the project and budget will be. If the project ends with one hospital and perhaps 30 participants, a freely available software package might suffice. However, if the scope goes beyond that and team members need to take advantage of more sophisticated capabilities, a more powerful package should be considered.

Self-Test 16.2

1. Interview with A codes might be: Fun, Best years, Risk, Work hard, Peers, Be open, Bad times, Hope you're lucky, Never could have planned, Opportunity, Take chances, Hard times, No bed of roses

   Interview with B codes might be: Influencing what happens, Don't have preconceptions of what is possible, Set the bar, Expend the energy, Look at history, I look to see what's the lesson, We're lucky, Colleagues, Open field, Everything is solvable, Can't plan for serendipity
2. How they are alike: Lucky, Hard work/expend the energy, Peers/colleagues, Open/open field, Never could have planned/can't plan for serendipity

   How they are different: A talks about opportunity, risk, taking chances, hard times and B mentions setting the bar, looking at history for lessons, and solving any problems

References

Ash JS, Sittig DF, Campbell E, Guappone K, Dykstra R. An unintended consequence of CPOE implementation: shifts in power, control, and autonomy. Proc Am Med Inform Assoc. 2006;2006:11–5.

Atlasti (n.d.) See www.atlasti.com. Accessed 8 June 2021.

Berg BL, Lune H. Qualitative research methods for the social sciences. 8th ed. Boston: Pearson; 2012.

Crabtree BF, Miller WL. Doing qualitative research. 2nd ed. Thousand Oaks, CA: Sage; 1999.

Dedoose (n.d.) See www.dedoose.com. Accessed 8 June 2021.

Denzin NK, Lincoln YS. Handbook of qualitative research. 2nd ed. Thousand Oaks, CA: Sage; 2000.

Geertz C. Interpretation of cultures. New York: Basic Books; 1973.

Kiyimba N, Lester JN, O'Reilly M. Using naturally occurring data in qualitative health research: a practical guide. Amsterdam: Springer; 2019.

Lupton D, editor. Doing fieldwork in a pandemic (crowd-sourced document); 2020. Available at https://docs.google.com/document/d/1clGjGABB2h2qbduTgfqribHmog9B6P0NvMgVuiH ZCl8/edit?ts=5e88ae0a#. Accessed 8 June 2021.

May C, Ellis NT. When procols fail: technical evaluation, biomedical knowledge, and the social production of "facts" about a telemedicine clinic. Soc Sci Med. 2001;53:989–1002.

May C, Gask L, Atkinson T, Ellis N, Mair F, Esmail A. Resisting and promoting new technologies in clinical practice: the case of telepsychiatry. Soc Sci Med. 2001;52:1889–901.

Miles MB, Huberman AM, Saldana J. Qualitative data analysis. 3rd ed. Thousand Oaks, CA: Sage; 2013.

National Institutes of Health, National Library of Medicine, Oral History Division. (n.d.) Medical Informatics Pioneers. https://lhncbc.nlm.nih.gov/LHC-research/LHC-projects/health-information/medical-informatics-pioneers.html. Accessed 8 June 2021.

Nielsen J, Mack RL, editors. Usability inspection methods. New York: Wiley; 1994.

Patton MQ. Qualitative evaluation methods. Thousand Oaks, CA: Sage; 1980.

Pope C, Mays N. Qualitative research in health care. 4th ed. Hoboken, NJ: Wiley; 2020.

QSR International (n.d.) See www.qsrinternational.com. Accessed 8 June 2021.

Rogers E. Diffusion of innovations. 5th ed. New York: Simon & Schuster; 2003.

Rogers R. Doing digital methods. Thousand Oaks, CA: Sage; 2016.

Stavri PZ, Ash JS. Does failure breed success: narrative analysis of stories about computerized physician order entry. Int J Med Inform. 2003;72:9–15.

Tolley EE, Ulin PR, Mack N, Robinson ET, Succop SM. Qualitative methods in public health: a field guide for applied research. Hoboken, NJ: Wiley; 2016.

Part IV
Special Study Types

Chapter 17
Mixed Methods Studies

Learning Objectives
The text, examples, Food for Thought questions, and self-tests in this chapter will enable the reader to:

1. Describe the relationship between triangulation and mixed methods.
2. Explain both the advantages and challenges of using mixed methods rather than quantitative or qualitative strategies alone.
3. Describe how different mixed methods research designs are depicted in a standard "typology of mixed methods designs."
4. Explain why data analysis for mixed methods studies is usually more difficult than for purely quantitative or purely qualitative studies.
5. Offer explanations about why usability and organizational systems-level studies are considered exemplars of mixed methods efforts.
6. Explain special ethical issues that exist in mixed methods evaluation that do not necessarily exist in either quantitative or qualitative projects that are separate.
7. Describe the future of mixed methods studies in informatics.

17.1 Introduction

Prior chapters have described the rationale and methods for conducting different types of quantitative and qualitative evaluation studies. The separation of the two strategies is traditional because, especially in biomedical and health fields, the majority of studies use either one or the other, but rarely both. This chapter will describe how the two strategies can be used in a well-planned sequence or in an intertwined manner, which is called a mixed methods approach, to take advantage of the benefits of both strategies. If mixed methods seem appropriate for addressing the study purpose and aims, and for producing the desired product, they offer many

© Springer Nature Switzerland AG 2022
C. P. Friedman et al., *Evaluation Methods in Biomedical and Health Informatics*, Health Informatics, https://doi.org/10.1007/978-3-030-86453-8_17

additional advantages over a single method from a scientific point of view. Because qualitative methods are inductive and quantitative methods are deductive, mixed methods can basically pursue a problem from both below and above in search of the truth. The approaches offer different views of the world, different paradigms, and together can be stronger than either of them alone. Together, they are ideal for helping to assure that the study team's approach is sociotechnical, considering a broad range of people and organizational issues as well as technology factors.

17.2   The Ultimate Triangulation

Earlier, the authors discussed the meaning of the term "triangulation" and why it is important in qualitative research. One method for assuring triangulation is to use a variety of methods, such as interviewing and observing, to seek the truth from different angles. Triangulation can go beyond qualitative methods, however, to include quantitative approaches in a meta-, or ultimate, version of triangulation. "Mixed methods" is a term that has been used in the literature to denote a variety of strategies along a continuum of the strength of mixing both qualitative and quantitative approaches. Figure 17.1 illustrates this continuum. Integration is a term often used to denote mixing of qualitative and quantitative data: the data can be more or less integrated at different stages in the study design (Teddlie and Tashakkori 2009). At the low end of the scale is the use of more than one method in either a qualitative or quantitative study. A qualitative study and a quantitative study can be separate and influence one another, but data may never actually be mixed. For example, a study team might be studying use of the EHR by medical assistants. The team might conduct a study that is ethnographic in nature, involving interviews and observations. The team could simultaneously be conducting a national survey of medical assistants about computer use utilizing an instrument developed for nurses. Different stakeholders might have sponsored the two studies, so the results are reported separately. One study is qualitative, one study is quantitative, and there is some information shared between the two simply because study team members learn about medical assistants' tasks from both studies. These are separate studies but can be considered at the lowest end of the continuum. At the highest end of the continuum in Fig. 17.1 is the blending of both qualitative and quantitative methods throughout the research process in multiple steps, from data gathering to reporting and interpretation. In this chapter, the authors are defining mixed methods as the use of a combination of both qualitative and quantitative methods to create synergized understandings of targeted phenomena. Mixed methods studies can be anywhere along the continuum outlined in Fig. 17.1. Simple sequential studies and more

Fig. 17.1 Continuum of extent of mixing qualitative and quantitative methods

complex studies involving more mixing of data will be described in sections below. Data from any of these models can be mixed during analysis to a greater or lesser degree. If not mixed at all, the design would be multimethod and not mixed method. If mixed during every step, it would be a "fully integrated design." (Nastasi and Hitchcock 2016).

The use of mixed methods for evaluation purposes has been increasing rapidly. Many published studies that claim to use mixed methods are not truly mixed, however. The term "mixed methods" has been misused not because evaluation study teams are trying to be misleading, but because it has been confused with the term "multimethod" and its definition has evolved gradually. Mixing means that both quantitative and qualitative data are gathered and the analysis involves data from both, thus influencing the results together. A multimethod study can collect data sets that are never mixed, whereas some degree of combined analysis is needed for a study to claim to use mixed methods. In addition, the term is often used for studies that are qualitative but use more than one qualitative method or that are quantitative but use different quantitative approaches. The stricter definition emphasizing mixing quantitative and qualitative methods has developed and become clearer over time. Because mixed methods studies have become more common and guidance has been developed for doing them properly (NIH Office of Behavioral and Social Sciences 2018), the quality of mixed methods studies has improved dramatically.

17.3   Why Use Mixed Methods

As noted in earlier chapters, quantitative methods can answer questions related to generalizability and breadth, while qualitative approaches offer depth and context. If a planned project needs both, then a mixed method strategy is a good fit.

Mixed methods are also most appropriate for producing certain "products."

- The Rapid Assessment Process (RAP), also called rapid ethnography or quick ethnography, uses mixed methods to produce an assessment of local culture before an intervention. For further information about RAP, please see Chap. 16.
- Case studies are often the result of mixed methods strategies. For further information, see Chap. 15.
- Human computer interaction (HCI) laboratory and usability studies that are semi-quantitative can be augmented with ethnography. Usability studies will be described below.
- Organizational studies likewise can benefit from the use of coordinated qualitative and quantitative strategies. These studies will be discussed below.

In addition, mixed methods are ideal for applied research (Creamer 2017). This is probably because they can evaluate both process and outcomes to guide implementation projects as they unfold over time. If they are used iteratively, they can capture a sociotechnical understanding in a longitudinal manner.

Mixed methods have some downsides as well, however (Cresswell and Plano Clark 2017). They often take longer to complete than one method of inquiry alone. They are time consuming because with more complex sequential designs, they should be iterative. For each step, strategies and tools must be modified based on what has been learned. The entire team needs to be involved as changes are made and most team members will have other projects demanding their time, so the intensity of effort at different stages must be carefully planned. Such studies must be conducted by an interdisciplinary team, ideally with at least some team members with methodological expertise in three areas, not just one: qualitative, quantitative and mixed methods themselves. The team members with mixed methods expertise should have training and experience in analyzing mixed methods data. This is because to be truly mixed in nature, the study's analysis phases are more challenging than with multimethod inquiry; mixed data analysis is challenging. Double the resources required by either quantitative or qualitative strategies, including multiple data analysis software products, may be needed. All team members need to understand all approaches even if they are not experts in all three. There can be disagreements among team members who have their favorite methodologies and have not yet reached the desired level of understanding about the benefits of mixed methods (O'Cathain et al. 2008). Another downside is that it may be harder to convince granting agencies to fund and journals to publish mixed methods research. Human subjects committees (Institutional Review Boards) may need educating about them as well. Mixed methods researchers always need to provide more extensive rationales for using these methods than if only one paradigm is involved. These researchers must defend the rigor of both sides, thus expending more effort on that score. Condensing explanations of methods and results is challenging within page limits of journals and grant proposals. As all stakeholders become more accustomed to mixed methods, however, these challenges should decrease.

17.4   Designing Mixed Methods Studies

Especially when teams are evaluating programs rather than conducting research with a goal of publication, they may not design the evaluation in sufficient detail. Often such evaluations unfurl almost haphazardly. For example, someone thinks a survey of users to determine their satisfaction with a new tool would be a good idea. Only after the results are in does it become obvious that respondents are anxious to share their views and a follow-up interview study would have been easier if the survey had asked for volunteers. Taking time at the beginning when a decision is made to do an evaluation to plan how to do it best over time can save effort in the future. By following recommendations in prior chapters, especially Chap. 5, readers should be able to succeed in crafting plans suited to the relevant evaluation strategy.

Suppose a team has decided that a research question requires an answer about the extent of the issue as well as a depth of knowledge about the issue itself. An example would be a research question with a focus on how individuals access a patient

portal: a qualitative study might aim to discuss the issue with patients to gain understanding of multiple ways to access a portal, while a follow-up quantitative study
might be done to discover how widespread those ways to access are available
nationwide. An evaluation team may decide that mixed methods can supply the
answer, and if team members have experience using these strategies, the members
also know that each new project needs to be designed differently. Teams usually
cannot simply duplicate a strategy they have used in the past. Like the use of qualitative methods alone, there is no recipe to follow. There exist a number of different
typologies of mixed methods studies based on the priority or importance of using
qualitative vs. quantitative techniques, where one is given more weight than the
other. The timing of one or the other is considered as well. Below in Fig. 17.2 is a
simplified typology that can be useful in making decisions about mixed methods
research designs. There exist many variations of this framework, (Leedy and
Ormrod 2016; Morgan 2014) and Fig. 17.2 reflects the authors' best effort to explain
in one diagram the different possibilities for designing these studies.

   This typology stresses that both timing and priority factor into decision making
about the sequence and emphasis given to either qualitative or quantitative parts of
a mixed methods study. Sequential designs, in which one approach follows the
other, would be at the lower end of the continuum of extent of mixing methods

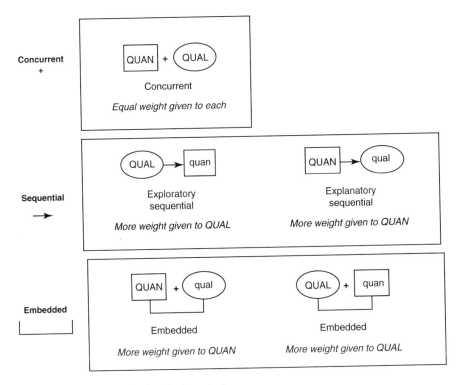

Fig. 17.2 Typology of mixed methods strategies

shown in Fig. 17.1 and would be considered fairly simple. A common and straightforward mixed methods design is called exploratory sequential, in which a qualitative design is done first, to explore an area about which little is known, followed by a survey that is developed based on the qualitative findings. If, in an exploratory sequential study, qualitative and quantitative methods are given equal effort in the extent of use of each approach, they are denoted as follows in capital letters: QUAL → QUAN. If one is emphasized more than the other, the lesser one is shown in small letters, qual → QUAN or QUAL → quan.

A second common design is one in which a quantitative study is done, followed by qualitative follow-up, perhaps to explain results of a survey, so it is called an explanatory design. As an example of an explanatory sequential study, one of the authors, with colleagues, developed a research question in 1998 when physician order entry (POE), now called computerized provider order entry (CPOE) was first available as a benefit of the EHR. The authors asked: what percent of hospitals have POE and how heavily is it used in those places? Results from a national survey indicated only 10% of hospitals claimed to have POE and there were few users (Ash et al. 1998). Informatics research indicated that POE could be very useful, so these results led to a research question asking why such a lauded tool was not widely disseminated. The answer could only be found by using qualitative methods to ask about and observe provider workflow. The team then crafted a research plan to conduct a follow-up qualitative study to determine both the upsides and downsides of CPOE (Ash et al. 2003). Unfortunately, the initial survey had not asked for volunteers to be interviewed, which would have expedited the qualitative phase of the study. This study sequence would be denoted as QUAN → QUAL. In another explanatory sequential study, Afable et al. analyzed survey results and then did a qualitative content analysis of student journal and instructor notes to assess technology-enhanced learning, (Afable et al. 2018) giving equal priority to each of the two phases of the study. Again, this was a QUAN → QUAL design, which is quite common. If emphasis is unequal in studies, the notation using the typology would be QUAN → qual or quan → QUAL.

Such sequential designs can become much more complex if both qualitative and quantitative methods are used in parallel, at the same time, and if several waves of one or the other occur as well. For example, in a study by Cohen et al. about patient entered data, a team first conducted a baseline survey and simultaneously analyzed data from the EHR and from the survey. Interviews followed. Interview data were converted to numbers, in a process called transformation of data, so that data could be compared to evaluate both process and outcomes measures quantitatively (Cohen et al. 2013). In another study that used multiple data sources, team members did pre- and post- surveys for an evaluation of a medical education mobile app, followed the pre-survey with focus groups to help interpret the results, and tracked usage data throughout the study (Davies et al. 2012). Another approach was used by a team that used mixed methods for developing case studies, in this instance to study processes using qualitative methods and outcomes using quantitative methods. They synthesized results using a matrix approach. This study team also very carefully outlined the rationale for using mixed methods (Sockolow et al. 2016).

An embedded design is another often-used design in evaluation efforts. A strategy with less emphasis is used simultaneously with one with greater emphasis. An example is a QUAN design with a qual design embedded within it. Often surveys of a primarily quantitative nature include open-ended questions designed to elicit qualitative text responses. Especially if the survey topic is one that stimulates respondents to answer open-ended questions with enthusiasm, the text answers can be extensive and analyzed qualitatively. An example is reported in a paper by Doberne et al. which explored pediatricians' satisfaction with EHRs (Doberne et al. 2017). Because the analysis process for embedded designs involves both kinds of data, allowing the qualitative comments to explain the quantitative answers, for example, embedded designs would be on the higher end of the continuum of complexity shown in Fig. 17.1.

Consideration needs to be given to the level of difficulty of the mixed methods: a simpler design will use fewer resources than a more complex design, and the budget will need to reflect the strategy. If only one type of qualitative and one type of quantitative strategy are selected, such as qualitative interviews followed by a survey, the study can be considered fairly simple. However, designs that involve a number of qualitative methods, perhaps interviews, observations, and content analysis of the minutes of meetings, in addition to several quantitative strategies such as user surveys and EHR data, all perhaps done more than once, the study will yield excellent data but will be quite challenging to manage. These are called cyclical or more advanced designs (Nastasi and Hitchcock 2016). As stressed in Chap. 5, a project manager will certainly be needed for such an ambitious project and intricate planning and evaluation of each step of the project must involve the entire team. For this kind of project, a flow diagram to explain the steps and a project management approach using project management tools can keep everyone informed of plans and changes.

Self-Test 17.1
With reference to Fig. 17.2 above:

1. Select a notation from the typology that would apply to a study in which a study team first conducted a national survey of hospitals to find out how many were using barcode medication administration systems and then the team selected ten of these hospitals to visit to collect data using interviews and observations.
2. What notation would you use to describe a study in which a team conducted four visits to sites in the U.S. using virtual scribe services abroad so the team could then design a large-scale survey about such services?
3. Three different surveys were administered to providers across a large hospital system before, during, and after a major EHR system upgrade. The evaluation team also conducted a series of ten interviews during each of those time periods as well. What notation in the typology would explain the design?

17.5   Data Collection Strategies

What is unique about data collection strategies for mixed methods studies is that, once again, planning must be especially careful and the planning requires input from experts in qualitative, quantitative, and mixed methods. If a survey is proposed as a first step, consider asking respondents to volunteer to be interviewed during a follow-up qualitative study. If a qualitative study about an intervention is to be first, followed by a survey to determine the breadth of use of the intervention, plan to have qualitative results influence development of survey questions. Iteratively check to assure that the study team's planned strategy still makes sense. For example, in team meetings, ask if preliminary results are answering your research questions or if you need to modify your strategy. When planning for data collection, carefully consider developing backup plans in case one of the steps becomes impossible to do. For example, if an evaluation plan is designed to assess the implementation of three successive and related informatics tools but the funding source for the third tool in the sequence is not forthcoming, the third phase of the evaluation will become impossible. The sponsors and team members will need to decide whether to halt their work, do further evaluation of the first two tools, and/or perhaps study what happened to derail the original implementation plans.

17.6   Data Analysis Strategies

Prior chapters have explained how to conduct data analysis for quantitative and qualitative evaluation efforts. For a mixed methods study, analysis should go beyond these separate analysis attempts, however. There must be an effort to mix data, to integrate data. This mixing is the strength of mixed methods, so team members should carefully plan how to do it. Once again, there is no recipe, so high degrees of creativity and judgement are needed. Quantitative analysis may be conducted separately from qualitative analysis and the results transformed so they can be blended. Numbers can be assigned to qualitative findings, by counting words, for example. Often, graphics can assist the research team in making sense of, for example, one set of qualitative results and one set of quantitative results. The two might be compared side by side in a table and the team members might evaluate how and why the results are alike or different. There are times when it makes most sense to transform qualitative data to numbers so the numbers can be integrated with quantitative data. For example, answers to interview questions could be categorized as strong vs. weak or high vs. low on a scale. The team studying patient-entered data in the study cited above by Cohen et al. conducted a baseline survey, then collected EHR data, and then did interviews and transformed the interview data (Cohen et al. 2013) so all phases could be compared. Another excellent fully integrated analysis was conducted by the team members who did the medical education mobile app study: they developed a matrix of both quantitative and qualitative aspects side by side (Davies et al. 2012).

There are no recipes for presenting the results of mixed methods studies. Possibilities include: tables such as those described above; quantitative results explained with examples from qualitative data; counts of the number of instances of patterns that arose in text data; and many other creative depictions depending on the purpose of the study. Such tables are often used as part of the analysis process, and then they are nearly always used when presenting results and contributing to discussions in papers.

Food for Thought

With reference to Fig. 17.2 above and the case example described in Sect. 15.1 about planning an evaluation of a patient portal implementation:

1. Design a mixed methods evaluation study that uses a QUAL → quan design. Why would you phase it this way?
2. Design a mixed methods evaluation study that uses a QUAL + QUAN design. How would you blend data during analysis?

17.7   Presenting Results of Mixed Methods Studies

Guidance is available for assessing mixed methods proposals and papers (Hong et al. 2018; Creswell 2016). In essence, strategies for rigor are similar to those for either qualitative or quantitative studies, but more emphasis is placed on the description of how the two will be integrated.

The first step in writing a report is to determine the audience, of course. If the evaluation was conducted for an internal quality improvement initiative, the stakeholders may include both clinical users and quality and administrative staff and only an internal report is needed. If publication in a research journal for broad dissemination is planned, different audiences and formats must be considered. Further guidance about reporting results is offered in Chap. 19.

When reporting a mixed methods study, a diagram of the research design is crucial. It needs to be made clear in the diagram where methods interfaced and how the mixing was done. The design of mixed methods displays is critical so that audience members can quickly visualize what was done and in what order (Guetterman et al. 2015). Figure 17.3 clearly indicates the phases in an exemplar hypothetical project, its iterative nature, and the points of analysis, both independent and mixed. If the focus of the evaluation was a patient portal, the study team might decide to conduct concurrent, equally weighted, quantitative and qualitative studies, gathering usage data from the system itself and interview data from both users and non-users of the portal, and mix the data to produce information on which to base development of a survey instrument. The survey analysis phase itself might then include mixing quantitative data from the primarily quantitative survey instrument with qualitative text data from open ended question responses that were included in the survey.

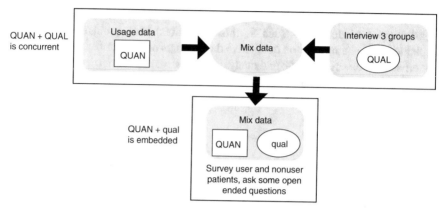

Fig. 17.3 Example of a mixed methods evaluation design diagram

17.8   Usability Studies as Exemplars of Mixed Methods

17.8.1   What Is Usability?

Usability is defined as "how useful, usable, and satisfying a system is for the intended users to accomplish goals in the work domain by performing certain sequences of tasks." (Johnson et al. 2011, p. 9). In usability testing, study teams gather varied types of data over time, often in novel and creative ways using mixed methods, with a goal of improving an information resource for the humans using it. Usability evaluation of information resources includes assessment of effectiveness, efficiency, and satisfaction.

Because study teams can assess usability using a broad array of methods, and because usability is affected by so many variables related to the environment within which an information resource is deployed, many disciplines contribute to the tradition and methods of usability studies. Usability rests on the field of human-computer interaction (HCI). Usability also rests on the discipline of human factors engineering, which is concerned with the design of systems taking into account the characteristics of humans.

This discussion will focus on usability assessment throughout the resource development life cycle. Usability studies must take into account the need for the resource, which ideally has been previously described. This is often done using mixed methods. Usability assessment then requires testing of different versions of resource prototypes, bringing into play methods that typically mix qualitative and quantitative, and are highly iterative as the results of usability studies contribute to rapid cycles of design, testing and revision. After deployment of a resource, usability testing remains important because a resource that is usable in the lab may perform differently in the field.

17.8.2 Importance of Usability

The history of informatics includes many examples of the development of resources before the importance of usability was recognized. In particular, many EHRs include vestiges of older systems designed for managing financial transactions or developed when programming capabilities were limited, so the technology drove design more than human needs did (Ash et al. 2003). Now that EHRs are widely deployed, many such systems continue to suffer from these historical limitations (Doberne et al. 2017). Usability, or lack thereof, has been blamed in large part for contributing to provider burnout. It has also been a cause of medical errors or near errors (Ash et al. 2003). The proliferation of consumer facing and mobile applications increases the potential for usability testing to become more widely used for both patient and health care professional-oriented systems. The maturation of informatics as a discipline will benefit as usability testing becomes an increasingly integral part of the selection, modification or development, implementation, and further assessment of all information resources.

17.8.3 Usability Testing Methods

Qualitative methods such as those described in Chap. 15 are frequently used in the early stages of resource development to inform design and validate the need for a system. Then, during the iterative design stage, some rather unique methods can be used, which include the following four general types of studies.

1. _Heuristic inspection_: Heuristic inspection, which engages a small group of reviewers who examine a system interface against a set of guidelines, can be a helpful early step in usability assessment. For this purpose, reviewers may employ Nielsen's heuristics, which include: visibility of system status; match between system and the real world; user control and freedom; consistency and standards; error prevention; recognition rather than recall; flexibility and efficiency of use; aesthetic and minimalist design; and promoting users' ability to recognize, diagnose, and recover from errors (Nielsen and Mack 1994). Heuristic inspection aligns closely with the qualitative professional review approach to evaluation introduced in Chap. 2.

2. _Laboratory studies_: Laboratory usability studies bring prospective users into a controlled environment where their interactions with a resource can be highly instrumented. These studies can address quantifiable metrics, assessing attributes such as: time to complete a task, percentage of each task completed, ratio of success to failure, number of errors, number of clicks, and number of backtracks. Data sources for these assessments include key stroke logs and other back-end tracking data generated by the resource itself. Eye tracking is useful for

understanding where the user's attention is focused when completing tasks. More qualitative methods include cognitive walkthrough, talk-aloud, or think-aloud techniques (Kushniruk et al. 1997). The results of all methods employed, some quantitative and some qualitative, can be mixed to reach conclusions about recommendations for improvements. A drawback to laboratory evaluation for health-related resources is the lack of contextual realism. The environment for testing, for example, is not the hectic and distracting environment where health care often takes place.

3. _Simulation:_ Simulations are an important variant on laboratory studies. While still undertaken outside of actual work environments, simulation studies seek to reproduce as much as possible the features of these environments that can profoundly affect resource usability (Kim and Park 2016). For example, usability of an information and communication resource to transmit clinical data and instructions between "first responders" in an ambulance and emergency department personnel can be tested with a mannikin of a patient in a real, but stationary, ambulance.

4. _Naturalistic settings:_ The fourth type of testing takes place in the setting where an information resource is deployed and where "real work" is being done. Many of the methods of laboratory studies, described above, can be used, with care taken to minimize disruption of the work process. A novel approach is a mobile usability study in which—with appropriate care taken—audio, video, and eye tracking follow cognitive walkthroughs in a real environment such as a hospital unit. (Kushniruk 2017) These data can be mixed with data collected by the information resource itself to provide a highly realistic picture of how users interact with a resource, the problems they encounter, and how these might be addressed.

Of course, this continuum from heuristic evaluation to evaluation in natural settings ranges from simple to much more complex studies. The point here is that the combination of methods requires mixing of different kinds of both qualitative and quantitative data in creative ways. More often than not, the quantitative data resulting from usability studies are not abundant enough to warrant the kind of statistical testing outlined in prior chapters. The selection of participants for usability testing, like that for purely qualitative studies, can be purposive and focused on including participants likely to provide the most valuable information rather than the often-large numbers required for statistical inference.

_17.8.4_ _Examples of Usability Studies in Informatics_

Informatics studies reporting results of usability evaluations in informatics range from fairly simple laboratory studies to complex arrays of studies conducted throughout the life cycle of development and implementation of an intervention.

What the following studies have in common is use of more than one method, the mixing of data during analysis and interpretation, and development of tables and figures that clearly report results. They are exemplars demonstrating the value of mixed methods in the usability domain.

An example of a relatively simple study combined data from eye tracking and task completion exercises with talk alouds to evaluate an HIV prevention mobile app, identifying potential sources of error (Cho et al. 2019). Also evaluating a mobile app, this one for stroke caregiver support, another study described a broad user centered design approach that began with focus groups and interviews, but then included usability studies during and after implementation (Caunca et al. 2020). The usability evaluation was conducted in the field with subjects using the app and completing multiple surveys. Data from the qualitative and quantitative phases were compared on multiple occasions and summarized in the report in matrices and graphs. Another study produced results evaluating a touch screen application for patient waiting room use so individuals could enter data into a questionnaire about asthma. The evaluation team did qualitative work and analyzed the data using both a top-down usability typology-focused template approach and a bottom up inductive more grounded approach focused on subjects' own phrasing as described in Chap. 16. This aspect was enhanced with laboratory testing of a prototype. As an example of the value of subjects first using the app and then being interviewed about it (which often raises issues that are not predicted by experts), the users indicated greater than expected concerns about the cleanliness of the tablets and touch screens (Cheung et al. 2020).

In summary, usability studies offer the opportunity to be creative with methods depending on the questions driving the evaluation, stage of the resource life cycle, and target population. While the basics have not changed since the recognition of the importance of usability studies in informatics (Nielsen and Mack 1994), newer technologies for recording, eye tracking, presenting simulations, and gathering data from the system itself in the background allow studies to be conducted in an increasingly more naturalistic manner.

Self-Test 17.2
1. What is the difference between usability, human-computer interaction, and human factors engineering?
2. Why do you think usability testing is not used more often before implementation of clinical information systems?
3. Which of the laboratory evaluation methods described above would produce text data that could be analyzed using qualitative data analysis methods?
4. Describe how a usability study, using simulated methods that are as realistic as possible, could be done for an information resource that supports critically ill patients in a cardiac intensive care unit.

17.9   Organizational Systems-Level Studies

17.9.1   What Are Organizational Systems-Level Studies?

Just as usability studies provide an excellent example of the use of mixed methods at the level of the individual and technology use, studies at the organizational level offer examples of the use of mixed methods in informatics to explore the use and impact of information resources when individuals work together as part of systems at multiple levels of scale. Levels of scale range from small groups, such the care team in an intensive care unit, to organizations such as hospitals. These organizations can be part of larger organizations, such as a healthcare delivery network (Friedman et al. 2015). Those systems can be part of even larger systems at national scale—the United Kingdom's National Health Service, for example—and these systems form a global network of health care interconnected through the World Health Organization. Also, all organizations routinely interact with other organizations. Teams that study organizations employ methods drawn from psychology, social psychology, sociology and anthropology (Robbins and Judge 2017).

As they apply to biomedical and health informatics, organizational studies can illuminate factors that explain the success or failure of information resources that are widely deployed, These studies can be focused at different levels of scale (Harrison and Shortell 2021) and at the group level of scale, studies might investigate effects of information resources on team cohesion, work processes, and communication. At the level of entire organizations, studies might address organizational structure and function, culture, policies, leadership, and change—either as predictors or of areas affected by the deployment of information resources.

Like usability studies, organizational and system studies make frequent use of mixed methods. Also, there are many distinctive characteristics of organization and system-level studies that study teams must address as part of the study design process. First and perhaps foremost, data to study phenomena at a particular level of scale are often collected at smaller levels of scale. For example, a study focused on teamwork might collect data at the individual level by individually interviewing the members of a group (a qualitative method) or asking individual members to complete attitude assessment questionnaires (a quantitative method). Of arguably equal importance, it is very difficult to manipulate large organizations in pursuit of controlled studies, so organizational and system level studies tend to employ correlational methods (as discussed in Chap. 13) on the quantitative side and the full range of qualitative methods discussed in Chaps. 14, 15, and 16.

17.9.2   Methods Used in Organizational and System-Level Studies

Many methods are used in conducting organizational system-level studies. The following are common methods, often used as parts of mixed methods studies (Robbins and Judge 2017).

1. *Field surveys*. These can be surveys of individuals responding with their own viewpoints, which are then aggregated to generate group or system level measures. On other occasions, a single respondent from an organization generates responses to a survey on behalf of the entire organization; for example a survey of EHR adoption across an entire hospital. When Ash et al. surveyed a sample of 1000 U.S. hospitals about CPOE, the study team sent a survey instrument to the chief medical information officer (CMIO), or someone in a similar position, at each hospital. The survey asked factual questions about commercial systems that were purchased by the hospital, rather than questions that sought the opinions of individual CMIOs (Ash et al. 1998)
2. *Field experiments*. These are interventional studies of the impact of informatics interventions at organizational scale. In ICT, these might entail implementing a system in some organizations and not others and comparing the two experiences. As noted above, these studies are challenging because it is difficult to allocate "treatments", and even more so to randomly allocate them, at the organizational level. The step-wedge design, described in Sect. 11.3.6, is one mechanism to conduct randomized field experiments.
3. *Meta-analyses*. These are comparisons of results of studies already conducted, as described in Chap. 5. Aggregations of studies at lower levels of scale can often lead to insights at higher levels.
4. *Analysis of existing data*: In informatics, studies using already existing data from the EHR fall into this category. Although EHR data are not always accurate and there are often issues related to aggregating the data, such data are becoming more available and more reliable. Additionally, the data from national surveys sponsored by government entities and foundations are often in the public domain and can be analyzed or re-analyzed to address important questions.
5. *Case studies*: Case studies at the level of one organization can be extremely useful if they are relevant to stakeholders within the organization and elsewhere. Massaro's classic case study describing what went wrong when CPOE for medications was implemented at the University of Virginia hospital had a profound impact on the field of informatics, not only because it was one of the first failure stories to be published but also because the analysis was so skillful and readers could relate to the issues (Massaro 1993).

Another example of a case study that was carefully planned as a mixed methods study was conducted by Spectrum Health when a new EHR system was being implemented across three hospitals in the healthcare system. User surveys were conducted before, during, and after implementation and the results were mixed with data from interviews done during multiple site visits. This was a longitudinal mixed methods study that influenced the implementation throughout the EHR life cycle (McMullen et al. 2015). In an example of a study at a higher organizational level, secondary data from an annual survey conducted by the American Hospital Association provided some information and primary data from a survey of hospital representatives with knowledge about the advanced capabilities of the hospitals' EHR systems were combined to find associations between organizational attributes, such as the size of ICT departments, and level of EHR capabilities (Holmgren et al. 2021).

In summary, evaluation studies of one organization can provide feedback to that organization to improve processes, whether they focus on implantation of informatics tools or of changes in patient care procedures. Evaluation studies at levels beyond one organization, to include organizations within a nation or even beyond, can discover important trends and best practices that might constructively influence policy. The methods described in prior chapters can be tailored to any organizational level. Mixed methods studies can be especially useful for answering evaluation questions about the complex issues involving informatics at the organizational and larger organizational system levels.

17.10   Special Ethical Considerations in Mixed Method Studies

Chapter 20 describes a number of ethical issues in evaluation studies in general. However, there are several special ethical issues related to mixed methods research in addition to those for either quantitative or qualitative designs conducted separately. If a survey is the first step in a mixed methods evaluation, the team may need to gather identifying information if qualitative follow-up is the next step. This means that the identity of survey subjects must become known, at least for those who volunteer. Mixed methods may place a higher burden of effort on subjects, taking more of their time, if they are to be queried more than once in a longitudinal study. Mixed methods study teams may need to provide additional justification to the IRB for the extent of the time commitment for involvement of human subjects over time, especially if monetary incentives are involved. Any time incentives are offered, they must be carefully planned so that they are effective in recruiting appropriate subjects but they cannot be construed as extortion. In mixed methods studies, in which subjects may be asked to expend quite a bit of time and effort, greater incentives may be needed and justified to sponsors and the IRB. Team members must also decide if they want to get approval for the whole study or submit approval requests in phases (e.g. do multiple submissions or modifications). Finally, qualitative team members will need to educate all of the stakeholders and all of the participants about human subjects issues related to all of the parts of the mixed methods study.

17.11   The Future of Mixed Methods Inquiry

The use of mixed methods is increasing quickly, along with guidance for conducting it skillfully (Hesse-Biber and Johnson 2015). Mixed methods studies will likely also be used at an increasing rate in health settings because of their ability to blend information collected using different strategies in a holistic and rigorous manner. They will be used increasingly in health informatics as well, since they are so well suited for evaluating this domain. They can cover all sociotechnical dimensions as defined by Sittig and Singh (Sittig and Singh 2010) from multiple points of view. They can provide ongoing feedback and evaluation so that the intervention under review can be modified and improved continuously. They can also produce a review of outcomes based on the most relevant variables as determined by the ongoing feedback. Because mixed methods are so critical in informatics for studying sociotechnical issues, more training and recognition of their importance are needed. Most informatics educational programs today do not include mixed methods training aside from superficial mention of it in qualitative and quantitative methods courses. While assuring that study teams include a combination of both qualitative and quantitative experts, this is no substitute for having team members well versed in strategies outlined in this chapter. It is especially important that someone on the team knows how to analyze the data in an integrated manner. Teamwork for evaluation efforts is becoming increasingly important as well, so training for effective teamwork can also strengthen mixed methods research.

Other trends include more attention to methods for managing data from both qualitative and quantitative data gathering approaches during the analysis process, for designing cyclical designs, for developing new types of graphical depictions (even using geographic information systems, for example), and for use of software that helps more with mixed methods. At present, the more sophisticated qualitative analysis software systems can accommodate some quantitative analysis and this trend will likely continue as mixed methods become even more popular.

Answers to Self-Tests

Self-Test 17.1
1. This would be a QUAN → QUAL study because it is sequential with equal emphasis given to each part.
2. This would be qual → QUAN since the study would be sequential with emphasis on the quantitative part.
3. This would be a QUAN + QUAL study because the parts were concurrent and given equal weight

Self-Test 17.2
1. The difference between usability, HCI, and human factors engineering is that usability evaluation methods can be used to assess human-computer interaction and human factors engineering concepts. In other words, usability evaluation provides strategies and techniques that are often used by study teams specializing in either HCI or human factors engineering.
2. Usability testing may not be used more often because there are not many experts available to promote it and the detailed methods are not generally taught in informatics teaching programs. Also, this kind of testing takes time and careful planning and developers and implementers may not want to take the necessary time.

3. Of the techniques that can be accomplished in the laboratory setting, cognitive walkthrough and think aloud methods seem like techniques that would produce text. If recordings were made while subjects explained what they were doing as they interacted with the computer system, the recording could be transcribed and the resulting text analyzed qualitatively.
4. This would be an excellent opportunity to use a mobile usability lab set-up, placed in the appropriate unit, to test how the environment might impact use of the information resource.

References

Afable MK, Gupte G, Simon SR, Shanahan J, Vimalananda V, Kim EJ, et al. Innovative use of electronic consultations in preoperative anesthesiology evaluation at VA medical centers in New England. Health Aff. 2018;37(2):275–82.

Ash JS, Gorman PG, Hersh WR. Physician order entry in U.S Hospitals. AMIA Proc. 1998;1998:235–9.

Ash JS, Gorman PG, Lavelle M, Payne TH, Massaro TA, Frantz GL, Lyman JA. A cross-site qualitative study of physician order entry. J Am Med Inform Assoc. 2003;10(2):188–200.

Caunca MR, Hartley SM, Wright CB, Czaja SJ. Design and usability testing of the stroke caregiver support system. A mobile-friendly website to reduce stroke caregiver burden. Rehab Nurs. 2020;45(3):166–77.

Cheung VLS, Kastner M, Sale JEM, Straus S, Kaplan A, Boulet LP. Development process and patient usability preferences for a touch screen tablet–based questionnaire. Health Inform J. 2020;26:233–47.

Cho H, Powell D, Pichon A, Kuhns LM, Garofalo R, Schnall R. Eye-tracking retrospective think-aloud as a novel approach for a usability evaluation. Int J Med Inform. 2019;129:366–73.

Cohen AN, Chinman MJ, Hamilton AB, Whelan F, Young AS. Using patient-facing kiosks to support quality improvement at mental health clinics. Med Care. 2013;51(3 Suppl 1):S13–20.

Creamer EG. Introduction to fully integrated mixed methods research. Thousand Oaks, CA: Sage; 2017.

Cresswell JW, Plano Clark VL. Designing and conducting mixed methods research. 3rd ed. Thousand Oaks, CA: Sage; 2017.

Creswell JW. Reflections on the MMIR: the future of mixed methods task force report. J Mixed Methods Res. 2016;10(3):215–9.

Davies BS, Rafique J, Vincent TR, Fairclough J, Packer MH, Richard Vincent R, et al. Mobile medical education (MoMEd) - how mobile information resources contribute to learning for undergraduate clinical students - a mixed methods study. BMC Med Ed. 2012;12:1.

Doberne JW, Redd T, Lattin D, Yackel TR, Eriksson CO, Mohan V, Gold JA, Ash JS, Chiang MF. Perspectives and uses of the electronic health record among United States pediatricians: a national survey. J Ambul Care Manag. 2017;40(1):59–68.

Friedman CP, Rubin J, Brown J, Buntin M, Corn M, Etheredge L, et al. Toward a science of learning systems: a research agenda for the high-functioning learning health system. J Am Med Inform Assoc. 2015;22:43–50.

Guetterman TC, Fetters MD, Creswell J. Integrating quantitative and qualitative results in health science mixed methods research through joint displays. Ann Fam Med. 2015;13:554–61.

Harrison MI, Shortell SM. Multi-level analysis of the learning health system: integrating contributions from research on organizations and implementation. Learn Health Sys. 2021;5:e10226.

Hesse-Biber SN, Johnson RB, editors. The Oxford handbook of multimethod and mixed methods research inquiry. Oxford, UK: Oxford University Press; 2015.

Holmgren AJ, Phelan J, Jha AK, Adler-Milstein J. Hospital organizational strategies associated with advanced EHR adoption. Health Serv Res. 2021. https://doi.org/10.1111/1475-6773.13655.

Hong QN, Pluye P, Fàbregues S, Bartlett G, Boardman F, Cargo M, et al. Mixed Methods Appraisal Tool (MMAT), version 2018. Registration of Copyright (#1148552), Canadian Intellectual Property Office, Industry Canada, 2018.

Johnson CM, Johnston D, Crowley PK, et al. EHR usability toolkit: a background report on usability and electronic health records (Prepared by Westat under Contract No. HHSA 290-2009-00023I). AHRQ Publication No. 11-0084-EF. Rockville, MD: Agency for Healthcare Research and Quality; 2011.

Kim J, Park JH. Effectiveness of simulation-based nursing education depending on fidelity: a meta-analysis. BMC Med Educ. 2016;16:152.

Kushniruk. The future of mobile usability, workflow, and safety testing. Stud Health Technol Inform. 2017;245:15–9.

Kushniruk AW, Patel VL, Cimino JJ. Usability testing in medical informatics: cognitive approaches to evaluation of information systems and user interfaces. AMIA Proc. 1997;1997:218–22.

Leedy PD, Ormrod JE. Practical research: planning and design. 11th ed. Boston, MA: Pearson; 2016.

Massaro TA. Introducing physician order entry at a major academic medical center. Acad Med. 1993;68:20–5.

McMullen C, Macey T, Pope J, Slot M, Lundeen P, Ash JS, Carlson N. How computerized provider order entry affects pharmacy: the Spectrum health experience. Am J Hosp Pharm. 2015;72(2):133–42.

Morgan DL. Integrating qualitative and quantitative methods: a pragmatic approach. Thousand Oaks, CA, Sage; 2014.

Nastasi BK, Hitchcock JH. Mixed methods research and culture-specific interventions: program design and evaluation. Thousand Oaks, CA: Sage; 2016.

Nielsen J, Mack RL, editors. Usability inspection methods. New York: Wiley; 1994.

NIH Office of Behavioral and Social Sciences. Best practices for mixed methods research in the health sciences. 2nd ed. Bethesda, MD: National Institutes of Health; 2018.

O'Cathain A, Murphy E, Nicholl J. Multidisciplinary, interdisciplinary, or dysfunctional? Team working in mixed-methods research. Qual Health Res. 2008;18:1574–85.

Robbins SP, Judge TA. Organizational behavior. 17th ed. Boston, MA: Pearson; 2017.

Sittig DF, Singh H. A new sociotechnical model for studying health information technology in complex adaptive healthcare systems. Qual Saf Health Care. 2010;19(Suppl 3):i68–74.

Sockolow P, Dowding D, Randell R, Favela J. Using mixed methods in health information technology evaluation. Stud Health Tech Inform. 2016;225:83–7.

Teddlie C, Tashakkori A. Foundations of mixed methods research: integrating quantitative and qualitative approaches in the social and behavioral sciences. Thousand Oaks, CA: Sage; 2009.

Chapter 18
Principles of Economic Evaluation and Their Application to Informatics

Learning Objectives
The text, examples, and self-tests in this chapter will enable the reader to:

1. Explain the purposes of economic studies as they apply to informatics.
2. Given an scenario requiring a cost study, determine which type of study is best suited to the study's purpose, which perspective(s) should be taken into account.
3. Distinguish between costs and charges, and between outcomes and benefits.
4. Given a table of costs and outcomes for a set of alternatives, determine which alternative yields the greatest cost-effectiveness.
5. Apply the concept of discounting to an economic analysis.
6. Describe the purpose of sensitivity analyses and the differing strategies for conducting them.
7. Describe the methods used to critique economic evaluations, and how these studies can be improved

18.1 Introduction to Economic Studies

This chapter returns to a primarily quantitative mode of thinking to discuss costs, and particularly how issues of resource cost and the various ways to represent outcomes can be factored into the logic of an evaluation. Although largely linked to quantitative evaluation approaches, this chapter has been placed toward the end of this volume because economic issues seem to span all evaluation approaches. Studies that are qualitative in approach could incorporate a formal cost analysis selected from the methods described below. While the focus and examples in this chapter address clinical issues because the the techniques described are mainly applied to clinical interventions and outcomes, the methods introduced apply across all of the application domains of biomedical and health information resources such as research or education.

© Springer Nature Switzerland AG 2022
C. P. Friedman et al., *Evaluation Methods in Biomedical and Health Informatics*, Health Informatics, https://doi.org/10.1007/978-3-030-86453-8_18

18.1.1 Motivation for Economic Analysis

The dramatic rise of health care expenditure over the past few decades has focused attention on health care resource allocation. For example, the United States spends close to 18% of its gross domestic product (GDP) on health care, which is $11,582 per person or just over $3.8 trillion, 2.7 times the figure quoted in the 2006 edition of this book (United States Centers for Medicare and Medicaid Services (CMS) 2021). Other nations spend from 5% to 14% of their GDP on health care, and the percentage of GDP directed toward health care expenditure in all countries continues to rise. These rising costs have prompted substantial efforts to reduce the ever-increasing component that health care represents in national budgets. However, technology and medical advances continue, and at a time when federal, state, and private insurance payers are cutting back reimbursement, there are continuing expansions in new pharmacological agents, diagnostic tests, medical devices and other technologies. At a more local level, hospitals, health care providers, and care-givers are faced with increasing costs and rising demand for new services and treatments, but have limited (and sometimes shrinking) budgets with which to meet those demands. Finally, there has been an increase in the role of patients in health care decisions and self-management actions, which may also require a more detailed understanding of costs and benefits and the response of consumers to changes in prices.

Therefore, at several levels there has been increasing interest in understanding the economic impact of decisions that administrators, clinicians, and health care systems make regarding which programs to institute and what services or interventions to offer. The overall goal of these investigations is to ensure that the financial resources expended at a national, regional, or local level are spent wisely, and that the decision maker is maximizing the benefit of the money spent to provide health care services. All financial resource expenditure decisions involve choices: monies spent in one area cannot be spent in another, and putting financial resources into a particular program (meaning any activity designed to maximize health and minimize ill health, rather than software development) implies that the program has more value than other potential programs that were not funded.

For example, a hospital must decide how to distribute limited financial resources between several competing demands for new services or programs. Should the hospital purchase a suite of new clinical dashboards that make it easier to carry out rapid learning cycles and implement a Learning Health System, in the hope of reducing error, and improving quality—or should it place those financial resources into equipment to conduct whole genome sequencing of bacterial isolates to improve the organization's ability to find hospital acquired infections? (Kumar et al. 2021) At a finer level of detail, departments within the hospital face similar decisions. Should the information systems division purchase new tablet computers and applications because of increasing complaints of long wait times for laboratory results in busy clinical areas, or spend its financial resources on improving the coverage of wireless networking?

The motivation for all economic analyses is to place decisions such as these in a rigorous, analytic framework. This will include all the relevant components important to the decision makers, to allow a pre-decision estimate of the consequences (in both financial resources and other outcomes) of the decision before the decision is actually taken. This chapter briefly introduces the principles of economic analysis as applied in health care, focusing on the components that are unique to biomedical and health informatics and clinical information systems.

18.2   Principles of Economic Analysis

Figure 18.1 illustrates a generic description of the choices facing a decision maker and describes the outcomes that must be simultaneously evaluated in the economic analysis of health care programs. The decision maker must choose whether to implement an innovation, which might be a clinical information resource: each choice implies a different stream of costs and outcomes. The purpose of any economic analysis is to make a quantitative comparison between the stream of costs and benefits that arise from each of the possible options being compared. The idea is to highlight the tension between costs and outcomes by calculating the costs required to achieve a given change in outcomes. It is important to emphasize that economic analyses can rarely, if ever, indicate the "correct" choice; they can only estimate the likely economic and clinical consequences of various choices. Whether a particular outcome gain is worth the costs involves political, ethical, and other concerns that are specific to the situation.

There has been increasing interest in the United States and abroad in the development of formal, consistent standards for the conduct and reporting of economic analyses in health care. In the United States, prompted by the realization that economic analyses were quite variable and showed differing adherence to even minimal standards of analysis and reporting (Udvarhelyi et al. 1992), the Public Health Service commissioned the Panel on Cost Effectiveness in Health and Medicine, which consisted of experts who developed a series of recommendations regarding the conduct of cost-effectiveness analyses in health care (Gold et al. 1996). Their recommendations, although not entirely complete, have become the *de facto*

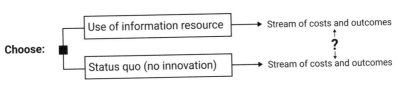

Fig. 18.1   Basic structure of an economic analysis. In general, all types of economic analyses follow this structure. A new program or strategy that can be introduced to replace an existing strategy produces a stream of costs and outcomes that are different from the stream of costs and outcomes produced by the alternative

standards for conducting, analyzing, and reporting cost-effectiveness analyses in the United States. They have now been supplemented by a second panel which met from 2012 and in 2016 recommended a set of standard methodological practices that all cost-effectiveness analyses should follow, to improve study quality and comparability: "*All cost-effectiveness analyses should report 2 reference case analyses: one based on a health care sector perspective and another based on a societal perspective. The use of an "impact inventory," which is a structured table that contains consequences (both inside and outside the formal health care sector), intended to clarify the scope and boundaries of the 2 reference case analyses is also recommended*" (Sanders et al. 2016).

In the United Kingdom, a similar effort and standards have been put forward by the National Institute for Clinical Excellence (NICE), which carries out cost-effectiveness studies on novel medical technologies using a procedures manual published on its Web site: www.nice.org.uk. Over the past 20 years the NICE organization has completed hundreds of effectiveness and cost-effectiveness analysis of medical procedures, which are used in coverage and reimbursement decisions by the U.K. National Health Service and other health services around the world. Some of these evaluate the cost effectiveness of information resources, such as online tools or applications to improve the mental health of patients or software that helps avoid repeated liver biopsies by assisting the interpretation of liver ultrasound scans (United Kingdom National Institute for Health and Care Excellence (NICE) 2020).

18.2.1   Types of Cost Studies

There are several different types of economic evaluation that are used to compare the effects of different choices and options in a health care setting. Table 18.1 provides a brief description of the various types of analysis and how they differ in terms of the measure of outcomes; all use the local currency to express costs. The characteristic that differentiates most economic analyses is the metric used to evaluate the benefits (Doubilet et al. 1986). In the simplest case, an analysis of the benefits goes no further than to assume the clinical effect of the possible strategies is the same; the only difference arises from difference in costs. This type of analysis is termed a cost-minimizing or cost minimization study, because with outcomes assumed equal, the only relevant goal would be to minimize costs. The most common clinical example of a cost-minimizing study would be an evaluation of a therapeutic drug substitution program in which the efficacy of the two drugs was equivalent and the analysis looks only at the costs of administering and monitoring the drug.

Cost-consequence studies are only slightly more complicated, as they list the costs (e.g. in dollars) and outcome in whatever units are appropriate for the specific situation being evaluated. For example, suppose new upgrades are being proposed for a wireless network to try to improve both network performance (perhaps measured in loading time for an electronic record of a standard size) and reliability (measured in expected number of minutes of unscheduled downtime per month). A cost-consequence study would simply generate lists, such as option A will cost

Table 18.1 Types of economic evaluation in health care

Type of analysis	Description	Measurement of outcome	Measurement of cost
Cost minimizing	Assumes outcomes are equivalent	None	Monetary amount; e.g. dollars
Cost consequence	Lists the costs and outcomes (even if multiple endpoints are used) of each option	Variable, possibly multiple	
Cost-effectiveness	Measures outcomes in clinical terms (lives, life expectancy, number of infections averted)	Clinical outcome	
Cost utility	Measures outcomes in utilities: measures of the preferences for the outcome state	Quality-adjusted life years (QALYs)	
Cost-benefit	Requires that outcomes (lives, quality of life) be given a monetary value	Monetary amount; e.g. dollars	

$238/clinical unit, will require 6.8 milliseconds to load a chart, and will have 35 min of unscheduled downtime per month; whereas option B will cost $312/clinical unit, will require 3.2 milliseconds to load a chart, and will have 41 min of unscheduled downtime each month. No attempt is made to equate the value of downtime and speed; the results are presented to allow the decision makers to place their own value on the relative worth of the various outcomes and the trade-off between them.

Cost-effectiveness analysis (CEA) is defined by the use of a clinically relevant outcome to quantify the benefits, and is consequently the most common type of economic analysis used in health care. The clinical effectiveness measure may span a wide range, from global outcomes such as lives saved to limited outcomes such as number of infections avoided. The critical aspect is that all the alternatives being evaluated are measured using the same outcome and the same metric. For example, a new component of a clinical information system may reduce the number of duplicate laboratory tests and x-rays ordered, but may cost financial resources to install and maintain. Assuming one believes the actual quality of care is unchanged, an appropriate metric might be the cost per number of duplicate tests avoided. For a clinical reminder system that warns clinicians when they are prescribing a drug to which a patient is allergic, the appropriate outcome might be cost per medication error avoided or even per anaphylactic reaction avoided. For interventions that affect the quality of care and have the potential to affect mortality, the appropriate outcome may be cost per life saved, or perhaps life-years saved, which would also incorporate the remaining life expectancy of a patient who benefited from the program.

Cost utility analysis (CUA) is an extension of cost-effectiveness analysis that differs by measuring the outcome as a "utility." This is a measure of the strength of preferences for the particular clinical outcome state, and is a measure of the quality of life attributable to living in a particular health state. It is intuitively obvious that people do not place the same value on various outcome states: a year of life after a stroke is worth less than a year of life without a stroke. There are multiple validated methods, such as SF-36 and EuroQOL, for measuring these preferences, as reviewed by Lara-Munoz and Feinstein

(Lara-Munoz and Feinstein 1999). There are several examples of scales throughout the literature (Feeny et al. 2002; Torrance and Feeny 1989). The debate about appropriate utility measures began five decades ago. The unique attribute of utilities is that they provide a quantitative assessment of the magnitude of difference between the perceived values of various health states. This allows the various outcome states resulting from different programs to be measured in quality-adjusted life years (QALYs), even if these programs are applied to people with different underlying diseases.

Cost-benefit analysis (CBA) is a form of economic evaluation that requires that both costs and benefits be valued in monetary terms. Therefore, to use CBA in health care, it is usually necessary to determine the value of the clinical outcome in currency terms, e.g. dollars. Because this means placing a value on the cost of human life, it is may sometimes be avoided as a mechanism for analysis in health care settings. However, if a policy maker needs to make a fair comparison between the value for money of different interventions or programs (for example, to compare a healthcare intervention with an educational or social care intervention), this is currently the most rationale way to frame the decision.

Each of the above analysis types is appropriate in a particular setting. For example, if outcomes are truly known to be equivalent between two different options, then ignoring outcomes and only analyzing costs (cost-minimizing study) may very well be an appropriate analysis. However, whichever analysis type is chosen, the costs (and benefits if included) need to be measured accurately, and in a manner comparable to other, similar studies. This chapter describes the techniques for conducting economic analysis, including the measurement of costs and outcomes; concentrates on their use in cost-effectiveness; and provides examples of cost-benefit analysis that have been used by health systems and hospital administrations to make various financial resource decisions in health care.

18.3   Conducting an Economic Analysis

When conducting an economic analyses, several decisions need to be made initially that set the stage for the type of costs and outcome data that will be used, the level of detail required in that data, and the sources and type of analysis that will be carried out. First, the study team must decide the perspective from which the analysis will be carried out: The patient? The hospital? The insurance company? Society? Second, the overall time frame of the analysis needs to be decided: does the study team care only about short-term effects, or will the technology choices alter the stream of resource use and benefits for many years to come?

18.3.1   The Perspective of the Analysis

The perspective of an analysis specifies from whose point of view the decision is being evaluated. The perspective of the analysis is important because different types of costs and benefits accrue to different components of the health care system, and therefore different costs and outcomes should be included in an analysis depending on who is making the decision and who accrues benefits. For example, consider a clinical decision rule in an information system that allows a nurse to triage symptomatic urinary tract infections and treat a portion of these women over the telephone rather than have everyone be seen and evaluated. Although this may reduce overall costs and improve the quality of care, whether or not it makes sense to a particular provider group depends on the reimbursement mechanism. Under a capitated system, where the health care provider is paid a monthly amount for all services provided, the benefit will accrue to the health care provider. Conversely, if providers operate under a fee-for-service insurance system, where they are paid only for people they see or tests they perform, they may lose financial resources that the insurance company gains.

The preceding example illustrates the critical importance of choosing an appropriate perspective from which the analysis is conducted. Different perspectives will include different types of costs and outcomes, depending on who bears the costs and who accrues the benefits. Four specific perspectives are considered below.

Societal Perspective   The societal perspective is by definition the broadest perspective, and therefore should include all costs and benefits, regardless of who pays the costs or to whom benefits accrue. Furthermore, in addition to a global inclusion of all monetary costs and clinical outcomes, a societal perspective should include many of the costs and benefits that an individual payer or insurance company might not choose to include, such as lost earnings or out-of-pocket expenses of the patients, or the time spent by family members in the care of an ill relative.

Health Care Payer or Insurance Company Perspective   Previous recommendations argued that the societal perspective should always be included in an analysis, as it was the broadest and most comparable. However, the inherent difficulty with including "everything" led the second panel (Sanders et al. 2016) to state that the healthcare payer perspective was a reasonable perspective to use for an analysis. The perspective of a large health care provider, system, or insurer makes substantial sense for many economic analyses because they are often making resource allocation decisions regarding what type of services to cover, and what financial resources to expend in the production and provision of those services.

Hospital Perspective   The hospital perspective is commonly used to evaluate the effects of information resources, as it is often the hospital that is expending financial resources. The hospital perspective is also usually straightforward: the cost structure of the institution is generally known, and the costs that the hospital must include in national cost reports are published. Furthermore, the financial responsibility of the hospital is generally straightforward, with nonprofit or government hospitals responsible for working within a global budget, while in private or for-profit hospitals there is the financial oversight of a board of directors or investors. The costs and benefits included are those that directly accrue to the bottom line of the hospital.

Patient Perspective Although arguably one of the most important perspectives from the individual patient's point-of-view, the patient's perspective is rarely used in comprehensive analyses. Part of the difficulty with the patient's perspective is that it does not include many costs simply because the patient doesn't pay them, the most obvious being costs paid by an insurance company.

18.3.2 Time Frame of the Analysis

Most decisions to implement a new program or introduce a new information resource set into motion a series of events (in terms of costs and benefits) that occur over time. It is not uncommon for these costs and benefits to occur at different times. When a hospital decides to implement a clinical information resource, the majority of the expenses may be required at the beginning (capital expenses to purchase hardware, software, and installation costs, financial resources required for training), whereas many of the benefits do not occur for several years, when clinical processes have been adapted to take advantage of the new information resource. The important concept is that the time frame used in an economic analysis must match the actual duration of time required to encompass all of the important costs and benefits of the program. The classic example of this is a comparison of the cost-effectiveness of a preventive intervention (e.g. using diet or drugs to lower cholesterol) with a therapeutic intervention (coronary artery bypass surgery) on mortality rates or quality-adjusted life years. Unless the data collection is carried out for a sufficient time for the cholesterol lowering strategy to have realized a benefit, the analysis will be biased in favor of the surgical intervention. However, in a different clinical scenario such as the treatment of acute urinary tract infection in otherwise healthy women which carries no risk of long-term events, a time frame of a few days could be appropriate when comparing different treatments.

Self-Test 18.1

1. For the following types of information resources and study needs, indicate which study methodology would be most appropriate for an economic analysis: cost minimizing, cost consequence, cost effectiveness, cost utility, or cost benefit.

    (a) You are a vendor of an electronic medical record, and want to show that installation of your product improves health care of hospitalized patients by decreasing length-of-stay and improving quality of care.
    (b) Your pharmacy department has proposed purchasing a new inventory control mechanism that will eliminate the need to carry large stocks of inventory on some items and decrease the number of pharmacy technicians that you need to hire. No changes in therapeutic outcome are expected.
    (c) A large employer wants to assess the impact of an electronic algorithm for detecting employees at high risk of work-related illness so that preventive healthcare measures can be offered that will decrease future illness burden,

absenteeism, and improve the efficiency and quality of life of its employee base.

(d) A pathology department is considering two digital pathology products that allow the remote viewing of slides by an off-site pathologist. The resolution quality is equivalent, but the hardware, software, and personnel costs appear different.

2. For the following proposed economic analysis, list the most appropriate perspective (payer, hospital, society):

(a) A study of the economic effects of instituting a therapeutic drug substitution program in a pharmacy.
(b) A study that investigates the overall impact of the digital transfer of clinical information between hospitals, home health agencies, nursing homes, private practitioners, and rehabilitation facilities on the overall quality of care of patients with orthopedic surgery in a region.

18.4   Definition and Measurement of Costs

For all types of economic analyses, the costs of each option must be clearly delineated and appropriately measured. Unfortunately, there is tremendous variability in the methods with which health care providers, hospitals, and insurance companies calculate costs. In addition, there are differences between certain definitions traditionally used by accountants from those used by health care economists. In this section, the differences between costs and charges are discussed, the various types of costs included in economic analyses in health care defined, and a general overview of the types of cost accounting systems typically used in hospitals provided. Finally, more generalized national or regional measures of costs that can be used in economic analyses are described.

18.4.1   Why Costs Do Not Equal Charges

One of the most important realizations over the past few decades with respect to health care costs is that there is little relationship between charges (the price that a hospital or provider asks the insurer or patient for payment) and the actual costs of that particular item or service to the provider (Finkler 1982). There are many reasons for this, from the market strength of many insurance companies that negotiate lower charges from providers to desires on the part of certain institutions to magnify the appearance of donated free care by having high prices for self-pay patients.

Direct Costs These are those costs that are an immediate consequence of the choice or decision being made. They typically include the costs of medical technology such as the information resource or service being studied and medical services (hospitalizations, medications, physician and other health care professional fees, durable medical equipment, etc.).

Time Costs These represent the amount of time required by providers and patients to use an information resource or participate in a treatment that should be included as a real cost to any program. If two health programs are equally effective and have equal monetary costs, the program that required less patient or provider time would be preferred. Details for measuring and evaluating different types of time expended in health care activities can be found in Gold et al. (1996). The fundamental concept is that time is a valuable commodity that should be included in the overall costs of a particular intervention. It is often true, however, that when considering clinical information system interventions, patient time considerations are ignored, often because the patient is in the hospital the entire time under either program, and the time-costs would likely cancel out. On the other hand, if a particular information system intervention had the effect of decreasing length-of-stay, the differential times of the patients could be included in the analysis (and should be, if the analysis is being conducted from the societal point of view).

Indirect Costs There are differences in the definition of indirect costs used by health care economists and accountants; many authors choose the term *productivity costs* to describe the types of costs being considered more directly. Indirect costs or productivity costs are the monetary values of lost productivity that occurs as a consequence of the intervention and the disease being treated. For example, if a patient treated under one specific strategy can return to work but a patient treated under an alternative strategy cannot, the values of the lost productivity should be charged to the first program. As noted above, this type of cost would be appropriately included only if the analysis were being conducted from a patient or societal perspective.

Intangible Costs (Pain and Suffering) There is no question that patients place a value on pain and suffering, as evidenced by the fact that patients spend financial resources (purchase pain relievers, accept operations that palliate symptoms) to eliminate or alleviate symptoms, even if the intervention has no effect on their length of survival. However, finding an appropriate value for this cost may be extremely difficult, and most investigators do not include pain and suffering as a cost. Instead, they include the values of that pain and suffering in the estimate of the quality of life of the particular health state. This is most commonly accomplished by using QALYs as the outcome measure, where the value of living in a state that includes pain and suffering is less than living in a state from which the pain and suffering are absent.

18.4.2 Mechanics of Cost Determinations

Once the perspective and time frame of the analysis is chosen, one must develop a method to assess and measure the costs that accrue from a particular intervention or program. There are, in general, two basic methods for determining costs: micro-costing methods and macro-costing methods. Each has benefits and difficulties, but can be used to develop accurate, robust cost analyses.

Micro-costing methods make use of detailed accounting principles to develop measures of how much each product, service, or option costs by breaking it down into individual parts and determining the cost of each component. For example, the cost of a chest x-ray includes the cost of the film, a small portion of the amortized cost of the equipment, staff time to take and process the x-ray, radiologist time to read it, a small amount of power to run the equipment, a small part of the cost of housekeeping to clean the radiology areas, etc. This process can be extremely complicated, but many cost-accounting systems keep track of inputs to the various products that they produce at that level of detail. Often, the calculation is accomplished at a slightly higher level of detail, in which the costs of an entire department (say radiology) are calculated based on what portion of the overall hospital budget is attributable to the department (salaries, capital equipment costs, overhead, etc.), and these costs compared to the total charges for the product produced (billed) by that department. This global *cost-to-charge ratio* is then applied to the charge for each individual product or service produced by that department to estimate the cost of that item. More detail regarding costing methodology is found in Drummond et al. (1997a), Chap. 4.

Macro-costing methods use truly global measures of the costs (or payments) for services. For example, in the United States, the Centers for Medicare and Medicaid Services (CMS) has calculated (through a very complicated resource-based analysis) the estimated average cost for every physician service from office visits to various procedures to the costs of hospitalization for all categories of diagnoses. For hospitals, these are called diagnostic related groups (DRGs). They not only represent what the federal government will pay for particular services, but also are designed to represent the average true cost of that service or procedure. In the U.S., for individual providers, CMS pays practitioners according to the Resource-Based Relative Value Scale (RBRVS), which represents a complex calculation of the education, training, difficulty, and risk of various services and procedures. More information on cost analyses are available in Drummond et al. (1997a), Gold et al. (1996), and at the NICE Web site: www.nice.org.uk.

Often, cost analyses do not need to be conducted at a level of detail that requires knowledge of the costs of every component of a health care provider or hospital's cost structure. This is especially true in determining the costs of information resources, when the costs may be dictated by market forces, where a particular vendor sells a record system for a particular price. Costs savings (in terms of decreased need for personnel, changes in pharmaceutical costs, or changes in maintenance fees) can also often be calculated from data derived from vendors, personnel files, and current hospital contracts. However, if a portion of the costs or benefits are measured as changes in the quantity of services, procedures, or clinical outcomes, a rigorous analysis of the true cost of those components needs to be accomplished.

18.5 Definition and Measurement of Outcomes and Benefits

Except for cost minimising studies, in which the clinically important outcomes are assumed to be equivalent, economic analyses must have a mechanism for measuring and quantifying the outcome of interest. One of the most difficult aspects of this task is making sure that the various outcomes are measured using the same metric, that is, that the units of measurement for all possible choices or strategies are the same. The simplest cases are those in which the study team can legitimately assume that the mortality and morbidity outcomes of a particular set of choices are the same, as then the outcome measure (such as changes in length-of-stay or duplicate tests avoided) are the appropriate choice. It is even possible to ignore the long-term benefits of a strategy if it has already been decided that the intermediate outcome that leads to a mortality benefit is desired. Consider a decision between different strategies to increase mammography screening for breast cancer in outpatients. Options to increase mammography rates might include a reminder and tracking alert system included in an electronic medical record, a text message reminder to patients incorporated into the scheduling software, or a publicity campaign such as wall posters within the physical confines of the practice. Each one of these options would have different costs and different impacts on the mammography rate for the practice. Provided that information was known (e.g., from literature reports from other practices that had tried these methods), estimates of the success rates could be determined, and estimates of the costs of each strategy could be derived as well. In this example, it would be appropriate to consider the cost-effectiveness outcome as costs per extra mammogram obtained, even though the long-run effect of the increasing mammography rate is to lower breast cancer incidence and mortality.

18.5.1 Matching the Correct Outcome Measure to the Problem

The choice of the appropriate outcome measure is dependent on the intended purpose of the economic analysis. If the analysis is designed to be used in regional or national resource allocation decisions across many different possible uses of the financial resources, it is necessary to measure outcomes in a quantity such as dollars, lives saved or QALYs that can be compared across the various different options.

However, economic analyses restricted to a much narrower outcome measure can be appropriate and valuable if their intended use is more local and restricted. For example, when considering the decision to purchase a new pharmacy module for a clinical information resource to prevent medication errors and interaction checking, a hospital administration may be satisfied with simple cost-consequence analyses, such as reporting the cost per medication error avoided by implementing a particular system. Again, this ignores the long-term health benefits of error reduction, but considers error reduction as an outcome in its own right.

18.5.2 Adjusting for Quality of Life

It is intuitively obvious and supported by innumerable quality-of-life studies that patients do not value all health outcomes similarly: a year of life after a stroke (cerebrovascular accident) is not valued as highly as a year of life in full health. Therefore, if the analysis is being conducted from society's perspective, the appropriate outcome is QALYs. However, the actual measurement of QALYs is not always straightforward: different assessment methods may arrive at different values, the assessed values may change over time, for some interventions that patients like no change in QALY can be measured, and the values may vary systematically with characteristics of the individuals responding to questionnaires. For these reasons, one should include an expert in utility or quality of life assessment if the study team intends to use QALYs as the outcome measure.

18.6 Cost-Minimizing Analysis

As noted above, the fundamental assumption in cost-minimizing analyses is that the clinical outcomes expected under each possible option are the same. The best example of such studies in health care are therapeutic substitution policies, where less expensive (usually generic) drugs are substituted for more expensive brand name drugs if the more expensive drug is ordered. Such therapeutic substitution policies are commonly embedded in clinical information systems that have prescribing components. As an example, simple case of an antibiotic substitution program will be examined.

Example 1: Antibiotic Substitution
The chair of the hospital formulary committee has requested that the pharmacy system institute an automatic therapeutic substitution program of a less expensive antibiotic (Cheapocillin) for the commonly ordered new antibiotic Cephokillumall. This rule is to be built directly into the pharmacy component of the hospital's clinical information system (CIS). The argument is that

Cheapocillin is substantially less expensive and by all reports equally effective for the treatment of infections. The infectious disease service does not object to the therapeutic substitution on clinical grounds. Therefore, the most important prerequisite for conducting a cost-minimization study has been met: there is good evidence and agreement that the two strategies have equivalent outcomes. This allows the economic analysis of the new information system to concentrate entirely on costs.

The most important concept in cost-minimizing studies is the appropriate identification and enumeration of the costs of the various strategies. Table 18.2 outlines the costs of the various therapeutic strategies. Cheapocillin has very low pharmaceutical costs per dose ($0.50), but comes as a powder that must be reconstituted with saline by the pharmacist. This procedure is calculated to cost $11.00 per dose. Because of its short half-life, Cheapocillin must be administered four times a day, resulting in a daily cost of $46.00. Cephokillumall is substantially more expensive for each dose ($22.00) but comes in a ready-to-administer vial, so preparation and administration costs for the pharmacy are reduced ($8.50). The recommended regimen is one dose per day, producing a daily cost of $30.50. However, because of potential side effects of Cephokillumall on the kidney, laboratory tests need to be obtained once on day 3 to assess kidney function, which cost $28.00. Since both drugs are given for 5 days, the total cost of Cheapocillin is $230.00, and the total cost of Cephokillumall is $152.50 plus $28.00 for the laboratory test, or $180.50.

Table 18.2 Example of a cost minimizing study of therapeutic substitution. This type of analysis assumes that the effectiveness of the two strategies (drugs in this case) is the same, and only compares the costs of each alternative

Cost	Cheapocillin	Cephokillumall
Cost per dose	$0.50	$22.00
Administration costs/dose	$11.00	$8.50
Doses per day	4	1
Total daily costs:	$46.00	$30.50
Cost of 5-day course	$230.00	$152.50
Laboratory costs (day 3)	$0.00	$28.00
Total costs	$230.00	$180.50

The point of this example is that even cost-minimizing studies may be more complicated than initial impressions, and the intuitive answer that Cheapocillin is obviously cheaper may not hold up after a complete accounting of all costs related to a particular strategy.

The analysis can be relatively straightforward as described or it can be more complicated, in several possible ways. For example, the costs of implementing a therapeutic substitution program will be substantially less in an information system that already has a pharmacy module with decision support functions built-in than would be the case if it had to be developed from a CIS that did not already have the pharmacy module. The costs of developing, implementing, and maintaining a particular function may also need to be included in the analysis, but the exact form and magnitude will be very dependent on the specific characteristics of the local information resources.

18.7   Cost-Effectiveness Analysis

The defining characteristic of cost-effectiveness is that the outcomes for each strategy are measured in the same, clinically meaningful outcome. The most general of these outcomes would be life expectancy (or its quality-adjusted companion, QALY). The purpose of this uniform measurement is that it allows various innovations (even if they are used in different diseases or patient groups) to be compared to each other and decisions about the efficient distribution of financial resources across various choices to be made. The economic principle that forms the foundation of CEA is that financial resources should be spent on the most cost-effective options first, and that new strategies or information resources added or funded in order of their cost-effectiveness. In theory, this ensures that the financial resources expended are purchasing the most "health" possible. So, when faced with a series of choices between possible health-improving innovations and a limited budget, economic principle would dictate that the analyst ranks the possible options in order of cost-effectiveness ratio (from the most cost-effective to the least cost-effective) and purchase innovations in decreasing cost-effectiveness order until the budget limit is reached. In practice, this rank ordering of options followed by spending the available health care resources is rarely explicitly done. Often financial resources directed in one area (e.g., information services) cannot be redirected to another area of public service (e.g., social work). This is true at all levels of decision making, with each level containing a series of political, social, and organizational barriers to the strict application of CEA. However, it remains useful to understand this as the underlying concept in the intended use of CEA, which is designed to achieve the highest quantity of the outcome (chosen and valued by the investigator) for the least cost.

Graphically, the comparisons used in cost-effectiveness analysis are illustrated in Fig. 18.2, which represents the cost-effectiveness space. For any new therapy or innovation that is being made, the costs and benefits need to be compared to the current strategy. For any new option, the new strategy can be more effective, less effective, or equivalent to the current strategy, and can be more expensive, less expensive, or equal in costs to the current strategy. This divides the cost-effectiveness space into four quadrants that have useful interpretations. The lower right quadrant would represent strategies that are both *cheaper* and *more effective* than the existing strategy: these programs should simply be implemented. Similarly, those strategies that fall in the upper left quadrant are both *more expensive* and *less effective* than the

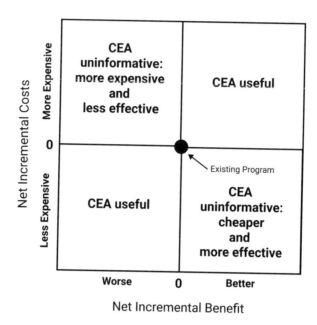

Fig. 18.2   The cost-effectiveness (CE) space. Compared to the costs and effects of a baseline program, a new strategy can be either more or less expensive or more or less effective that the current strategy. This divides the CE space into four quadrants. The upper left quadrant represents those strategies that are both less effective and more expensive that the existing program; these programs require no choice and are clearly dominated by the current strategy. Similarly, no analysis is required for strategies that fall in the right lower quadrant. Projects in this quadrant are cheaper and better than the existing strategy, and should simply be adopted. The real benefit of cost-effectiveness analysis (CEA) is in the right upper quadrant (where strategies are more expensive but produce better outcomes) and in the left lower quadrant (where strategies are cheaper, but do not produce equivalent outcomes). It is in these areas that a trade-off exists between cost and effectiveness, the condition required for a CEA. *Note:* This type of analysis assumes that the effectiveness of the two strategies (drugs in this case) is the same, and only compares the costs of each alternative

current strategy: these should be avoided. It is only in the two remaining quadrants (the upper right quadrant, where strategies are *more expensive* and *more effective*, and the lower left, where strategies are *less expensive* but *not as effective*) that the use of CEA is appropriate. It is in these areas, where there is a *trade-off* between costs and benefits, that CEA is most useful.

18.7.1   The Incremental Cost-Effectiveness Ratio

A crucial aspect of a CEA is that something cannot be cost-effective in and of itself; it is only cost-effective (or not) compared to an alternative. Consequently, the statement "This app is cost-effective" is nonsensical; it must be accompanied by a description of what the app is being compared to. In fact, CEA is most useful when the strategies being examined represent a range of possibilities, each with different costs and effectiveness. Typical CEA studies provide the results of each possible strategy, and compare each to the next least effective or expensive.

The global outcome measure that defines cost-effectiveness analysis is the incremental cost-effectiveness ratio (ICER). This represents the ratio of the net costs that will be expended by implementing a particular innovation divided by the net benefit, measured as an appropriate clinical outcome. It is defined as:

$$\frac{Cost_B - Cost_A}{Effectiveness_B - Effectiveness_A}$$

or the net costs of moving to strategy B from strategy A divided by the net benefits (measured as a clinical outcome) of choosing strategy B over A. The units of the ICER are dollars per unit outcome, which represents the cost of an additional unit of the particular outcome measure. For example, if the outcome of an intervention is measured in lives saved, the ICER would be in units of dollars per life saved.

To illustrate, a completely generic example, with hypothetical programs that both cost financial resources and save lives is examined. Although this may not be the most common outcome measure used in CEA studies in biomedical and health informatics, it represents the most common application of CEA in health care and is a reasonable starting point to develop an understanding of the technique.

Example 2: A Series of Lifesaving Therapies
A simple CEA is illustrated in Table 18.3 and displayed graphically in Fig. 18.3. In the figure, only the upper right quadrant of Fig. 18.2 is represented for simplicity; all three strategies evaluated are more expensive and better for patients than the current strategy. Assume that there are three possible strategies for treating a particular disease, and they have different costs and effects, as shown in the table. Current therapy is described by strategy A, which costs $10,000 and produces a mean of 3.5 years of survival. Two new therapies have been devised: strategy B costs $20,000 (an additional $10,000 compared to strategy A) but produces a longer survival of 4.5 years (an additional year compared to strategy A). A third treatment (strategy C) is even more expensive at $35,000, but does produce improved results, with patients living 5.0 years after receiving that therapy.

To calculate the ICER, the first step is to calculate the net costs and effects of moving from the base strategy to the next best strategy. From strategy A to B, the net cost is $10,000, the net effectiveness is 1 year: the ICER is therefore $10,000/life year gained. Then, the use of strategy C instead of strategy B costs an additional $15,000 ($35,000–$20,000), and gains another 0.5 life years (5.0–4.5): the ICER of moving from strategy B to strategy C is $30,000/ life year. The calculations are illustrated graphically in Fig. 18.3. The net costs and effects (in terms of life years) are plotted in the cost-effectiveness space. The slope of the line between each possible strategy is the ICER of that strategy.

Table 18.3 Example of cost-effectiveness (CE) analysis of several lifesaving interventions. For each strategy, the table provides the cost of the strategy and the outcome in life-years gained. The net costs are calculated as the difference in costs from the previous strategy, and the net effects are the difference in outcomes for a strategy compared to the next least effective. The incremental cost-effectiveness ratio (ICER) is simply the net cost divided by the net effects

Strategy	Cost	Outcome (year of life)	Net cost	Net effect	Incremental CE ratio
Strategy A	$10,000	3.5	Baseline strategy		
Strategy B	$20,000	4.5	$10,000	1 year	$10,000/year vs. strategy A
Strategy C	$35,000	5.0	$15,000	0.5 years	$30,000/year vs. strategy B

Fig. 18.3 The cost-effectiveness space for Example 2. Assume the baseline strategy is A; the graph depicts the net (incremental) costs required to purchase the next best strategies B and then C, as well as the net or incremental gains that the strategy would provide. *Note:* For each strategy, the table provides the cost of the strategy and the outcome in life-years gained. The net costs are calculated as the difference in costs from the previous strategy, and the net effects are the difference in outcomes for a strategy compared to the next least effective. The incremental cost-effectiveness ratio (ICER) is simply the net cost divided by the net effects

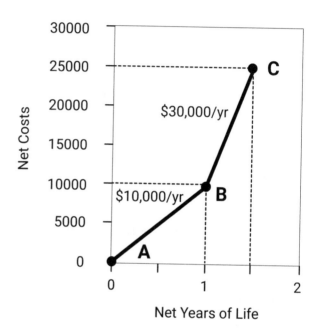

It is left to health policy makers to decide how they will use such figures. For example, if controlling health expenditure is important, they may decide to sanction widespread use of strategy B but reserve strategy C for those patients most likely to benefit, in view of its much higher ICER.

18.7.2   Cost-Benefit Analyses: The Cost-Benefit Ratio

As noted in Sect. 18.2.1 on types of cost analyses, cost-benefit analyses (CBAs) are distinguished by the valuation of both the costs and outcomes of a strategy in monetary terms. The cost-benefit ratio then simply measures the ratio between the incremental costs of choosing a strategy over the benefits (measured in monetary units) of each strategy:

$$\frac{Cost_B - Cost_A}{Monetary\,Benefit_B - Monetary\,Benefit_A}$$

The advantage of CBA, and one of the reasons for its use in many fields other than health care, is that the interpretation of the cost-benefit (CB) ratio is straightforward: any strategy with a CB ratio less than 1 means that the benefits are valued more than the costs, and instituting that strategy produced a net benefit. Cost-benefit ratios of greater than 1 indicate that the costs are greater than the benefits: such projects should not usually be undertaken. This applies even for projects undertaken for the public good since if something has a value, for the purpose of CBA it is necessary to quantify that value in dollars. The value of the public good is then contained in the benefit side of the equation. The problem often is trying to agree on a monetary value for commodities like health status or increased security resulting from police protection, military strength, etc. It is important to note that authors sometimes report the benefit-cost ratio (that is, benefits divided by costs, the inverse of cost-benefit ratio), in which case options with ratios over 1 are favored and those with a ratio under 1 are ruled out.

Example 3: Cost-Benefit Analysis
A hospital needs to upgrade its billing and patient accounts system because the current version of the system is no longer supported by the vendor. There are two options, A and B, which differ in their startup and maintenance costs. These options also have different effects on the ability to produce accurate bills, their personnel staffing requirements, and the speed with which bills are submitted (which decreases the financial resources caught up in accounts receivable). Table 18.4 details the costs and benefits (both measured in dollars) for each option. For example, system A is more expensive to purchase than system B, and has higher costs of personnel training and higher maintenance fees. However, it also is a more accurate system, leading to increased billing revenue, as well as requiring fewer personnel to operate and maintain, which may save the salary of a part time data-entry clerk ($14,000). Assume that all these costs have been properly spread over the expected life of the product. Although system A is more expensive, it returns more benefit for every dollar spent. The CB ratio indicates that it costs only 44 cents for each dollar it saves, whereas system B costs 63 cents for each dollar it returns. Although any project with a CB ratio of less than 1 produces more benefit than it costs, choices can also be ranked, based on which options produces the highest return.

Self-Test 18.2

1. Consider the following costs of deploying alternative components of an information resource and the number of lives each component is expected to save. If the

Table 18.4 Simple cost-benefit analysis. In this analysis, two replacement billing systems are compared with respect to their system and maintenance costs, as well as the effects they have on personnel needs and their effect on billing revenue and debt service from decreased time in accounts receivable

	Information resource A	Information resource B
Costs		
Cost of system	$16,000	$12,500
Cost of personnel training	$2850	$1000
Maintenance contract	$1850	$1250
Total cost:	$20,700	$14,750
Benefits		
Increased billing revenue	$32,000	$22,000
Decreased staffing requirements	$14,000	$0
Decreased debt service	$1340	$1340
Total benefit:	$47,340	$23,340
Cost-benefit ratio	0.44	0.63

stakeholder group making this decision feels that it can spend no more than $75,000 per life year saved, which components strategy should it pursue? (Hint: calculate the ICER of each strategy, ranked by effectiveness.)

Strategy	Cost	Outcome (life years gained)
Strategy A	$80,000	3.5
Strategy B	$50,000	3
Strategy C	$30,000	2.5
Strategy D	$90,000	3.6

2. Currently, a health plan is spending $150,000 on a health prevention program that the plan feels increases the life expectancy of the patient population by 10 life-years (averaged over the entire population). A proposal has been made for a new program that costs $130,000, and is known elsewhere to save 15 life years. Is a formal CEA of this program necessary?

18.7.3 Discounting Future Costs and Benefits

It is clear from common sense and basic economic analysis that dollars spent today are not directly equivalent to those spent in the past or in the future. The changing value of a dollar over time has two components: inflation and the real rate of return. Inflation represents the change in prices in an economy, and requires that prices (and therefore costs) be standardized to a common base year if expenditures from multiple different years are to be added, compared, or analyzed. Dollars spent in different years can be adjusted to a base year through the consumer price index (CPI) (United States Bureau of Labor Statistics (BLS) 2021). There is also a health care specific price index, the Medical Price Index (MPI), which is used to adjust the prices of medical and health care goods and services (United States Bureau of Labor Statistics (BLS) 2021). In addition, however, even after adjusting for inflation, there is a need to account for the fact that dollars can be invested and produce greater value in the future. Even if the inflation rate were zero, the real rate of return (interest rate) in society would return more dollars next year to an investment made today; therefore, a dollar spent today is not worth the same as a dollar spent tomorrow.

The *present value* (PV) of an expenditure of X dollars t years in the future with an interest rate of r is

$$PV = \frac{X}{\left(1+r\right)^{t}}$$

So if a particular strategy incurs a cost of $100 next year, and the interest rate is 5%, the value today of that $100 expenditure next year is only $95.23 [100/(1 + 0.05)^{1}] This is because $95.23 invested today at 5% interest produces the $100 needed for the program a year from now. For a stream of costs $(X_1, X_2, \ldots, X_n)$ made over time under an interest rate of r, the present value is

$$PV = \frac{X_1}{\left(1+r\right)^{1}} + \frac{X_2}{\left(1+r\right)^{2}} + \frac{X_3}{\left(1+r\right)^{3}} + \frac{X_4}{\left(1+r\right)^{4}} \ldots + \frac{X_n}{\left(1+r\right)^{n}}$$

For all resource use that occurs generally more than a year in the future, it is both reasonable and appropriate to discount those costs to their present value equivalents.

Discounting Benefits   There is a natural intuition about the reasons analysts must discount costs: it is clear that a dollar today is not worth the same as a dollar tomorrow. Although the same intuition regarding benefits does not exist, there are equally persuasive reasons why analysts must discount the value of future benefits (years and/or quality of life) as well. One of the consequences of discounting costs but not discounting benefits is that the analysis would produces a change in the relative valuation of outcomes with respect to financial resources over time. In other words, by discounting costs but not benefits, the amount of monetary value attached to a particular outcome would change over time, even if the societal value for those outcomes were constant. If a life-year is worth $50,000 today, discounting costs, but not benefits, implies that the optimal ICER next year is less than $50,000/life year. The problem of discounting benefits is described in more detail in Gold et al. (1996), Drummond et al. (1997a), and in a landmark technical paper by Keeler and Cretin (1983). However, virtually all methodology groups that have examined the problem agree that costs and benefits should be discounted at the same rate.

Choosing a Discount Rate   The choosing of an appropriate discount rate is not trivial, however. It is important to remember that the discount rate is quite different from the inflation rate, which represents the "price" of money. In economic analyses, the discount rate refers to the real growth rate of money, after the effects of inflation have been removed. There are several recommendations for choosing a discount rate; they differ by country and are sometimes dependent on what type of resource (public or private) is being used. Most recommendations are between 3% and 5%, and discussions regarding these choices can be found in Drummond et al. (1997a).

18.8 Sensitivity Analysis

It is virtually impossible to conduct an economic analysis without making several assumptions regarding a particular component of the costs or benefits. One of the most powerful benefits of using economic analysis is the ability to test the effect of different assumptions on the outcome. However, such "sensitivity analyses" have many other uses. During the design and development of an economic analysis, the investigator can use sensitivity analysis to ensure that the components of the analysis represent the situation as designed. For example, it is often possible to know ahead of an analysis what the expected effect of a particular change in a particular parameter should have on the answer. If one of the inputs in a CEA is the effectiveness of a particular therapy or information resource, then the cost-effectiveness (CE) of an option should increase as the predicted effectiveness of that particular treatment increases. To make sure an analytic model is working as expected, a series of these analyses should be conducted: an option's CE ratio should increase as the cost of that option increases, and should decline as the cost decreases. Although it seems obvious, when the analytic model that has been constructed to evaluate a problem is complex, this is the easiest mechanism for checking that the model does represent the actual problem well enough.

There are two major types of sensitivity analysis: structural sensitivity analysis, in which the actual components of the analysis are varied to understand the effects of various different assumptions regarding the structure of the model, and parameter sensitivity analysis, in which the values of the variables in the model (costs, probabilities of various outcomes, etc.) are varied over reasonable ranges to understand the effect of variability in the parameter estimates.

18.8.1 Structural Sensitivity Analyses

The goal of structural sensitivity analysis is to help ensure that the important and correct components of the problem have been included and are related in an appropriate manner. For example, an analysis of an antibiotic interaction and allergy-checking program in a pharmacy system might be analyzed only considering the effect on error reduction (number of allergic reactions, number of medication side effects). However, the new procedures may have an unpredictable effect on pharmacists' time, and a complete analysis would generally include what happens to the workloads of the pharmacists. Although a common assumption is that it would save pharmacist time, it is possible that a whole new set of activities (for example, responding to complaints from clinicians regarding the medication interaction checker) may actually increase pharmacist time. Also, it is possible that the interaction program would teach the clinicians, and improve the accuracy and efficiency of their prescribing behavior over time. Therefore, a simple analysis would not include pharmacist time or the effect on long-term clinical behavior, and there is no direct way of including that in a simple model by just changing a probability or value in the analysis. Including pharmacist time and long-term effects on clinician behavior would require a model with entirely new components in it, and would represent a structural sensitivity analysis since the structure of the model representing the strategies is extended to include a different level of detail concerning the potential outcomes of each strategy.

18.8.2 Parameter Estimate Sensitivity Analyses

Virtually every number involved in an economic analysis is a point estimate of a quantity that could, in reality, be different from the specific number included in the model. For example, even in the situation in which the estimate of a particular cost is the result of an economic analysis conducted alongside a randomized controlled trial, the cost estimate is only accurate within some confidence limits. Had the trial been conducted in a different set of patients or a different setting, undoubtedly the estimated costs would be (at least slightly) different. More commonly, the estimate may come from the literature, and is accompanied by a measure of its accuracy given by the 95% confidence interval or standard deviation. Although the accuracy with which parameters for a particular analysis are known may vary, it is rarely (if ever) the case that a parameter is known exactly. Therefore, an analysis of the model outputs using slightly different estimates for each parameter is necessary.

Traditionally, the most common methods of describing the variability of results in an economic analysis was to provide *baseline*, *best-case*, and *worst-case* calculations. These are illustrated in Figs. 18.4 and 18.5. Assume that a CEA has been done as described in Fig. 18.4. There are two strategies, A and B, each with different costs in several categories (hospital costs, ambulatory costs, pharmacy costs, and time costs) and strategy B is both more expensive and more effective (it adds 5 years of life expectancy for this particular disease). The straightforward calculation of the ICER is to calculate the net change in costs incurred by moving from the standard therapy A to strategy B, and the net change in benefits. As shown, strategy B produces 5.5 additional years of life for a cost of $53,000, giving an ICER of $9655/year.

A common method to evaluate the sensitivity of a result is to combine all of the worst (least favorable) assumptions and calculate a worst-case CE ratio, and repeat the analysis with all of the best (most favorable) assumptions and calculate a best-case CE ratio, as shown in Fig. 18.5. More sophisticated approaches to sensitivity analysis are now used frequently but require expert support from a health economist, so are outside the scope of this book. These methods include carrying out a series of one-way or multi-way sensitivity analyses and then displaying the results

Strategy A

Variable	Baseline
Hospital costs	$60,000
Ambulatory costs	$4,600
Drug costs	$900
Time costs	$2,800
Total costs:	$68,300

Outcomes: 5.0 years

Strategy B

Variable	Baseline
Hospital costs	$102,000
Ambulatory costs	$9,400
Drug costs	$2,000
Time costs	$8,000
Total costs:	$121,400

Outcomes: 10.5 years

$$\frac{\text{NET Costs: Strategy B} - \text{Strategy A}}{\text{NET Benefits: Strategy B} - \text{Strategy A}} = \frac{\$121,400 - 68,300 = \$53,100}{10.5 \text{ years} - 5.0 \text{ years} = 5.5 \text{ years}} = \$9.655/\text{year}$$

Fig. 18.4 Sample cost-effectiveness analysis, describing the comparison between two strategies, A and B, in which B is both more expensive and more effective than A, and calculating a cost-effectiveness (CE) ratio. *Note:* In this analysis, two replacement billing systems are compared with respect to their system and maintenance costs, as well as the effects they have on personnel needs and their effect on billing revenue and debt service from decreased time in accounts receivable

Net Difference in Costs

Variable	Lowest	Baseline	Highest
Hospital costs	$30,000	$42,000	$85,000
Ambulatory costs	$3,900	$4,800	$6,800
Drug costs	$900	$1,100	$1,300
Time costs	$4,000	$5,200	$6,400
Total costs:	$38,800	$53,100	$99,500

Net Difference in Benefits

Life expectancy	4.8 years	5.5 years	8.2 years

Best Case:
Least Expensive
Most Effective

Baseline Case:
Best estimate of
Costs and effects

Worse Case:
Most Expensive
Least Effective

$$\frac{\$38,800}{8.2 \text{ years}} = \$4732/\text{year} \qquad \frac{\$53,100}{5.5 \text{ years}} = \$9655/\text{year} \qquad \frac{\$99,500}{4.8 \text{ years}} = \$20729/\text{year}$$

Fig. 18.5 An example of a best-case/worst-case sensitivity analysis, combining all of the worst (least favorable) assumptions to calculate a worst-case CE ratio, and repeating the analysis with all of the best (most favorable) assumptions to calculate a best-case CE ratio

in tables or a "tornado diagram" and probabilistic sensitivity analysis using Monte Carlo or other stochastic methods.

Self-Test 18.3

1. The implementation of a particular information resource in your health system will produce the following streams of projected costs and benefits (all defined in terms of dollars).

	Year 1	Year 2	Year 3	Year 4	Year 5
Costs	$100,000	$40,000	$10,000	$10,000	$10,000
Benefits	0	$30,000	$40,000	$50,000	$60,000

(a) Without considering discounting, is investing in this project a good strategy?
(b) Also without discounting future revenue and cost streams, what is the cost-benefit ratio of this project?
(c) Assume a discount rate of 5% and that the hospital considers all costs and benefits as occurring at the end of each year. What are the present values of the cost and benefit streams?
(d) What is the cost benefit ratio of this project with the inclusion of discounting at 5%?

18.9 Special Characteristics of Cost-Effectiveness Studies in Biomedical and Health Informatics

Although the principles of economic analysis are the same regardless of the content area of the problem being addressed, there are often special circumstances surrounding the specific nature of a particular problem that a study team needs to incorporate in economic analyses of that particular area. There are several such special circumstances in informatics that need to be understood in order to conduct accurate and appropriate economic analyses.

18.9.1 System Start Up and Existing Infrastructure Costs

The deployment of a particular feature or use of a clinical information resource may appear very different depending on what infrastructure or components of the resource are already in place. For example, the decision to introduce order entry into a clinical information system will appear very different if the existing clinical information system already has components that can accept orders and relay them to the pharmacy, radiology system, etc. Similarly, consider the decision to introduce an allergy-checking and medication interaction system after a high-profile error has occurred in a hospitalized patient. The costs of such a component could look very different if it required purchasing an entire pharmacy system, as compared to the situation where adding a new module to an existing system will accomplish the task.

Similarly, infrastructure costs that have already been expended (whether the decision was appropriate or not) should *not* be included in the costs of a new strategy, even if that strategy uses those components. Ignoring such "sunk costs" builds on the incremental concepts that have already been stressed. For example, in the decision to implement computerized physician order entry in a clinical record system that is already scheduled for deployment, the cost of the record system installation, development, maintenance, etc., is not charged to the physician order entry component as it is a sunk cost: the decision to install those components had already been made. The only reasonable costs to include in the decision regarding physician order entry are the additional costs that will now be incurred to install, manage, and update the order entry component.

18.9.2 Sharing Clinical Information System Costs

The idea of sharing costs arises because the costs and benefits of information resources are often distributed across many departments. The hospital can't exist without billing, so if the billing system is used for an additional purpose, how much of the billing system cost should be attributed to the new purpose? As argued above, the basic cost of the billing system should not be included. However, any additional costs required to modify it or interface it with the clinical information system should be included, as should some component of the maintenance contracts, upgrades, and other ongoing costs. However, it is sometimes very difficult to understand how to distribute those costs across the various components that use them.

Information resources usually have life cycles running on a different time scale than the processes they support. For example, a particular clinical system that is installed to improve patient safety and decrease fatal drug errors may have a system life of 5 to 8 years, and require multiple upgrades. At the end of the life span of the product, a new system or comprehensive upgrade may have to be purchased. If the benefits of this medication interaction system have long-term effects, the evaluation from a societal point-of-view would require including the continuing costs of system replacement, etc. However, these distant future costs are often not included in individual hospital analyses.

18.10 Critically Appraising Sample Cost-Effectiveness Studies in Biomedical and Health Informatics

This section provides a critical appraisal or critique of two sample studies in which the effects of a clinical information system were economically evaluated. The guides to the evaluation of economic analyses by Drummond et al. (1997b) and O'Brien et al. (1997) provide a useful template for the critique of such published studies. There are four important components when critiquing the validity of economic study results (from Drummond et al.):

1. Did the analysis provide a full economic comparison of health care strategies?
2. Were the costs and outcomes properly measured and valued?

3. Was an appropriate allowance made for uncertainties in the analysis?
4. Are the estimates of the costs and outcomes related to the baseline risk in the treatment population?

The two articles are each reviewed using these questions. It is useful to have a copy of the full article available while reading the critiques.

The study by Wang et al. (2003) examines the financial costs and benefits of instituting a hypothetical electronic medical record. No attempt is made to associate the effects with clinical outcomes; the study is only concerned with financial results such as increased charge capture, decreased drug utilization, and decreased costs in maintaining and pulling paper charts. The article uses data from the published literature, the authors' own electronic medical record system, and expert opinion to make assumptions regarding the costs and benefits of system implementation. The setting of the study is a US practice in which some of the patients are fully capitated (the practice only receives a set amount per month to care for these individuals, and must pay for health care expenditures out of that pool) and some of the patients have fee-for-service insurance, in which the practice is paid according to a bill that is generated by the practice. The practice, therefore, accrues benefit if it reduces expensive medication use and laboratory tests in the capitated group, and accrues benefit if it can improve and enhance billing practices in the fee-for-service group. The analysis is carried out for 5 years, with setup and initialization costs only accruing in year 1, and maintenance costs, license fees, and other recurring costs charged in the years that they occur. The costs and expected changes in expenditure for various laboratory tests and the increase in billing accuracy for patients and other benefits are listed, with ranges placed on those estimates. The authors calculate that, on average, the net benefit from instituting an electronic medical record was over $86,000 per healthcare provider over a 5-year period, or $13,000 per provider per year.

With respect to the first question posed by Drummond et al., the answer is mixed: the authors provide only a single alternative (electronic medical record, EMR). However, this is the relevant comparison, and multiple possible EMRs could be evaluated through the extensive sensitivity analyses. With respect to the second point, although extensive analysis was completed regarding the various costs and benefits, the values represent only the financial gains: no clinical gains are included, making the analysis undervalue the EMR (assuming that the EMR does provide clinical benefits). The analysis clearly incorporated uncertainties in the estimates of input parameters and conducted multiple one-way and several multiway sensitivity analysis of critical estimates to determine how robust the analysis was to variability in assumptions. Finally, the last criterion regarding validation asks whether the outcomes and costs are related to the baseline risks of the population: this requirement relates much more directly to the economic analysis of a particular disease process, and is not particularly relevant in this study. On the whole, this study adheres to formal methodological recommendations quite well.

The second study by Evans et al. (1998) details the effects of the implementation of a computer program to help clinicians choose the proper antibiotic regimen for seriously ill individuals in the intensive care unit at a large academic hospital, using a pre–post design where the outcomes are measured for a year before implementation and then for a year after implementation. The program has multiple clinical algorithms that evaluate the type of infection, the current, local resistance patterns of bacterial strains at that hospital, and the level of illness of the individual, as well as many other clinical factors. The program not only recommends whether or not a particular patient needs an antibiotic, but also makes recommendations on the particular drug to use, the dose, and the interval. Over the course of the hospitalization, as clinical conditions and laboratory results change, the program makes recommendations for the modification of the patient's antibiotic regimen. Benefits of the program are measured in terms of the number of antibiotics ordered, the duration of dose, the number of days of excess antibiotics provided, and the length and costs of hospital and intensive care unit stays.

Turning to the critique, the analysis clearly does not provide a comprehensive analysis of the economic outcomes of the program. Many of the outcomes are clinical, and there is no attempt to include evidence that the program improved life expectancy, etc. Length of stay and avoidance of allergic reactions are the only two clinical outcomes considered. Furthermore, with respect to whether costs and outcomes were correctly valued, again the answer must be no for a comprehensive analysis. There are no costs of developing and implementing the information resource included in the analysis, yet the authors indicate that they have been working on this for over a decade, which must represent a large amount of financial resources. As already noted, clinical outcomes are not valued, and benefits are essentially only calculated as a pre–post change in these outcome variables; long-term effects (survival, etc.) are not included. There is very little attention to sensitivity analysis, although ranges for the outcomes found in the trial are provided. Finally, since the analysis takes place in an environment identical to the expected use, the incidence of various infections, antibiotic uses, etc., are exactly what would be expected to be seen in clinical practice at a similar facility.

From an economic point of view, this study is clearly more a cost-consequence than a true cost-effectiveness or cost-benefit analysis. Because there was no attempt to include the costs of development and implementation, it is not possible to evaluate the benefits of duplicating this effort elsewhere. However, the study does provide a remarkable amount of useful information, indicating that, if well conducted, any type of economic analysis may be useful. It provides an excellent estimate of the amount of wasted and inappropriate care that can be removed from the use of antibiotics in seriously ill hospitalized patients through the use of computer-based treatment algorithms.

18.11   Conclusion

This chapter introduces the basic methods used in economic analysis, but necessarily only scrapes the surface of what is a complex and developing field, especially when coupled with rapid developments in the software used to develop economic models, methods for eliciting and valuing quality of life and preferences for different outcome states, and emerging new approaches to using economic models to inform decision making. The intention of this chapter was not to provide all the skills necessary to design and carry out full scale economic analyses, but rather to provide the skills to understand: what is involved, the benefits of the approach, and if it is wise to discuss the problem with a health economist, and the language to prime that discussion. A brief section summarizing and critiquing two published economic evaluations was included to demonstrate the kind of studies being published in health and biomedical informatics and to illustrate how these can be improved. Overall, it is hoped that providing study teams with this brief introduction will encourage more economic studies to be carried out in our field, with the benefits that these bring to decision and policy makers. There is evidence that this is already happening, with the increasing number of economic studies being carried out in health and biomedical informatics enabling some recent systematic reviews of economic evaluations to be published (Ghani et al. 2020; Sadoughi et al. 2018; Murphie et al. 2019; Jacob et al. 2017).

Acknowledgments This chapter is closely based on a longer chapter from the second edition of this book written by Professor Mark Roberts of the University of Pittsburgh. The authors would like to sincerely acknowledge Professor Roberts' significant contribution to this new version.

Answers to Self-Tests

Self-Test 18.1

1. (a) Cost-effectiveness (if quality of life were included, could be cost-utility as well).
   (b) Cost-minimizing (or could be cost-consequence).
   (c) Cost-utility.
   (d) Cost-minimizing.

2. (a) Hospital.
   (b) Society.

Self-Test 18.2

1. First, rank the strategies by effectiveness (C, B, A, D), and calculate the incremental costs and benefits of moving to the next best strategy. Then calculate the ICER by dividing the incremental costs by the incremental effectiveness for each strategy:

	Cost	Incremental costs	Life-years	Incremental effectiveness	ICER
Strategy C	$30,000		2.5		
Strategy B	$50,000	$20,000	3	0.5	$40,000
Strategy A	$80,000	$30,000	3.5	0.5	$60,000
Strategy D	$90,000	$10,000	3.6	0.1	$100,000

Since the decision makers feel that they can only spend up to $75,000 year, the optimal decision is strategy A as it yields the greatest gain in life years while remaining within the spending limit; strategy D is too expensive for the small extra benefit it provides.

2. No, this is a program that falls into the "cheaper and better" quadrant of the CEA plane. No further analysis is necessary.

Self-Test 18.3

(a) Undiscounted, the value of the streams simple $170,000 in costs and $180,000 in benefits, so it is a good investment.

	Year 1	Year 2	Year 3	Year 4	Year 5	Present value
Costs	$100,000	$40,000	$10,000	$10,000	$10,000	$170,000
Benefits	0	$30,000	$40,000	$50,000	$60,000	$180,000

(b) The undiscounted cost-benefit ratio is $170,000/$180,000 or 0.94, indicating that it costs you 94 cents for each dollar of return.

(c) To calculate the present value of a discounted stream, each year's costs and benefits must be discounted by

$$PV = \frac{X}{(1+r)^t}$$

which produces the following table (for the first entry, $0.9524 \times 100,000 = \$95,239$, excluding rounding error.

	Year 1	Year 2	Year 3	Year 4	Year 5	Present value
Discount multiplier	0.9524	0.9070	0.8638	0.8227	0.7835	
Costs	$95,238	$36,281	$8638	$8227	$7835	$156,220
Benefits	$0	$27,211	$34,554	$41,135	$47,012	$149,911

(d) The cost-benefit ratio is now $156,220/$149,911, or 1.04, indicating an unfavorable CB ratio.

This example shows that, when the majority of costs occur early and benefits occur late in a project, discounting can change the conclusion about cost benefit.

References

Doubilet P, Weinstein MC, McNeil BJ. Use and misuse of the term "cost-effective" in medicine. N Engl J Med. 1986;314:253–6.

Drummond MF, O'Brien B, Stoddart GL, Torrance GW. Methods for the evaluation of health care programmes. 2nd ed. Oxford, UK: Oxford Medical Publications; 1997a.

Drummond MF, Richardson WS, O'Brien BJ, Levine M, Heyland D. Users' guides to the medical literature XIII. How to use an article on economic analysis of clinical practice A. Are the results of the study valid?. Evidence-Based Medicine Working Group. JAMA. 1997b;277:1552–7.

Evans RS, Pestotnik SL, Classen DC, Clemmer TP, Weaver LK, Orme JF Jr, et al. A computer-assisted management program for antibiotics and other antiinfective agents. N Engl J Med. 1998;338:232–8.

Feeny D, Furlong W, Torrance GW, Goldsmith CH, Zhu Z, DePauw S, et al. Multiattribute and single-attribute utility functions for the health utilities index mark 3 system. Med Care. 2002;40:113–28.

Finkler SA. The distinction between costs and charges. Ann Intern Med. 1982;96:102–9.

Ghani Z, Jarl J, Sanmartin Berglund J, Andersson M, Anderberg P. The cost-effectiveness of mobile health (mHealth) interventions for older adults: systematic review. Int J Environ Res Pub Health. 2020;17:5290.

Gold MR, Siegel JE, Russel LB, Weinstein MC. Cost effectiveness in health and medicine. New York: Oxford University Press; 1996.

Jacob V, Thota AB, Chattopadhyay SK, Njie GJ, Proia KK, Hopkins DP, et al. Cost and economic benefit of clinical decision support systems for cardiovascular disease prevention: a community guide systematic review. J Am Med Inform Assoc 2017;24:669–676. Available from https://doi.org/10.1093/jamia/ocw160. Accessed 3 July 2021.

Keeler EB, Cretin S. Discounting of life-saving and other nonmonetary effects. Manag Sci. 1983;29:300–6.

Kumar P, Sundermann AJ, Martin EM, Snyder GM, Marsh JW, Harrison LH, et al. Method for economic evaluation of bacterial whole genome sequencing surveillance compared to standard of care in detecting hospital outbreaks. Clin Infect Dis 2021;73:e9–e18. Available from https://doi.org/10.1093/cid/ciaa512. Accessed 3 July 2021.

Lara-Munoz C, Feinstein AR. How should quality of life be measured? J Investig Med. 1999;47:17–24.

Murphie P, Little S, McKinstry B, Pinnock H. Remote consulting with telemonitoring of continuous positive airway pressure usage data for the routine review of people with obstructive sleep apnoea hypopnoea syndrome: a systematic review. J Telemed Telecare. 2019;25:17–25.

O'Brien BJ, Heyland D, Richardson WS, Levine M, Drummond MF. Users' guides to the medical literature XIII. How to use an article on economic analysis of clinical practice B. What are the results and will they help me in caring for my patients?. Evidence-Based Medicine Working Group. JAMA. 1997;277:1802–6.

Sadoughi F, Nasiri S, Ahmadi H. The impact of health information exchange on healthcare quality and cost-effectiveness: a systematic literature review. Comput Methods Programs Biomed 2018;161:209–232. Available from https://doi.org/10.1016/j.cmpb.2018.04.023. Accessed 3 July 2021.

Sanders GD, Neumann PJ, Basu A, Brock DW, Feeny D, Krahn M, et al. Recommendations for conduct, methodological practices, and reporting of cost-effectiveness analyses: second panel on cost-effectiveness in health and medicine. JAMA 2016;316:1093–1103. Available from https://doi.org/10.1001/jama.2016.12195. Accessed 3 July 2021.

Torrance GW, Feeny D. Utilities and quality-adjusted life years. Int J Technol Assess Health Care. 1989;5:559–75.

Udvarhelyi IS, Colditz GA, Rai A, Epstein AM. Cost-effectiveness and cost-benefit analyses in the medical literature: are the methods being used correctly? Ann Intern Med. 1992;116:238–44.

United Kingdom National Institute for Health and Care Excellence (NICE). Medical Technologies Guidance (MTG27): virtual touch quantification to diagnose and monitor liver fibrosis in chronic Hepatitis B and C. London: United Kingdom National Institute for Health and Care Excellence (NICE); 2020. Available from https://www.nice.org.uk/guidance/mtg27. Accessed 3 July 2021.

United States Bureau of Labor Statistics (BLS). Economic News Release: Consumer Price Index (CPI). Washington, DC: Division of Consumer Prices and Price Indexes, United States Bureau of Labor Statistics (BLS), United States Department of Labor (DOL); 2021. Available from https://www.bls.gov/news.release/cpi.toc.htm. Accessed 3 July 2021.

United States Centers for Medicare & Medicaid Services (CMS). The National Health Expenditure Accounts (NHEA): Historical Data. Baltimore, MD: United States Centers for Medicare & Medicaid Services (CMS); 2021. Available from https://www.cms.gov/Research-Statistics-Data-and-Systems/Statistics-Trends-and-Reports/NationalHealthExpendData/NationalHeal thAccountsHistorical#:~:text=U.S.%20health%20care%20spending%20grew,spending%20 accounted%20for%2017.7%20percent. Accessed 3 July 2021.

Wang SJ, Middleton B, Prosser LA, Bardon CG, Spurr CD, Carchidi PJ, et al. A cost-benefit analysis of electronic medical records in primary care. Am J Med. 2003;114:397–403.

Part V
Ethical, Legal and Social Issues in Evaluation

Chapter 19
Proposing Studies and Communicating Their Results

Learning Objectives

The text, examples, and Food for Thought questions in this chapter will enable the reader to:

1. Write evaluation study proposals using a structured format.
2. Given a study proposal, referee (critique) it using the criteria presented in this chapter.
3. Frame reporting of evaluation study results as a process of communication.
4. Suggest alternatives to the traditional written report as means of communicating study results.
5. Contrast the methods that are customarily used to report quantitative studies with those used to report qualitative studies.
6. Identify the benefits and potential risks associated with "exiting the train" at different points in the disengagement process.

19.1 Introduction

This chapter addresses a set of critical issues for evaluation. These are the often "hidden" but important considerations that can determine if a study receives the financial support that make its conduct possible, if a study in progress encounters procedural difficulties, or if a completed study leads to settled decisions that might shape the improvement, adoption, or possibly the rejection of an information resource. Whether a study is funded—either internally by the organization developing or deploying the resource, or externally by a separate funding agency—often depends on how well the study plan is represented in a proposal. Whether a study encounters procedural difficulties often depends on the study team's adherence to general ethical standards as well as more specific stipulations built into an

© Springer Nature Switzerland AG 2022
C. P. Friedman et al., *Evaluation Methods in Biomedical and Health Informatics*, Health Informatics, https://doi.org/10.1007/978-3-030-86453-8_19

evaluation contract. Whether a study leads to settled decisions depends on how well and convincingly the study findings are represented in various reports.

Studies can succeed or fail to make an impact for reasons other than the technical soundness of the evaluation design: the considerations that have occupied so much of this book. Conducting an evaluation study is a complex and time-consuming effort, requiring negotiation skills and the ability to compromise between conflicting interests. Members of a study team conducting an evaluation must be communicators, managers, and politicians—in addition to methodologists.

This chapter provides a glimpse into many of these less technical issues. The focus is on proposals that express study plans, the related process of refereeing other people's proposals and reports, and the know-how to communicate the study results via many possible mechanisms. The treatment of each of these issues here includes just the rudiments of what a fully accomplished study team must know and be able to do. Readers should access additional resources to amplify what is presented here, including the references listed at the end of the chapter.

19.2   Writing Evaluation Proposals

19.2.1   Why Proposals Are Necessary and Difficult

A proposal is a plan for a study that has not yet been performed (Gerin and Kepelewski 2011; Miner and Ball 2019). A proposal usually also makes a case that a study should be performed and, often as well, a case that the recipient of the proposal—the prospective funder—should make available the financial and other resources needed to conduct the study. In many circumstances, evaluation studies must be represented in formal proposals before a study begins. This is required for several reasons. First, the negotiations about the scope and conduct of a study require a clear representation of the study plan. Second, if the study team is seeking resources from an external agency to conduct the study, funding agencies almost always require a formal proposal or application. Third, students conducting evaluation studies as part of their thesis or dissertation research must propose this work to their committees, with formal approval required before the work can begin. Fourth, committees overseeing research on human subjects—also called ethics committees or institutional review boards (IRBs)—require written advance plans of studies to ensure that these plans comply with ethical standards for the conduct of a study. Field studies of clinical information resources usually involve protected health information about care recipients, which requires that these studies obtain ethics approval. Laboratory studies of information resources supporting clinical work, research, or education may also require human subjects review if data are collected directly from members of the general public, practitioners, study teams, or students who are participants in these studies.

When studies are nested within information resource development projects that are funded by the organization developing the resource, a formal proposal for the study may not technically be required. Nonetheless, creation of a written description of the study plan is still a good idea. Sound evaluation practice always includes a process of negotiation with important members of the client group for whom the study is conducted. These negotiations cannot occur properly without some written representation of the study plan. An evaluation "contract" based on an unwritten understanding of how a study will be conducted, absent any kind of written proposal, can lead to significant misunderstandings. With no written anchor for the conduct of the study, these misunderstandings can be difficult to resolve. A written evaluation plan, even when not required to secure funding, is also an important resource to support study planning and execution. Conducting a study without a written plan is like building a house without a blueprint. The study team is always feeling its way along. Changes in a plan are always possible, but it is helpful for the study team to be keenly aware of the changes being made in the originally conceived plan. Although they are described somewhat differently, qualitative studies can be reflected in a written study plan just as readily as quantitative studies.

Evaluation studies are particularly difficult to describe, and several helpful resources are available (Miller et al. 1989; Wilce 2021; United States Centers for Disease Control and Prevention (CDC) 2011). Writing a study proposal is challenging largely because it requires the author to describe events and activities that have not yet occurred. In general, writing a plan is intrinsically more difficult than describing, in retrospect, events that have occurred and have been experienced by the people who would describe them. Writers of proposals must portray their plans in ways that are logical and comprehensible. Uncertainty about what ultimately will happen when the study is undertaken must be acknowledged but not overstated in a way that will make readers of the plan lack confidence in the study team. In addition to having a clear idea of what they want to do, proposal writers must know what constitutes the complete description of a plan (what readers of the plan will expect to be included), the format these descriptions are expected to take, and the style of expression considered appropriate.

Writing a persuasive proposal is part science and part art. Although this assertion is impossible to confirm, it is likely that many studies, that would have been very valuable had they been conducted, were never performed because of the study team's inability to describe the proposed study satisfactorily in writing.

19.2.2   Format of a Study Proposal

One format to describe a study is embodied in the U.S. Public Health Service Form 398 (PHS 398) (United States National Institutes of Health (NIH) 2021).[1] Even if the study team is not planning to apply to the U.S. government for funds, this format provides a sound, proven generic structure for articulating a study plan. (Writing a proposal is difficult enough; having to invent a format is yet one more challenging thing for the study team to do.) Another reason to use the format of PHS 398 is that many, perhaps most, readers have grown accustomed to study plans in this format and indeed may be writing their own proposals using it. They then tacitly or overtly expect the details of the study plan to unfold in a particular sequence. When the plan unfolds in the expected sequence, it is easier for referees and other readers to understand.

A complete proposal using PHS 398 has multiple parts; proper completion of all of them is important if a team is applying to a U.S. government agency for research funding. This chapter focuses on the parts of the form that express the study plan itself:

Specific Aims
Research/Study Strategy

1. Significance
2. Innovation
3. Approach

Bibliography and References Cited
For proposals that are submitted for funding, study teams usually find themselves challenged to make their proposals terse enough to comply with the page length restrictions. Page limits vary from funding agency to funding agency; and within agencies they can vary. Writing proposals is thus usually an exercise in editing and selective omission. Rarely are study teams groping for things to say about their proposed study. In many cases a single proposal describes a large development project of which evaluation is one component. Section 19.2.4 of this chapter explores how to manage that situation.

19.2.3   Suggestions for Expressing Study Plans

This section provides guidance for expressing study designs using the format of Form PHS 398, and includes specific suggestions for describing the study's "Approach". A checklist for writing proposals, and to support refereeing them as discussed later, is included at the end of this section.

---

[1] Details of the official version of Form 398 change from time to time, as do other agencies' required formats. Study teams planning to apply to a funding agency should consult and follow that agency's detailed prescriptions for format and content.

Specific Aims In this section of the proposal the study team describes what it hopes to achieve in the proposed work. The format of this section should consist of a pre-amble, which provides a general rationale for the study, followed by an expression of the specific aims as discrete entities. *For evaluation studies, the kind of study questions described in Sect. 3.2 can serve to express the specific aims.* It is best to number the discrete aims (e.g., Aim 1, Aim 2 or Question 1, Question 2, etc.), so that later in the proposal these can be referenced by number. As a general rule of thumb, a study should have three to seven specific aims/questions. If the study team finds itself expressing the study with one or two aims, the aims may be too general and can be subdivided. Correspondingly, if a study is expressed with eight or more aims, the study itself may be too broad or the aims may be stated with excessive granularity. Even though specific investigative questions might change, particularly in a qualitative study, the general purposes or "orienting" questions that guide the study from the outset can be stated here.

Significance This section should establish the need for this particular study/project, not a general need for studies of this type. After finishing this section, the reader should be able to answer this question: "How will stakeholders be better off if the aims of this study are accomplished?" Although it is not solely a literature review, this section must substantiate its major points with appropriate citations to the literature, using the methods of Evidence-Based Health Informatics described in Chap. 5. For evaluation studies, the pertinent literature might include unpublished documents or technical reports about the information resource under study. In general, it is not a good idea for the study team to cite too much of its own prior work in this section.

Innovation The innovation section of a proposal is, in general, more important to research studies than evaluation studies. For research studies, the section should address any aspects of the study, including the methods employed, that are novel or "firsts of their kind". In informatics evaluation studies, some caution is required. The prospective funders may be primarily interested in the likelihood that the study will successfully and informatively address the questions that guide it, and less interested in the novelty of the methods and approach. Study teams should tie any innovation to its benefit in addressing the study questions, and not to the study team's own interests in seeing how useful a novel method will turn out to be.

Approach This section contains a description of the study being proposed. It includes the following:

- *Restatement of the study aims*: Even though the study aims were expressed earlier in proposal, this repetition helps the reader bring the study back into focus.
- *Overview of the study design*: To give the reader the "big picture," this should establish the type of study being proposed, as discussed in Chap. 3. If a field study is proposed, it is important to explain how the study will fit into its health care, research, or educational environment. If the study is quantitative, explain

whether the design is descriptive, comparative, or correlational—and why this choice was made. Provide an overview of the study groups and the timing of the intervention. If the study is qualitative, include an overview of the data collection strategies and analytic procedures that will be employed

- *Preliminary Studies:* This section describes previous relevant work undertaken by the study team and their collaborators. This section emphasizes results of this previous work and how the proposed study builds on these results. This will serve to reassure prospective funders that the team can actually carry out successfully everything that is proposed. If measurement studies have been performed previously, for example, this section describes the methods and results of these studies. Any pilot data and their implications are included here. Depending on the funding agency, this may be the place for study team members to describe their own backgrounds and qualifications, although in applications for U.S government grants, this is inappropriate because there are separate sections of the application requiring team members to include their professional biographical sketches.

- *Study method details*: For quantitative studies, this must include specific information about participants and their sampling/selection/recruitment; investigative procedures with a clear description of the intervention (the information resource and who will use it, in what forms); description of the independent and dependent variables; how each of the variables will be measured (the instrumentation, with reliability/validity data if not previously reported); a data analysis plan (what statistical tests in what sequence); and a discussion of sample size, which in many cases will include a formal power analysis. Samples of any data collection forms, or other instruments, should be provided in an appendix to the proposal, if appendices are allowed.

  For qualitative studies, the study details include the kinds of data that will be collected (who is anticipated to be interviewed, the types of documents that will be examined, the types of activities that will be observed); how study documents will be maintained and by whom; and the plan for consolidating and extracting patterns and themes from the data. The reader of the proposal, if conversant with qualitative methods, will understand that many of the detailed ideas expressed in this section may change as the study unfolds. For the benefit of readers who are unfamiliar with the nature of qualitative methods, it is wise to provide a convincing rationale and more details about the strategies being proposed than one would for a quantitative study.

  In most circumstances, and especially when a study plan contains four or more aims/questions, it is best to describe the study method for each aim/question individually and in sequence.

- *Project management plan:* For evaluations, it is important to describe the study team and its relation to the resource development team and how decisions to modify the study, should that be necessary, will be made. The "role" figure (Fig. 2.2), and related concepts introduced in Chap. 2 of this book, may help determine the content of this section.

- *Communication/reporting plan:* For evaluations, it is important to explain the report(s) to be developed, by whom, and with whom they will be shared in draft and final form. Communication is discussed in detail in Sect. 19.4 below.
- *Timeline:* The proposal should include, preferably in chart form, a timeline for the study. The timeline should be as detailed as possible given page length limitations.

Proposal Quality Checklist

A. Specific aims/questions

  1. Establishes a numbering system (Aim/Question 1, Aim/Question 2, ...)
  2. Includes preamble followed by a list of numbered aims/questions

B. Significance

  1. Establishes the need for this study/project (not a general need for studies of this type)
  2. States how we will be better off if we know the answers to these questions
  3. Uses the literature extensively (30+ references)
  4. Does not cite too much of the study team's own work

C. Approach

  1. Does the proposal use the structure of the aims to organize the approach? Are the following included?

    (a) (Re)statement of aims/study questions
    (b) Overview of design
    (c) Management plan
    (d) Reporting plan
    (e) Timeline in as much detail as possible

  2. Progress report/preliminary studies

    (a) Describes relevant previous work of principal study team or collaborators
    (b) Emphasizes results of this work and how proposed study builds on these results
    (c) Does not paraphrase study team's curriculum vitae
    (d) Reports pilot data

  3. For quantitative studies

    (a) Participants and their selection/recruitment
    (b) Experimental procedures/intervention
    (c) Independent and dependent variables

(d) How variables will be measured (instruments and any reliability/
validity data not previously reported)
(e) Data analysis plan (which statistical tests in what sequence)
(f) Power analysis and discussion of sample size

4. For qualitative studies

(a) Kinds of data that will be collected
(b) From whom data will be collected
(c) How study documents will be maintained
(d) Plan for consolidating and generating themes from data

D. In general

1. Does the format/layout help the reader understand the project?
2. If there is an unsolved problem, does principal study team show awareness of the issues involved and how others have addressed them?
3. Is the cascade problem (if any) adequately addressed?
4. Are specimen data collection forms included in the appendix?

19.2.4 Special Issues in Proposing Evaluations

Studies Nested in Larger Projects Many evaluations are proposed not in a free-
standing manner but rather as part of a larger development project. In this case the
evaluation is best expressed as one specific aim of the larger study, possibly with
sub-aims for each study question. The background and significance of the evaluation is then discussed as part of the "Background and Significance" section of the
proposal; the same would be true for the "Preliminary Studies" section of the proposal. The evaluation methods would be described in detail as a major part of the
"Approach" section. Under these circumstances, the specific evaluation plans must
be described in a highly condensed form. Depending on the scope of the evaluation,
this may not be a problem. If sufficient space to describe the evaluation is unavailable in the main body of the proposal, the study team might consider including one
or more technical appendices to provide further detail about the evaluation.[2]

---

[2] Some caution is required here if the proposal is to be submitted for U.S. federal funding. The
guidelines for Form 398 specifically state that the appendix should not be used "to circumvent the
page limitations of the research plan." It may, for example, be acceptable to include specimen data
collection forms in the appendix and also to provide in the appendix more technical information
about the forms. However, if the main body of the proposal says that "the evaluation design is
described in Appendix A" and the appendix in fact contains the entire description of the design, the
proposal will likely be considered unacceptable by the funding agency and will be returned for
modification. Study teams applying for U.S. federal funding should read the applicable policy on
appendices very carefully.

Currently Irresolvable Design Issues At the time they write a proposal, study teams often find themselves with a design issue that is heavily dependent on information that is currently unavailable or otherwise difficult to resolve. In this case, the best strategy is *not* to try to hide the problem. An expert, careful reader will probably detect the unmentioned problem and consider the study team naive for not being aware of it. Hence, the study team should expose an unsolved problem, show awareness of the issues involved, and, above all, how they or others in comparable situations have addressed issues of this type. This strategy often succeeds in convincing the reader/reviewer of the proposal that although the study team understandably does not know what to do now, it will make a good decision when the time comes during the execution of the study.

The "Cascade" Problem A related issue is the so-called cascade problem, which occurs when the plan for *Stage N* of a study depends critically on the outcome of *Stage N – 1*. There is no simple solution to this problem when writing a proposal. The best approach is to describe the dependencies, how the necessary decisions about *Stage N* will be made, and possibly describe in some detail the plan for *Stage N* under what the study team considers the most likely outcome of *Stage N – 1*. Some proposal writers consider the existence of a cascade problem to indicate the boundaries of a study. If the outcome of *Stage N* depends critically on *Stage N – 1*, *Stage N* becomes a different study and is described in a separate proposal that is written later, after the work on the previous stage has started and progressed.

19.3 Refereeing Evaluation Studies

After a study proposal is submitted to a funding body, it is usually refereed—that is, reviewed for merit by one and usually more individuals experienced in evaluation design and methods. It is therefore useful for those writing a proposal for an evaluation study also to understand the refereeing process and some its own challenges (Wieczorkowska and Kowalczyk 2021; Pier et al. 2018). In addition, once a study team has succeeded in obtaining funding for several research and evaluation projects, funding organizations are quite likely to ask members of the team to referee other study proposals. Therefore, this section briefly discusses how one goes about reviewing a proposed study. Many of these concepts also apply to reviewing a completed study that has been submitted for formal presentation or publication.

Many funding organizations or journals provide referees with a checklist of review criteria they would like addressed, which obviously take precedence over the generic advice that follows. In general, the questions that referees can ask themselves, when refereeing an evaluation proposal, include the following:

- Is there a study question and specific aims, and are they clearly formulated? Often there is more than one question per study.

- Is the study question important and worth answering, or is the answer banal or already well-established from other studies?
- Are the investigative methods described in sufficient detail to determine what is being proposed or what was already done?
- Are these methods appropriate to answer the study question, given the potential biases and confounding factors; that is, is the study design likely to result in work that is internally valid?
- Is the study setting sufficiently typical to allow useful conclusions to be drawn for those working elsewhere; that is, is the study externally valid? (This point may not always be crucial for an evaluation done to satisfy a "local" need.)
- Is it feasible for the study teams to carry out the methods described within the resources requested?
- Does the proposal address the criteria or any special instructions provided by the funding agency? (Sometimes, the funding agency staff will provide special instructions to a particular referee, based on that referee's background and experience.)

For completed studies submitted as a report, or for more formal presentation or publication, the following additional criteria may apply:

- Does the interpretation of the data reflect the sources of the data, the data themselves, and the methods of analysis used?
- Are the results reported in sufficient detail? In quantitative studies, do all summary statistics, tables, or graphs faithfully reflect the conclusions that are drawn? In qualitative studies, is there a clear and convincing argument? Is the writing sufficiently crisp and evocative to lend both credence and impact to the portrayal of the results?
- Are the conclusions valid, given the study design, setting and results, and other relevant literature?

Some ethical issues related to refereeing are important to emphasize. Prospective referees who believe they have conflicts of interest that could affect their judgment about the proposed work should of course decline the assignment. In some cases, individuals might perceive the potential of appearance of a conflict, but still believe they can be unbiased in their assessments. In such cases, it is wise still to decline the assignment, as the appearance of a potential bias can be as erosive as the bias itself. Potential referees should also decline the assignment if they do not have direct experience in the evaluative methods being proposed. For example, persons experienced only with quantitative evaluations should not referee proposals of largely qualitative work, and vice versa.

19.4 Communicating the Results of Completed Studies

19.4.1 What Are the Options for Communicating Study Results?

Once a study is complete, the results must be communicated to the stakeholders and others who with an interest in the study and its results. In many ways, *communication* of evaluation results, a term preferred over *reporting* in this book, is the most challenging aspect of evaluation. Several works addressing the challenges of reporting in informatics (Brender et al. 2013; Sepucha et al. 2018), and more generally (Creswell et al. 2011; Torres et al. 2005) are available for further reference.

Basic theory (Shannon 1949) tells us that successful communication requires a sender, one or more recipients, and a channel linking them, along with a message that travels along this channel, and a clear understanding of the message by the receiver (Table 19.1). Seen from this perspective, successful communication is challenging in several respects. Most of all, success requires that the recipient of the message attend to the information the message contains. That is, for evaluations, the recipient must read the report if it is in writing or participate in a meeting intended to convey evaluation results. This requirement challenges the study team to create a report the stakeholders will want to read or to choreograph a meeting they will be motivated to attend. As noted, successful communication also requires that the recipient understand the message, which challenges study teams to draft written documents at appropriate reading levels, with audience-appropriate technical detail. This may require different versions of the written report, developed with care to convey the same messages, to appeal to several different audiences.

Table 19.1 Reporting as a communication process

Elements of communication	Equivalent in evaluation studies
A sender of the message	The study team or investigative team
A message	The results of the study
A communication channel	A report, conversation, meeting, Web site, journal, newspaper or newsletter article, broadcast radio or television, etc.
Recipient(s) of the message	The stakeholders and other audiences for the study results

Overall, study teams should recognize that their obligation to communicate does not end with the submission of a written document comprising their technical evaluation report. The report is one means or channel for communication, not an end in itself. Depending on the nature, number, and location of the recipients, there is a large number of options for communicating the results of a study:

- Written reports

  - Document(s) prepared for specific audience(s)
  - Internal newsletter article
  - Published journal article, with appropriate permissions
  - Monograph, picture album, or book

- One-to-one or small group meetings

  - With stakeholders or specific stakeholder groups
  - With general public, if appropriate

- Formal oral presentations

  - To groups of project stakeholders
  - Conference presentation with poster or published paper in proceedings
  - To external meetings or seminars

- Internet

  - Project Web site
  - Online preprint
  - Internet based journal

- Other

  - Video describing the study and information resource
  - Interviews with journalists

A written report is not the sole medium for communicating evaluation results. Verbal, graphical, or "multimedia" approaches can be helpful as ways to enhance communication with specific audiences. Another useful strategy is to hold a "town hall meeting" to discuss a written report after it has been released. Photographs or videos of the work setting for a study, the people in the setting, and the people using the resource can be effective. If appropriate permissions are obtained, these images—whether included as part of a written report, shown at a town hall meeting, or placed on a Website—can be worth many thousands of words. The same may be true for recorded statements of resource users. If made available, with permission, as part of a multimedia report, the voices or even video of the participants can convey the feelings behind the words that can enhance the credibility of the study team's conclusions.

19.4.2   What Is the Role of the Study Team: Reporter or Change Agent?

In addition to the varying formats for communication described above, study teams have other decisions to make after the data collection and analysis phases of a study are complete. One key decision is what personal role they will adopt after these formal investigative aspects of the work are complete. They may elect only to communicate the results, or they may also choose to persuade stakeholders to take specific actions in response to the study results, and perhaps even assist in the implementation of these actions. This raises a key question: Is the role of a study team to simply record and communicate study findings or does it include engagement with the study stakeholders to help them change how they work as a result of the study?

Answering this question about the role of a study team requires recognition that an evaluation study, particularly a successful one, has the potential to trigger a series of events. The study team's potential role in this cascade is depicted below:

Conduct study: data collection and analysis
Communicate the results as a neutral report compliant with the contract

- Interpret the results and communicate the meaning of these to stakeholders
- Recommend actions for stakeholders to take
- Suggest to stakeholders how to implement the recommended actions
- Participate as a change agent in the implementation process

Viewing the aftermath of a study in this way is most important when a study is conducted for a specific audience that needs to make decisions and then take specific actions requiring careful planning, but it also can assist the study team when the intended consequences of the evaluation are less clear.

Some study teams—perhaps enthused by the clarity of their results and an opportunity to use them to improve health care, biomedical research, or education—prefer to go beyond reporting the results and conclusions to making recommendations, and then helping the stakeholders to implement them. Figure 19.1 illustrates the

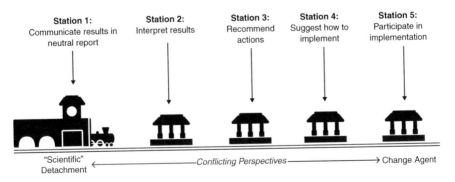

Fig. 19.1 Scientific detachment vs. change agent: when to get off the train?

dilemma often faced by study teams about whether to retain what might be seen as their scientific detachment and merely report the study results—metaphorically leaving the "train" at the first or second "stations"—or stay engaged somewhat longer. Study teams who choose to remain may become engaged in helping the stakeholders interpret what the results mean, guiding them in reaching decisions, and perhaps even in implementing the actions decided upon. The longer they stay on the train, the greater the extent to which study teams must leave behind their scientific detachment and take on a role more commonly associated with implementers or change agents. Some confounding of these roles is inevitable when the study is performed by individuals within the organization developing the information resource under study. There is no hard-and-fast rule for deciding when to leave the train; the most important realization for study teams is that the different stations exist and that a decision about where to exit is inevitable. As discussed in the following section, the evaluation contract can provide guidance for this decision.

19.4.3 Role of the Evaluation Contract

In all studies, the evaluation contract assumes a central role in shaping what will happen after the data collection and analysis phases of the study are completed. A possible dilemma arises if an audience member, perhaps the director of the resource development project under study, disagrees with the conclusions of a draft evaluation report. The contract, if properly written, protects the study team and the integrity of the study, often, but not always, by making the study team the final authority on the content of the report. Under these circumstances, the contract stipulates that reactions to draft evaluation reports have the status of advice, and the study team is under no obligation to modify the report in accord with reactions. In practice, the reactions to draft evaluation reports usually do not raise these kinds of dilemmas but rather provide crucial information to the study team that improves the report. Nonetheless, these kinds of problematic situations can arise, and the contract offers a measure of guidance in managing through them.

The evaluation contract should also help study teams decide what to include in their initial report. In particular, the contract could help study teams decide to what degree they should pass judgment on the information resource—for example, by declaring a project successful—or leave such judgments to the stakeholders. If the contracted role of the study team does include passing judgment, these judgments should be specific, justified by reference to study results, and not unduly generalized beyond the settings and information resources studied. The contract should also specify the circumstances under which the study results could be published in a professional journal and thus contribute, for the benefit of others, to the evidence base of informatics.

19.5   Specific Issues of Report Writing

As a practical matter, almost all evaluations result in a written report, irrespective of whatever other communication modes they employ. Deciding what to include in a written evaluation report is often difficult. As a study is nearing completion, whether the study is quantitative or qualitative in primary orientation, much data will have been collected and many analyses performed. Interpretations of the data usually have been fueled by reports of other studies in the literature and the study team's previous personal experiences. The key question for reporting, as it is when deciding what to study (Miller and Sittig 1990), lies in distinguishing what is necessary, in contrast to what might be interesting, to include for the various stakeholders.

Because evaluations are carried out for specific groups with specific interests, the task of report writing benefits from attention to what these groups need to know, and may be shaped by the evaluation contract. If the report is for a single person with a specific decision to make, that individual's interests guide the contents of the report. More typically, however, the study team is writing for a range of audiences, perhaps including all members of the organization deploying the resource, the specific resource users, the lay public, biomedical informaticians and ICT personnel, policy makers, and others. Members of each audience expect more detail in the areas of most interest to them. There are inevitable challenges to make the report sufficiently brief to be read by busy people, to include enough detail to convincingly address the study questions, and to meet the needs of specific stakeholder groups. One strategy, best suited to a large study, is to produce modular reports. The study team could describe in an overview document the details of the information resource, the problem it is addressing, and the setting where it has been deployed, and then refer to this overview in subsequent evaluation component reports. Each component report would then address a specific aspect of the report and/or the information needs of a particular audience.

19.5.1   The IMRaD Format for Writing Study Reports

The IMRaD (Introduction, Method, Results, and Discussion) acronym describes a generic format for reporting any study based on empirical data (Sollaci and Pereira 2004). This would include evaluation studies. While best suited for communicating evaluation study results to technical/scientific audiences, this model encourages study teams to be clear about the evaluation questions that were addressed and the data that were used or collected to answer the questions—helping the audiences determine if the inferences drawn from the data are justified. An IMRaD report includes the following components:

1. Introduction to the problem, review of relevant literature, and statement of study aims or questions.

2. Methods employed, ideally described in enough detail to allow another study team to replicate the study.
3. Results of data analyses, often summarized in tables or figures.
4. Discussion of the results and potential limitations of the study, and conclusions drawn in the context of other studies.

IMRaD implies a linear flow in execution of a study that aligns very well with the quantitative approaches to evaluation. With some allowances for distinctive features of qualitative studies, as discussed in the following section, this general structure of reporting can be a useful guide for qualitative and mixed methods studies as well.

19.5.2 Writing Qualitative Study Reports

The goals of reporting a qualitative study may include describing the resource; how it is used; how it is "seen" by various groups; and its effects on people, their relationships, and organizations. To these ends, the qualitative study team will typically include direct quotations, interesting anecdotes, revealing statements, lessons learned, and examples of the insights, prejudices, fears, and aspirations that study participants expressed—all with due regard to confidentiality and the contract or memorandum of understanding negotiated at the study outset.

Reporting of qualitative studies raises a number of special issues:

• In comparison with a quantitative study, writing a qualitative report is less formulaic and often more challenging to the written communication skills of the study team. Conveying the feelings and beliefs, and often the hopes and dreams, of people in their work environment in relatively few words can require expressive skills that transcend those required in quantitative studies. Reports typically require numerous drafts before they communicate as intended.

• As in all evaluation studies, it is essential to respect the confidentiality of study subjects. In qualitative studies, fieldwork directly exposes the study subjects to the study team, and the use of quotations and images in reports can readily reveal identities. Measures to be taken to protect subjects should be laid out in the evaluation contract, also recognizing that the stipulations may need to be altered to address difficult problems or conflicts as they emerge. Before distributing an evaluation report, the study team must show each participant any relevant passages that might allow them to be identified and allow each participant to delete or modify the passage, if there are any privacy concerns.

• The study report is typically an evolving document, written in consultation with the client group. Version control is important, and it is often unclear when the report is finished. Here again, the evaluation contract may be helpful for determining when the report is sufficiently mature for release.

• The report can itself change the environment being studied by formalizing and disseminating insights about the information resource. Thus, study teams must adopt a responsible, professional approach to its writing and distribution.

- It can be difficult to summarize a qualitative study in a relatively small number of pages without losing the richness of the personal experiences that qualitative studies strive to convey. There is a danger that reports describing such studies can be unpersuasive or come across as more equivocal than the underlying data and analytic results can actually justify. To counteract this problem, authors can use extracts from typical statements by subjects or brief accounts of revealing moments to illustrate and justify their conclusions in the same way quantitative study teams summarize a mass of data in a set of tables or statistical metrics.

19.6 Conclusion

This chapter called particular attention to issues of communication. Before a study even begins, and often to secure the resources needed to carry out the study, communicating the planned intent of the study team in a proposal is an essential step. As discussed, it is challenging for many reasons to peer into the future and describe something that has not yet occurred. The structured formats and guidelines offered in this chapter can, in this regard, be of significant assistance.

Communication challenges also arise as the end of a study approaches. Evaluation studies require extended, arduous, and exacting work. It would be tempting for a study team to complete their data analyses, write up the results, submit them, and then move on to the next study: Mission completed! This chapter sought to clarify that, particularly for studies in service to stakeholders who want to use the study results to inform major decisions, significant work often lies ahead. Just as there was a myriad of choices in designing and conducting the study itself, there is a myriad of choices for communicating the results. Most important, the results of a study must be communicated to stakeholders in ways they can understand and put to beneficial use.

Food for Thought

1. With regard to Fig. 19.1, why might study teams performing qualitative studies be inclined to "stay on the train" longer than teams performing quantitative studies?
2. Refer to the *YourView* portal resource at fictitious Nouveau Community Hospital, as introduced in Self-test 2.1 in Sect. 2.4. Suppose that a completed study pertaining to the privacy aspects of *YourView* found that the system was poorly designed and that patients had good reason to be concerned about their privacy. When the study team presented these results to the senior management of the hospital, they asked the team to submit the report only to them and destroy all other copies. What do you think the study team should do and what would be the rationale for the team's action?

3. A study team has completed a quantitative study of the effects of an app on blood pressure control and finds it to be highly effective. The study uses very sophisticated statistics. The team has been asked by the organization that paid for the study to give a report of the results to low-literacy patient groups in order to motivate them to use the app. What strategies might work well to communicate these results?

References

Brender J, Talmon J, de Keizer N, Nykänen P, Rigby M, Ammenwerth E. STARE-HI - statement on reporting of evaluation studies in health informatics. Appl Clin Inform. 2013;4:331–58.

Creswell JW, Klassen AC, Plano Clark VL, Smith KC. Best practices for mixed methods research in the health sciences. Bethesda, MD: United States National Institutes of Health (NIH), Office of Behavioral and Social Sciences Research; 2011.

Gerin W, Kepelewski CH. Writing the NIH Grant proposal. 2nd ed. Thousand Oaks, CA: Sage; 2011.

Miller PL, Sittig DF. The evaluation of clinical decision support systems: what is necessary versus what is interesting. Med Inf. 1990;15:185–90.

Miller RA, Patil R, Mitchell JA, Friedman CP, Stead WW, Blois MS, et al. Preparing a medical informatics research grant proposal: general principles. Comput Biomed Res. 1989;22:92–101.

Miner JT, Ball KC. Proposal planning and writing. 6th ed. Greenwood, ABC-CLIO, LLC: Santa Barbara, CA; 2019.

Pier EL, Brauer M, Filut A, Kaatz A, Raclaw J, Nathan MJ, et al. Low agreement among reviewers evaluating the same NIH grant applications. Proc Nat Acad Sci 2018;115:2952–2957. Available from https://www.pnas.org/content/115/12/2952. Accessed 17 June 2021.

Sepucha KR, Abhyankar P, Hoffman AS, Bekker HL, LeBlanc A, Levin CA, et al. Standards for UNiversal reporting of patient decision aid evaluation studies: the development of SUNDAE checklist. BMJ Qual Saf. 2018;27:380–8.

Shannon CE. The mathematical theory of communication. Urbana, IL: University of Illinois Press; 1949.

Sollaci LB, Pereira MG. The introduction, methods, results, and discussion (IMRAD) structure: a fifty-year survey. J Med Libr Assoc. 2004;92:364–7.

Torres RT, Preskill H, Piontek ME. Evaluation strategies for communicating and reporting: enhancing learning in organizations. 2nd ed. Thousand Oaks, CA: Sage; 2005.

United States Centers for Disease Control and Prevention (CDC). Developing an effective evaluation plan: setting the course for effective program evaluation. Atlanta, GA: United States Centers for Disease Control and Prevention (CDC), National Center for Chronic Disease Prevention and Health Promotion, Office on Smoking and Health, Division of Nutrition, Physical Activity, and Obesity; 2011. Available from https://www.cdc.gov/obesity/downloads/cdc-evaluation-workbook-508.pdf. Accessed 14 June 2021.

United States National Institutes of Health (NIH). Application for a Public Health Service Grant: Grant Application Form PHS 398 (Revised 03/2020). Washington, DC: United States Department of Health & Human Services, Public Health Service; 2021. Available from https://grants.nih.gov/grants/funding/phs398/phs398.html. Accessed 14 June 2021.

Wieczorkowska G, Kowalczyk K. Ensuring sustainable evaluation: how to improve quality of evaluating grant proposals? Sustainability. 2021;13:28–42.

Wilce M. CDC evaluation plan template. Atlanta, GA: United States Centers for Disease Control and Prevention (CDC); 2021. Available from https://www.cdc.gov/tb/programs/Evaluation/Guide/PDF/Evaluation_plan_template.pdf. Accessed 14 June 2021.

Chapter 20
Ethics, Safety, and Closing Thoughts

Learning Objectives

The text, examples, Food for Thought question, and self-tests in this chapter will enable the reader to:

1. Describe the role of Institutional Review Boards (IRBs) for ensuring ethical evaluation studies that are conducted for research purposes.
2. Explain how gaining IRB approval can be of benefit to an evaluation study team.
3. Describe the difficulties that can be encountered when a study team needs to gain IRB approval.
4. Discuss what is meant by ICT safety, how it relates to ethics, and how evaluation studies can help to ensure the safe implementation and use of ICT.
5. Describe what ethical issues might be encountered during each phase of the evaluation process.
6. Summarize the major trends in the discipline of informatics that will likely impact the work of informatics evaluation study teams in the future.

20.1 Introduction

This final chapter introduces a number of thought-provoking challenges that need to be addressed if ICT evaluation efforts are to succeed in helping to improve ICT and its impact of the quality of healthcare. It first describes ethical challenges related specifically to informatics evaluation projects, including unique human subjects issues and assuring the safety of ICT itself. It then focuses on ethical challenges during each phase of an evaluation project. The chapter concludes with a description of the authors' closing thoughts about the future of evaluation methods and efforts in informatics.

© Springer Nature Switzerland AG 2022
C. P. Friedman et al., *Evaluation Methods in Biomedical and Health Informatics*, Health Informatics, https://doi.org/10.1007/978-3-030-86453-8_20

20.2   Broad Ethical Issues in ICT Evaluation

Ethics in general means "the moral principles that govern a person's behavior or the conducting of a behavior" (Seroussi et al. 2020, p. 7). There are many resources about health sciences research ethics that are applicable to informatics, but resources about the ethics of the field of informatics such as the codes of ethics of the American Medical Informatics Association (Petersen et al. 2018) and International Medical Informatics Association (IMIA n.d.) are especially important reading for informaticians. Here, we will discuss issues even more specific to informatics *evaluation* studies, many of which involve protection of human subjects. An especially challenging issue is that standard practices for managing ethical issues in controlled trials may not be suitable for a qualitative informatics evaluation. The fine line between research and operations studies can be hard to distinguish in informatics, which complicates ethical considerations. And finally, while informatics studies hold promise for decreasing inequities, they can also increase them.

20.2.1   Standard Practices for the Protection of Human Subjects

Many informatics evaluation projects involve human subjects and raise the same ethical issues that arise in all human subjects research in the health care domain. However, there exist several special issues for informatics evaluation study teams. First, there is debate about ethical issues related to newer types of studies, such as analysis of social media platform data or secondary use of data, that have still not been resolved. Second, there is at times an unclear boundary between evaluation studies for internal purposes and for research-oriented studies, which means that the required oversight for studies is unclear as well. Finally, there exists great potential for increasing inclusion of diverse populations in informatics evaluation projects.

Suitability of IRB Oversight for Informatics Committees called Institutional Review Boards (IRBs) exist in organizations that conduct human subjects research studies to assure that researchers are adhering to federal regulations concerning human subjects. Studies must be proposed to IRBs and approved by them before the studies can commence. Studies that have higher risks to subjects need to go through a full review by an IRB. Other types of studies that are considered low risk may be designated as exempt from review, or may qualify for expedited review, but must be seen by the IRB regardless. An application must be submitted for both types of studies. Most IRBs process exempt or expedited proposals fairly quickly, depending on their workload. Studies that are more invasive and therefore of higher risk to subjects, must be fully reviewed, and this process can take some time. Studies of information technology applied to health care, research, and education each invoke human subjects' considerations in slightly different ways, so it is always wise to consult with an IRB as early as possible in planning.

When they are engaged, IRBs will offer study teams specific instructions regarding, for example, from whom informed consent must be obtained and how. These instructions make the lives of the study team members easier in the long run by removing these difficult considerations from the sphere of the team's sole judgment, allowing the study to proceed free from concerns that appropriate ethical procedures are not being followed. There is often a fine line between quality improvement evaluations of operational activities, which may not be considered research and therefore might not need IRB approval, and projects considered to be research. These distinctions are subtle and are usually made on a case-by-case basis. When in doubt, study teams should request a determination from the relevant IRB whether or not its review and approval is needed. Many informatics evaluation projects are considered of such low risk to subjects that the IRB will expeditiously approve them in the exempt or expedited categories.

In the U.S., IRBs differ from organization to organization, usually not in major ways since they all adhere to federal regulations, but in minor ways that can complicate planning for studies. For example, if a study has three senior investigators at three different institutions, each of them will need approval from their home institutions even if none of the subjects will be located there. If the subjects are located elsewhere, at five different sites, for example, the IRB at each site will need to approve the study. Some organizations will cede oversight to another IRB, but others will not. Some have very different requirements for gaining consent from subjects.

For example, one of the authors was part of a study that had three co-investigators, so each of their organizations approved the study first. All three IRBs approved the ability to consent subjects, none of whom were patients, verbally. One of the site IRBs required that both subjects and any patients nearby (though none were subjects) sign a three-page consent form and a local employee needed to be present during the signing process. In another study, one of the authors, who had IRB approval from the home site to do verbal consents for a focus group, arrived to find that the site IRB not only required written consent, but that each page of the six-page consent form had to be signed and include the subject's social security number on each page. Twenty minutes of the one-hour focus group was spent gaining consents. The point here is that during the planning stage of a study, time must be factored in for gaining IRB approval(s). Time must also be allocated for each study team member to undergo the training required by each different IRB as well. Many organizations are now outsourcing human subjects training, so it is possible the same training modules might be required for several sites, which would save team members' time. In addition, some organizations are consolidating their IRB activities so that if all study sites belong to a certain group, only one IRB approval may be needed. In summary, it may be getting easier to manage IRB issues for multi-site studies related to informatics.

Ethical review mechanisms are only somewhat standardized in the U.S. and they certainly are not internationally. Countries differ in their requirements, especially for biomedical and social science research. Since many informatics projects are a combination of both, they are not clear cut and often generate discussion.

Quality Improvement vs. Research There is a fine line between studies that are planned to improve the quality of care within an organization and studies that are for research purposes that, once results are disseminated, can add to the scientific body of knowledge. See Chap. 5 for more information about evidence-based informatics. Quality improvement seeks to discover if changes made in practice have an influence on patient care. Research seeks to develop evidence to guide practice. The dilemma for informatics is that many studies in this discipline include aspects of both. If organizations are to become Learning Health Systems, the two types of studies need to be done together, however (Faden et al. 2013). Faden et al. contend that there is concern that because of the "fuzziness of the distinction" and the "oversight burdens" of research with IRB approval, some worthwhile studies of operations are not being done (Faden et al. 2013). In a number of organizations, informaticians and quality improvement specialists in hospital environments measure quality for reporting purposes while academic informatics departments belonging to the same organization conduct grant funded research completely separately because their goals are different. There is now a recognition that the quality improvement and academic staff could be benefiting from collaborating, because their ultimate goal, improving care, is shared. For research and practice in informatics to move forward, a different ethics paradigm or framework is needed, according to ethical scholars (Faden et al. 2013; Selby and Krumholz 2013; Goodman 2020). They believe that the level of risk should be much more finely assessed before it is compared to the risk of harm.

Other Media As discussed in Chap. 16, analysis of other media such as social media posts, raises new and difficult ethical issues. Again, it is the fuzziness, like that between operational studies and research studies, that causes difficulties. For example, in one published study, a study team had approval to gather and analyze data from a group of online discussions if moderators of those groups gave permission. The team could not get permission from many of the moderators since the moderators thought that announcing to the discussants that a researcher was doing a study about them would have a negative impact on the discussions and would make discussants uncomfortable. This presented an ethical dilemma for the team members: should they follow moderators' suggestions and be unobtrusive or adhere to IRB policies and gather inadequate data from more welcoming sites (Sugiura et al. 2017). This is a difficult issue from a regulatory point of view because it is an example of the kind of research that can be stultified by oversight.

Inclusion of Diverse Populations Another difficult ethical issue is inclusion of diverse populations in informatics studies. Evaluations of informatics interventions that might improve the health of vulnerable populations are becoming more common as awareness of inequities has increased. For quantitative studies, the existing data may not be inclusive or surveys may not reach these people, and for qualitative studies, there may be issues such as language barriers. These issues should be identified and addressed during the research design planning phase of any project (Bekemaier et al. 2019; DeCamp and Lindvall 2020).

20.2.2 The Safety of Implementing ICT

Informaticians have a moral imperative to attempt to do more good than harm with their implementations. Beginning in 1999 when the Institute of Medicine report To Err is Human (Corrigan et al. 1999) highlighted the potential of clinical information systems to increase patient safety, leaders within the U.S. government and countries around the world became interested in promoting the adoption of electronic health records and clinical decision support. In the U.S. a federal budgetary stimulus plan provided incentives for implementing such systems for both hospitals and ambulatory health care organizations (ARRA 2009). The widespread and fast rollout of systems and research within the informatics community about the unintended consequences of ICT (Ash et al. 2007) precipitated discussion of strengthening regulations in this area. The Office of the U.S. National Coordinator for HIT and similar agencies in other countries began certification of the technology itself as well as helping develop incentives for the "meaningful use" of HIT (ARRA 2009). An Institute of Medicine report about health information technology and patient safety published in 2011 offered strongly worded recommendations for reporting mechanisms for harms caused by ICT because until more data became available about the nature of the risks, it would be difficult addressing them. Some, such as problems with usability, were obvious, but studies cited in that report generally concluded that causes of harm were extremely complex and when ICT was a factor, it was one of many (IOM study Committee on Patient Safety and Health Information Technology 2011). Therefore, the safety of ICT itself was recognized as a critical focus of evaluation studies.

Unfortunately, decision makers still do not know how large the problems are, either those related to patient safety or with errors caused by information systems (Bates and Singh 2018). Better methods are needed for studying the joint problem. Incident reporting has progressed, especially in Europe, Australia, and the U.S., but it is still voluntary and hard to recognize how much ICT contributes to total errors (Singh and Sittig 2020). In the U.S., a group of Patient Safety Organizations gathers these voluntary data, allowing researchers to "deep dive" into the reports to study ICT issues, but of course the data are incomplete (Bates and Singh 2018). In the European Union, Medical Device Directive 2007/47/EC1 defines ICT software as a medical device (EU Medical Device Regulation 2017/745 2017) and data are therefore collected more broadly than in the U.S. where only limited types of software associated with medical devices are regulated in this way. The Food and Drug Administration (FDA) database of reported errors called Manufacturer and User

Device Experience (MAUDE) (MAUDE n.d.) has been made available to researchers around the world, but data on only these limited types of software are included. Much more data are needed: a national coordination center was proposed in the Institute of Medicine HIT safety report, but there has been little progress on that effort (Bates and Singh 2018). Without standard metrics for reporting and without universal reporting, trustworthy evaluation cannot take place. Such evaluations are sorely needed if reporting organizations are destined to truly become Learning Health Organizations that can continually improve. It is an ethical imperative that ICT safety should become one of the most important initiatives of the informatics community.

Important smaller steps have fortunately been taken to assess ICT safety within individual organizations. The U.S. Office of the National Coordinator funded development of a series of guides/checklists that organizations can use for evaluating ICT safety and for planning for improvement. The Safety Assurance Factors for EHR Resilience (SAFER) guides are freely available and their use is increasing because CMS is now mandating it (SAFER n.d.), but progress on reaching higher scores has been limited (Sittig et al. 2018). The Leapfrog test that uses simulation within an organization's EHR to measure progress in medication safety and the use of CPOE shows that improvements were made, but that particular problem still exists (Classen et al. 2020).

Usability continues to be a large safety-related factor because, although measurement and methods have improved, knowledge is not being applied—some say because EHR vendors are not making changes, but also because organizations are not necessarily changing configurations at their end based on usability research. See Chap. 17. Progress has been slow and much more needs to be done to translate what has been learned through relevant usability studies into practice (Carayon and Hoonakker 2019).

Food for Thought

Since ICT will continue to be used even though it might at times cause patient harm, what are some ways that you, as an aspiring evaluation study team member, might want to study ICT safety?

20.2.3   Encouraging Diversity

In the case described earlier (Sugiura et al. 2017) about the study team investigating social media discussion groups, it might have been the most vulnerable individuals whose voices most needed to be heard who would have been blocked from providing data if groups were disenfranchised because moderators were protecting them from harm in a manner different from the IRB's way.

Selection of subjects in quantitative studies has ethical considerations in that strategies need to include seeking out outliers or the underserved. Researchers need to carefully assure that the sample is truly representative of the population.

A colleague was studying subjects with a specific medical condition and found that many of them were illiterate. She developed an oral interview method for gaining their consent and assuring that they understood the study that her IRB approved. Otherwise, these interviewees could not have been included. Language is a large barrier to studying representative populations, either using surveys, interviews, or observations.

20.3 Ethical Issues During Each Phase of a Study

Ethical issues exist for all informatics evaluations. Those specific to each phase of evaluation for both quantitative and qualitative studies are described below.

20.3.1 Planning for Ethical Evaluation

Careful consideration of the ethics of a project begins during the planning phase so that thoughtful decisions made early can help avoid setbacks later. Ethical dilemmas can surprise a study team and arise during every phase of a study. Some dilemmas can be predicted, but some do not occur until the team is deeply into the project and the issues might threaten to undermine the study. For example, for one of the author's studies the original plan was to conduct observations in addition to interviews at five sites. After gaining entry to a site, getting their IRB approval, scheduling appointment times for eight study team members for the three consecutive days of on-site work, and securing airline reservations and lodging for Oregon, Texas, and Massachusetts-based research team members, the site administration decided 1 week in advance that observations would not be allowed. This would leave a gap in the data, of course, so cancelling the site visit was considered. The team would have missed gathering any data from a site that could offer a rural perspective, which was important to the study's representativeness of underserved populations. For this reason, they believed it was an ethical imperative to conduct the visit even though their plans were radically disrupted. The investigators instead adjusted: several team members volunteered to stay home and interview guides were modified so that appropriate interview participants were asked to describe clinical areas and processes since team members could not observe them. Team members had to note in publications about the project that observation data were not available from this one site, which was somewhat embarrassing. However, the site visit was extremely useful, despite the setback.

Decisions need to be made early in planning about _who should do the evaluating_. The goal from an ethics point of view is to reduce bias. Especially in qualitative studies, bias is recognized and managed rather than eliminated. For any study, however, it is useful to confront bias openly. Informaticians are generally champions of new technology, so whether they are conducting quantitative or qualitative studies, they need to be aware of this bias. Multidisciplinary teams are ideal because their diversity provides continuous assessments of team member biases. Even better, teams with outside assistance can solicit help in assessing team member bias (Reid et al. 2018).

Mixed teams of insiders and outsiders are ideal. The integrity and professionalism of study team members raise important ethical considerations throughout all the phases of the study. Study teams are in a strong position to bias the collection, interpretation, and reporting of study data in such a way as to favor—or disfavor—the information resource, its developers, its users, and the organization that deploys it. One mechanism to address this concern would restrict the pool of potential study teams to independent agents, commissioned by an independent organization with no strong predispositions toward or profit to be made from specific outcomes of the evaluation. While there is a role for evaluations conducted with this extreme level of detachment, it is impractical and actually suboptimal as a general strategy, since multidisciplinary teams of insiders and outsiders are actually more ideal.

Whether team members are qualitative or quantitative, multidisciplinary, insiders or outsiders in the organization being studied, or outside experts, they should be practicing reflexivity. For qualitative studies, attention to reflexivity, a process to improve awareness of bias, can assist in managing unwanted bias throughout the life cycle of the project. When deciding whom to involve, team leaders must think carefully about the makeup of the research team. Especially in qualitative studies, the instruments for gathering data are the researchers and all data gathered are filtered through their brains. While this works well for gaining multiple perspectives, it is also an invitation for bias. Besides guarding against some bias by comparing perspectives, experienced study teams use reflexivity as a safeguard. Researchers identify their biases during the planning stage and throughout the study so they are aware of their own personal ethical issues. Reflexivity can clear the "foggy lenses" of prejudice, such as that held by a developer for an application self-produced (Adler-Milstein 2019).

It is not always obvious _what to evaluate_. Selecting the specific aspect of the project that needs to be evaluated is not trivial and it is tempting to plan based on what stakeholders predict will give the most positive results. In determining what aspects of a project to evaluate, remember to consider the potential for bias built into the application that is the focus of study. There is need at this point to evaluate the extent of potential bias inherent in the application. For example, sometimes the bias is built in to artificial intelligence (AI) systems through their algorithms (DeCamp 2020). In other words, algorithms built using faulty data sets that do not include vulnerable population data could lead to misleading AI systems and suboptimal system-generated recommendations. A serious discussion among all team members, some of whom might want to study risks and some of whom what to study successes, will likely result in studying both, despite individual preferences.

Many decisions about _how to evaluate_ have ethical implications. A basic consideration at early stages should be an assessment of risk to subjects, either physical or psychological. One would hope that benefits to the knowledge base (if a study is designed to produce generalizable published results, for example) or to the study teams trying to improve a system would outweigh risks to subjects. There are many kinds of risk, ranging from the possibility that the data might fall into the wrong hands to taking too much of the subjects' time or perhaps causing some mental discomfort while discussing informatics interventions. Informaticians should be especially adept at assuring confidentiality and privacy of the data, which are generally considered risks in any project. Despite new EU data protection privacy regulations (EU Medical Device Regulation 2017/745 2017) and HIPAA in the U.S., patients are still wary about privacy. In some countries, social science research has far less stringent oversight than biomedical research. When planning how to conduct an evaluation, the study team must also be reflexive about power dynamics. The study team members must not only be careful to avoid coercing patients into taking part in studies, but they must conduct themselves professionally so they do not bias participants (Reid et al. 2018).

Related to coercing participants is the topic of incentivizing participants. The study team should think carefully and perform an assessment of any direct monetary costs to the subjects associated with their participation, including but not restricted to transportation, that is necessary for studies that might include in-person interviews or focus groups, or participation in usability studies conducted in a physical laboratory. Compensation for subjects' time is considered to be ethical, though there is a fine line between modest incentives and overly generous compensation that might engender in the participants an obligation to respond in a way that is shaped by their perception of what the study team expects.

Particularly in qualitative studies, decisions about how to evaluate are often made iteratively as more is learned during a project. At times large changes in strategy may be needed, so study teams need to be flexible and creative. The changes may require new ethical considerations not present in the original plans, and for studies conducted under IRB regulations, major modifications must be submitted to the IRB for review and approval. For example, when the beginning of the pandemic struck in the spring of 2020, ethnographers around the world were affected when they could no longer conduct onsite fieldwork. Many developed innovative methods such as asking local collaborators to video record events occurring inside study sites, or by conducting interviews and focus groups over Zoom. Video recording raises human subjects concerns that the on-site presence of an observer does not. For example, confidentiality is more of an issue with video than audio because participants can be more easily recognized than by voice alone. Because new issues could arise with changing study techniques, especially those of additional risk, an IRB modification should be sought in those circumstances.

Other ethical considerations related to moving from face-to-face fieldwork to virtual fieldwork involve the whole arena of digital ethnography studies (Sugiura et al. 2017). This includes virtual ethnography, netnography, mobile device methods for survey research, and a growing toolbox of methods that use secondary or preexisting data. For example, a study team may decide to observe largescale public events, but attendees cannot give consent, so researchers must be aware of the sensibilities, safety and security of these people. Researchers may want to analyze social media discussions, but they might need permission from moderators first.

Decisions might include whether or not to tell discussants that researchers are present because such disclosure might cloud their discussions.

If the project already has IRB approval and a change in strategy is needed, a request for a modification of the original approval must be made. For low-risk projects, this is usually a simple process and the review provides a skilled assessment by IRB analysts about ethical issues you may not have considered.

From the beginning, the study team must make multiple decisions at different points in time about _whom to involve_. There are numerous ethical considerations when planning what stakeholder groups to involve and how participants might play additional roles, such as by becoming collaborators, beyond being passive subjects.

Informatics projects should include multiple stakeholders, as discussed in Chap. 22. Study teams have a responsibility towards stakeholders such as funders, employers, sponsors, colleagues, and sometimes industry. It takes time, energy, and decision making during the planning stage to identify the most important stakeholder groups and how to involve them. Different stakeholder groups may not always agree, making decisions more difficult. On the other hand, the evaluation cannot be conducted without their cooperation. One area of critical decision making is agreeing on ethical issues such as what might be considered proprietary and what can be published. If interviews and observations are involved, there must be agreement about what study teams will and will not share with, for example, administrators of the organization. A team may have to struggle with wanting to share with collaborators, such as CMIOs within an organization, criticisms staff members have expressed. Such decisions should be discussed in each instance about how to handle sensitive issues and should be memorialized in the evaluation contract discussed in Chaps. 2 and 3.

As an example, if the team finds that training has been minimal and users complain because they do not know what the capabilities of the system are, recommendations about training rather than criticisms of management should be made. During debriefings with insider collaborators, understand that defensiveness will be a natural reaction. These collaborators may not want to hear what the study team needs to report to them. These debriefings need to be prepared carefully so that they are viewed as helpful and are not casting blame. In qualitative studies, subjects or informants should be considered teachers because they are instructing study teams how to view their culture. It is important that these insiders are treated with respect and appreciation.

20.3.2   The Data Gathering and Storage Phase

For qualitative studies, attention to reflexivity, a process to improve awareness of bias, can assist in managing unwanted bias throughout the life cycle of the project, especially during data gathering when participants are volunteering their valuable time. During the _storage phases,_ both quantitative and qualitative studies must adhere to ethical principles about access to data, and these principles are often unclear.

Early planning must include decision making about how to locate pre-existing data or how to gather new data. If pre-existing data are to be used, arrangements to access them can often take longer than anticipated.

Getting and using quantitative data must follow a predetermined process within any organization in order to protect *data privacy and confidentiality*. The recurring theme of confidentiality is of special concern when data must be processed or stored off-site or include items likely to be sensitive to patients (e.g., their HIV status or psychiatric history) or professionals (e.g., their performance or work load). Study team members must negotiate inter-agency agreements before sharing will be permitted. One approach is to "anonymize" the data by removing obvious identifiers, although, especially in the case of rare diseases or professionals with unique skill sets, it may still be possible to identify the individuals concerned by a process of triangulation. There are many strategies for keeping data secure. Physical security measures (locking up servers in a secure room) are effective methods of restricting access to confidential data. Software access controls on databases and encryption of data sent over the Internet are also useful safeguards.

Selection of sites to study in qualitative studies raises additional ethical issues. Many informatics evaluations are conducted at the home organizations of the study teams. This is so the organization benefits, the researchers have knowledge of the culture and professional contacts for gaining entrée, and travel is not needed. However, there can be inherent bias and an inability to see different perspectives at a home site. Selecting other sites is more difficult and time consuming, but is more likely to yield new knowledge. Sites willing to volunteer might not be representative, however, since they are probably proud of their successes. Study teams should make an extra effort to seek out sites that represent the normal or that are outliers. Outliers can be excellent sources: an organization that has rejected purchasing a new informatics tool will employ stakeholders who have given that decision deep thought.

Selection of subjects in qualitative studies is done purposively, meaning that you deliberately seek out those who have knowledge about your focus, and both the normal and outlier groups. A decision usually must be made whether to include non-users of an intervention as well as users. Sometimes they have thought a good deal about the topic and can eloquently describe their rejection of the tool. It is unwise to target subjects who have no knowledge, however, and it is wrong to waste time and resources pursuing them. Ethics requires careful consideration during subject selection and avoidance of a biased group, such as champions only.

Selection of subjects who are patients raises different ethical questions from those concerning professional subjects. Often extra steps, sometimes time consuming, must be taken to get permission to approach different groups in an effort to be inclusive, though it is the right thing to do.

Subject recruitment and consent are processes that must be scrutinized carefully to assure safe, fair, and equitable treatment of participants in the study. Recruitment for informatics studies involves many ethical dilemmas, but most are similar to those encountered in other fields. Individual IRBs have their own guidelines about recruitment, all designed to assure the subject is well informed before consenting.

There is a fine line between giving so much information that the subject is either too well prepared and lacks spontaneity, or that the subject becomes so guarded that the information shared is not helpful. It may seem misleading to explain that you are studying best practices for implementing an information system instead of sharing that you are investigating unintended consequences and safety risks, but while you study one, you learn about the other, and the vagueness of your purpose statement is warranted. Some IRBs allow subjects to give verbal consent if they have agreed to information offered on an information sheet summarizing the study. Others require written consent, sometimes with an internal staff member present.

When recruiting, be aware of power differentials and avoid any semblance of coercion. For example, if your inside collaborator is the chief medical informatics officer, subjects in information technology may feel obligated to be part of the study because of that person's power in the organization. Students may feel pressure to take part if a professor asks and patients may feel obligated if their providers ask them. Most professionals actually want to talk about information resources and most patients want to help improve health care, so many are willing to take part.

Data gathering is another process with important ethical implications. Most studies are performed in conjunction with informatics and information technology professionals and other stakeholders within an organization. The goal is ultimately to improve health and health care, through a quality improvement effort or by disseminating generalizable results. Study team members may be asked to continue to play a role in the change process, deciding to "stay on the train," to continue the analogy presented in Chap. 19. Therefore, many informatics projects are similar to action research, in that they can be designed to empower communities to make change. Because these action research-oriented projects may be somewhat invasive and not merely observational, there are additional ethical considerations, especially during the data gathering process. For example, if patients are interviewed about their use of a patient portal, they will likely discuss their medical issues during the interviews, and if they are observed, they may use the portal for reasons related to their personal health issues. This information must be carefully protected.

Clinicians are often engaged as study team members and their involvement raises unique ethical issues as a result. They need to be reflexive about their biases if they observe medical practices that make them uncomfortable. They need to balance their professional ethics with those of the need to gather data. It is not unusual for a study team member who is also a clinician to see a mistake about to be made and must quickly decide how to handle the situation. As a fly on the wall observer, the clinician should not interfere, but as a clinician, the person should speak up. A pharmacist on one team faced such a dilemma and gently suggested to a subject that they might want to rethink an order. A nurse on another team was asked by a doctor to help him roll over a patient. The nurse knew that, as an observer, the nurse should not get involved, but the nurse was trying to establish rapport, so the nurse decided, in this low-risk circumstance, the nurse should do it. Study team members who are clinicians are often asked medical questions by other clinicians, especially if a subject has chatted with the observer and knows the observer's specialty. The observer needs to be ready with a response indicating that the observer is in the ethnographer role at the moment and not a clinical role. If the situation were urgent and the observer's professional ethics mandated intervention, the observer would need to step in, however.

Self-Test 20.1
1. What could the study team do before finalizing the evaluation study question to avoid marginalizing study participants?
2. During an interview for the Nouveau *YourView* study, an interviewee who is well known as an astute informatics researcher asks for a copy of the transcript of their interview to use as the basis for a future publication. This person thinks that because their responses to questions, which took a good bit of preparation ahead of time, were quite eloquent and poetic, they would be publishable. How should the interviewer respond to this request?
3. A study team member who is a pharmacist observes a medication mistake about to happen. The study team has promised that it would be unobtrusive during observations. Should the pharmacist intervene or allow it to happen?
4. A study team member who is an ophthalmologist is observing in a primary care setting and the provider being observed asks for an opinion about a patient with an eye issue. What should the team member do?

20.3.3 Analysis and Interpretation Phase

This phase can be fraught with ethical dilemmas if decisions are made that allow bias to infiltrate results. In qualitative fieldwork, it is common for inexperienced study team members to blend descriptions of observations with their own interpretations so that when the fieldnotes are analyzed, it is unclear which is fact and which is interpretation. Separate fieldnotes or memos should hold interpretation off to the side, to be considered later. It is often tempting to over-interpret results in order to make a point or justify recommendations, but this is not justifiable. One way to guard against the temptation is to plan on member checking, which means sharing results with subjects and insiders to find out if they appear to be true to participants.

20.3.4 Reporting Results Phase

Results of most evaluation studies will eventually be disseminated. The original contract should specify exactly how the study team will share results. Chapter 19 offers more detail. If an internal report is to be written by a study team, it should describe in detail the methods for the study and report results without identifying participants, if that is the agreement. For publication more broadly, it is not just a matter of preserving privacy and confidentiality of subjects and organizations as appropriate, but also being honest about how your results lead to whatever recommendations you make. The following are suggestions about what to include. Select the most important findings that suit your audience and that you can thoroughly justify based on your results. Avoid false conclusions. Be open and clear about the limitations of the study. Be careful about the timing of publication. For example, when preprints came into broad use during the pandemic, papers were at times published before results were complete because there was an urgency to add to the knowledge base quickly, though this was risky. Consider lessons learned from negative results and the need to publish even those results. Publish about the downsides as well as the upsides.

Self-Test 20.2

An evaluation team has completed two phases of a study, one qualitative and the other a quantitative study of administrative data. They are trying to find out if the perceptions of providers that they are not closing out their charts as fast as they did 5 years ago is correct. During qualitative work, they learn that providers are almost unanimous in sensing that their chart completion time is longer and they explain it is because of the EHR, which has become even more complex over the past 5 years. However, the administrative data indicate that across all specialties, chart completion time is faster by an average of 2 days.

1. What paper or papers should the team plan on publishing about these results?
2. What should the team do to try to reconcile these disparate results?

20.4   Ethical Considerations for the Future

Informatics evaluation methods are rapidly becoming more sophisticated. They are using more complex study designs. Mixed methods that truly blend findings from different data gathering strategies are a goal more often being reached. That means that there are different ethical considerations during the life cycle of each piece of the project. Often there are simultaneous phases, so a study might have six parts, each with its own series of phases, each with its own ethical dilemmas, as discussed above. In addition, the qualitative phases are always in flux, changing as new knowledge is gained. If a study team plans on using the same subjects for multiple parts of the study, the burden on their time is far increased. Not only that, but they cannot remain anonymous if they are to be contacted multiple times.

Evaluation studies are entering into uncharted territory. They are using new kinds of data, like EHR data and large data sets, geospatial data, and virtual sources for qualitative data in exciting new ways. There are many unanswered questions about intellectual property ownership and data ownership that the legal system has not yet addressed. The IRB can provide guidance, and if you think its members do not understand the challenges of informatics evaluation research, it pays to take the time to educate them.

20.5 Closing Thoughts About the Future of Informatics Evaluation

In this section, the authors will offer their personal views regarding trends in the development and deployment of information resources that will guide future informatics evaluation. The discussion will focus on issues and types of interventions that teams will be studying, on evolving methods for evaluating the interventions, and on the future of informatics evaluation more generally.

20.5.1 What Will Be Studied

Interesting and innovative new technologies will need to be studied before and after they are deployed for use by health care professionals, and the exact nature of these innovations is hard to predict. However, there are broad categories of tools and foci which are emerging as important trends deserving the attention of everyone engaged in informatics and creating important needs for evaluation. These foci follow below in no particular order.

EHR Safety The risk to patient safety from implementation of ICT grows commensurate with adoption of new tools. The rapid rate of EHR adoption, and the many applications that are ancillary to EHRs as their use becomes routinized, such as use of machine learning models in clinical decision support, will continue to instigate unintended consequences that create potential for patient harm. It is likely that future regulation will require evaluation of the safety and effectiveness of this class of information resources.

Health Information Exchange The challenges of interoperability raise the possibility that protected health information can be lost or corrupted when moving among systems. The potential value of interoperability increases geometrically as more digital data become available and standards-based exchange becomes increasingly routine. Greater study of its benefits and challenges will become a necessity over time.

Telemedicine and Telehealth While evaluation studies of telemedicine and telehealth efforts have been growing in number over the past decade, the COVID-19 pandemic forced a sudden burgeoning of these alternatives to in-person face to face communication. The safety, accuracy, efficiency and effectiveness of these technologies require increased study in light of their widespread and largely unanticipated influence on health care.

Usability This continues to be a large factor because, while information resources of all types become more sophisticated, they can become more difficult to use, but excellence in design can mitigate this effect. For this reason, usability studies of sophisticated information resources that are under development, and as they are deployed, will become increasingly important in future years. It will be important that usability studies are inclusive across age and education, and considerate of users' differing cultural backgrounds.

Patient Centeredness People worldwide are taking more active roles in their health care, in many cases in partnership with clinicians and making use of increasingly sophisticated information resources. These trends require increased study of how "patients" make use of these resources and the degree to which they promote better health.

Disparities Interactive resources that provide valuable information, guidance, and advice—using inexpensive and easy to maintain technology—hold enormous potential to reduce health disparities. However, if deployed in ways that do not respect the needs of underserved populations, they could have the opposite effect. Evaluation studies can help avoid this most unfortunate outcome.

Closed-Loop Resources There is understandably enormous excitement about the potential of closed-loop systems that operate routinely without direct human control and intervention, making use of models developed through artificial intelligence and machine learning. It is imperative that evaluation of such resources extend beyond software verification and validation. That these resources "do what they're supposed to do" is necessary but not sufficient to their success. It is essential to also study safety and human elements, such as the degree to which people trust these resources.

Robotics Some robotics applications will operate autonomously while others will engage robots in support of people who will interact with them. The many implications of these applications, holding great potential to improve health, will be an important area for future evaluation study.

Resources Addressing Non-clinical Domains This book has addressed clinical informatics applications to a significant extent because this domain is where evaluation studies have concentrated and methods have developed. Applications supporting personal and public health, education, and research will inevitably proliferate. This proliferation will accelerate as extremely valuable new kinds of data, most notably geo-spatial and social media data, become omnipresent. When this happens, study teams should—and will, in the authors' estimation—increasingly devote their attention to these domains.

Quality Improvement Often quality improvement professionals are quite separate within organizations from informatics and health information professionals, both organizationally in different departments and therefore in physical space as well.

However, quality improvement efforts depend on informatics tools and informaticians should be involved with quality improvement staff in a closer partnership. The two-way dependency is likely to grow, and the evaluation specialists within both areas will broaden their expertise as a result.

Health Services Research   This is another area that is separate yet complementary and sometimes overlapping with informatics. Especially as more population level studies are conducted, their results will be of greater use to health services researchers and evaluation study teams (Adler-Milstein 2019).

New Kinds of Data and More Data   This is another quickly growing and exciting change largely precipitated by public health informatics. EHR data, geospatial data, virtual/social media data, data supporting the addition of social determinants of health data to records will all need new kinds of data infrastructures which must be evaluated (Murray 2016).

Bioinformatics Implementations   There are few studies being published about evaluations of bioinformatics implementations. This is likely because many are within the commercial sector, which is not incentivized to either conduct or publish them. In academia, bioinformatics study team members do not have a history of conducting evaluations of the development and implementation of their tools, but they, like their colleagues in the clinical disciplines, will likely be helping their organizations become Learning Health Systems, in which case evaluation studies will become more common in bioinformatics.

The Role of Relevant Theories to Support Evidence-Based Health and Biomedical Informatics   New health or biomedical information resource are often based loosely on previous resources with a liberal dose of new components inspired by current technical possibilities. As a result, we accept that failure of new versions of information resources to achieve their desired impact is common. However, engineers working in more mature disciplines such as aeronautics or civil engineering follow a much more conservative approach, using materials and mechanisms that have a strong track record of success, and theories that reliably predict which components or configurations will achieve the desired outcome. For informatics to become a mature, professional discipline we need a commitment to identifying (through well-designed evaluation studies) what works, when and for whom, and then adopting and disseminating that material to all who work the field. This evidence-based informatics approach requires the help of study teams to identify and test relevant theories for their ability to improve the usability, accuracy or impact of information resources (Wyatt 2016). A recent open access textbook on evidence based informatics includes a number of case studies of this approach (Ammenwerth and Rigby 2016), while a more recent book on theory in informatics explores the types of theory that should help put our discipline onto a sounder footing (Scott et al. 2019). All this implies that in the future, some evaluation studies will be designed to test the applicability of a new or well known theory in a specific use case for which insufficient evidence exists about what kind of information resource will help solve the problem. Note that the results of these studies then need to be disseminated - whether positive or negative—to help others improve their practice.

20.5.2 Future Trends in Evaluation Methods

In this volume, the authors have described a broad range of methodologies that are in present use for studying biomedical and health information resources. Looking to the future, and driven by many of the trends in information resource development described above, evaluation methods will follow suit. Study teams in informatics will likely adopt several of the following emerging enhancements to existing qualitative and quantitative methods.

Mixed Methods Mixed methods studies will proliferate in informatics because they are suitable for studies that seek to investigate contextual factors and because by using both inductive and deductive approaches, a more thorough study can be done with both depth and breadth.

Teamwork A team approach to evaluation will become more commonplace. Although teamwork sometimes takes extra effort, experience is teaching that a team approach has many advantages. Of utmost importance is that it aids the rigor or trustworthiness of a study. The different backgrounds and perceptions of the team members help to assure that different views are considered and that bias is minimized.

Rapid Assessment Rapid assessment procedures have been shown to be efficient and rigorous and they provide a package of methods that can guide study development. They include the two attributes listed above: they depend on mixed methods to gain a contextual view of what is being studied, and they involve a multidisciplinary team approach.

Attributing Causality From Routine or Real-World Data Chapter 13 described many of the challenges and some of the solutions to inferring whether an intervention such as an information resource was responsible for observed changes using only routine data. Several novel techniques based on instrumental variables, acyclic graphs and other methods are emerging but are as yet rarely used in biomedicine, let alone in health and biomedical informatics. We anticipate that over the next decade these methods will gradually become more widely known and used, and that future developments will allow them to be applied in a wider range of evaluation scenarios.

Importing Evaluation Methods from Other Disciplines Several of the new methods being applied to attribute causality from routine data have been imported to biomedicine from other disciplines, such as econometrics. As discussed at an international workshop on eHealth evaluation held in London in 2016 (Scott et al. 2019), we anticipate that in future a range of methods will be imported from engineering, public health, psychology and other disciplines that will make the task of the informatics study team easier and the results more robust. Some examples include multiphase optimization, fractional factorial experiments, and sequential multi assignment randomized trials.

More Rigor to Qualitative and Mixed Methods Studies, The pursuit of rigor in all studies, but especially in qualitative and mixed methods studies, has been stimulated partly by quantitative researchers disparaging their use. To counter attacks from quantitative experts who have called qualitative work too subjective, qualitative specialists have developed techniques for establishing rigor that are being widely adopted. In summary, informatics evaluation has matured commensurate with the maturation of informatics itself. Advances in the practice of qualitative research and mixed methods research have been especially robust over the past decade. Too often evaluation is not conducted when it should be, however, so there is a need for enlightenment of stakeholders to support evaluation. Even when evaluation is conducted, its results do not seem to impact practice in the field or development of better tools, with usability studies as an example.

For more evaluation to be conducted, more resources from organizations that are implementing systems should be earmarked for evaluation purposes. If they are aiming to become Learning Health Systems, they will need a steady feedback process for learning from their experiences. The exciting trends in support for evaluation, development of new methods, and the progress towards Learning Health Systems will require a workforce that is trained and committed to these activities. The goal of this volume is to present readers with lessons from the past and with foundations for advancing useful methods for the future of evaluation methods in health informatics.

Answers to Self-Tests

Self-Test 20.1

1. What could the study team do before finalizing the evaluation study question to avoid marginalizing study participants?

   Involve participants. Purposively select representatives of the study population and invite them to voice their concerns and desires about what the study question should include.

2. During an interview for the Nouveau YourView study, an interviewee who is well known as an astute informatics researcher asks for a copy of the transcript of their interview to use as the basis for a future publication. This person thinks that because their responses to questions, which took a good bit of preparation ahead of time, were quite eloquent and poetic, they would be publishable. How should the interviewer respond to this request?

This is a common situation and as long as the transcript does not include any information that should not be divulged, the interviewee should be given a copy when it is available.

3. A study team member who is a pharmacist observes a medication mistake about to happen. The study team has promised that it would be unobtrusive during observations. Should the pharmacist intervene or allow it to happen?

The pharmacist should not allow it to happen and should therefore intervene since it is his ethical duty as a professional. A gentle intervention would be appropriate, such as suggesting the observed person reconsider the decision.

4. A study team member who is an ophthalmologist is observing in a primary care setting and the provider being observed asks for an opinion about a patient with an eye issue. What should the team member do?

The team member would need to assess the situation. If the answer were to be quick and a nod would suffice, it would be better for the team member to do that than to create an awkward situation that might alienate the person being observed. If the medical situation was a complex one, the team member might suggest that they discuss it later. The team member is there to observe and could also say in a kindly manner that right now was observation time but later there could be a discussion.

Self-Test 20.2

1. What paper or papers should the team plan on publishing about these results?

This is a dilemma that must be discussed by the team. They could publish the results of each part of the study separately, but that would be quite self-serving of them. They would likely get the papers accepted, but basically they would be withholding information from readers. They could write one paper and try to explain reasons for discrepancies, but they would be inferring important findings without adequate data. A better option would be to further explore a new evaluation question about why they had such intriguing results.

2. What should the team do to try to reconcile these disparate results?

They could look further at their data, but if answers were not there, they should gather more data to try to reconcile the results. They might do some member checking, showing the results of the quantitative study to some purposively selected experts who might help interpret them.

References

Adler-Milstein J. Health informatics and health services research: reflections on their convergence. J Am Med Inform Assoc. 2019;26:903–4.

Ammenwerth E, Rigby M, editors. Evidence based informatics. Studies in Health Technology and Informatics. Amsterdam: IOS Press; 2016.

ARRA. American Recovery and Reinvention Act of 2009 (ARRA) (Pub. L. 111-5); 2009.

Ash JS, Sittig DF, Dykstra RH, Guoppone G, Carpenter JD, Seshadri V. Categorizing the unintended sociotechnical consequences of computerized provider order entry. Int J Med Inform. 2007;76S:S21–7.

Bates DW, Singh H. Two decades since to err is human: an assessment of progress and emerging priorities in patient safety. Health Aff. 2018;37(11):1736–43.

Bekemaier B, Park S, Backonja U, Ornelas I, Turner AM. Data, capacity-building, and training needs to address rural health inequities in the Northwest United States: a qualitative study. J Am Med Inform Assoc. 2019;26:825–34.

Carayon P, Hoonakker P. Human factors and usability for health information technology: old and new challenges. Yearb Med Inform. 2019;28(1):71–7.

Classen DC, Holmgren AJ, Co Z, Newmark LP, Seger D, Danforth M, Bates DW. National trends in the safety performance of electronic health record systems from 2009-2018. JAMA Netw Open. 2020;3(5):e205547.

Corrigan JM, Kohn LT, Donaldson MS, editors. To err is human: building a safer health system. Washington, DC: National Academies Press; 1999.

DeCamp M, Lindvall C. Latent bias and the implementation of artificial intelligence in medicine. J Am Med Inform Assoc. 2020;27:2020–3.

EU Medical Device Regulation 2017/745. 2017. https://eumdr.com. Accessed 17 June 2017.

Faden RR, Kass NE, Goodman N, Pronovost P, Tunis S, Beaushamp TL. An ethics framework for a Learning Health System: a departure from traditional research ethics and clinical ethics. Ethical Oversight of Learning Health Care Systems, Hastings Center Report Special Report 43, no. 1. 2013;S16–S27.

Goodman KW. Ethics in health informatics. Yearb Med Inform. 2020;29:26–31.

IMIA. (n.d.). https://imia-medinfo.org/wp/imia-code-of-ethics. Accessed 21 June 2021.

IOM study Committee on Patient Safety and Health Information Technology. Health IT and patient safety: building safer systems for better care. Washington, DC: National Academies Press; 2011.

MAUDE (n.d.). https://www.accessdata.fda.gov/scripts/cdrh/cfdocs/cfmaude/search.cfm. Accessed 21 June 2021.

Murray E, Hekler EB, Anderson G, Collins LM, Doherty A, Hollis C, et al. Evaluating digital health interventions: key questions and approaches. Am J Prev Med. 2016;51:843–51.

Petersen C, Berner ES, Embi PJ, Hollis KF, Goodman KW, Koppel R, et al. AMIA's code of professional and ethical conduct 2018. J Am Med Inform Assoc. 2018;25:1579–82.

Reid AM, Brown JM, Smith JM, Cope AC, Jamieson S. Ethical dilemmas and reflexivity in qualitative research. Perspect Med Educ. 2018;7:60–75.

SAFER guides. (n.d.). https://www.healthit.gov/topic/safety/safer-guides. Accessed 21 June 2021.

Scott P, De Keizer N, Georgiou A. Applied interdisciplinary theory in health informatics: a knowledge base for practitioners. Amsterdam: IOS Press; 2019.

Selby JV, Krumholz HM. Ethical oversight: Serving the best needs of patients. Ethical Oversight of Learning Health Care Systems, Hastings Center Report Special Report 43, no. 1. 2013;S34–36.

Seroussi B, Hollis KF, Soualmia LF. Transparency of health informatics processes as the condition of healthcare professionals' and patients' trust and adoption: the rise of ethical requirements. Yearb Med Inform. 2020;29:7–10.

Singh H, Sittig DF. Framework for safety-related electronic health record research reporting: the SAFER reporting framework. Ann Int Med. 2020;172:S92–S100.

Sittig DF, Salimi M, Aiyagari R, Banas C, Clay B, Gibson KA, et al. Adherence to recommended electronic health record safety practices across eight health care organizations. J Am Med Inform Assoc. 2018;25:913–8.

Sugiura L, Wiles R, Pope C. Ethical challenges in online research: public/private perceptions. Res Ethics. 2017;13:184–99.

Wyatt JC. Evidence-based health informatics and the scientific development of the field. Stud Health Technol Inform. 2016;222:14–24.

Glossary

Accuracy (1) Extent to which the measured value of some attribute of an information resource, or other object, agrees with the accepted value for that attribute or "gold standard"; (2) extent to which a measurement in fact assesses what it is designed to measure (roughly equivalent to "validity").

Action research A disciplined method for intentional learning from experience characterized by intervention in real world systems followed by close scrutiny of the effects. The aim of action research is to improve practice and it is typically conducted by a combined team of practitioners and researchers. Originally formulated by social psychologist Kurt Lewin. [Adapted from Wikipedia definition, www.wikipedia.org].

Alerting resource Resource that monitors a continuous signal or stream of data and generates a message (an alert) in response to patterns or items that may require action on the part of the care provider.

Allocation concealment Ensuring that those recruiting participants to a randomized trial have no knowledge of the group to which each participant will be allocated. Failure to conceal allocation has been shown (see meta regression techniques) to be a major cause of bias in such studies, and is best avoided by blinding those recruiting participants and recruiting them only by communication with a central trials office.

Analysis of variance (ANOVA) General statistical method for determining the statistical significance of effects in experimental studies. The F test is the basic inferential test statistic for analysis of variance. (See Chap. 12).

Art criticism approach An evaluation approach where an experienced "critic", who is a connoisseur of information resources, interacts with a resource and writes a review of it.

Assessment bias Occurs when anyone involved in a demonstration study allows their own personal beliefs, whether positive or negative, to affect the results.

Association Association tells you whether two variables are related.

Atheoretic Without a theory.

© Springer Nature Switzerland AG 2022
C. P. Friedman et al., *Evaluation Methods in Biomedical and Health Informatics*, Health Informatics, https://doi.org/10.1007/978-3-030-86453-8

Attenuation The "silent" effect of measurement error that reduces the observed magnitude of the correlation between the two attributes in a demonstration study.

Attribute Specific property of an object that is measured, similar to "construct."

Audit trail A record of the study, one sufficiently detailed enough to allow someone else to follow the study's history and determine if the investigation and the resulting data provided an adequate basis for the argument and conclusions.

Baseline study Study undertaken to establish the value of a variable of interest prior to an intervention such as the deployment of an information resource.

Before-after study A comparative study in which something is measured during a baseline period and then again after an intervention has occurred, e.g., an information resource is installed.

Bias (1) *Measurement bias:* Any systematic deviation of a set of measurements from the truth. (2) *Cognitive bias:* A set of consistent tendencies of all humans to make judgments or decisions in ways that are less than optimal. (3) Bias in demonstration studies—see confounding.

Bioinformatics implementations The use of tools for collecting and analyzing biological data at molecular and cellular levels.

Biomedical and health informatics Addresses the collection, management, processing, and communication of information related to health care, biomedical research, personal health, public and population health, and education of health professionals. Informaticians focus much of their work on "information resources" that carry out these functions.

Blinding In a comparative study, ensuring that participants in a study and those making measurements on them are unaware of the group to which the participant has been allocated.

Bug report User's report of an error in a program. The rate of bug reports over time may provide a measure of improvement in an information resource.

Calibration (1) Extent to which human participants' estimates of the probability of an event agree with the frequency with which the event actually occurs. (2) Extent to which appraisals by judges actually agree, as opposed to being correlated.

Carryover effect A **carryover** effect is an effect that "carries over" from one experimental condition to another. Whenever subjects perform in more than one condition (as they do in within-subject designs) there is a possibility of **carryover** effects.

Cascade problem occurs when a component of a study plan for *Stage N* of a study depends critically on the outcome of *Stage N – 1*.

Case-control study A study where participants are assigned to a group (a value of a discrete independent variable) based on some event or action that happened in the past; for example people who did or did not have a particular disease.

Case crossover design In the **case-crossover design**, the study population consists of subjects who have experienced an episode of the health outcome of interest.

Causation *Causation* indicates that an event affects an outcome. In *statistics*, correlation doesn't necessarily imply *causation*. In *statistics*, causation means that one thing will cause the other.

Central tendency A descriptive summary of a dataset through a single value that reflects the center of the data distribution.

Chains of effect The links between introducing a biomedical or health information resource and achieving improvements in the outcomes of interest.

Champion Someone innovative and forward thinking, or who advocates for adoption of an information resource.

Checklist effect The improvement observed in performance due to more complete and better-structured data collection about a case or problem when paper- or computer-based forms are used.

Chi-Square Test of significance to find a relationship between two categorical variables.

Child codes Sub-codes in a code book.

Clients of users Individuals and groups served by those for whom the resource was designed.

Clinical trial Prospective experimental study where a clinical intervention (e.g., an information resource) is put to use in the care of a selected sample of patients. Clinical trials almost always involve a control group, formed by random allocation, which receives either no intervention or a contrasting intervention.

Cluster randomized design A study design where randomization occurs at a group level, usually when measurements will be made at a lower level of scale.

Code book Code books include definitions of themes and sub-themes that are used as terminology for the coding of narrative text.

Codes Labels or terms used in both quantitative and qualitative studies to characterize elements of data.

Cognitive walkthrough Usability testing protocols which are effective in understanding users' thought patterns and strategizing as they work with a resource.

Cohen's kappa (κ) Compares the agreement between the variables in a contingency table against that which might be expected by chance.

Cohorts A **cohort** is a group of subjects who share a defining characteristic (typically subjects who experienced a common event in a selected time period, such as a birthday).

Cohort study Prospective study where two or more groups (not randomly selected) are selected for the presence or absence of a specific attribute, and are then followed forward over time, in order to explore associations between factors present at the outset and those developing later.

Communication A process including a message, a sender of the message, one or more message recipients, and a channel linking them along which the message travels. Successful communication results in a clear understanding of the message by the receiver.

Comparative study Experimental demonstration study where the values of one or more dependent variables are compared across discrete groups corresponding to values of one or more independent variables. The independent variables are typically manipulated by the investigator, but may also reflect naturally occurring groups in a study setting.

Complete Means that all possible conditions (combinations of each level of each independent variable) are included in the design.

Comparison-based approach An approach to evaluation in which the information resource under study is compared to one or more contrasting conditions, which may include other resources or the absence of the resource entirely.

Confidence interval (CI) Belief that the true value of the statistic being estimated lies within a certain identified level of confidence.

Confounding Problem in experimental studies where the statistical effects attributable to two or more independent variables cannot be disaggregated. Also, the "hidden" effects of a bias or a variable not explicitly included in an analysis, that threatens internal validity.

Confounding by indication A term used when a variable is a risk factor for a disease among nonexposed persons and is associated with the exposure of interest in the population from which the cases derive, without being an intermediate step in the causal pathway between the exposure and the disease.

Control strategies Strategies in interventional demonstration studies that help study teams credibly attribute a change in a dependent variable to one or more independent variables.

Construct validity This approach to measurement validation addresses the question: Does the measurement of the attribute under study correlate with measurements of other attributes (also known as constructs) in ways that would be expected theoretically?

Constructivism Belief in the social construction of reality, that people build their views of reality based on their own beliefs.

Constructs Attributes are sometimes referred to as "constructs" in part as a reminder that attributes of objects, that are measured in evaluation studies, are products of human ingenuity.

Consultation system Decision support system that offers task- and situation-specific advice when a decision-maker requests it.

Content analysis Technique widely used with narrative data to assign elements of verbal data to specific categories (see Chap. 10). Usually, the categories are defined by examining all or a specific subset of the data.

Content validity This is the most basic notion of measurement validity, also known as "face" validity. Content validity addresses whether the measurement process, by inspection or observation, appears to measure what it is intended to measure.

Context of use Setting in which an information resource is situated. It is generally considered important to study a resource in the context of use as well as in the laboratory. Synonym: "the field."

Contingency table Cross-classification of two or more nominal or ordinal variables. The relation between variables in a contingency table can be tested using chi-square or many other statistics. When only two variables, each with two levels, are classified, this is called a "two by two table (2×2)." (See Chap. 8).

Contract An agreement that will identify the deliverables that are required, who has interests in or otherwise will be concerned about the study results, where the study personnel will be based, the resources available, the timeline for reports and other deliverables, any constraints on what can be studied or where, and budgets and funding arrangements

Control (control group) In experimental studies, the intervention(s) specifically engineered to contrast with the intervention of interest. It can be no treatment other than the normal treatment, an accepted alternative treatment, or no treatment disguised as a treatment (placebo).

Controlled before-after study A kind of before-after study in which either external or internal controls, or both, are used to reduce confounding. (See Chap. 7).

Convenience sampling Less deliberate sampling of study participants based largely on availability and happenstance.

Correlation A measure of strength and direction of the association between two variables.

Correlational study Non-experimental demonstration study, conducted in a setting in which manipulation is not possible, that establishes correlations or statistical associations among independent and dependent variables.

Cost benefit analysis Economic analysis in which both costs and outcomes are measured in terms of money. It requires methods to value clinical benefit in terms of financial resources. The result is a statement of the type "running the reminder system costs $20,000 per annum but saves $15 per patient in laboratory tests."

Cost-benefit ratio Measures the ratio between the incremental costs of choosing a strategy over the benefits (measured in monetary units) of each strategy.

Cost effectiveness analysis Economic analysis that measures costs in dollars, and outcomes in a single health care outcome (such as life expectancy, number of infections averted) that is consistent across options. The result is a statement of the type "running the reminder system costs $20,000 per annum but saves one laboratory test per patient."

Cost minimizing analysis Economic analysis that chooses the lowest cost strategy out of several options. A fundamental assumption is that the outcomes are equivalent.

Cost to charge ratio The ratio of the overall costs a department or hospital spends related to the global measure of charges for the services it provides. It is used to develop individual cost measures for specific services or items by assuming that the same cost-charge ratio found for the organization as a whole applies to each component of the organization.

Cost-consequence analysis Economic analysis that simply lists the costs in terms of money and the outcomes in whatever measure is appropriate for the particular condition. The number of outcomes may be single or multiple, and no attempt is made to analytically compare costs and outcomes.

Cost utility analysis (CUA) Is an extension of cost-effectiveness analysis that differs by measuring the outcome as a "utility."

Credibility Believability. The extent to which audiences will trust the results of a study.

Criterion-related validity This approach to measurement validation addresses the question: Do the results of a measurement process correlate with some external standard or predict an outcome of particular interest?

Critical appraisal This is the process of carefully and systematically examining research to judge its trustworthiness and its value and relevance in a particular context.

Critiquing system Decision support system in which the decision maker describes the task (such as a patient) to the system then specifies his or her own plan to the system. The system then generates advice—a critique—which explores the logical implication of those plans in the context of the task data and the resource's stored knowledge.

Cyclical mixed methods studies Term used for complex mixed methods studies that include numerous phases.

Data Observations of phenomena occurring in the world which can be captured quantitatively as numbers or qualitatively using language and images.

Data analysis plan A *Data Analysis Plan* (DAP) involves putting thoughts about data analysis into a plan of action.

Data cleaning The process of removing incorrect, duplicate or ambiguous data prior to carrying out the analysis.

Data controller The organization legally responsible for controlling access to and preservation of person-level data.

Data mining The practice of analyzing large databases of previously collected data, typically without pre-stated hypotheses about relationships among the variables in the dataset.

Data readiness level Term used to characterize whether data are suitable for deployment. These measures allow managers and decision makers to understand project status, and where investment is needed without intimate familiarity with the data itself.

Data saturation In qualitative studies, this means closure or convergence, at which point study team members are seeing and hearing about what they already know.

Data warehouse Central repositories of integrated data from one or more disparate sources.

Decision support system (decision-aid) Information resource that compares at least two task characteristics with knowledge held in computer-readable form and then guides a decision maker by offering task-specific or situation-specific advice.

Decision-facilitation approach An evaluation approach that seeks to resolve issues important to developers and managers, so these individuals can make decisions about the future of the resource.

Demonstration study Study that establishes a relation—which may be associational or causal—between a set of measured variables. (See Chaps. 7 and 8).

Dependent variable In a correlational or experimental study, the main variable of interest or outcome variable, which is thought to be affected by or associated with the independent variables. (See Chap. 7).

Descriptive study A one-group demonstration study that seeks to measure the value of a variable in a sample of participants. A study with no independent variable.

Design validation studies Studies focusing on the quality of the processes of information resource design and development, most likely by asking an expert to review these processes.

Development team Those who designed, created, and maintain an information resource.

Diffusion of Innovations Theory Diffusion of Innovations is a theory that seeks to explain how, why, and at what rate new ideas and technology spread.

Direct costs In health care analyses, direct costs represent the actual purchase of goods and services related to a particular chosen strategy. (See also Indirect costs).

Discounting future costs and benefits Takes into account that a dollar spent today is not worth the same as a dollar spent tomorrow.

Diverse populations In health care, groups such as patients comprised of multiple races, ages, genders, ethnicities, and orientations.

Domain knowledge Understanding of the concepts pertaining to a particular aspect of health or disease.

Dose-response relationship Describes the magnitude of the response of an organism, as a function of exposure (or doses) to a stimulus or stressor (usually a chemical) after a certain exposure time.

Double-blind study Clinical trial in which neither patients nor care providers are aware of the treatment groups to which participants have been assigned.

Duplicate record An unwanted **record** that has the same key as another **record** in the same file.

EBHI See Evidence-based health informatics.

Ecological context The surroundings that impact an information resource development or implementation project.

Economic Analysis The purpose of any economic analysis is to make a quantitative comparison between the stream of costs and benefits that arise from each of the possible options being compared.

Editing style Investigators develop codes as they review the data, making notes as they read and reread the various texts they have assembled.

Effect size A measure of the degree of association between the dependent variable and each of the independent variables.

Embedded design A mixed methods strategy with both qualitative or quantitative methods used simultaneously with one given greater emphasis.

Emergent design Study where the design or plan of research can and does change as the study progresses. This is a characteristic of qualitative studies.

Empirical methods Study or discovery methods based on data derived from observation or experience.

Errors of commission (analogous to type I error, false-positive error) Generically, when an action that is taken turns out to be unwarranted or an observed positive result is, in fact, incorrect. In statistical inference, a type I error occurs when an investigator incorrectly rejects the null hypothesis.

Errors of omission (analogous to type II error, false-negative error) Generically, when an action that should have been taken is not taken or a negative test result is incorrect. In statistical inference, a type II error occurs when an investigator incorrectly fails to reject the null hypothesis.

Ethics The moral principles that guide the behavior of individuals and groups.

Ethnography Set of research methodologies derived primarily from social anthropology. The basis of many of the qualitative evaluation approaches.

EQUATOR Enhancing the QUAlity and Transparency Of health Research network.

Evaluation (of an information resource) There are many definitions of evaluation, including (1) the process of determining the extent of merit or worth of an information resource; and (2) a process leading to a deeper understanding of the structure, function, and/or impact of an information resource.

Evaluation machine An evaluation machine would somehow enable all stakeholders to see how a work or practice environment would appear if an information resource had never been introduced.

Evaluation mindset A way of thinking that embraces the idea that studies should meet the needs of stakeholders.

Evidence-based health informatics (EBHI) Use of the current best evidence in the field of health informatics.

Experimental design Plan for a study that includes the specification of the independent and dependent variables, the process through which participants will be assigned to groups corresponding to specific combinations of the independent variables, and how and when measurements of the dependent variables will be taken.

Experimental study A comparative study purposefully designed by an investigator to explore cause-and-effect relations through such strategies as the use of control, randomization, and analytic methods of statistical inference.

Explanatory design Qualitative methods follow quantitative methods in a mixed methods design

Exploratory data analysis An approach of analyzing data sets to summarize their main characteristics, often using statistical graphics and other data visualization methods.

Exploratory sequential Quantitative methods follow qualitative methods in a mixed methods design.

External validity The extent to which results can justify conclusions about other contexts (that is, the extent to which results can be generalized).

Eye tracking Technology that is useful for understanding where the user's attention is focused when completing tasks

False negative error, or a false negative A result that incorrectly indicates the absence of a condition when the condition is present.

False positive error (FP) or a false positive A result that incorrectly indicates the presence of a condition when the condition is not present.

Feasibility study Preliminary "proof-of-concept" evaluation demonstrating that an information resource's design can be implemented and will provide reasonable output for the input it is given. Similar to a pilot study.

Field study Study of an information resource where the information resource is used in a real-life context such as ongoing health care. Study of a deployed information resource (Compare with Laboratory study).

Field user effect studies Studies focusing on how an information resource affects the activities of health promotion and care, research, public health, or education.

Field-function studies Variants of laboratory-function studies, field-function studies focus on an information resource that is not yet ready for use in the real world with a goal of testing it using real world problems or challenges.

Fisher's exact test A test of significance used in the analysis of contingency tables when cell sizes are small and chi-square is not the appropriate test.

Formal interviews Occasions where both the study team member and interviewee are aware that the answers to questions are being recorded for direct contribution to the evaluation study.

Formative study Study with the primary intent of improving the information resource under study by answering developer questions rapidly enough to allow the results to influence decisions they take. An example would be providing the developers with regular feedback or user comments during a pilot (Compare with Summative study).

Fully integrated design Qualitative and quantitative mixed methods study data are mixed during each phase of the study.

Fully structured interview There is a schedule of questions that are always presented in the same words and in the same order; similar to a verbal survey.

Fundamental theorem of biomedical and health informatics The theorem suggests that the goal of informatics is to deploy information resources that help persons—clinicians, students, scientists, and, increasingly, all members of society—do whatever they do "better" than would be the case without support from these information resources.

Generalizability theory An extension of the classical theory of measurement. Using this approach, study teams can conduct measurement studies taking into account the effects of multiple types of observations.

Goal-free approach An evaluation approach where those conducting the evaluation are purposefully blinded to the intended effects of an information resource and collect whatever data they can gather to enable them to identify all the effects of the resource, regardless of whether or not these are intended.

Gold standard Expression of the state of the art in the application domain or the "truth" about the condition of a task (such as the diagnosis of a patient) against which performance of an information resource can be compared in practice, gold standards are usually not knowable, so studies often employ the best approximation to the "truth" that is available to the investigator.

Grey literature Unpublished sources of information that are often difficult to locate.

Grounded theory A theory which seeks to develop new theory entirely from the newly collected data.

Groups of participants See Study groups of participants

Hawthorne effect An effect that can bias the results of studies: the notion that the act of studying human performance changes performance.

Health care payer or insurance company perspective The perspective of a large health care provider, system, or insurer makes substantial sense for many economic analyses because they are often making resource allocation decisions regarding what type of services to cover, and what financial resources to expend in the production and provision of those services.

Health services research Inquiry about how individuals or populations get health care and what can be done to improve health care.

Heterogeneity In **statistics** means that your populations, samples or results are different.

Heuristic inspection Engaging a small group of reviewers who examine a system interface against a set of guidelines can be a helpful early step in usability assessment.

Hierarchical (or "nested") approach in measurement studies When tasks are the observation of primary interest, specific subsamples of tasks are assigned to specific subsamples of measurement objects.

Histogram Continuous data can be graphed as a histogram by plotting the number of respondents with each value.

Hospital perspective The hospital perspective is commonly used to evaluate the effects of information systems, as it is often the hospital that is expending financial resources. The hospital perspective is also usually straightforward: the cost structure of the institution is generally known, and the costs that the hospital must include in national cost reports are published.

Human factors engineering Concerned with the design of systems taking into account the characteristics of humans.

Human factors Those aspects of the design of an information resource that relate to the way users interact with the information resource, primarily addressing the issues involved in a user interface design (related to ergonomics and human-computer interaction).

Human subject A person who is included in a research or evaluation project as a participant.

Human-computer interaction Human-computer interaction is a field which focuses on the interfaces between people and computers.

Hunches In qualitative studies, personal theories based on a person's experience in the field.

ICER Incremental cost-effectiveness ratio. Represents the ratio of the net costs that will be expended by implementing a particular innovation divided by the net benefit, measured as an appropriate clinical outcome.

ICT Information and communications technology.

Immersion/crystallization style The least structured of Crabtree and Miller's analysis styles with study team members spending extended periods of time reading and interpreting the text and gaining an intuitive sense of the data prior to writing a description of their interpretation

Impact Effect of an information resource on an application area such as health care, usually expressed as changes in the actions or procedures undertaken by workers or as client outcomes such as patient morbidity and mortality.

Implementation bias Extra effort is taken beyond what would be done in service to the typical user or organization that deploys the resource in order to generate new information.

IMRaD An "Introduction, Method, Results, and Discussion" acronym that describes a generic format for reporting any study based on empirical data.

In situ **studies** Conducted in an ongoing clinical, research, or educational environment.

Incremental costs In economic analysis in health care, these are the costs of implementing the next logical "option", irrespective of the magnitude of effect on costs and benefits. In general, the incremental costs are defined as the total costs that will be incurred by choosing one strategy over another.

Independent observations One independent element of measurement data. "Independent" means that, at least in theory, the result of each observation is not influenced by other observations.

Independent variable In a correlational or interventional study, a variable thought to determine or be associated with the value of the dependent variable.

Indirect costs In economic analyses in health care, indirect costs are those that result from the choice of a strategy on the individuals who are treated, such as lost productivity, missed days of work, etc. This is quite different from the accounting definition of indirect costs. (See also Direct costs).

Inductive reasoning One seeks to go beyond simple documentation and description of phenomena that are observed to make broader generalizations.

Informal interviews Spontaneous discussions between the study team and persons in the field.

Informants One of several terms for the individuals who are being studied.

Information or measurement bias Concerns problems with determining either the outcome or the "exposure".

Information resource Generic term for a system that seeks to enhance information management or communication in a biomedical domain by providing task-specific information directly to workers (often used equivalently with "system"). Usually but not always computer-based.

Informed consent A process through which a study team gets agreement from a participant after communicating the risks and benefits to the person.

Infrastructure costs Infrastructure is the set of fundamental facilities and systems that support the sustainable functionality of households and firms.

Institutional Review Boards (IRBs) A committee that oversees research ethics to ensure the research follows U.S. ethical guidelines.

Instrument Technology employed to make a measurement, such as a paper or digitally administered questionnaire. The instrument encodes and embodies the procedures used to determine the presence, absence, or extent of an attribute in an object.

Instrumental variable analysis In statistics, econometrics, epidemiology and related disciplines, the method of instrumental variables (IV) is used to estimate causal relationships when controlled experiments are not feasible or when a treatment is not successfully delivered to every unit in a randomized experiment.

Intangible costs (pain and suffering) Patients place a value on pain and suffering, as evidenced by the fact that patients spend financial resources (purchase pain relievers, accept operations that palliate symptoms) to eliminate or alleviate symptoms, even if the intervention has no effect on their length of life or survival.

Internal validity Internal validity is determined by how well a study can rule out alternative explanations for its findings (usually, sources of <u>systematic error</u> or 'bias').

Interpretive One seeks to go beyond simple documentation and description of phenomena that are observed to increase understanding of them.

Interrupted time series study A comparative study design in which several measurements are made before and several after the intervention. The analysis attempts to show that a step change in the dependent variable is statistically more likely to have occurred during the interval associated with the intervention than during any other interval.

Interval variable A continuous variable in which meaning can be assigned to the differences between values, but there is no real zero point so it lacks ratio properties. (Compare with Ratio variables).

Intervention In an experimental study, the activity, information resource, treatment etc., that distinguishes the study groups.

Interventional studies The study team prospectively creates a contrasting set of conditions that enable the exploration of the questions of interest.

Intuitionist–pluralist worldview The philosophical underpinning of qualitative evaluation approaches, emphasizing that there exist many legitimate viewpoints about any phenomenon under study.

IRBs See Institutional Review Boards.

Items The individual elements of a form, questionnaire, or test that are used to record ratings, knowledge, attitudes, opinions, or perceptions.

Iterative loop of qualitative processes A loop in a figure showing how fluid the process is, always aimed at answering the evaluation questions. At any point in the process, a decision can be made to loop back to revisit steps and even to start over with new or revised research questions, however.

Judge Human, usually a domain expert, who, through a process of observation, makes an estimate of the value of an attribute for an object or set of objects.

Key informant Sponsor, someone inside the organization who knows well the people involved and the resources being studied.

Knowledge-based system Class of information resource that provides advice by applying an encoded representation of knowledge within a biomedical domain to the state of a specific patient or other task.

Knowledge In an informatics context, understanding of a health problem by people or machines that is necessary to inform decisions about that problem.

Laboratory function studies Studies that go beyond usability of an information resource to explore whether whatever the information resource does has the potential to be beneficial.

Laboratory study Study that explores important properties of an information resource in isolation from the application setting. (Compare with Field study.)

Laboratory usability studies Brings prospective users into a controlled environment where their interactions with a resource can be highly instrumented.

Laboratory-user effect studies Studies where the intended users of an information resource—practitioners, patients, researchers, students, or others—interact with the resource and are asked what they *would do* with the advice or other output that the resource generates, but no actions are taken as a result.

Learning Health Systems In health care, organizations continuously learn from data so that they can continuously improve.

Linking datasets Used to bring together information from different sources in order to create a new, richer dataset. This involves identifying and combining information from corresponding records on each of the different source datasets.

Logical-positivist approach The philosophical underpinning of quantitative evaluation approaches, emphasizing that entities have properties that can be measured and that all observers should agree about the measurement result.

Logistical factors Many measurement processes are strongly influenced by procedural, temporal, or geographic factors, such as the places where and times when observations take place.

Macro-costing methods These use truly global measures of the costs (or payments) for services. For example, in the United States, the Centers for Medicare and Medicaid Services (CMS) has calculated (through a very complicated resource-based analysis) the estimated average cost for every physician service from office visits to various procedures to the costs of hospitalization for all categories of diagnoses.

Mann Whitney Test A nonparametric statistical test, for randomly selected values X and Y from two populations, indicating that the probability of X being greater than Y is equal to the probability of Y being greater than X.

Marginal costs In economic analyses, the marginal cost is defined as the cost of a single extra unit of output. Often in health care this measure has little meaning, as many activities are bundled: for example, many components of a clinical information resource are bundled, and there is no option of dividing either the inputs or expected outputs of their use, so true marginal costs cannot be calculated. (See also Direct costs, Indirect costs).

Matching Means finding one (or more) control patient for every intervention patient who resembles them in all relevant features.

Measurement study Study to determine the extent and nature of the errors with which a measurement is made using a specific instrument. (Compare with demonstration study.)

Measurement The process of assigning a value corresponding to the presence, absence, or degree of presence, of a specific attribute in a specific object.

Medical device An instrument, apparatus, implement, machine, contrivance, implant, in vitro reagent, or other similar or related article used in healthcare or health promotion.

Member checking In qualitative investigation, the process of reflecting preliminary findings back to individuals in the setting under study; one way of confirming that the findings are truthful.

Meta regression The use of systematic review, meta-analysis and regression techniques on a large body of primary studies to uncover significant associations between aspects of the study design, intervention etc., and a single outcome variable.

Meta-analysis A set of statistical techniques for combining quantitative study results across a set of completed studies of the same phenomenon to draw conclusions more powerful than those obtainable from any single study of that phenomenon. Used in many systematic reviews or overviews.

Meta-narrative review This is the qualitative equivalent of meta-analyses, with a focus on describing how key concepts in a domain emerged and changed over time.

Micro-costing methods These make use of detailed accounting principles to develop measures of how much each product, service, or option costs by breaking it down into individual parts and determining the cost of each component

Mixed methods Study team members collect and analyze both quantitative and qualitative data within the same study.

Multimethod Using qualitative and quantitative methods but data are not mixed.

Multi-level modelling Analyses that simultaneously take into account data that exist hierarchically at various levels of scale; e.g. people, groups of those people, or organizations that form from the groups.

Multiple co-occurring observations Multiple observations conducted at approximately the same time. The observations can be crafted in ways that are different enough for each observation to create a unique challenge for the object yet similar enough that they all measure essentially the same attribute.

Multiple perspectives theory Based on general systems theory, an approach that uses technical, organizational, and personal perspectives to understand a complex system.

Narrative analysis Frames informants' observations as "stories" and addresses both the words and flow of the narrative.

Naturalistic Performed without purposeful manipulation of the environment under study.

Naturally occurring text Online sources that include online discussion groups, Facebook groups, YouTube, podcasts, and even digital mapping and geospatial technologies.

Needs assessment studies Studies that seek to clarify the information problem the resource is intended to solve.

Nested (hierarchical or multilevel) designs Tests to see if there is variation between groups, or within **nested** subgroups of the attribute variable.

Nominal variable Variable that can take a number of discrete values but with no natural ordering or interval properties.

Non-parametric tests Class of statistical tests (such as the chi squared and Mann Whitney U tests) that requires few assumptions about the distribution of values of variables in a study (e.g., that the data follow a normal distribution).

Normal distribution or the Gaussian distribution A probability distribution that is symmetric about the mean. Observations near the mean are more frequent in occurrence than data far from the mean. In graph form, it is a bell curve.

NP-hard problem To verify a program using brute-force methods requires application of every combination of possible input data items and values for each in all possible sequences.

Null hypothesis In inferential statistics, the hypothesis that an intervention will have no effect: that there will be no differences between groups and no associations or correlations among variables.

Number needed to treat (NNT) The NNT in epidemiology is the average number of patients who need to be treated to prevent one additional bad outcome (e.g. the number of patients that need to be treated for one of them to benefit compared with a control in a clinical trial). NNT is equal to the reciprocal of the absolute difference in event rates between the two groups.

Object (of measurement) Entity on which a measurement is made and to which a measured value of a variable is assigned.

Object classes Information resources, persons (who can be patients, care providers, students, and researchers), groups of persons, and organizations (healthcare and academic) on which measurements are made.

Objective (1) Noun: state of practice envisioned by the designers of an information resource, usually stated at the outset of the design process. Specific aims of an information resource. (2) Adjective: a property of an observation or measurement such that the outcome is independent of the observer (cf. Subjective).

Objectives-based approach An evaluation approach that seeks to determine if a resource meets its design or performance objectives.

Objects-by-observations matrix A representation of the results of multiple observations made for each of a set of objects in a measurement process.

Observation of primary interest The classical theory of measurement allows consideration of only one type of observation (judges, tasks, or items) in a measurement process. The observation of primary interest is the type of observation included in a measurement study.

Observational study (naturalistic study) Approach to study design that entails no experimental manipulation. Can be descriptive (dependent variable only) or correlational, in which the study team typically draws conclusions by carefully observing users with or without an information resource.

Opinion leaders Respected individuals in a social system. Others in the system will follow their lead.

Optical character recognition The use of technology to distinguish printed or handwritten text characters inside digital images of physical documents, such as a scanned paper document.

Oral history A technique for obtaining personal stories and perspectives about factors of historical significance.

Ordinal variable Variable that can take a number of discrete values which have a natural order (compare with Nominal variable).

Organizational and system studies Information resources are deployed into a larger context. These studies focus on how the effects of these resources shape and are shaped by the environment in which they function. Also, use of mixed methods in informatics to explore the use and impact of information resources when individuals work together as part of systems at multiple levels of scale.

Outcome variable Similar to "dependent variable," a variable that captures the end result of a health care, research or educational process; for example, long-term operative complication rate, citation rate of an article or mastery of a subject area.

Overfitting When a statistical model that powers an information resource is carefully adjusted to achieve maximal performance on training data, this adjustment may worsen its accuracy on a fresh set of data or "test set" due to a phenomenon called overfitting.

Panel study Study design in which a fixed sample of respondents provides information about a variable, often at different time periods.

Parameter estimate sensitivity analyses Sensitivity analysis provides an opportunity to determine the level of accuracy needed in the estimation of the parameter.

Parent codes Main topics in a code book in qualitative studies.

Part-whole correlation The strength and direction of the relationship between one observation in a measurement process and the total score of each object.

Participant In an evaluation study, the entities on which observations are made. Although persons are often the participants in informatics studies, information resources, groups, or organizations can also be the participants in studies.

Patient centeredness Citizens worldwide are being encouraged to play greater roles in their health care, partnering with the health care team.

Person-level anonymized data The result of the process of removing personally identifiable information from data sets, so that the people whom the data describe remain anonymous.

PICO format The PICO model, as adapted for informatics studies, expresses four elements comprising a complete study question: P = Participants in the study; I = Information resource or other Intervention being studied; C = Comparison between the information resource under study and an alternative resource or the absence of the resource; and O = Outcomes of interest to the stakeholders.

Pilot study Trial version of a study (often conducted with a small sample) to ensure that all study methods will work as intended or to explore if there is an effect worthy of further study. (See also Feasibility study).

Positive predictive value The probability that participants with a positive prediction truly have the predicted outcome.

Power Statistical term describing the ability that a study, given its sample size, will detect an effect of a specified magnitude. The power of a study equals one minus the probability of making a type II error.

Practical significance Difference or effect due to an intervention that is large enough to affect professional practice. With large sample sizes, small differences can be statistically significant but may not be practically significant. In health care, similar to "clinical significance."

Precision In measurement studies, the extent of unsystematic or random error in the results. High precision implies low measurement error. Similar to Reliability.

Problem impact studies Studies focusing on the actual outcomes resulting from deployment of an information resource; for example, the health status of individuals and populations.

Process variable Variable that measures what is done by staff in an evaluation study, such as accuracy of diagnosis or number of tests ordered by health care workers.

Processes Donabedian defines *Process* as the things done to and for the patient (e.g. defaulter tracing and hospital referrals).

Product liability When a developer or distributor of an information resource is held legally responsible for its effects on staff decisions, regardless of whether they have taken due care when developing and testing the information resource.

Professional review approach The well-known "site visit" approach to evaluation where a group travels to the site(s) where an information resource is deployed and collects qualitative data from the stakeholders.

Prophecy formula A formula providing a way to estimate the effect on reliability of adding equivalent observations to or deleting observations from a measurement process.

Prospective studies Studies designed before any data are collected. A cohort study is a kind of prospective study, while most data mining activities are retrospective studies.

Propensity score A remedy for confounding by indication: The propensity score is the probability of treatment assignment conditional on observed baseline characteristics.

Purposive selection Deliberate selection is based on the purpose of the study, and the selection strategy can evolve as the progress of the study reveals initially unanticipated needs for subjects who bring potentially novel viewpoints.

QALY (Quality Adjusted Life Years) QALYs are a method for adjusting measures of the quantity of life/survival by the quality of being in a particular state. Utility weights (between 0 and 1) are assigned to health states, and the value of being in that health state is defined as the length of time in that state times the utility of being in that state.

Qualitative data analysis software (QDA) Allows the study team to share data, store, organize, count, search and retrieve, code, group codes into themes, and visualize patterns across codes, themes, and documents.

Qualitative methods A family of study methods that is not dependent on numerical data and that seek to provide an deep understanding of phenomena and the meaning individuals give to them.

Quantitative approaches Evaluation methods that use the results of measurements expressed as numbers to generate results that inform decisions.

Quasi-legal approach An evaluation approach that establishes a mock trial, or other formal adversarial proceeding, to judge the merit or worth of an information resource.

Random sample A method used in any kind of study for picking participants to ensure that the study findings will be generalizable to the entire population. Entails obtaining a complete listing of all participants and randomly selecting a subset as a sample.

Randomized allocation A method used in comparative or interventional demonstration studies for reliably determining whether an intervention, such as an information resource to support researchers, causes a change in some dependent variable, such as research productivity.

Randomized studies Experimental studies in which all factors that cannot be directly manipulated by the investigator are controlled through random allocation of participants to groups.

Ratio variable Continuous variable in which meaning can be assigned to both the differences between values and the ratio of values (compare with Interval variable). Ratio variables support both division and multiplication, in addition to addition and subtraction.

RE-AIM Framework with five components: Reach, Effectiveness, Adoption, Implementation, and Maintenance. Using this framework, evaluation teams can study the impact of innovations at both the individual (i.e., end-user) and organizational (i.e., delivery agent) levels.

Real world data Data that derive from actual practice without intervention or manipulation.

Reference standard See Gold standard.

Reflexivity The conscious recognition of one's own biases and the equally conscious design of the study to address them.

Regression, linear Statistical technique, statistically equivalent to analysis of variance, in which a continuous dependent variable is modeled as a linear combination of one or more continuous independent variables.

Regression, logistic Statistical technique in which a dichotomous dependent variable is modeled as an exponential function of a linear combination of one or more continuous or categorical independent variables.

Regression discontinuity study design (RDD) A quasi-experimental correlational study design that aims to determine the causal effects of interventions by comparing outcomes in groups to which an intervention has been assigned or not due to them lying just above or just below a cutoff or threshold in a continuous variable.

Regression to the mean The statistical tendency for any single measurement that lies far from the mean to, when measured again, "regress" or revert to a value that is closer to the mean.

Reliability Extent to which the results of measurement are reproducible or consistent (i.e., are free from unsystematic error).

Report A vehicle for communicating the methods and results of a study.

Research Systematic inquiry that seeks to develop evidence to guide practice.

Responsive/illuminative approach An evaluation approach that seeks to represent the viewpoints of those who are users of the resource or who are part of the environment where the resource operates.

Retrospective studies Studies in which existing data, often generated for a different purpose, are reanalyzed to explore a question of interest to the study team. A case-control study is a kind of retrospective study.

Reverse causation A direct causal effect in a direction opposite to what is hypothesized.

ROC analysis Receiver operating characteristic analysis. First used in studies of radar signal detection, it is typically used with a test that yields a continuous value but is interpreted dichotomously. ROC analysis documents the trade-off between false-positive and false-negative errors across a range of threshold values for taking the test result as positive and results in a curve, the area under which gives a good estimate of the performance of a predictive model.

Sample size calculation Estimate of the number of participants required in a study.

Sampling strategy Method for selecting a sample of participants used in a study. The sampling strategy determines the nature of the conclusions that can be drawn from the study.

Saturation The point when a qualitative study has gathered enough data because nothing new is being learned.

Secular trends Systematic changes, not directly related to the resource under study, that take place over time.

Semi-structured interview Study team members specify in advance a set of topics they would like to address but are flexible as to the order in which these topics are addressed and open to discussion of topics not on the prespecified list.

Sensitivity (1) Performance measure equal to the true positive rate. In an alerting information resource, for example, the sensitivity is the fraction of cases requiring an alert for which the information resource actually generated an alert. (2) In information resource design, it is the extent to which the output of the information resource varies in response to changes in the input variables.

Sensitivity analysis In health economics, the extent to which the results of the economic modeling exercise vary as key input variables or assumptions are varied. Tests the effect of different assumptions on the outcome.

Sentiment analysis A method of qualitative data analysis that can help evaluate tone, intent, and emotions.

Sequential designs One approach follows the other in a mixed methods study.

Simulation studies Seek to reproduce as much as possible the features of environments that can profoundly affect resource usability.

Single-blind study Study in which the participants are unaware of the groups to which they have been assigned.

Simpson's Paradox *Simpson's Paradox* is a statistical phenomenon where an association between two variables in a population emerges, disappears or reverses when the population is divided into subpopulations.

Societal perspective The societal perspective in economic evaluation is by definition the broadest perspective, and therefore should include all costs and benefits, regardless of who pays the costs or to whom benefits accrue.

Socio-technical 8-dimensional model Describes eight areas that must be included in any thorough context-driven assessment.

Software validation Validation studies explore whether the "right" information resource was built, which involves both determining that the original specification was fit to the problem it was built to address and that the resource as built is performing to specification.

Software verification Verification means checking whether the resource was built to specification.

Sources of variation Factors that cause the results of measurements on a given object to vary from observation to observation.

Specificity Performance measure equal to the true negative rate. In a diagnostic information resource, for example, specificity is the fraction of cases in which a disease is absent and in which the information resource did not diagnose the disease.

Spurious correlation A connection between two variables that appears causal but is not.

Stakeholders Actors who may be affected by a biomedical or health information resource, and each may have a unique view of what constitutes benefit. Also, those with a need to know about the study and its results.

Standard error of measurement The uncertainty added to the result of measurement on each object, as a result of random (also called unsystematic) error.

Standard error of the mean An indicator of the uncertainty in the value of the mean result of measurements performed on a sample of objects.

Statistical disclosure *Statistical disclosure* control (SDC), also known as **statistical disclosure** limitation (SDL) or **disclosure** avoidance, is a technique used in data-driven research to ensure no person or organization is identifiable from the results of an analysis of survey or administrative data, or in the release of microdata.

Statistical significance An observed difference or effect that, using methods of statistical inference, is unlikely to be due to chance alone.

Step-wedge design The collection of observations across a group of clusters during a baseline period in which no clusters are exposed to the intervention. Following this, at regular intervals, or steps, one or more clusters are randomized to receive the intervention and all participants are once again measure.

Structural sensitivity analysis A type of economic sensitivity analysis in which the goal is to help ensure that all the important and correct components of the problem have been included and are related in an appropriate manner.

Structure validation studies Studies that address the static form of the software, often after the first prototype has been developed.

Study design Designing a study of any kind to generate results likely to be useful to stakeholders and which will fit within the available envelope of time, staff, financial and other resources.

Study groups of participants A scientifically interesting **group** of **participants**, such as **participants** who have a certain condition, medical history, exposure to the information resource, or demographic property.

Study power Sometimes called sensitivity, this means how likely it is that the **study** will distinguish an actual effect from one of chance.

Study setting The physical, social, or experimental site within which research is conducted.

Study setting bias An inappropriate context is selected for a study.

Subject See Participant.

Subjective probability Individual's personal assessment of the probability of the occurrence of an event of interest.

Subjective Property of observation or measurement such that the outcome depends on the observer (compare with Objective).

Summative study Study designed primarily to demonstrate the value of a mature information resource (compare with Formative study).

Sunk costs In economics and business decision-making, a sunk cost is a cost that has already been incurred and cannot be recovered, so should not be taken into account in an economic evaluation study.

Systematic review A secondary research method that attempts to answer a pre-defined question using an explicit, rigorous approach based on exhaustive searches for published and unpublished studies of the appropriate study design.

Systems analysis Breaking down the steps for systematically developing a system such as information resource.

Talk-aloud, or think-aloud These protocols are effective in understanding users' thought patterns and strategizing as they work with a resource

Task domain The task domain, chosen by a study team, defines and limits the set of tasks that will be included in a measurement process where tasks are the observation of interest.

Tasks Health-related activities that are conducted by resource users in their professional or personal lives. Test cases against which the performance of human participants or an information resource is studied.

Technology Acceptance Model (TAM) The technology acceptance model is an information systems theory that models how users come to accept and use a technology.

Template style Qualitative analysis that uses a *preexisting* list of terms, called codes in the language of qualitative analysis.

Tendency effects When judges are the observation of primary interest, some judges are consistently overgenerous or lenient; others are consistently hypercritical or stringent. Others locate all of their ratings in a narrow region.

Test-retest reliability A method to estimate measurement reliability by repeating the same measurement over short time intervals.

Test set A *test set* in machine learning is a secondary (or tertiary) data set that is used to test a machine learning program after it has been trained on an initial training set.

Theory Plausible or scientifically acceptable general principle or body of principles offered to explain phenomena.

Thick description In qualitative work, those conducting an evaluation attempt to describe something in great detail to promote deeper understanding of a phenomenon.

Time costs These represent the amount of time required by providers and patients to use an information resource or participate in a treatment and should be included as a real cost to any program.

Time frame When used in an economic analysis the time frame must match the actual duration of time required to encompass all of the important costs and benefits of the program.

Training data Data derived from a large dataset collected in one or more settings, often at some time in the past (and sometimes many years ago), or in a different setting from that in which a predictive model or classifier will be used.

Training set bias An exaggeration of model perfomance when the study does not take account of the many issues that may reduce performance in a new setting.

Transformation of data Applying a mathematical function to an entire dataset (such as taking the square root or logarithm) to make the distribution of non-normal data more closely resemble a normal distribution. Usually carried out to enable standard statistical tests or regression to be used without violating assumptions of normality. Also, conversion of qualitative text data in a mixed methods study into numeric form.

Trusted third party An entity who manages a warehouse or safe haven for the analysis of person-level data, which receives datasets for linkage from two or more parties who all trust the third party.

Triangulation Drawing a conclusion from multiple sources of data that address the same issue. A method used widely in subjectivist research.

Two by two (2 × 2) table Contingency table in which only two variables, each with two levels, are classified.

Type I error In statistical inference, a type I error occurs when an investigator incorrectly rejects the null hypothesis, typically inferring that a study result is positive when it is in fact negative.

Type II error In statistical inference, a type II error occurs when an investigator incorrectly fails to reject the null hypothesis, typically inferring that a study result is negative when it is in fact positive.

Underrepresentation Fewer participants than needed in studies of participants from certain age groups, ethnic groups, genders or with what is known in many countries as legally protected characteristics.

Unintended disclosure The malicious or accidental re-identification individuals from anonymised data to yield confidential or sensitive information. Often this entails exposure of the data to individuals outside your organization, but it can also mean exposure to unauthorized individuals inside your organization.

Unintended effects Consequences that were unexpected and unplanned.

Unit of analysis error Error in analysis that results when clustering based on a naturally occurring unit (e.g., the provider, practice, or health care institution) to which the intervention assigned is ignored in the analysis of individual (often patient) level data.

Unmeasured confounders Confounding occurs when the effect of the exposure of interest (for example, use of an information resource) on the outcome variable mixes with the effects of other variables that can change the outcome variable. They are unmeasured if they are not taken into account.

Unstructured interview There are no or few predetermined questions.

Usability How useful, usable, and satisfying a system is for the intended users to accomplish goals in the work domain by performing certain sequences of tasks.

Usability studies Studies that address how well intended users can operate or navigate an information resource to help meet their needs.

Usage logs Detailed data generated automatically by an information resource of the "moves" a user makes during their interaction with the resources. This might include text entered, links followed, and choices made from menus as well as time spent and frequency of use.

Users of the resource Individuals and groups who interact with an information resource.

Validation (1) In software engineering: the process of determining whether software is having the intended effects (similar to evaluation). (2) In measurement: the process of determining whether an instrument is measuring what it is designed to measure. (See Validity).

Validity (1) In demonstration studies or experimental designs: internal validity is the extent to which a study is free from design biases that threaten the interpretation of the results; external validity is the extent to which the results of the experiment generalize beyond the setting in which the study was conducted. (2) In measurement: the extent to which an instrument that measures what it is intended to measure. Validity is of three basic kinds: content, criterion-related, and construct.

Variable Quantity measured in a study. Variables can be measured at the nominal, ordinal, interval, or ratio levels.

Verification Process of determining whether software is performing as it was designed to perform (i.e., according to the specification).

Virtual data enclave or safe haven The virtual data enclave (VDE) provides access for trained, authorized users to restricted-use data and is a virtual machine launched from the researcher's own desktop but operating on a remote server, similar to remotely logging into another physical computer.

Visual analog scales A graphical scale such as a line with a beginning point and an end point that allows a respondent to mark an answer, corresponding that to person's perception or belief, anywhere along a continuum.

Volunteer effect People who volunteer as participants, whether to complete questionnaires, participate in psychology experiments, or test-drive new cars or other technologies, are atypical of the population at large.

Vulnerable individuals Those whose voices most need to be heard; those at greater risk for getting less or lower quality health care.

Wilcoxon signed rank test Is a non-parametric statistical **test** used to compare two related samples, matched samples, or repeated measurements on a single sample to assess whether their population mean **ranks** differ (i.e., it is a paired difference **test** for non-normal data).

Index

A

Absolute difference, 271, 280
Accuracy, 263, 264
Accurate, 231
Action research, 375, 486
Advantages and disadvantages of correlational
    studies using routine data, 293–295
Age, ethnic group, gender and other
    biases, 218
Allocation, 315
Allocation and recruitment biases, 242–243
Analysis methods for descriptive
    studies, 221–224
Analysis of variance (ANOVA), 272
Analyzing studies with continuous dependent
    variables and discrete independent
    variables, 270–277
Analyzing studies with continuous
    independent and dependent
    variables, 277–280
Anatomy of all evaluation studies, 27–28
Applying for permissions, 298
Approach, 461
Art criticism approach, 35
Assessment bias, 214, 305
Association, 233
Atheoretic, 351
Attenuation, 167
Attribute, 106
Attribute–object class pairs, 107
Audit trail, 357

B

Before-after studies, 234

Biases associated with data
    collection, 243–244
Bioinformatics implementations, 491
Biomedical and health informatics, 3
Bonferroni method, 251
Box and whisker plot, 224

C

Calibration, 155, 231
Carryover effect, 244
Cascade problem, 465
Case crossover design, 314
Case-control, 306
Case studies, 374, 405, 417
Causation, 230, 291
Central tendency, 222
Chains of effect, 17–18
The Challenge of inferring causation from
    association, 301–304
Challenges when designing interventional
    studies, 230–233
Champions, 392
Charges, 431
Checklist effect, 246
Child codes, 389
Chi-square, 265–266
Choice of effect size metrics, absolute change,
    relative change, number needed to
    treat, 280–281
Classical theory of measurement, 131–132
Classifier, 267
Clients of those users, 29
Closing thoughts, 475, 477–493
Cluster randomized, 238–239

Code book, 389
Codes, 383
Cognitive walkthrough, 414
Cohen's *d*, 271
Cohen's Kappa, 266
   a useful effect size index, 266–267
Cohorts, 296
Common cause, 302, 305
Communication, 467
Comparison-based approach, 34
Conceptual diagram of framework, 395
Conditions, 208
Confidence interval, 216, 221, 222, 271
Confounder, 236
Confounding, 305, 309–312
Confounding by indication, 305, 312
Constructive decisions, 82
Constructivist, 330
Constructs, 107
Construct validity, 148–149
Content analysis, 382
Content validity, 146–147
Context, 89
Contingency tables, 262–263
Continuous, 221
Continuum of extent of mixing qualitative and
   quantitative methods, 404
Contract, 45
Control, 231
Control for confounding, 313–314
Control strategies, 233
Control strategies for interventional
   studies, 233–241
Convenience sampling, 360
Correcting the correlation for
   attenuation, 171–172
Correlation, 155, 291
Correlational studies, 123, 208, 211, 259, 290
Cost-benefit analysis (CBA), 428
Cost-benefit ratio, 440
Cost-consequence studies, 426
Cost-effectiveness, 437
Cost-effectiveness analysis (CEA), 427
Cost-minimizing/cost minimization study, 426
Costs, 431
Cost utility analysis (CUA), 427
Credibility (believability), 339
Criterion-related validity, 147–148
Critically appraising, 304
Crude accuracy, 263
C statistic, 270
Cut-point, 267
Cyclical, 409

D
Data analysis plan, 297
Data availability and quality, 18–19
Data cleaning, 299
Data controllers, 295, 298
Data dredging, 251
Data extract, 296
Data mining, 208, 290
Data readiness level, 294
Data saturation, 356
Data warehouse, 294
Decision-facilitation approach, 34
Decision making processes, 18
Definition of the qualitative
   approach, 330–332
Definitions of evaluation, 26–27
Degrees of freedom, 274
Demonstration studies, 104, 115, 120, 259
Dependent variables, 120, 209
Descriptive design, 207
Descriptive and interventional demonstration
   studies, 165–166
Descriptive, interventional and correlational
   studies, 206
Descriptive studies, 122, 210
Design considerations in measurement studies
   to estimate reliability, 141–144
Designing measurement processes, 178
Design validation studies, 50
Development team, 29
Diagnose measurement problems, 159–164
Diana Forsythe, 330
Differential recall bias, 305
Diffusion of innovations theory, 392
Direct costs, 431
Discounting, 442–444
Discrete, 221
Discrimination, 263
Distinguishing evaluation and research, 11
Diverse populations, 478
Domain knowledge, 16–17
Dose-response, 304
Duplicate records, 297

E
Ecological context, 338
Economic analysis, 425
Editing style, 383
Effectiveness, 248–249
Effect size, 230, 248–249, 260
Effect studies, 69–70
EHR safety, 489

Elevator speech, 371
Embedded design, 409
Emergent, 337
Empirical methods, 26, 337
Engagement, 233, 291
EQUATOR website, 297
Estimate the size and direction of potential
    biases, 300
Ethics, 475–488, 490–493
Ethnographic observation, 368
Ethnography, 331
Evaluation machine, 7, 231
Evaluation mindset, 80
Evaluation paradox, 252
Evidence-based health informatics (EBHI),
    80, 90–91, 352, 491
Expert panels, 367
Exploratory data analysis, 299
Explanatory design, 408
Exploratory sequential, 408
External validity, 206, 212
Eye tracking, 413

F
Factors leading to false inference of
    causation, 304–314
Fallacy of case-control studies, 239–240
False negative (FN), 249, 264
False positive (FP), 249, 264
Feedback effect, 247
Field, 341
Field-function studies, 52
Field notes, 370
Field studies, 331
Field user effect studies, 53, 68–70
Fisher's exact test, 265–266
Focus groups, 367
Formal interviews, 363
Formative decisions, 82
Frameworks, 92–93
Fully integrated design, 405
Fully structured interview, 363
Function study, 66–68
Fundamental theorem, 6
Future of informatics evaluation, 489–493
The future of qualitative data gathering
    methods, 377

G
General linear models, 280
Generalizability theory, 151–152

Goal-free approach, 35
Gold standards, 19
Grand strategy for analysis of study
    results, 260–262
Graphical portrayal of the results, 222
Graphing results as a way to visualize
    statistical interactions, 276
Grey literature, 94
Grounded theory, 351
Groups of participants, 208

H
Hawthorne effect, 14, 215
Health care payer/Insurance company
    perspective, 429
Health information exchange, 489
Health services research, 491
Heterogeneity, 304
Heuristic inspection, 413
Hierarchical approach, 191
Histogram, 221, 222
Historically controlled, 234
Hospital perspective, 429
House's typology, 30
Human-computer interaction (HCI), 412
Human factors engineering, 412
Human subjects, 476
Hunches, 350

I
Illuminative/responsive study type, 329
Immersion/crystallization style, 383
Immortal time bias, 305
Implementation bias, 219
Improving measurements that use items as
    observations, 196–199
Improving measurements that use tasks as
    observations, 188–191
IMRaD, 471
In situ studies, 82
In the field, 349
In vitro studies, 82
Incremental cost-effectiveness ratio
    (ICER), 439
Independent observations, 109–110
Independent variables, 120, 209
Indices of variability, 222–224
Indirect costs, 432
Inductive, 330
Inductive thinking, 329, 331, 335, 336
Informal interviews, 363

Informants, 360
Information bias, 304, 307–309
Information/measurement bias, 307
Information resources, 9
Informed consent, 477
Innovation, 461
Institutional Review Boards (IRBs), 476
Instrument, 108, 315
Instrumental variable, 315
Instruments, 191, 200
Intangible costs (pain and suffering), 432
Integration, 404
Interactions, 274
Internal and external validity, 206
Internal validity, 206
Internal validity of interventional
    studies, 241–248
Internally and externally controlled before-
    after studies, 236–237
Internally valid, 212
Interpretation, 391
Interpretive, 330
Interquartile range, 224
Interval, 113, 221
Interventional studies, 123, 208, 211,
    229–231, 240, 244, 249, 250,
    252, 259
Intuitionist–pluralist, 31
Items, 112, 177
Items are the primary observation, 191–200
Iterative loop of qualitative
    processes, 349–350
Iterative qualitative loop, 341–342

J
Jottings, 370
Judges, 112, 177
Judges' opinions are the primary
    observations, 180–184

K
Key informant, 362
Key object classes, 110–111
Knowledge, 19

L
Laboratory function studies, 51–52
Laboratory usability studies, 413
Laboratory-user effect studies, 52–53
Learning health system, 292, 478

Level of evaluation, 87–88
Levels of measurement, 112–114, 209
Life cycle, 88–89
Linking datasets, 296
Literature searching, 91
Logical–positivist, 30
Logistic regression, 280
Logistical factors, 112

M
Machine learning, 267
Macro-costing methods, 433
Macroethnography, 368
Main effects, 274
Mann Whitney test, 271
Matched controls, 239–240
Matching, 239, 313
Matrix of sites and subjects, 394
Maturity of the information resource
    employed, 55
Mean, median, or mode, 221
Measurement, 106
Measurement errors, 137–140, 165–166
Measurement instruments, 108
Measurement noise, 305
Measurement process, 105–110
Measurement studies, 104, 115, 117–119
Medical device, 15
Member checking, 342, 356
Membership bias, 305
Memos, 370, 390
Mendelian randomization, 315
Meta-analysis, 95, 304
Meta-narrative reviews, 95
Methods to estimate reliability, 135–137
Micro-costing methods, 433
Microethnography, 368
Missing data, 215–216, 297
Mixed methods, 403, 404
Mixed methods evaluation design
    diagram, 412
Mixed methods studies, 403
Multi-level modelling, 210
Multimethod, 405
Multiple co-occurring observations, 135
Multiple perspectives framework, 352
Multiple regression, 278
Multivariate modelling, 209, 313

N
Narrative analysis, 392

National Institute for Clinical Excellence (NICE), 426
Natural history of a qualitative study, 340–343
Naturalistic, 330
Naturally occurring data, 373
Naturally occurring text, 393
Needs assessment studies, 50, 64–66
Negative predictive value, 264
Negotiation of the "ground rules" of the study, 340
Nominal, 112, 221
Non-parametric, 222, 271
Non-response bias, 305
Normal distribution, 222
NP-hard problem, 20
Null hypothesis, 249
Number needed to treat, 281
Number of judges needed, 182
Number of tasks needed, 187

O
Object, 106
Object classes, 106
Objectives-based approach, 34
Objects-by-observations matrix, 129, 159–162
Observation details, 371
Observation of primary interest, 177
Opinion leaders, 392
Optical character recognition, 298
Options for communicating study results, 467–468
Oral history, 366
Ordinal, 112, 221
Organization and system studies, 54, 70–72
Organizational systems-level studies, 416–418
Overfitting, 219–221, 281–282
Oversight, 476

P
Paired sample t-test, 271
Parent codes, 389
Participants, 208, 209, 360
Participatory action research (PAR), 375
Part-whole correlations, 155
Patient centeredness, 490
Person-level data, 292, 295
Perspective of an analysis, 428
Philosophical bases of evaluation, 30–32
PICO formalism, 219
PICO format, 47
Placebo effect, 241–242

Planning, 79–98
Positive predictive value, 264
Precision-recall (P-R) curve, 270
Predictive model, 231
Problem impact, 70–72
Problem impact studies, 54
The process of designing and conducting a demonstration study, 213
Process of measurement studies, 156–158
Professional practice and culture, 14–15
Professional review approach, 36
Project advisory groups, 97
Project management, 96–98
Propensity analysis, 312
Propensity matching, 312
Propensity score, 312
Prophecy formula, 129, 142
Proposing studies, 457–466, 468–470, 472–474
Prospective studies, 121, 230
Purposive selection, 360
Pyramid of challenges affecting the three types of demonstration studies, 207

Q
Qualitative approaches, 25
Qualitative data analysis, 381–383, 385–394, 396, 397
Qualitative data analysis software (QDA), 386
Qualitative evaluation approaches, 329, 331, 333, 334, 336, 337, 339–345
Qualitative methods, 331
Qualitative objectivity, 339
Qualitative research, 331
Qualitative study design, 347, 349–355, 357, 358, 360–374, 376–379
Quality-adjusted life years (QALYs), 428
Quality improvement, 478, 490
Quantitative approaches, 25
Quantitative objectivity, 339
Quasi-legal approach, 35

R
Random allocation, 218
Randomization by group, 239
Randomized controlled study, 237
Random selection, 218
RAP fieldwork manual, 376
Rapid assessment, 492
Rapid assessment process (RAP), 375, 405

Rating item with a graphical response
    scale, 192
Ratings paradox, 199–200
Ratio, 113, 221
RE-AIM, 351
Real world evidence, 290, 293
Real world data, 213
Reasons for performing evaluations, 12–13
Receiver operating characteristic (ROC), 268
Refereeing evaluation studies, 465–466
Reflexivity, 353–354, 482
Regression analysis, 277
Regression discontinuity study design, 316
Regression to the mean, 232
Relative difference, 271, 281
Reliability, 129
Reliability estimate, 171
Report, 471
Reporting guidelines, 297
Research, 478
Responsive/illuminative approach, 36
Result communication, 467, 471
Retrospective studies, 122, 211
Reverse causation, 302, 311
Rigor, 353–359
ROC analysis, 267–270
Role of the evaluation contract, 470
Role of the study team, 469–470
Roles in evaluation studies, 28–30

S
Safe haven, 292
Safety, 475, 477–488, 490–493
Safety and regulation, 15–16
Safety of ICT, 475, 479
Sample size, 230
Sample size calculation, 217, 218
Sampling of users and tasks, 58–59
Saturation, 338
Secular trends, 231
Selection bias, 304, 306
Semi-structured interview, 363
Sensitivity, 264
Sensitivity (detection rate), 264
Sensitivity analyses, 444
Sentiment analysis, 385
Sequential designs, 407
Setting in which the study takes place, 55
Significance, 461
Simpson's paradox, 305, 310
Simulation, 414
Simultaneous external controls, 235–236

Simultaneous randomized controls, 237–238
Skewed, 222
Social media, 292
Social media posts, 478
Societal perspective, 429
Socio-technical 8-dimensional model, 336
Software validation, 10
Software verification, 10
Sources of variation, 181–182
Sources of variation among items, 194
Sources of variation among tasks, 185–187
Special issues in proposing
    evaluations, 464–465
Special issues with ANOVA, 275–277
Specific aims, 461
Specific issues of report writing, 471–473
Specificity, 264
Spurious correlation, 302–303
Stakeholder needs, 85–87
Stakeholders, 28, 339
Standard deviation, 221, 222
Standard error, 221
Standard error of measurement, 139–140, 155
Standard error of mean, 155, 216, 224
Statistical disclosure risk, 301
Statistical disclosure rules, 298
Statistical inference, 248–251
Statistical power, 250
Statistically significant, 261
Step-wedge design, 238–239
Stratification, 313
Structure of hypercritic study, 169
Structure, processes, or outcomes, 206
Structure validation studies, 50–51
Study design, 206
Study design scenarios and examples, 63–77
Study groups, 211
Study power, 230, 249
Study proposal, 460
Study questions, 45–46
Study setting, 219
Study setting bias, 219
Subjects, 360
Summative decisions, 82
Sunk costs, 447
Surrogate outcome, 250
Systematic reviews, 95
Systems analysis, 334–336

T
Talk-aloud, 414
Task domain, 177

Tasks, 57, 111, 177
Tasks are the primary observation, 184–191
Teams, 96
Teamwork, 492
Technology acceptance model (TAM), 351
Telemedicine, 489
Template style, 383
Tendency effects, 181
Terminology for demonstration
    studies, 208–210
Test-retest reliability, 135
Test set, 219, 282
Theory, 350
Thick description, 338
Think-aloud, 414
Threats to internal validity, 214–217
Threats to external validity, 217–221
Threshold, 267
Time costs, 432
Time frame, 430
Top-down and bottom-up approaches, 384
Tower model, 4
Training data, 219
Training set, 282
Training set bias, 219–221
Transferable, 231
Transformation of data, 299, 408
Tree of evaluation, 43
Triangulation, 354, 404–405
TRIPOD reporting guideline, 220
True negative (TN), 264
True positive (TP), 264
Trusted third party, 294, 298
Trusted virtual data enclave, 292
t-test, 270
Type I error, 249, 261
Type II error, 249
Types of data used in correlational
    studies, 292–293
Types of demonstration studies, 207–208
Typology of mixed methods strategies, 407

U
Uncontrolled study, 233–234
Underrepresentation, 218
Unintended disclosure, 296
Unintended effects, 81
Unit of analysis error, 239
Univariate, 209
Unmasking bias, 305
Unmeasured confounders, 310
Unstructured interview, 363
Usability, 66–68, 224, 412, 489
Usability studies, 51
Usability testing methods, 413–414
Usage logs, 292
Users of the resource, 29
Using Contingency (2 × 2) Tables: Indices of
    Effect Size, 263–265
Utility, 427
Utility (usefulness), 339

V
Validity, 129
Validity and its estimation, 144–149
Validity of demonstration studies, 212–213
The value of different approaches, 343–344
Variables, 209
Virtual data enclave, 298
Virtual ethnography techniques, 376
Visual analog scale, 192
Volunteer effect, 218
Vulnerable individuals, 480

W
Well-behaved observation, 162
When are qualitative studies
    appropriate, 336
Wilcoxon signed rank test, 271
Writing evaluation proposals, 458–465
Writing qualitative study reports, 472–473

Printed in the United States
by Baker & Taylor Publisher Services